Of The Elements

18

13	14	15	16	17	2 4.215 0.95 **He** 0.179¤ Helium 4.002 602 ±2
5 +3 4275 2300 **B** 2.34 Boron 10.811 ±5	6 ±4,2 4470 4100 **C** 2.62 Carbon 12.011	7 ±3,5,4,2 77.35 63.14 **N** 1.251¤ Nitrogen 14.006 74 ±7	8 -2 90.18 50.35 **O** 1.429¤ Oxygen 15.999 4 ±3	9 -1 84.95 53.48 **F** 1.696¤ Fluorine 18.998 403 2 ±9	10 27.10 24.55 **Ne** 0.901¤ Neon 20.179 7 ±6
13 +3 2793 933.25 **Al** 2.7 Aluminum 26.981 539 ±5	14 +4 3540 1685 **Si** 2.33 Silicon 28.085 5 ±3	15 ±3,5,4 550 317.3 **P** 1.82 Phosphorus 30.973 762 ±4	16 ±2,4,6 717.8 388.4 **S** 2.07 Sulfur 32.066 ±6	17 ±1,3,5,7 239.1 172.2 **Cl** 3.17¤ Chlorine 35.452 7 ±9	18 87.3 83.81 **Ar** 1.784¤ Argon 39.948

10	11	12	13	14	15	16	17	18
+2,3 **Ni** kel 4 ±2	29 +2,1 2836 1358 **Cu** 8.96 Copper 63.546 ±3	30 +2 1180 693 **Zn** 7.14 Zinc 65.39 ±2	31 +3 2478 303 **Ga** 5.91 Gallium 69.723	32 +4 3107 1210 **Ge** 5.32 Germanium 72.61 ±2	33 ±3,5 876 **As** 5.72 Arsenic 74.921 59 ±2	34 -2,4,6 958 494 **Se** 4.80 Selenium 78.96 ±3	35 ±1,5 332.25 265.90 **Br** 3.12 Bromine 79.904	36 119.80 115.78 **Kr** 3.74¤ Krypton 83.80
+2,4 **Pd** dium .42	47 +1 2436 1234 **Ag** 10.5 Silver 107.868 2 ±2	48 +2 1040 594 **Cd** 8.65 Cadmium 112.411 ±8	49 +3 2346 430 **In** 7.31 Inidum 114.818 ±3	50 +4,2 2876 505 **Sn** 7.30 Tin 118.710 ±7	51 ±3,5 1860 904 **Sb** 6.68 Antimony 121.760	52 -2,4,6 1261 723 **Te** 6.24 Tellurium 127.60 ±3	53 ±1,5,7 458 387 **I** 4.92 Iodine 126.904 47 ±3	54 165 161 **Xe** 5.89¤ Xenon 131.29 ±2
+2,4 **Pt** num 8 ±3	79 +3,1 3130 1338 **Au** 19.3 Gold 196.966 54 ±3	80 +2,1 630 234 **Hg** 13.5 Mercury 200.59 ±2	81 +3,1 1746 577 **Tl** 11.85 Thallium 204.383 3 ±2	82 +4,2 2023 601 **Pb** 11.4 Lead 207.2	83 ±3,5 1837 545 **Bi** 9.8 Bismuth 208.980 37 ±3	84 +4,2 1235 527 **Po** 9.4 Polonium (~210)	85 ±1,3,5,7 610 575 **At** Astatine (~210)	86 211 202 **Rn** 9.91¤ Radon (~222)

+3 **Gd** Minium 25 ±3	65 +3 3496 1630 **Tb** 8.27 Terbium 158.925 34 ±3	66 +3 2835 1682 **Dy** 8.54 Dysprosium 162.50 ±3	67 +3 2968 1743 **Ho** 8.80 Holmium 164.930 32 ±3	68 +3 3136 1795 **Er** 9.05 Erbium 167.26 ±3	69 +3,2 2220 1818 **Tm** 9.33 Thulium 168.934 21 ±3	70 +3,2 1467 1097 **Yb** 6.98 Ytterbium 173.04 ±3	71 +3 3668 1936 **Lu** 9.84 Lutetium 174.967
+3 **Cm** ium 47)	97 +4,3 **Bk** Berkelium (247)	98 +3 900 **Cf** Californium (251)	99 **Es** Einsteinium (254)	100 **Fm** Fermium (257)	101 **Md** Mendelevium (260)	102 **No** Nobelium (259)	103 **Lr** Lawrencium (262)

Exploring Chemical Analysis

["The Experiment" by Sempé. Copyright C. Charillon, Paris.]

Exploring Chemical Analysis

Daniel C. Harris

Michelson Laboratory
China Lake, California

W. H. Freeman and Company
New York

Acquisitions Editor: Deborah Allen
Project Editor: Mary Louise Byrd
Text and Cover Designer: Blake Logan
Illustration Coordinator: Bill Page
Illustrator: Network Graphics
Production Coordinator: Julia DeRosa
Composition: York Graphic Services
Manufacturing: RR Donnelley & Sons Company

Library of Congress Cataloging-in-Publication Data

Harris, Daniel C., 1948–
    Exploring chemical analysis / Daniel C. Harris.
      p.   cm.
    Includes bibliographical references (p.   –) and index.
    ISBN 0-7167-3042-1
    1. Chemistry, Analytic.   I. Title.
  QD75.2.H368   1996
  543—dc20

96-10937
CIP

Printed in the United States of America

First printing, 1996

Exploring Chemical Analysis

Exploring Chemical Analysis

My goal for this book was to provide a *short, engaging, elementary* introduction to analytical chemistry for students whose primary interests generally lie outside of chemistry. You might use this book in your analytical chemistry course or in lower-level courses such as freshman laboratory, when a brief introduction to quantitative analysis is required. You will find a distinct flavor of biological and environmental subject matter that I hope will capture your interest and give analytical chemistry a vital place in your studies.

Exploring Chemical Analysis covers modern analytical chemistry, not just wet chemical analysis. Major topics include statistics, volumetric analysis, spectrophotometry, acid-base equilibrium, electrochemistry and potentiometry, chromatography and electrophoresis, gravimetric and combustion analysis, and atomic spectroscopy. Laboratory experiments are provided for most major analytical methods, with emphasis on environmental and biological samples.

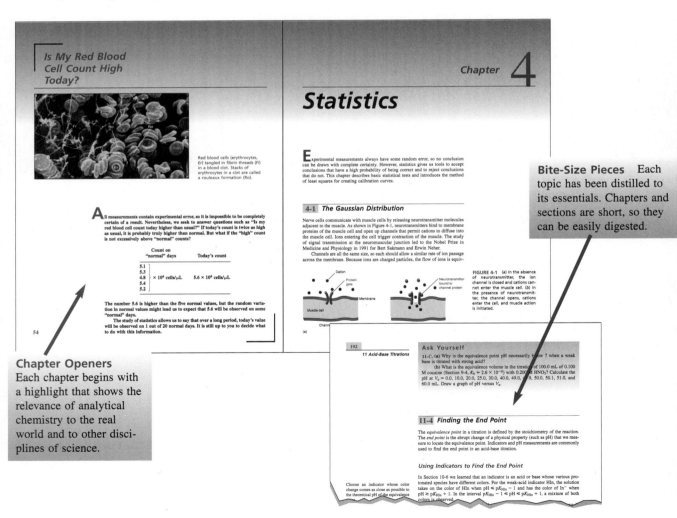

Is My Red Blood Cell Count High Today?

Red blood cells (erythrocytes, Er) tangled in fibrin threads (Fi) in a blood clot. Stacks of erythrocytes in a clot are called a rouleaux formation (Ro).

All measurements contain experimental error, so it is impossible to be completely certain of a result. Nevertheless, we seek to answer questions such as "Is my red blood cell count today higher than usual?" If today's count is twice as high as usual, it is probably truly higher than normal. But what if the "high" count is not excessively above "normal" counts?

Count on "normal" days	Today's count
5.1	
5.3	
4.8 $\times 10^6$ cells/μL	5.6 $\times 10^6$ cells/μL
5.4	
5.2	

The number 5.6 is higher than the five normal values, but the random variation in normal values might lead us to expect that 5.6 will be observed on some "normal" days.

The study of statistics allows us to say that over a long period, today's value will be observed on 1 out of 20 normal days. It is still up to you to decide what to do with this information.

54

Chapter 4

Statistics

Experimental measurements always have some random error, so no conclusion can be drawn with complete certainty. However, statistics gives us tools to accept conclusions that have a high probability of being correct and to reject conclusions that do not. This chapter describes basic statistical tests and introduces the method of least squares for creating calibration curves.

4-1   The Gaussian Distribution

Nerve cells communicate with muscle cells by releasing neurotransmitter molecules adjacent to the muscle. As shown in Figure 4-1, neurotransmitters bind to membrane proteins of the muscle cell and open up channels that permit cations to diffuse into the muscle cell. Ions entering the cell trigger contraction of the muscle. The study of signal transmission at the neuromuscular junction led to the Nobel Prize in Medicine and Physiology in 1991 for Bert Sakmann and Erwin Neher.

Channels are all the same size, so each should allow a similar rate of ion passage across the membrane. Because ions are charged particles, the flow of ions is equiv-

FIGURE 4-1 (a) In the absence of neurotransmitter, the ion channel is closed and cations cannot enter the muscle cell. (b) In the presence of neurotransmitter, the channel opens, cations enter the cell, and muscle action is initiated.

192

11 Acid-Base Titrations

Ask Yourself

11-C. (a) Why is the equivalence point pH necessarily below 7 when a weak base is titrated with strong acid?

(b) What is the equivalence volume in the titration of 100.0 mL of 0.100 M cocaine (Section 9-4, $K_b = 2.6 \times 10^{-6}$) with 0.200 M HNO_3? Calculate the pH at V_a = 0.0, 10.0, 20.0, 25.0, 30.0, 40.0, 49.0, 49.9, 50.0, 50.1, 51.0, and 60.0 mL. Draw a graph of pH versus V_a.

11-4   Finding the End Point

The *equivalence point* in a titration is defined by the stoichiometry of the reaction. The *end point* is the abrupt change of a physical property (such as pH) that we measure to locate the equivalence point. Indicators and pH measurements are commonly used to find the end point in an acid-base titration.

Using Indicators to Find the End Point

In Section 10-6 we learned that an indicator is an acid or base whose various protonated species have different colors. For the weak-acid indicator HIn, the solution takes on the color of HIn when pH ≤ pK_{HIn} − 1 and has the color of In− when pH ≥ pK_{HIn} + 1. In the interval pK_{HIn} − 1 ≤ pH ≤ pK_{HIn} + 1, a mixture of both colors is observed.

Choose an indicator whose color change comes as close as possible to the theoretical pH of the equivalence

Bite-Size Pieces   Each topic has been distilled to its essentials. Chapters and sections are short, so they can be easily digested.

Chapter Openers
Each chapter begins with a highlight that shows the relevance of analytical chemistry to the real world and to other disciplines of science.

Demonstrations and Color Plates I can't come to your classroom to present chemical demonstrations, but I can tell you about some of my favorites and show you color photos of how they look. Color photos are located near the center of the book.

Ask Yourself Sections end with a brief, straightforward problem or question entitled "Ask Yourself," which reinforces the main theme of the section. These are the first problems that you should do as you work your way through this course. Complex material is broken down into elementary steps in these questions. Complete solutions to "Ask Yourself" questions can be found at the back of this book.

Spreadsheets Spreadsheets are introduced as an optional tool. They are useful in this course and will serve you well in many ways outside this course.

Demonstration 13-2

The Human Salt Bridge

A salt bridge is an ionic medium through which ions diffuse to maintain electroneutrality in each compartment of an electrochemical cell. One way to prepare a salt bridge is to heat 3 g of agar (the stuff used to grow bacteria in a Petri dish) with 30 g of KCl in 100 mL of water until a clear solution is obtained. The solution is poured into a U-tube and allowed to gel. The bridge is stored in saturated aqueous KCl.

To conduct this demonstration, set up the galvanic cell in Figure 13-2 with 0.1 M $ZnCl_2$ on the left side and 0.1 M $CuSO_4$ on the right side. You can use a voltmeter or a pH meter to measure voltage. If you use a pH meter, the positive terminal is the socket to which the glass electrode goes and the negative terminal is the reference electrode socket.

You should be able to write the two half-reactions for this cell and use the Nernst equations (13-6 and 13-7) to calculate the theoretical voltage. Measure the voltage with a conventional salt bridge. Then replace the salt bridge with one made of filter paper freshly soaked in NaCl solution and measure the voltage again. Finally, replace the filter paper with two fingers of the same hand and measure the voltage again. Your body is really just a bag of salt housed in a *semipermeable membrane* (skin) through which ions can diffuse. Small differences in voltage observed when the salt bridge is replaced can be attributed to the junction potential discussed in Section 14-2.

Challenge One hundred and eighty students at Virginia Polytechnic Institute and State University made a salt bridge by holding hands. (Their electrical resistance was lowered by a factor of 100 by wetting everyone's hands.) Can your school beat this record?

Line Notation

A shorthand notation employing just two symbols is commonly used to describe electrochemical cells:

| phase boundary || salt bridge

Figure 13-2 is represented by the *line diagram*.

$$Zn(s) \mid ZnCl_2(aq) \parallel CuSO_4(aq) \mid Cu(s)$$

boundary is indicated by a vertical line. The electrodes are shown at the - and right-hand sides of the diagram. The contents of the salt bridge fied.

Interpreting Line Diagrams of Cells

diagram for the cell in Figure 13-3. Write an anode reaction (an oxi-e left half-cell, a cathode reaction (a reduction) for the right half-cell, ell reaction.

235

160

9 Monoprotic Acid-Base Equilibria

Ask Yourself

9-D. (a) What is the pH and fraction of association of a 0.100 M solution of a weak base with $K_b = 1.00 \times 10^{-5}$?

$$\text{fraction of association} = \frac{[BH^+]}{[BH^+] + [B]} = 0.020$$

(b) A 0.10 M solution of a base has pH = 9.28. Find K_b.
(c) A 0.10 M solution of a base is 2.0% associated. Find K_b.

9-5 Fractional Composition Equations

We wish to find expressions for the fraction of an acid in each form (HA and A^-). These equations will be useful in future problems involving acids and bases. Consider an acid with formal concentration F:

$$HA \xrightleftharpoons{K_a} H^+ + A^- \qquad K_a = \frac{[H^+][A^-]}{[HA]}$$

We know that the concentrations of HA and A^- must add up to F because these are the only two species containing the group A. This relationship is called a *mass balance*:

The mass balance would more appro- Mass balance: $F = [HA] + [A^-]$

e gives $[A^-] = F - [HA]$, which can be plugged into

$$K_a = \frac{[H^+](F - [HA])}{[HA]}$$

$$[HA] = \frac{[H^+]F}{[H^+] + K_a} \qquad (9\text{-}14)$$

es in the form HA is called α_{HA}.

$$_{HA} = \frac{[HA]}{[HA] + [A^-]} = \frac{[HA]}{F}$$

F gives

$$_{HA} = \frac{[HA]}{F} = \frac{[H^+]}{[H^+] + K_a} \qquad (9\text{-}15)$$

tion in the form A^-, designated α_{A^-}, can be obtained:

$$_{A^-} = \frac{[A^-]}{F} = \frac{K_a}{[H^+] + K_a} \qquad (9\text{-}16)$$

	A	B	C
1	Constants:	Temp (°C)	Density (g/mL)
2	a0 =	5	0.99997
3	0.99989	10	0.99970
4	a1 =	15	0.99911
5	5.3322E-05	20	0.99821
6	a2 =	25	0.99705
7	-7.5899E-06	30	0.99565
8	a3 =	35	0.99403
9	3.6719E-08	40	0.99223
10			
11	Formula:		
12	C2 = $A$3+($A$5*B2)+($A$7*B2^2)+($A$9*B2^3)		

Ask Yourself

3-E. Can you reproduce the spreadsheet in Figure 3-3 on your computer? The boiling point of nitrogen (N_2) at 1 atm pressure is −196°C. Use your spreadsheet to find the kelvin and degrees Fahrenheit equivalents of −196°C. Check your answers with your calculator.

3-6 Spreadsheet + Graph = Power

Although a spreadsheet has great power for crunching numbers, humans require a visual display to appreciate the relationship between two columns of numbers. Therefore, results of a spreadsheet calculation are plotted directly by the spreadsheet program or can be pasted into a graphing program for display.

Figure 3-4 shows a spreadsheet for computing the density of water as a function of temperature (°C) with the equation

$$\text{density (g/mL)} = a_0 + a_1 \cdot T + a_2 \cdot T^2 + a_3 \cdot T^3 \qquad (3\text{-}8)$$

where $a_0 = 0.999\,89$, $a_1 = 5.332\,2 \times 10^{-5}$, $a_2 = -7.589 \times 10^{-6}$, and $a_3 = 3.671\,9 \times 10^{-8}$. The constants a_0 to a_3 are entered in column A. Column B is labeled "Temp (°C)" and column C is labeled "Density (g/mL)". Enter values of temperature in column B. In column C2, type the formula "=$A$3+($A$5*B2)+($A$7*B2^2)+($A$9*B2^3)", which uses the exponent symbol ^ to compute T^2 and T^3. When you enter the formula, the number 0.999 97 is computed in cell C2. The remainder of column C is completed with a FILL DOWN (or COPY) command. The spreadsheet is not finished until it is documented by entering text in cells A11 and A12 to show what formula was used in column C.

To see the relationship between the input in column B and the output in column C we need a graph. At this point, you will need instruction for using your spreadsheet software or a graphing program to plot column C on the y-axis (the ordinate) and column B on the x-axis (the abscissa). Results displayed in Figure 3-5 show that the change in density from 5° to 15°C is small relative to the change in density between 30° and 40°C.

3-6 Spreadsheet + Graph = Power

FIGURE 3-4 Spreadsheet for computing the density of water as a function of temperature.

FIGURE 3-5 Density of water computed with the spreadsheet in Figure 3-4.

Equation 3-8 is accurate to five decimal places over the range 4° to 40°C.

A spreadsheet is not complete until it is documented.

Marginal Notes   Don't miss them! This learning tool provides emphasis and amplification of specific points in the text, as well as occasional questions for you to answer as you read this book.

Boxes   *Explanation* boxes expand or explain key points from the text. *Informed Citizen* boxes present interesting supplementary descriptive material that helps you to see how analytical chemistry is applied to real problems of importance to real people.

Vocabulary and Glossary   Important terms are in **bold** type and listed at the end of each chapter. Definitions are found in the Glossary at the back of the book.

Hydrolysis is the reaction of anything with water. Specifically, the reaction $L^- + H_2O \rightleftharpoons HL + OH^-$ is called hydrolysis.

L^- can be treated as monobasic with $K_b = K_{b1}$.

A tougher problem.

K_{b1} tells us that L^- will not **hydrolyze** (react with water) very much to give HL. Furthermore, K_{b2} tells us that the resulting HL is such a weak base that hardly any further reaction to make H_2L^+ will occur.

We therefore treat L^- as a monobasic species, with $K_b = K_{b1}$. The results of this (fantastic) approximation can be outlined as follows:

$$\underset{\underset{0.050-x}{L^-}}{H_2NCHCO_2^-} + H_2O \xrightleftharpoons{K_{b1} = 5.59 \times 10^{-5}} \underset{\underset{x}{HL}}{H_3NCHCO_2^-} + \underset{x}{OH^-}$$

$$\frac{x^2}{F-x} = 5.59 \times 10^{-5} \Rightarrow x = [OH^-] = 1.64 \times 10^{-3}\ M$$

$$[HL] = x = 1.64 \times 10^{-3}\ M$$

$$[H^+] = K_w/[OH^-] = 6.08 \times 10^{-12}\ M \Rightarrow pH = 11.22$$

$$[L^-] = F - x = 4.84 \times 10^{-2}\ M$$

The concentration of H_2L^+ can be found from the K_{b2} equilibrium (12-6).

$$HL + H_2O \xrightleftharpoons{K_{b2}} H_2L^+ + OH^- \quad K_{b2} = \frac{[H_2L^+][OH^-]}{[HL]} = \frac{[H_2L^+]x}{x} = [H_2L^+]$$

We find that $[H_2L^+] = K_{b2} = 2.13 \times 10^{-12}\ M$, and the approximation that $[H_2L^+]$ is insignificant relative to [HL] is well justified. In summary, if there is any reasonable separation between K_1 and K_2 (and, therefore, between K_{b1} and K_{b2}), *the fully basic form of a diprotic acid can be treated as monobasic, with $K_b = K_{b1}$.*

The Intermediate Form, HL

A solution prepared from leucine, HL, is more complicated than one prepared from either H_2L^+ or L^-, because HL is both an acid and a base.

$$HL \rightleftharpoons H^+ + L^- \qquad K_a = K_2 = 1.79 \times 10^{-10} \qquad (12\text{-}7)$$
$$HL + H_2O \rightleftharpoons H_2L^+ + OH^- \qquad K_b = K_{b2} = 2.13 \times 10^{-12} \qquad (12\text{-}8)$$

...ule that can both donate and accept a proton is said to be **amphiprotic**. The ...sociation reaction (12-7) has a larger equilibrium constant than the base ...is reaction (12-8), so we expect the solution of leucine to be acidic. ...ever, we cannot simply ignore Reaction 12-8, even if K_a and K_b differ by ...orders of magnitude. Both reactions proceed to a nearly equal extent, ...uced in Reaction 12-7 reacts with OH^- from Reaction 12-8, thereby driving ...12-8 to the right. ...reat this case correctly, we resort to writing a *charge balance* (Section 11-7), ...ys that the sum of positive charges in solution must equal the sum of neg-...arges in solution. The procedure is applied to leucine, whose intermediate form ...L) has no net charge. However, the results apply to the intermediate form ...iprotic acid, regardless of its charge.

Box 10-1 *Explanation*

Strong Plus Weak Reacts Completely

A strong acid reacts with a weak base "completely" because the equilibrium constant is large:

$$\underset{\substack{\text{Weak}\\ \text{base}}}{B} + \underset{\substack{\text{Strong}\\ \text{acid}}}{H^+} \rightleftharpoons BH^+ \quad K = \frac{1}{K_a} \quad (\text{for } BH^+)$$

If B is tris, then the equilibrium constant for reaction with HCl is

$$K = \frac{1}{K_a} = \frac{1}{10^{-8.075}} = 1.2 \times 10^8 \leftarrow \text{A big number}$$

A strong base reacts "completely" with a weak acid because the equilibrium constant is, again, very large:

$$\underset{\substack{\text{Strong}\\ \text{base}}}{OH^-} + \underset{\substack{\text{Weak}\\ \text{acid}}}{HA} \rightleftharpoons A^- + H_2O \quad K = \frac{1}{K_b} \quad (\text{for } A^-)$$

If HA is acetic acid, then the equilibrium constant for reaction with NaOH is

$$K = \frac{1}{K_b} = \frac{K_a \text{ for HA}}{K_w} = 1.7 \times 10^9 \leftarrow \text{Another big number}$$

The reaction of a strong acid with a strong base is even more complete than a strong plus weak reaction:

$$\underset{\substack{\text{Strong}\\ \text{acid}}}{H^+} + \underset{\substack{\text{Strong}\\ \text{base}}}{OH^-} \rightleftharpoons H_2O \quad K = \frac{1}{K_w} = 10^{14} \leftarrow \underset{\text{Ridiculously big!}}{}$$

If you mix a strong acid, a strong base, a weak acid, and a weak base, the strong acid and base will neutralize each other until one is used up. The remaining strong acid or base will then react with the weak base or weak acid.

Question   Does the pH change in the right direction when HCl is added?

$$pH = pK_a + \log\left(\frac{\text{moles of B}}{\text{moles of BH}^+}\right) = 8.075 + \log\left(\frac{0.090\ 6}{0.041\ 6}\right) = 8.41$$

...e of solution is irrelevant.

...receding example illustrates that *the pH of a buffer does not change very* ...*n a limited amount of a strong acid or base is added.* Addition of 12.0 ...M HCl changed the pH from 8.61 to 8.41. Addition of 12.0 mL of 1.00 ...1.00 L of unbuffered solution would have lowered the pH to 1.93. ...hy does a buffer resist changes in pH? **It does so because the strong acid** ...**consumed by B or BH⁺.** If you add HCl to tris, B is converted to BH⁺. ...NaOH, BH⁺ is converted to B. As long as you don't use up the B or ...dding too much HCl or NaOH, the log term of the Henderson-Hassel-...ation does not change very much and the pH does not change very much. ...tion 10-1 illustrates what happens when the buffer does get used up. The ...its maximum capacity to resist changes of pH when pH = pK_a. We will ...his point later.

...ourself

...) What is the pH of a solution prepared by dissolving 10.0 g tris plus ...ris hydrochloride in 0.250 L water? ...) What will the pH be if 10.5 mL of 0.500 M $HClO_4$ is added to (a)? ...) What will the pH be if 10.5 mL of 0.500 M NaOH is added to (a)?

Calibration curve

You should be able to use Equation 4-16 to find the uncertainty in a result derived from a calibration curve.

Important Terms

average	Gaussian distribution	slope	t test
blank	mean	standard deviation	y-intercept
calibration curve	method of least squares	standard solution	
confidence interval	Q test	Student's t	

Problems

4-1. What is the relation between the standard deviation and the precision of a procedure? What is the relation between standard deviation and accuracy?

4-2. What fraction of observations in an ideal Gaussian distribution lies within $\mu \pm \sigma$? Within $\mu \pm 2\sigma$? Within $\mu \pm 3\sigma$?

4-3. The ratio of the number of atoms of the isotopes ^{69}Ga and ^{71}Ga in samples from different sources is listed in the following table:

Sample	$^{69}Ga/^{71}Ga$	Sample	$^{69}Ga/^{71}Ga$
1	1.526 60	5	1.528 94
2	1.529 74	6	1.528 04
3	1.525 92	7	1.526 85
4	1.527 31	8	1.527 93

(a) Find the mean value of $^{69}Ga/^{71}Ga$.

(b) Find the standard deviation and relative standard deviation.

(c) Sample 8 was analyzed seven times, with $\bar{x} = 1.527\ 93$ and $s = 0.000\ 07$. Find the 99% confidence interval for sample 8.

4-4. What is the meaning of a confidence interval?

4-5. For the numbers 116.0, 97.9, 114.2, 106.8, and 108.3, find the mean, standard deviation, and 90% confidence interval for the mean.

4-6. The calcium content of a mineral was analyzed five times by each of two methods. Are the mean values significantly different at the 95% confidence level?

Method	Ca (percent composition)
1	0.027 1, 0.028 2, 0.027 9, 0.027 1, 0.027 5
2	0.027 1, 0.026 8, 0.026 3, 0.027 4, 0.026 9

4-7. Find the 95 and 99% confidence intervals for the mean mass of nitrogen from chemical sources, Table 4-3.

4-8. Two methods were used to measure the specific activity (units of enzyme activity per milligram of protein) of an enzyme. One unit of enzyme activity is defined as the amount of enzyme that catalyzes the formation of one micromole of product per minute under specified conditions.

Method	Enzyme activity (five replications)				
1	139	147	160	158	135
2	148	159	156	164	159

Is the mean value of method 1 significantly different from the mean value of method 2 at the 95% confidence level?

4-9. Students measured the concentration of HCl in a solution by various titrations in which different indicators were used to find the end point.

Indicator	Mean HCl concentration (M) (± standard deviation)	Number of measurements
1. Bromothymol blue	0.095 65 ±0.002 25	28
2. Methyl red	0.086 86 ±0.000 98	18
3. Bromocresol green	0.086 41 ±0.001 13	29

Worked Examples These appear frequently throughout the book to explain, amplify, and reinforce most types of calculations.

EXAMPLE Finding pK_a of a Weak Acid

A 0.100 M solution of hydrazoic acid is found to have pH = 2.83. Find pK_a for this acid.

$$:\ddot{N}=\overset{..}{N}=\overset{..}{N}-H \ \overset{K_a}{\rightleftharpoons}\ :\ddot{N}=N=\ddot{N}:^- \ +\ H^+ $$

Hydrazoic acid   Azide (N_3^-)
0.100 − x   x   x

SOLUTION We know that $[H^+] = 10^{-pH} = 10^{-2.83} = 1.4_8 \times 10^{-3}$ M. Because $[N_3^-] = [H^+]$ in this solution, $[N_3^-] = 1.48 \times 10^{-3}$ M and $[HN_3] = 0.100 - (1.4_8 \times 10^{-3}) = 0.098\ 5$ M. From these concentrations, we calculate K_a and pK_a:

$$ K_a = \frac{[N_3^-][H^+]}{[HN_3]} = \frac{(1.48 \times 10^{-3})^2}{0.098\ 5} = 2.2_2 \times 10^{-5} $$
$$ \Rightarrow pK_a = -\log(2.2_2 \times 10^{-5}) = 4.65 $$

Ask Yourself

9-C. (a) What is the pH and fraction of dissociation of a 0.100 M solution of the weak acid HA with $K_a = 1.00 \times 10^{-5}$?
(b) A 0.045 0 M solution of HA has a pH of 2.78. What is pK_a for this acid?
(c) A 0.045 0 M solution of HA is 0.60% dissociated. What is pK_a for this acid?

9-4 **Weak-Base Equilibrium**

Problems and Answers Each chapter has a set of problems to provide practice and build understanding of analytical chemistry. Short answers to numerical problems can be found at the back of the book. My philosophy is that you are entitled to immediate feedback after struggling through a homework problem. "Did I get it right or have I missed something?"

Problems   17

in (a) watts; (b) joules per second; (c) calories per second; (d) calories per hour.

1-8. (a) Refer to Table 1-4 and calculate how many meters are in 1 inch. How many inches are in 1 m?
(b) A mile is 5 280 feet and a foot is 12 inches. The speed of sound in the atmosphere at sea level is 345 m/s. Express the speed of sound in mi/s and mi/hr.
(c) There is a delay between lightning and thunder in a storm, because light reaches us almost at once, but sound is slower. How many meters, kilometers, and miles away is lightning if the sound reaches you 3.00 s after the light?

1-9. Define the following measures of concentration:
(a) molarity   (e) volume percent
(b) molality   (f) parts per million
(c) density   (g) parts per billion
(d) weight percent   (h) formal concentration

1-10. What is the formal concentration (expressed as mol/L = M) of NaCl when 32.0 g is dissolved in water and diluted to 0.500 L?

1-11. If 0.250 L of aqueous solution with a density of 1.00 g/mL contains 13.7 μg of pesticide, express the concentration of pesticide in (a) ppm and (b) ppb.

1-12. The concentration of the sugar glucose ($C_6H_{12}O_6$) in human blood ranges from about 80 mg/dL before meals up to 120 mg/dL after eating. Find the molarity of glucose in blood before and after eating.

1-13. (a) How many grams of perchloric acid, $HClO_4$, are contained in 100.0 g of 70.5 wt % aqueous perchloric acid?
(b) How many grams of water are in 100.0 g of solution?
(c) How many moles of $HClO_4$ are in 100.0 g of solution?

1-14. How many grams of boric acid [$B(OH)_3$, MW 61.83] should be used to make 2.00 L of 0.050 0 M solution?

1-15. Water is fluoridated to prevent tooth decay.
(a) How many liters of 1.5 M KF should be added to a reservoir with a diameter of 100 m and a depth of 20 m to give 1.2 ppm F$^-$?
(b) How many grams of solid KF should be added to the same reservoir to give 1.2 ppm F$^-$?

1-16. How many grams of 50 wt % NaOH (FW 40.00) should be diluted to 1.00 L to make 0.10 M NaOH?

1-17. A bottle of concentrated aqueous sulfuric acid, labeled 98.0 wt % H_2SO_4, has a concentration of 18.0 M.
(a) How many milliliters of reagent should be diluted to 1.00 L to give 1.00 M H_2SO_4?
(b) Calculate the density of 98.0 wt % H_2SO_4.

1-18. How many grams of methanol (CH_3OH, MW 32.04) are contained in 0.100 L of 1.71 M aqueous methanol?

1-19. A dilute aqueous solution containing 1 ppm of solute has a density of 1.00 g/mL. Express the concentration of solute in g/L, μg/L, μg/mL, and mg/L.

1-20. The concentration of $C_{20}H_{42}$ (MW 282.55) in winter rainwater in Figure 1-4 is 0.2 ppb. Assume that the density of rainwater is close to 1.00 g/mL; find the molar concentration of $C_{20}H_{42}$.

1-21. A 95.0 wt % solution of ethanol (CH_3CH_2OH, MW 46.07) in water has a density of 0.804 g/mL.
(a) Find the mass of 1.00 L of this solution and the grams of ethanol per liter.
(b) What is the molar concentration of ethanol in this solution?

1-22. (a) How many grams of nickel are contained in 10.0 g of a 10.2 wt % solution of nickel sulfate hexahydrate, $NiSO_4 \cdot 6H_2O$ (FW 262.85)?
(b) The concentration of this solution is 0.412 M. Find the density.

1-23. A 500.0-mL solution was prepared by dissolving 25.00 mL of methanol (CH_3OH, density = 0.791 4 g/mL) in chloroform. Find the molarity of the methanol.

1-24. Describe how to prepare exactly 100 mL of 1.00 M HCl from 12.1 M HCl reagent.

1-25. Cesium chloride is used to prepare dense solutions required for isolating cellular components with a centrifuge. A 40.0 wt % solution of CsCl (FW 168.36) has a density of 1.43 g/mL.
(a) Find the molarity of CsCl.
(b) How many milliliters of the concentrated solution should be diluted to 500 mL to make 0.100 M CsCl?

1-26. Protein and carbohydrates provide 4.0 Cal/g, whereas fat gives 9.0 Cal/g. (Remember that 1 Calorie, with a capital C, is really 1 kcal.) The weight percents of these components in some foods are

Food	Protein wt %	Carbohydrate wt %	Fat wt %
shredded wheat	9.9	79.9	—
doughnut	4.6	51.4	18.6
hamburger (cooked)	24.2	—	20.3
apple	—	12.0	—

Calculate the number of calories per gram and calories per ounce in each of these foods. (Use Table 1-4 to convert grams into ounces, remembering that there are 16 ounces in 1 pound.)

om disso-
one BH$^+$

(9-12)

A weak-base problem has the same algebra as a weak-acid problem, except $K = K_b$ and $x = [OH^-]$.

Ask Yourself

12-D. Consider the titration of 50.0 mL of 0.050 0 M malonic acid with 0.100 M NaOH.
(a) How many milliliters of titrant are required to reach each equivalence point?
(b) Calculate the pH at V_b = 0.0, 12.5, 25.0, 37.5, 50.0, and 55.0 mL.
(c) Put the points from (b) on a graph and sketch the titration curve.

Key Equations

Diprotic acid equilibria	$H_2A \rightleftharpoons HA^- + H^+$ $K_{a1} = K_1$
	$HA^- \rightleftharpoons A^{2-} + H^+$ $K_{a2} = K_2$
Diprotic base equilibria	$A^{2-} + H_2O \rightleftharpoons HA^- + OH^-$ K_{b1}
	$HA^- + H_2O \rightleftharpoons H_2A + OH^-$ K_{b2}
Relation between K_a and K_b	Monoprotic system $K_aK_b = K_w$ Triprotic system $K_{a1}K_{b3} = K_w$
	Diprotic system $K_{a1}K_{b2} = K_w$ $K_{a2}K_{b2} = K_w$
	$K_{a2}K_{b1} = K_w$ $K_{a3}K_{b1} = K_w$
pH of H_2A (or BH_2^{2+})	$H_2A \overset{K_{a1}}{\rightleftharpoons} H^+ + HA^-$ $\dfrac{x^2}{F-x} = K_{a1}$
	$F-x$ x x

(This gives [H$^+$], [HA$^-$], and [H$_2$A]. You can solve for [A^{2-}] from the K_{a2} equilibrium.)

pH of HA$^-$ (or diprotic BH$^+$)	$pH = \frac{1}{2}(pK_1 + pK_2)$
pH of A^{2-} (or diprotic B)	$A^{2-} + H_2O \overset{K_{b1}}{\rightleftharpoons} HA^- + OH^-$ $\dfrac{x^2}{F-x} = K_{b1} = \dfrac{K_w}{K_{a2}}$
	$F-x$ x x

(This gives [OH$^-$], [HA$^-$], and [A^{2-}]. You can find [H$^+$] from the K_w equilibrium and [H$_2$A] from the K_{a1} equilibrium.)

Diprotic buffer	$pH = pK_1 + \log\left(\dfrac{[HA^-]}{[H_2A]}\right)$ $pH = pK_2 + \log\left(\dfrac{[A^{2-}]}{[HA^-]}\right)$

Both equations are always true and either can be used, depending on which set of concentrations you happen to know.

Titration of H$_2$A with OH$^-$		
$V_b = 0$	Find pH of H_2A	$V_b = \frac{3}{2}V_{e1}$ $pH = pK_2$
$V_b = \frac{1}{2}V_{e1}$ $pH = pK_1$		$V_b = V_{e2}$ Find pH of A^{2-}
$V_b = V_{e1}$ $pH \approx \frac{1}{2}(pK_1 + pK_2)$		$V_b > V_{e2}$ Find concentration of excess OH$^-$

Key Equations Key equations are highlighted in the text and are collected at the end of each chapter. If you understand how to use these equations, you have probably mastered much of the material in the chapter.

Contents

Chapter 4   *Statistics   55*

Chapter 5   *A Skirmish with Equilibrium 73*

Chapter 6   *Good Titrations   95*

What Am I Doing in This Course?

Linda Hughes

A recurring theme in this textbook is the use of analytical chemistry to study the ecology of a saltwater aquarium. Interactions among fish, plants, and bacteria are reflected by the concentrations of nitrogen compounds, such as urea, nitrite, nitrate, and ammonia in the water. The introduction of marine ecology into the analytical chemistry classroom was pioneered by Kenneth Hughes at Georgia Tech.

For most of its users, analytical chemistry is a tool, not an end in itself. If you are interested in biology, medicine, environmental science, geology, oceanography, archeology, or paleontology, you might rely on analytical chemistry to study how a nerve cell communicates with its neighbors, to find out if your patient's blood enzyme concentrations are elevated, or to chart the temperature of the ocean over the past 50 000 years

It is possible that you will make analytical measurements yourself later in your career, but it is more likely that you will just be a consumer of analytical results. One purpose of this course is for you to understand how analysts derive the results that you will use. Another purpose is to understand the limitations (the uncertainties) of the results. A third purpose is for you to develop laboratory skills and careful work habits that you can apply in your own field.

Scientists cannot prove theories. We can only disprove them. For example, we cannot be 100% sure that greenhouse gases created by burning fossil fuels are raising the temperature of the atmosphere and will have significant effects on the climate. Nonetheless, politicians and engineers must make decisions every day to take actions in the absence of certainty. This course will help you, as an informed citizen, to appreciate how technical information is obtained and interpreted. This course should help give you a basis for positive actions in your own field of interest.

Preface

This book is intended to provide a *short, interesting, elementary* introduction to analytical chemistry for students whose primary interests generally lie outside of chemistry. I selected topics that I think should be covered in introductory courses and refrained from going into more depth than necessary.

Mastering the material in this course requires that you work problems and try laboratory experiments to gain experience and develop skills. After each section is a question entitled "Ask Yourself," which helps you begin assimilating what you have read. Solutions to these questions are at the back of the book, as are short answers to most end-of-chapter problems. Complete solutions are found in the *Solutions Manual*.

This book benefited greatly from suggestions and advice of Deborah Allen and John Haber at W. H. Freeman and Company. Jodi Simpson's copy editing smoothed my prose, corrected mistakes, and brought to life a unicorn you will meet. Mary Louise Byrd is primarily responsible for converting my manuscript into the handsome book you are reading. Blake Logan created the design and Andrew P. Zutis laid out the pages. I am grateful to the following people who read all or part of the manuscript and provided detailed comments and suggestions: David White (University of Illinois at Urbana), J. S. Johar (Wayne State College), Chu-Ngi Ho (East Tennessee State University), Herman Stein (Bronx Community College), Joseph J. Topping (Towson State University), and Russ Wiley (University of Northern Iowa). Paul Weber (Briar Cliff College) and Ed Koubek (U.S. Naval Academy) provided helpful comments on experiments that I adapted from their publications. Comments from Prof. Stein and piercing questions from my son David helped focus this book on its intended audience. Initial assessments were provided by Kenneth Hughes (Georgia Tech), Lee C. Coombs (California Polytechnic State University at San Luis Obispo), and Michael D. Hampton (University of Central Florida). Helpful discussions were held with Kenneth Hughes and, at Michelson Laboratory, with Eric Erickson, Mike Seltzer and Greg Ostrom. I thank the many people who provided original photographs. Sources of illustrations and sources of some ideas are listed in the Credits.

My wife, Sally, worked on every aspect of this book and contributed greatly to its accuracy and readability. My son Doug checked the solutions to problems, corrected mistakes in the text, and urged me repeatedly to simplify the problems.

All large analytical chemistry texts, including my own *Quantitative Chemical Analysis,* are alligators that could overwhelm you with their size and strength. *If you don't like to wrestle with alligators, this book was written for you.* I truly appreciate comments, criticisms, and suggestions from students and teachers. You can reach me at the Chemistry and Materials Branch, Research and Technology Division, Michelson Laboratory, China Lake CA 93555.

Dan Harris
September 1996

Chemical
Measurements

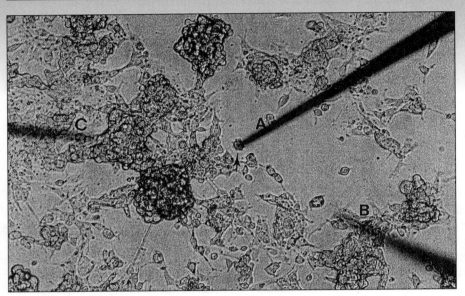

A. G. Ewing, Pennsylvania State University

Optical micrograph showing microelectrode (A) and two micropipets (B and C) positioned near a single nerve cell. The electrode measures release of neurotransmitter chemicals from the cell when the cell is stimulated with nicotine from a pipet.

Every phenomenon in science is described in terms of a small set of units that enumerate quantities such as mass, length, time, temperature, electric current, and chemical concentration. To encompass the enormous range of measurements—from the size of a quark inside a proton to the distance between galaxies—we use prefixes such as *zepto* (10^{-21}) and *tera* (10^{+12}) to multiply the basic units.

When stimulated by nicotine (yes, the stuff in tobacco), the nerve cell discharges approximately 20 bursts per second, each burst containing approximately 130 zeptomoles (130×10^{-21} moles or 80 000 molecules) of the neurotransmitter dopamine. Each discharge lasts about 10 milliseconds (10×10^{-3} seconds). To make such measurements, the electrode must respond to picoamperes (10^{-12} amperes, pronounced PEE-ko-amperes) of electric current. One of the many goals of this first chapter is to acquaint you with units and prefixes.

Chapter 1

Getting Down to Basics

Most people who practice analytical chemistry do not consider themselves to be analytical chemists. For example, chemical analysis is an essential tool used by biologists to understand how organisms function and by doctors to diagnose disease and monitor the response of a patient to treatment. Environmental scientists measure chemical changes in the atmosphere, water, and soil in response to the activities of both man and nature. Whether or not you aspire to be an analytical chemist, you are taking this course because you will either make chemical measurements yourself or you will need to understand analytical results reported by others.

Qualitative analysis is the process of identifying *what* is in a sample. **Quantitative analysis** is the process of measuring *how much* of the substance is in a sample. Most of this book deals with quantitative analysis, but many of the techniques we study also play a role in qualitative analysis. The species (chemical substance) to be identified or measured, such as chlorophyll in a leaf or iron in blood, is called the **analyte.**

This chapter begins with an overview of the analytical process. We then introduce units and prefixes needed to study analytical chemistry. You should learn to convert a measurement from one set of units to another. We define measures of chemical concentration and review calculations necessary for preparing solutions in the laboratory.

Terms to learn are written in **bold** type and listed in the Glossary.

Chemists refer to any molecule or ion as a *species.* Species is both singular and plural.

1-1 Steps in a Chemical Analysis

The object to be analyzed—such as a tree, a fossil, sediment from a lake, or a train-load of coal—is called the *lot,* or *gross sample.* Most materials of interest are **heterogeneous,** a term meaning that the composition varies from one part to another. By contrast, a **homogeneous** material is uniform throughout.

Heterogeneous: composition varies from place to place within a material
Homogeneous: composition is the same throughout the material

1

Before we can perform a meaningful chemical analysis, we must obtain a smaller *bulk sample* whose composition is representative of the lot (Figure 1-1). This process, called **sampling,** is described in Box 1-1. The bulk sample is then converted to a homogeneous *laboratory sample* from which we take small portions, called **aliquots,** for replicate chemical analyses. The purpose of repeating the analysis is to obtain an estimate of the uncertainty in the result, which is as important as the result itself. **Sample preparation** refers to the process of converting the heterogeneous bulk sample into a homogeneous laboratory sample. Sample preparation also refers to the processes in which a very dilute sample is concentrated or species that interfere with the analysis of the intended analyte are eliminated.

Interference occurs when a species other than analyte increases or decreases the response of the analytical method and makes it appear that there is more or less analyte than is actually present. **Masking** refers to the transformation of an interfering species into a form that is not detected by the analytical method. For example, Ca^{2+} in natural waters can be measured with a reagent called EDTA. Al^{3+} interferes with this analysis because it also reacts with EDTA. Therefore Al^{3+} is masked by treating the sample with excess F^- to form AlF_6^{3-} that does not react with EDTA.

$$Ca^{2+} + EDTA^{4-} \longrightarrow Ca(EDTA)^{2-}$$
$$Al^{3+} + EDTA^{4-} \longrightarrow Al(EDTA)^{-}$$

However,

$$AlF_6^{3-} + EDTA^{4-} \longrightarrow \text{no reaction}$$

Most chemical analyses have the following steps:

Sampling	**1.** Obtain a representative bulk sample from the lot.
Sample Preparation	**2.** Extract from the bulk sample a homogeneous laboratory sample.
	3. Convert the laboratory sample into a form suitable for analysis, a process that usually involves dissolving the sample. Samples with a low concentration of analyte may need to be concentrated prior to analysis.
	4. Remove or *mask* species that interfere with the chemical analysis.
Analysis	**5.** Measure the concentration of analyte in several aliquots. In some cases, different analytical methods are used with similar samples to make sure that each method gives the same result.
Interpretation	**6.** Interpret your results and draw conclusions.

Most of this book deals with analytical methods used in step 5. The chapter on statistics is vital to step 6—interpreting experimental results. Sampling and sample preparation in steps 1–4 are treated in occasional examples.

FIGURE 1-1 *Sampling* is the process of selecting a representative bulk sample from the lot. *Sample preparation* is the process that converts a bulk sample into a homogeneous laboratory sample. Sample preparation also refers to steps that eliminate interfering species or that concentrate the analyte.

Ask Yourself

1-A. (a) What is the difference between a *heterogeneous* and a *homogeneous* material?

(b) Read Box 1-1 and ask yourself: What is the difference between a *random* heterogeneous material and a *segregated* heterogeneous material?

(c) Again refer to Box 1-1 and decide what the difference is between a *random* sample and a *composite* sample. When would each be used?

Box 1-1 *Explanation*

Constructing a Representative Sample

To construct a representative bulk sample from a heterogeneous lot, you first visually divide the material into segments. A **random sample** is collected by taking portions from the desired number of segments chosen at random. If you want to measure the magnesium content of the grass in the 10-meter × 20-meter field in panel (*a*), you could divide the field into 20 000 small patches that are 10 cm on a side. After assigning a number to each small patch, you could use a computer program to pick 100 numbers at random from 1 to 20 000. The grass in each of these 100 patches is then harvested and combined to construct a representative bulk sample for analysis.

For **segregated** materials (in which different regions have different compositions), a representative **composite sample** is constructed. For example, the field in panel (*b*) has three different types of grass segregated into regions A, B, and C. You could draw a map of the field on graph paper and measure the area in each region. In this case, 66% of the area lies in region A, 14% lies in region B, and 20% lies in region C. To construct a representative bulk sample from this segregated material, take 66 of the small patches from region A, 14 from region B, and 20 from region C. You could do this by drawing random numbers from 1 to 20 000 to select patches until you have the desired number from each region. Once you have the desired number of patches in a region, you would disregard further numbers from that region.

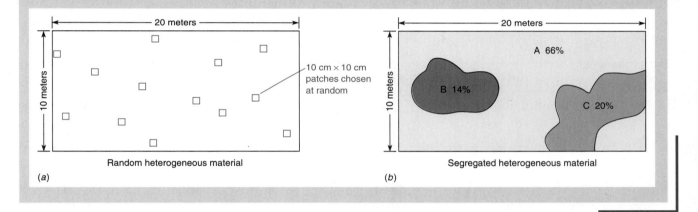

(*a*) Random heterogeneous material (*b*) Segregated heterogeneous material

1-2 *SI Units and Prefixes*

SI units of measurement derive their name from the French *Système International d'Unités*. *Fundamental units* (base units) from which all others are derived are defined in Table 1-1. Standards of length, mass, and time are the *meter* (m), *kilogram* (kg), and *second* (s), respectively. Temperature is measured in *kelvins* (K), amount of substance in *moles* (mol), and electric current in *amperes* (A). Table 1-2 lists other quantities that are defined in terms of the fundamental quantities. For example, force is measured in *newtons* (N), pressure is measured in *pascals* (Pa), and energy is measured in *joules* (J), each of which can be expressed in terms of the more fundamental units of length, time, and mass.

It is sometimes convenient to use the prefixes in Table 1-3 to express large or small quantities. For example, the pressure of dissolved oxygen in arterial blood is

Pressure is force per unit area. 1 pascal (Pa) = 1 N/m². The pressure of the atmosphere is about 100 000 Pa.

3

TABLE 1-1   **Fundamental SI units**

Quantity	Unit (symbol)	Definition
Length	meter (m)	The meter is the distance light travels in a vacuum during $\frac{1}{299\ 792\ 458}$ of a second.
Mass	kilogram (kg)	One kilogram is the mass of the prototype kilogram kept at Sèvres, France.
Time	second (s)	The second is the duration of 9 192 631 770 periods of the radiation corresponding to a certain atomic transition of ^{133}Cs.
Electric current	ampere (A)	One ampere of current produces a force of 2×10^{-7} newtons per meter of length when maintained in two straight, parallel conductors of infinite length and negligible cross section, separated by one meter in a vacuum.
Temperature	kelvin (K)	Temperature is defined such that the triple point of water (at which solid, liquid, and gaseous water are in equilibrium) is 273.16 K, and the temperature of absolute zero is 0 K.
Luminous intensity	candela (cd)	Candela is a measure of luminous intensity visible to the human eye.
Amount of substance	mole (mol)	One mole is the number of particles equal to the number of atoms in exactly 0.012 kg of ^{12}C (approximately $6.022\ 136\ 7 \times 10^{23}$).
Plane angle	radian (rad)	There are 2π radians in a circle.
Solid angle	steradian (sr)	There are 4π steradians in a sphere.

TABLE 1-2   **SI-derived units with special names**

Quantity	Unit	Symbol	Expression in terms of other units	Expression in terms of SI base units
Frequency	hertz	Hz		1/s
Force	newton	N		$m \cdot kg/s^2$
Pressure	pascal	Pa	N/m^2	$kg/(m \cdot s^2)$
Energy, work, quantity of heat	joule	J	$N \cdot m$	$m^2 \cdot kg/s^2$
Power, radiant flux	watt	W	J/s	$m^2 \cdot kg/s^2$
Quantity of electricity, electric charge	coulomb	C		$s \cdot A$
Electric potential, potential difference, electromotive force	volt	V	W/A	$m^2 \cdot kg/(s^3 \cdot A)$
Electric resistance	ohm	Ω	V/A	$m^2 \cdot kg/(s^3 \cdot A)$

approximately 1.3×10^4 Pa. Table 1-3 tells us that 10^3 is assigned the prefix k for "kilo." We can express the pressure in multiples of 10^3 as follows:

$$1.3 \times 10^4 \ \cancel{Pa} \times \frac{1 \text{ kPa}}{10^3 \ \cancel{Pa}} = 1.3 \times 10^1 \text{ kPa} = 13 \text{ kPa}$$

The unit kPa is read "kilopascals." It is a good idea to write units beside each number in a calculation and to cancel identical units in the numerator and denominator. This practice ensures that you know the units for your answer. If you intend to calculate pressure and your answer comes out with units other than pascals, something is wrong.

Question In the past, medical technologists referred to one microliter (μL) as one lambda (λ). How many microliters are in one milliliter?

EXAMPLE	Counting Neurotransmitter Molecules with an Electrode

The photograph at the beginning of this chapter shows an electrode used to measure neurotransmitter molecules discharged from a nerve cell. Box 1-2 describes the process by which packets of neurotransmitters are released in discrete bursts. Each neurotransmitter molecule that diffuses to the surface of the electrode releases two electrons. The average burst of charge measured at the electrode is 37 fC (femtocoulombs, 10^{-15} C). One coulomb of charge corresponds to 6.24×10^{18} electrons. How many molecules are released in an average burst?

SOLUTION Table 1-3 tells us that 1 fC equals 10^{-15} C. Therefore 37 fC corresponds to

$$37 \text{ fC} \times \left(\frac{10^{-15} \text{ C}}{\text{fC}}\right) = 37 \times 10^{-15} \text{ C}$$

The number of electrons in 37 fC is

$$(37 \times 10^{-15} \text{ C}) \times \left(\frac{6.24 \times 10^{18} \text{ electrons}}{\text{C}}\right) = 2.3 \times 10^5 \text{ electrons}$$

Because each neurotransmitter molecule releases two electrons, the number of molecules in one burst is

$$(2.3 \times 10^5 \text{ electrons}) \times \left(\frac{1 \text{ molecule}}{2 \text{ electrons}}\right) = 1.2 \times 10^5 \text{ molecules}$$

TABLE 1-3 Prefixes

Prefix	Symbol	Factor
yotta	Y	10^{24}
zetta	Z	10^{21}
exa	E	10^{18}
peta	P	10^{15}
tera	T	10^{12}
giga	G	10^{9}
mega	M	10^{6}
kilo	k	10^{3}
hecto	h	10^{2}
deca	da	10^{1}
deci	d	10^{-1}
centi	c	10^{-2}
milli	m	10^{-3}
micro	μ	10^{-6}
nano	n	10^{-9}
pico	p	10^{-12}
femto	f	10^{-15}
atto	a	10^{-18}
zepto	z	10^{-21}
yocto	y	10^{-24}

Ask Yourself

1-B. What are the names and abbreviations for each of the prefixes from 10^{-24} to 10^{24}? Which abbreviations are capitalized?

1-3 Conversion Between Units

Although SI is the internationally accepted system of measurement in science, other units are encountered. Conversion factors are found in Table 1-4. For example, common non-SI units for energy are the *calorie* (cal) and the *Calorie* (with a capital C, which represents 1 000 calories, or 1 kcal). Table 1-4 states that 1 cal is exactly 4.184 J (joules).

You require approximately 46 Calories per hour (h) per 100 pounds (lb) of body mass to carry out basic functions required for life, such as breathing, pumping blood, and maintaining body temperature. This minimum energy requirement for a conscious person at rest is called the *basal metabolism*. A person walking at 2 miles per hour on a level path requires approximately 45 Calories per hour per 100 pounds of body mass beyond basal metabolism. The same person swimming at 2 miles per hour consumes 360 Calories per hour per 100 pounds.

1 *calorie* of energy will heat 1 gram of water from 14.5° to 15.5°C.
1 000 *joules* will raise the temperature of a cup of water by about 1°C.
1 cal = 4.184 J

1 pound (mass) $\approx$ 0.453 6 kg
1 mile $\approx$ 1.609 km

Box 1-2 *Informed Citizen*

Exocytosis of Neurotransmitters

Nerve signals are transmitted from the *axon* of one nerve cell (a *neuron*) to the *dendrite* of a neighboring neuron across a junction called a *synapse*. A change in electric potential at the axon causes tiny membrane-enclosed compartments called *vesicles* to fuse with the cell membrane. This process, called *exocytosis*, releases neurotransmitter molecules stored in the vesicles. When neurotransmitters bind to receptors on the dendrite, gates open up to allow cations to cross the dendrite membrane. Cations diffusing into the dendrite change its electric potential, thereby transmitting the nerve impulse into the second neuron.

The cell in the photograph at the beginning of this chapter releases the neurotransmitter dopamine when stimulated by nicotine. Each molecule of dopamine that diffuses to the microelectrode gives up two electrons.

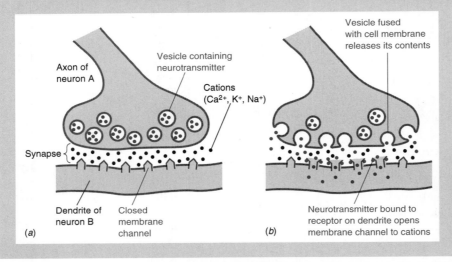

Action of neurotransmitters at synapse. Neurotransmitters are released by *exocytosis* when vesicles merge with the outer cell membrane. Cells can take molecules in by the reverse process, called *endocytosis*.

TABLE 1-4 Conversion factors

Quantity	Unit	Symbol	SI equivalent[a]
Volume	liter	L	$*10^{-3}$ m^3
	milliliter	mL	$*10^{-6}$ m^3
Length	angstrom	Å	$*10^{-10}$ m
	inch	in.	*0.025 4 m
Mass	pound	lb	*0.453 592 37 kg
Force	dyne	dyn	$*10^{-5}$ N
Pressure	atmosphere	atm	*101 325 Pa
	bar	bar	$*10^5$ Pa
	torr	1 mm Hg	133.322 Pa
	pound/in.^{2}	psi	6 894.76 Pa
Energy	erg	erg	$*10^{-7}$ J
	electron volt	eV	$1.602\ 177\ 33 \times 10^{-19}$ J
	calorie, thermochemical	cal	*4.184 J
	Calorie (with a capital C)	Cal	*1 000 cal = 4.184 kJ
	British thermal unit	Btu	1 055.06 J
Power	horsepower		745.700 W
Temperature	centigrade (= Celsius)	°C	$*K - 273.15$
	Fahrenheit	°F	$*1.8(K - 273.15) + 32$

a. An asterisk (*) indicates that the conversion is exact (by definition).

The figure below shows pulses of electric current measured at the microelectrode, each arising from the dopamine from an individual vesicle. The number of electrons in each pulse tells us how many dopamine molecules were released from one vesicle.

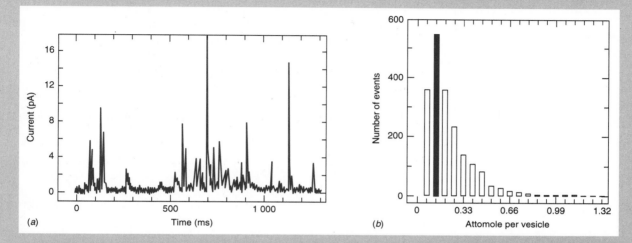

(a) Bursts of electric current arise from exocytosis of individual vesicles. The vertical axis (y-axis, called the **ordinate**) is measured in units of picoamperes (pA, 10^{-12} A). The horizontal axis (x-axis, called the **abscissa**, pronounced ab-sis-a) is measured in milliseconds (ms, 10^{-3} s). (b) Histogram showing relative frequencies of various sizes of packets of neurotransmitter. Each bar represents a 0.066 amol (= 66 zmol) interval. The second bar (= 2 × 66 = 132 zmol) represents the most common packet size. The average packet size is greater than 132 zmol, however, because the distribution is not symmetric.

EXAMPLE Unit Conversions

Express the rate of energy use for a walking person (46 + 45 = 91 Calories per hour per 100 pounds of body mass) in joules per hour per kilogram of body mass.

SOLUTION We will convert each non-SI unit separately. First note that 91 Calories equals 91 kcal. Table 1-4 states that 1 cal = 4.184 J, or 1 kcal = 4.184 kJ, so

$$91 \text{ kcal} \times \frac{4.184 \text{ kJ}}{1 \text{ kcal}} = 381 \text{ kJ}$$

Table 1-4 also says that 1 lb is 0.453 6 kg; so 100 lbs = 45.36 kg. The rate of energy consumption is, therefore,

$$\frac{91 \text{ kcal/h}}{100 \text{ lb}} = \frac{381 \text{ kJ/h}}{45.36 \text{ kg}} = 8.4 \frac{\text{kJ/h}}{\text{kg}}$$

Correct use of significant figures is discussed in Chapter 3.

You could have written this expression as one calculation, with appropriate unit cancellation:

$$\text{rate} = \frac{91 \text{ kcal/h}}{100 \text{ lb}} \times \frac{4.184 \text{ kJ}}{1 \text{ kcal}} \times \frac{1 \text{ lb}}{0.453 \text{ 6 kg}} = 8.4 \frac{\text{kJ/h}}{\text{kg}}$$

1 W = 1 J/s

One watt is 1 joule per second. The person in the previous example expends 8.4 kilo-joules per hour per kilogram of body mass while walking. **(a)** How many watts per kilogram does the person use? **(b)** If the person's mass is 60 kg, how many watts does he expend?

SOLUTION **(a)** The person expends 8.4×10^3 J/h/kg. Because an hour contains 3 600 s (60 s/min $\times$ 60 min/h), the required power is

$$8.4 \times 10^3 \ \frac{J}{h \cdot kg} \times \frac{3\ 600\ s}{1\ h}$$

Oops! The units didn't cancel out. I guess we need to use the inverse conversion factor:

$$8.4 \times 10^3 \ \frac{J}{h \cdot kg} \times \frac{1\ h}{3\ 600\ s} = 2.33 \ \frac{J}{s \cdot kg} = 2.33 \ \frac{J/s}{kg} = 2.33 \ \frac{W}{kg}$$

(b) Our intrepid walker has a mass of 60 kg. Therefore his power requirement is

$$2.33 \ \frac{W}{kg} \times 60\ kg = 140\ W$$

Ask Yourself

1-C. A 120-pound woman working in an office expends about 2.2×10^3 kcal/day, whereas the same woman climbing a mountain needs 3.4×10^3 kcal/day.
    (a) How many joules per day does the woman expend in each activity?
    (b) How many seconds are in one day?
    (c) How many joules per second (= watts) does the woman expend in each activity?
    (d) Which requires more power (watts), the office worker or a 100-W light bulb?

1-4   *Chemical Concentrations*

The minor species in a solution is called the **solute** and the major species is the **solvent.** In this text, most discussions concern *aqueous* solutions, in which the solvent is water. **Concentration** refers to how much solute is contained in a given volume or mass.

$$\text{molarity (M)} = \frac{\text{moles of solute}}{\text{liters of solution}}$$

The liter is named after the French-man Claude Litre (1716–1778), who named his daughter Millicent. Do you suppose her friends called her Millie Litre?
    "Mole Day" is celebrated at 6:02 A.M. on October 23 (10/23) at many schools.

Molarity and Molality

Molarity (M) is the number of moles of a substance per liter of solution. A mole is Avogadro's number of atoms or molecules or ions (6.022×10^{23} mol^{-1}). A **liter** (L) is the volume of a cube that is 10 cm on each edge. Because 10 cm = 0.1 m, we can say that 1 L = $(0.1$ m$)^3 = (10^{-1})^3$ m$^3 = 10^{-3}$ m^3. In Figure 1-2, chemical con-

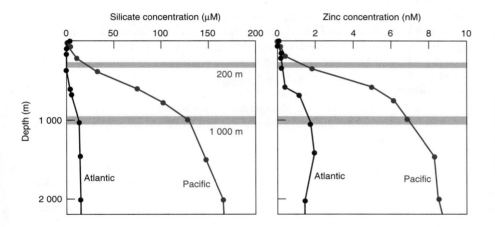

FIGURE 1-2 Concentration profiles of dissolved silicate and zinc in the north Atlantic and north Pacific Oceans. The oceans are *heterogeneous:* Samples from depths of 200 or 1 000 m do not have the same concentrations of each species. Living organisms near the ocean surface deplete seawater of both silicate and zinc.

centrations in the ocean are expressed in micromoles per liter (10^{-6} mol/L = μM) and nanomoles per liter (10^{-9} mol/L = nM). The molarity of a species is usually designated by square brackets, as in [Cl$^-$].

The **atomic weight** of an element is the number of grams containing Avogadro's number of that element's atoms. The **molecular weight** (MW) of a compound is the sum of atomic weights of the atoms in the molecule. It is the number of grams containing Avogadro's number of molecules of the compound.

EXAMPLE **Molarity of Salts in the Sea**

(a) Typical seawater contains 2.7 g of salt (sodium chloride, NaCl) per deciliter (= dL = 0.1 L). What is the molarity of NaCl in the ocean? **(b)** MgCl$_2$ has a concentration of 0.054 M in the ocean. How many grams of MgCl$_2$ are present in 25 mL of seawater?

SOLUTION (a) The molecular weight of NaCl is 22.99 (Na) + 35.45 (Cl) = 58.44 g/mol. The moles of salt in 2.7 g are

You can find atomic weights in the periodic table on the inside front cover of this book.

$$\text{moles of NaCl} = \frac{(2.7 \ \cancel{g})}{\left(58.44 \ \dfrac{\cancel{g}}{\text{mol}}\right)} = 0.046 \ \text{mol}$$

so the molarity is

$$[\text{NaCl}] = \frac{\text{mol NaCl}}{\text{L of seawater}} = \frac{0.046 \ \text{mol}}{0.1 \ \text{L}} = 0.46 \ \text{M}$$

(b) The molecular weight of MgCl$_2$ is 24.30 (Mg) + [2 × 35.45 (Cl)] = 95.20 g/mol, so the number of grams in 25 mL is

$$\text{grams of MgCl}_2 = 0.054 \ \frac{\cancel{\text{mol}}}{\cancel{L}} \times 95.20 \ \frac{g}{\cancel{\text{mol}}} \times 25 \times 10^{-3} \ \cancel{L} = 0.13 \ g$$

An *electrolyte* dissociates into ions in aqueous solution. Magnesium chloride is a **strong electrolyte,** which means that it is completely dissociated into Mg^{2+} and 2 Cl$^-$ in water. The concentration of MgCl$_2$ molecules in seawater is close to zero,

Strong electrolyte: completely dissociated into ions in solution
Weak electrolyte: partially dissociated in solution

because there are very few $MgCl_2$ molecules. Sometimes the molarity of a strong electrolyte is referred to as **formal concentration** (F), to indicate that the substance is really converted to other species in solution. When we say that the "concentration" of $MgCl_2$ is 0.054 M in seawater, we are really referring to its formal concentration (0.054 F). The "molecular weight" of a strong electrolyte is more properly called the **formula weight** (FW), because it refers to the sum of atomic weights of the atoms in the formula, even though there are no such molecules with that formula.

A **weak electrolyte** such as acetic acid, CH_3CO_2H, is partially split into ions in solution:

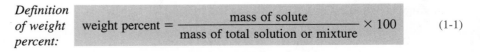

	Formal concentration	Percent dissociated
	0.1 F	1.3%
	0.01	4.1
	0.001	12.4

A solution prepared by dissolving 0.010 00 mol of acetic acid in 1.000 L has a formal concentration of 0.010 00 F. The actual molarity of CH_3CO_2H is 0.009 59 M because 4.1% is dissociated into $CH_3CO_2^-$ and 95.9% remains as CH_3CO_2H. Nonetheless, we customarily say that the solution is 0.010 00 M acetic acid and understand that some of the acid is dissociated.

Molality (m) is a designation of concentration expressing the number of moles of substance per kilogram of solvent (not total solution). Unlike molarity, molality is independent of temperature. Molarity changes with temperature because the volume of a solution usually increases when it is heated.

Confusing abbreviations:

mol = moles

$$M = \text{molarity} = \frac{\text{mol solute}}{\text{L solution}}$$

$$m = \text{molality} = \frac{\text{mol solute}}{\text{kg solvent}}$$

Percent Composition

The percentage of a component in a mixture or solution is usually expressed as a **weight percent** (wt %):

Definition of weight percent:
$$\text{weight percent} = \frac{\text{mass of solute}}{\text{mass of total solution or mixture}} \times 100 \qquad (1\text{-}1)$$

A common form of ethanol (CH_3CH_2OH) is 95 wt %, which has 95 g of ethanol per 100 g of total solution. The remainder is water. Another common expression of composition is **volume percent** (vol %):

Definition of volume percent:
$$\text{volume percent} = \frac{\text{volume of solute}}{\text{volume of total solution}} \times 100 \qquad (1\text{-}2)$$

Although units of weight or volume should always be expressed to avoid ambiguity, weight is usually implied when units are absent.

$$\textbf{density} = \frac{\text{mass}}{\text{volume}} = \frac{\text{g}}{\text{mL}}$$

A closely related dimensionless quantity is

specific gravity =
$$\frac{\text{density of a substance}}{\text{density of water at 4°C}}$$

Because the density of water at 4°C is very close to 1 g/mL (Figure 3-5), specific gravity is nearly the same as density.

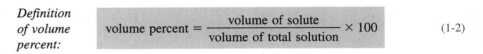

EXAMPLE **Converting Weight Percent to Molarity**

Find the molarity of HCl in a reagent labeled "37.0 wt % HCl, density = 1.188 g/mL." The **density** of a substance is the mass per unit volume.

SOLUTION We need to find the moles of HCl per liter of solution. To find moles of HCl, we need to find the mass of HCl. The mass of HCl in 1 L is 37.0% of the mass of 1 L of solution. The mass of 1 L of solution is $(1.188 \text{ g/mL})(1\,000 \text{ mL/L}) = 1\,188$ g/L. The mass of HCl in 1 L is

$$HCl\left(\frac{g}{L}\right) = 1\,188 \; \frac{\text{g solution}}{L} \times 0.370 \; \frac{\text{g HCl}}{\text{g solution}} = 439.6 \; \frac{\text{g HCl}}{L}$$

↑
This is what 37.0 wt % HCl means

The molecular weight of HCl is 36.46 g/mol, so the molarity is

$$\text{molarity} = \frac{\text{mol HCl}}{\text{L solution}} = \frac{439.6 \text{ g HCl/L}}{36.46 \text{ g HCl/mol}} = 12.1 \; \frac{\text{mol}}{L} = 12.1 \text{ M}$$

Parts per Million and Parts per Billion

Concentrations of trace components of a sample can be expressed as **parts per million** (ppm) or **parts per billion** (ppb), terms that mean grams of substance per million or billion grams of total solution or mixture.

Definition of parts per million:

$$ppm = \frac{\text{mass of substance}}{\text{mass of sample}} \times 10^6 \tag{1-3}$$

Definition of parts per billion:

$$ppb = \frac{\text{mass of substance}}{\text{mass of sample}} \times 10^9 \tag{1-4}$$

Question What do you suppose the definition of parts per trillion is?

Masses must be expressed in the same units in the numerator and denominator.

The density of a dilute aqueous solution is close to 1.00 g/mL, so *we frequently equate 1 g of water with 1 mL of water,* although this equivalence is only approximate. Therefore 1 ppm corresponds to 1 μg/mL (= 1 mg/L) and 1 ppb is 1 ng/mL (= 1 μg/L).

1 ppm ≈ 1 μg/mL
1 ppb ≈ 1 ng/mL
The symbol ≈ is read "is approximately equal to."

EXAMPLE   **Converting Parts per Billion to Molarity**

Hydrocarbons are compounds containing only hydrogen and carbon. Plants manufacture hydrocarbons (called fats or lipids) as components of the membranes of cells and vesicles. The biosynthetic pathway leads mainly to compounds with an odd number of carbon atoms. Figure 1-3 shows the concentrations of hydrocarbons washed from the air by rain in the winter and summer. The preponderance of odd-number hydrocarbons in the summer suggests that the source is mainly from plants. The more uniform distribution of odd- and even-number hydrocarbons in the winter suggests a man-made origin. The concentration of $C_{29}H_{60}$ in summer rainwater is 34 ppb. Find the molarity of this compound in nanomoles per liter (nM).

SOLUTION A concentration of 34 ppb refers to 34 ng of $C_{29}H_{60}$ per gram of rainwater, which we equate to 34 ng/mL = 34×10^{-9} g/mL. To find moles per liter, we first find grams per liter:

$$34 \times 10^{-9} \; \frac{g}{mL} \times \frac{1\,000 \text{ mL}}{L} = 34 \times 10^{-6} \; \frac{g}{L}$$

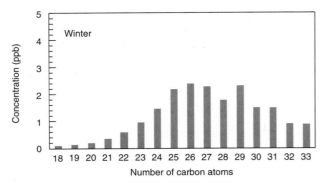
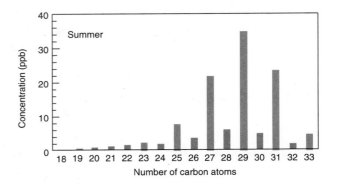

FIGURE 1-3 Concentrations of alkanes (hydrocarbons with the formula C_nH_{2n+2}) found in rainwater in Hannover, Germany, in the winter and summer in 1989 are measured in parts per billion (= μg hydrocarbon/L of rainwater). Summer concentrations are higher and compounds with an odd number of carbon atoms predominate.

Because the molecular weight of $C_{29}H_{60}$ is 408.8 g/mol, the molarity is

$$\text{molarity of } C_{29}H_{60} \text{ in rainwater} = \frac{34 \times 10^{-6} \text{ g/L}}{408.8 \text{ g/mol}} = 8.3 \times 10^{-8} \text{ M}$$
$$= 83 \times 10^{-9} \text{ M} = 83 \text{ nM}$$

With reference to gases, ppm usually indicates volume rather than mass. For example, 8 ppm carbon monoxide in air means 8 μL of CO per liter of air. Always write the units to avoid confusion. Figure 1-4 shows gas concentration measured in *parts per trillion* by volume (picoliters per liter).

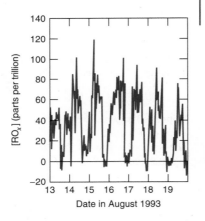

FIGURE 1-4 Concentration of RO_x (parts per trillion by volume = pL/L) measured outside the S. G. Mudd Building at the University of Denver. RO_x refers to the combined concentrations of HO (hydroxyl radical), HO_2 (hydroperoxide radical), RO (alkoxy radical, where R is any organic group), and RO_2 (alkyl peroxide radical). These reactive species are created mainly by photochemical reactions driven by sunlight. Concentrations peak around 2:00 P.M. each day and fall close to zero during the night.

Ask Yourself

1-D. The density of 70.5 wt % aqueous perchloric acid is 1.67 g/mL. Note that grams refers to grams of *solution* (= g $HClO_4$ + g H_2O).
  (a) How many grams of solution are in 1.00 L?
  (b) How many grams of $HClO_4$ are in 1.00 L?
  (c) How many moles of $HClO_4$ are in 1.00 L? This is the molarity.

1-5 Preparing Solutions

To prepare a solution with a desired molarity, weigh out the correct mass of pure reagent, dissolve it in solvent in a *volumetric flask* (Figure 1-5), dilute with more solvent to the desired final volume, and mix well by inverting the flask 20 times.

EXAMPLE Preparing a Solution with Desired Molarity

Cupric sulfate is commonly sold as the pentahydrate, $CuSO_4 \cdot 5H_2O$, which means that there are five moles of H_2O for each mole of $CuSO_4$ in the solid crystal. The formula weight of $CuSO_4 \cdot 5H_2O$ ($= CuSO_9H_{10}$) is 249.69 g/mol. How many grams of $CuSO_4 \cdot 5H_2O$ should be dissolved in a 250-mL volumetric flask to make a solution containing 8.00 mM Cu^{2+}?

SOLUTION An 8.00 mM solution contains 8.00×10^{-3} mol/L. Because 250 mL is 0.250 L, we need

$$8.00 \times 10^{-3} \frac{\text{mol}}{\cancel{L}} \times 0.250\,\cancel{L} = 2.00 \times 10^{-3} \text{ mol } CuSO_4 \cdot 5H_2O$$

The required mass of reagent is

$$(2.00 \times 10^{-3}\,\cancel{\text{mol}}) \left(249.69\,\frac{\text{g}}{\cancel{\text{mol}}}\right) = 0.499 \text{ g}$$

The procedure is to weigh 0.499 g of solid $CuSO_4 \cdot 5H_2O$ into a 250-mL volumetric flask, add about 200 mL of distilled water, and swirl to dissolve the reagent. Then dilute with distilled water up to the 250-mL mark and invert the flask 20 times to ensure complete mixing. The solution contains 8.00 mM Cu^{2+}.

Dilute solutions can be prepared from concentrated solutions. Typically, a desired volume or mass of the concentrated solution is transferred to a volumetric flask and diluted to the intended volume with solvent. The number of moles of reagent in V liters containing M moles per liter is the product $M \cdot V = (\text{mol}/\cancel{L})(\cancel{L}) = \text{mol}$. When a solution is diluted from a high concentration to a low concentration, the number of moles of solute is unchanged. Therefore we equate the number of moles in the concentrated (conc) and dilute (dil) solutions:

Dilution formula:

$$\underbrace{M_{\text{conc}} \cdot V_{\text{conc}}}_{\substack{\text{Moles taken from} \\ \text{concentrated solution}}} = \underbrace{M_{\text{dil}} \cdot V_{\text{dil}}}_{\substack{\text{Moles placed in} \\ \text{dilute solution}}} \qquad (1\text{-}5)$$

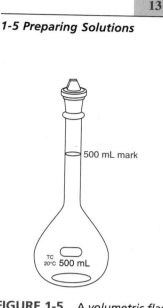

FIGURE 1-5 A *volumetric flask* contains a specified volume when the liquid level is adjusted to the middle of the mark in the thin neck of the flask. Use of this flask is described in the next chapter.

500 mL mark

TC 20°C 500 mL

EXAMPLE	Preparing 0.1 M HCl

The molarity of "concentrated" HCl purchased for laboratory use is ~12.1 M. How many milliliters of this reagent should be diluted to 1.00 L to make 0.100 M HCl?

The symbol ~ is read "approximately."

SOLUTION The required volume of concentrated solution is found with Equation 1-5:

$$M_{\text{conc}} \cdot V_{\text{conc}} = M_{\text{dil}} \cdot V_{\text{dil}}$$
$$(12.1 \text{ M})(x \text{ mL}) = (0.100 \text{ M})(1\,000 \text{ mL}) \implies x = 8.26 \text{ mL}$$

The symbol $\implies$ is read "implies that."

It is all right to express both volumes in mL or both in L. The important point is to use the same units for volume on both sides of the equation so that the units cancel. To make 0.100 M HCl, place 8.26 mL of concentrated HCl in a 1-L volumetric flask and add ~900 mL of water. After swirling to mix, dilute to the 1-L mark with water and invert the flask 20 times to ensure complete mixing.

EXAMPLE **A More Complicated Dilution Calculation**

A solution of ammonia in water is called "ammonium hydroxide" because of the equilibrium

$$NH_3 \quad + H_2O \rightleftharpoons \quad NH_4^+ \quad + \quad OH^-$$

Ammonia   Ammonium   Hydroxide
                              ion   ion

The density of concentrated ammonium hydroxide, which contains 28.0 wt % NH_3, is 0.899 g/mL. What volume of this reagent should be diluted to 500 mL to make 0.250 M NH_3?

SOLUTION To use Equation 1-5, we need to know the molarity of the concentrated reagent. The density tells us that the reagent contains 0.899 grams of solution per milliliter of solution. The weight percent tells us that the reagent contains 0.280 grams of NH_3 per gram of solution. To find the molarity of NH_3 in the concentrated reagent, we need to know the moles of NH_3 in one liter:

$$\text{grams of } NH_3 \text{ per liter} = 899 \ \frac{\text{g solution}}{L} \times 0.280 \ \frac{\text{g } NH_3}{\text{g solution}} = 252 \ \frac{\text{g } NH_3}{L}$$

$$\text{molarity of } NH_3 = \frac{252 \ \frac{\text{g } NH_3}{L}}{17.03 \ \frac{\text{g } NH_3}{\text{mol } NH_3}} = 14.8 \ \frac{\text{mol } NH_3}{L} = 14.8 \ M$$

Now we use Equation 1-5 to find the volume of 14.8 M NH_3 required to prepare 500 mL of 0.250 M NH_3:

$$M_{conc} \cdot V_{conc} = M_{dil} \cdot V_{dil}$$

$$14.8 \ \frac{\text{mol}}{L} \times V_{conc} = 0.250 \ \frac{\text{mol}}{L} \times 0.500 \ L$$

$$\Rightarrow V_{conc} = 8.45 \times 10^{-3} \ L = 8.45 \ \text{mL}$$

The correct procedure is to place 8.45 mL of concentrated reagent in a 500-mL volumetric flask, add about 400 mL of water, and swirl to mix. Then dilute to exactly 500 mL with water and invert the flask 20 times to mix well.

EXAMPLE **Preparing a Parts per Million Concentration**

Drinking water usually contains 1.6 ppm fluoride (F^-) for prevention of tooth decay. Consider a reservoir with a diameter of 450 m and a depth of 10 m. **(a)** How many liters of 0.10 M NaF should be added to produce 1.6 ppm F^-? **(b)** How many grams of solid NaF could be used instead?

SOLUTION **(a)** Assuming that the density of water in the reservoir is close to 1.00 g/mL, 1.6 ppm F^- corresponds to 1.6×10^{-6} g F^-/mL or

$$1.6 \times 10^{-6} \ \frac{\text{g } F^-}{mL} \times 1\,000 \ \frac{mL}{L} = 1.6 \times 10^{-3} \ \frac{\text{g } F^-}{L}$$

The atomic weight of fluorine is 19.00, so the desired molarity of fluoride is

$$\text{desired [F}^-\text{] in reservoir} = \frac{1.6 \times 10^{-3} \; \frac{\text{g F}}{\text{L}}}{19.00 \; \frac{\text{g F}}{\text{mol}}} = 8.42 \times 10^{-5} \; \text{M}$$

The volume of the reservoir is $\pi r^2 h$, where r is the radius and h is the height:

$$\text{volume of reservoir} = \pi \times (225 \text{ m})^2 \times 10 \text{ m} = 1.59 \times 10^6 \text{ m}^3$$

Volume of cylinder $= \pi r^2 h$

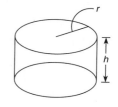

To use the dilution formula, we need to convert m³ to liters. There are 1 000 L in one cubic meter (see Table 1-4). Thus the volume of the reservoir in liters is

$$\text{volume of reservoir (L)} = 1.59 \times 10^6 \text{ m}^3 \times 1\,000 \; \frac{\text{L}}{\text{m}^3} = 1.59 \times 10^9 \text{ L}$$

Finally, we are in a position to use the dilution formula 1-5:

$$M_{\text{conc}} \cdot V_{\text{conc}} = M_{\text{dil}} \cdot V_{\text{dil}}$$

$$0.10 \; \frac{\text{mol}}{\text{L}} \times V_{\text{conc}} = \left(8.42 \times 10^{-5} \; \frac{\text{mol}}{\text{L}}\right) \times (1.59 \times 10^9 \; \text{L})$$

$$\Rightarrow V_{\text{conc}} = 1.3 \times 10^6 \text{ L}$$

We require 1.3 million liters of 0.10 M F⁻. Note that our calculation assumed that the final volume of the reservoir is 1.59×10^9 L. Even though we are adding more than 10^6 L of reagent, this is small compared to 10^9 L. Therefore the approximation that the reservoir volume remains 1.59×10^9 L is pretty good.

(b) Even if we use solid NaF instead, we still require 8.42×10^{-5} M F⁻. The number of moles of F⁻ in the reservoir is $(1.59 \times 10^9 \text{ L}) \times (8.42 \times 10^{-5} \text{ mol/L}) = 1.34 \times 10^5$ mol F⁻. Because one mole of NaF provides one mole of F⁻, we need $(1.34 \times 10^5 \text{ mol NaF}) \times (41.99 \text{ g NaF/mol NaF}) = 5.6 \times 10^6$ grams of NaF.

Ask Yourself

1-E. A 48 wt % solution of HBr in water has a density of 1.50 g/mL.
   (a) How many grams of solution are in 1.00 L?
   (b) How many grams of HBr are in 1.00 L?
   (c) What is the molarity of HBr?
   (d) How much solution is required to prepare 0.250 L of 0.160 M HBr?

Key Equations

Molarity (M) $[A] = \dfrac{\text{moles of solute A}}{\text{liters of solution}}$

Weight percent $\text{wt \%} = \dfrac{\text{mass of solute}}{\text{mass of solution or mixture}} \times 100$

Volume percent $\qquad$ vol % = $\dfrac{\text{volume of solute}}{\text{volume of solution or mixture}} \times 100$

Density $\qquad$ density = $\dfrac{\text{grams of substance}}{\text{mL of substance}}$

Parts per million $\qquad$ ppm = $\dfrac{\text{mass of substance}}{\text{mass of sample}} \times 10^6$

Parts per billion $\qquad$ ppb = $\dfrac{\text{mass of substance}}{\text{mass of sample}} \times 10^9$

Dilution formula $\qquad$ $M_{conc} \cdot V_{conc} = M_{dil} \cdot V_{dil}$

M_{conc} = concentration (molarity) of concentrated solution

M_{dil} = concentration of dilute solution

V_{conc} = volume of concentrated solution

V_{dil} = volume of dilute solution

Important Terms

abscissa	heterogeneous material	ordinate	segregated material
aliquot	homogeneous material	parts per billion	SI units
analyte	interference	parts per million	solute
atomic weight	liter	qualitative analysis	solvent
composite sample	masking	quantitative analysis	strong electrolyte
concentration	molality	random sample	volume percent
density	molarity	sample preparation	weak electrolyte
formal concentration	molecular weight	sampling	weight percent
formula weight			

Problems

1-1. What is the difference between *qualitative* and *quantitative* analysis?

1-2. List the steps in a chemical analysis.

1-3. What does it mean to *mask* an interfering species?

1-4. (a) List the SI units of length, mass, time, electric current, temperature, and amount of substance. Write the abbreviation for each.

  (b) Write the units and symbols for frequency, force, pressure, energy, and power.

1-5. Write the name and number represented by each symbol. For example, for kW you would write kW = kilowatt = 10^3 watts.

(a) mW $\quad$ **(c)** kΩ $\quad$ **(e)** TJ $\quad$ **(g)** fg
(b) pm $\quad$ **(d)** μC $\quad$ **(f)** ns $\quad$ **(h)** dPa

1-6. Express the following quantities with abbreviations for units and prefixes from Tables 1-1 through 1-3:

(a) 10^{-13} joules
(b) $4.317\,28 \times 10^{-8}$ coulombs
(c) $2.997\,9 \times 10^{14}$ hertz
(d) 10^{-10} meters
(e) 2.1×10^{13} watts
(f) 48.3×10^{-20} moles

1-7. Table 1-4 states that 1 horsepower = 745.700 watts. Consider a 100.0-horsepower engine. Express its power output in (a) watts; (b) joules per second; (c) calories per second; (d) calories per hour.

1-8. (a) Refer to Table 1-4 and calculate how many meters are in 1 inch. How many inches are in 1 m?

(b) A mile is 5 280 feet and a foot is 12 inches. The speed of sound in the atmosphere at sea level is 345 m/s. Express the speed of sound in mi/s and mi/hr.

(c) There is a delay between lightning and thunder in a storm, because light reaches us almost at once, but sound is slower. How many meters, kilometers, and miles away is lightning if the sound reaches you 3.00 s after the light?

1-9. Define the following measures of concentration:

(a) molarity (e) volume percent
(b) molality (f) parts per million
(c) density (g) parts per billion
(d) weight percent (h) formal concentration

1-10. What is the formal concentration (expressed as mol/L = M) of NaCl when 32.0 g is dissolved in water and diluted to 0.500 L?

1-11. If 0.250 L of aqueous solution with a density of 1.00 g/mL contains 13.7 μg of pesticide, express the concentration of pesticide in (a) ppm and (b) ppb.

1-12. The concentration of the sugar glucose ($C_6H_{12}O_6$) in human blood ranges from about 80 mg/dL before meals up to 120 mg/dL after eating. Find the molarity of glucose in blood before and after eating.

1-13. (a) How many grams of perchloric acid, $HClO_4$, are contained in 100.0 g of 70.5 wt % aqueous perchloric acid?

(b) How many grams of water are in 100.0 g of solution?

(c) How many moles of $HClO_4$ are in 100.0 g of solution?

1-14. How many grams of boric acid [$B(OH)_3$, MW 61.83] should be used to make 2.00 L of 0.050 0 M solution?

1-15. Water is fluoridated to prevent tooth decay.

(a) How many liters of 1.5 M KF should be added to a reservoir with a diameter of 100 m and a depth of 20 m to give 1.2 ppm F^-?

(b) How many grams of solid KF should be added to the same reservoir to give 1.2 ppm F^-?

1-16. How many grams of 50 wt % NaOH (FW 40.00) should be diluted to 1.00 L to make 0.10 M NaOH?

1-17. A bottle of concentrated aqueous sulfuric acid, labeled 98.0 wt % H_2SO_4, has a concentration of 18.0 M.

(a) How many milliliters of reagent should be diluted to 1.00 L to give 1.00 M H_2SO_4?

(b) Calculate the density of 98.0 wt % H_2SO_4.

1-18. How many grams of methanol (CH_3OH, MW 32.04) are contained in 0.100 L of 1.71 M aqueous methanol?

1-19. A dilute aqueous solution containing 1 ppm of solute has a density of 1.00 g/mL. Express the concentration of solute in g/L, μg/L, μg/mL, and mg/L.

1-20. The concentration of $C_{20}H_{42}$ (MW 282.55) in winter rainwater in Figure 1-4 is 0.2 ppb. Assume that the density of rainwater is close to 1.00 g/mL; find the molar concentration of $C_{20}H_{42}$.

1-21. A 95.0 wt % solution of ethanol (CH_3CH_2OH, MW 46.07) in water has a density of 0.804 g/mL.

(a) Find the mass of 1.00 L of this solution and the grams of ethanol per liter.

(b) What is the molar concentration of ethanol in this solution?

1-22. (a) How many grams of nickel are contained in 10.0 g of a 10.2 wt % solution of nickel sulfate hexahydrate, $NiSO_4 \cdot 6H_2O$ (FW 262.85)?

(b) The concentration of this solution is 0.412 M. Find the density.

1-23. A 500.0-mL solution was prepared by dissolving 25.00 mL of methanol (CH_3OH, density = 0.791 4 g/mL) in chloroform. Find the molarity of the methanol.

1-24. Describe how to prepare exactly 100 mL of 1.00 M HCl from 12.1 M HCl reagent.

1-25. Cesium chloride is used to prepare dense solutions required for isolating cellular components with a centrifuge. A 40.0 wt % solution of CsCl (FW 168.36) has a density of 1.43 g/mL.

(a) Find the molarity of CsCl.

(b) How many milliliters of the concentrated solution should be diluted to 500 mL to make 0.100 M CsCl?

1-26. Protein and carbohydrates provide 4.0 Cal/g, whereas fat gives 9.0 Cal/g. (Remember that 1 Calorie, with a capital C, is really 1 kcal.) The weight percents of these components in some foods are

Food	Protein wt %	Carbohydrate wt %	Fat wt %
shredded wheat	9.9	79.9	—
doughnut	4.6	51.4	18.6
hamburger (cooked)	24.2	—	20.3
apple	—	12.0	—

Calculate the number of calories per gram and calories per ounce in each of these foods. (Use Table 1-4 to convert grams into ounces, remembering that there are 16 ounces in 1 pound.)

Weighing Femtomoles of DNA

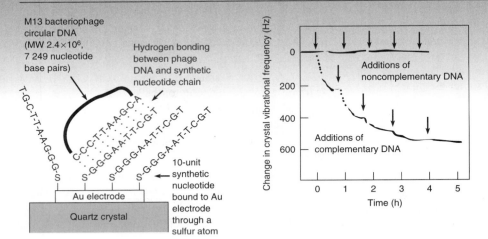

T he vibrating quartz crystal that keeps time in your wristwatch and computer can be used as an extremely sensitive mass detector. When a substance is *adsorbed* (bound) on the crystal's surface, the vibrational frequency of the crystal changes in proportion to the adsorbed mass.

The figures show how a quartz crystal can detect nanogram quantities of one specific *deoxyribonucleic acid* molecule (the genetic material, DNA) from a *bacteriophage* (a virus that attacks bacteria). The crystal is coated with a synthetic chain of DNA containing 10 nucleotides (designated C, G, A, and T) chosen to bind to a complementary sequence on the bacteriophage DNA: C and G always bind to each other, and A and T always bind to each other. When stimulated by an oscillating electric field applied through a thin gold coating on its surface, the crystal vibrates at a precisely defined characteristic frequency.

To measure DNA, the crystal is immersed in a solution containing many different DNA molecules. If the complementary DNA molecule with the correct sequence of nucleotides is present, it binds to the synthetic DNA and lowers the vibrational frequency of the crystal. Arrows in the graph indicate additions of equal portions of DNA to the solution. The lower curve shows the change in crystal frequency when DNA with the correct complementary sequence is added. Successive additions give smaller responses as the electrode becomes saturated with DNA. The upper curve illustrates the absence of response when noncomplementary DNA is added.

A frequency change of 560 Hz corresponds to 590 ng (0.25 pmol) of DNA. The smallest quantity that could be detected by this system was approximately 25 ng—or about 11 fmol.

Chapter 2

Tools of the Trade

Analytical chemistry extends from simple "wet" chemical procedures (such as titrations) to elaborate instrumental methods. In this chapter we describe basic laboratory apparatus and manipulations associated with chemical measurements.

2-1   Safety with Chemicals and Waste

The primary safety rule is not to do something that you (or your instructor) consider to be dangerous. If you believe that an operation is hazardous, discuss it with your instructor and do not proceed until sensible procedures and precautions are in place. If you still consider an activity to be too dangerous, don't do it.

Before beginning work, you should be familiar with safety precautions appropriate to your laboratory. Goggles or safety glasses with side shields (Figure 2-1) are worn at all times to protect you from flying chemicals and glass, which visit labs from time to time. (Although you may be very careful, one of your neighbors will not be so careful.) A flame-resistant lab coat, long pants, and shoes that cover your feet help protect you from spills and flames. Rubber gloves protect you when pouring concentrated acids, but organic solvents can penetrate rubber gloves. Food and chemicals should not mix: Don't bring food or drink into the lab.

Treat chemical spills on your skin *immediately* by flooding the affected area with water and then seeking medical attention. Clean up spills on the bench, floor, or reagent bottles immediately to prevent accidental contact by the next person who comes along.

Solvents and concentrated acids that produce harmful fumes should be handled in a fume hood that sweeps vapors away from you and out through a vent on the roof. The hood is not meant to transfer toxic vapors from the chemistry building to the cafeteria; it is designed to dilute low levels of pollutants to even lower levels. Never generate large quantities of toxic fumes that are allowed to escape through the hood.

FIGURE 2-1   Goggles or safety glasses with side shields are required in every laboratory.

Report any accidents immediately to your instructor, who should have advice on first aid and cleanup procedures.

Box 2-1 *Informed Citizen*

Disposal of Chemical Waste[†]

Many chemicals that we commonly use are harmful to plants, animals, and people if carelessly discarded. For each experiment, your instructor should establish safe procedures for waste disposal. Options include (1) pouring solutions down the drain and diluting with tap water, (2) saving the waste for disposal in an approved landfill, (3) treating waste to decrease the hazard and then pouring it down the drain or saving it for a landfill, and (4) recycling the chemical. *Chemically incompatible wastes should never be mixed with each other, and each waste container must be labeled to indicate the quantity and identity of its contents.*

A few examples illustrate different approaches to managing lab waste. Dichromate ($Cr_2O_7^{2-}$) is reduced to Cr^{3+} with sodium hydrogen sulfite ($NaHSO_3$), treated with hydroxide to make insoluble $Cr(OH)_3$, and evaporated to dryness for disposal in a landfill. Waste acid is mixed with waste base until nearly neutral (as determined with pH paper) and then poured down the drain. Waste iodate (IO_3^-) is reduced to I^- with $NaHSO_3$, neutralized with base, and poured down the drain. Waste silver or gold is chemically treated to recover the metals. Toxic gases used in a fume hood are bubbled through a chemical trap or burned in a flame to prevent escape from the hood.

[†]Handling and disposing of chemicals is described in *Prudent Practices for Handling Hazardous Chemicals in Laboratories* (1983) and *Prudent Practices for Disposal of Chemicals in Laboratories* (1983), both available from National Academy Press, 2101 Constitution Avenue N.W., Washington, DC 20418. See also G. Lunn and E. B. Sansone, *Destruction of Hazardous Chemicals in the Laboratory* (New York: Wiley, 1994), and M. A. Armour, *Hazardous Laboratory Chemical Disposal Guide* (Boca Raton, FL: CRC Press, 1991).

Preservation of a habitable planet demands that we minimize waste production and responsibly dispose of chemical waste (Box 2-1). Recycling of chemicals is practiced in industry for economic as well as ethical reasons; it should be a component of pollution control in your lab.

All vessels should be labeled immediately to show what they contain. If you don't do this, you may (no, you *will*) forget what is in some of your own containers and then you won't know what to do with them. Unlabeled waste is extremely expensive to discard, because you must analyze the contents before you can legally dispose of it.

The lab notebook must

1. State what was done
2. State what was observed
3. Be understandable to others

Without a doubt, somebody reading this book today is going to make an important discovery in the future and will seek a patent. The lab notebook is your legal record of your discovery. Therefore, each notebook page should be signed and dated. Anything of potential importance should also be signed and dated by a second person.

2-2 *Your Lab Notebook*

The critical functions of your lab notebook are to state *what you did* and *what you observed,* and it should be *understandable by a stranger.* The greatest error, made even by experienced scientists, is writing ambiguous notes. After a few years, even the author of a notebook cannot interpret what was done or seen. Writing in *complete sentences* is an excellent way to reduce this problem.

The measure of scientific "truth" is the ability to reproduce an experiment. A good lab notebook will allow you or anyone else to duplicate an experiment in the exact manner in which it was conducted the first time.

Beginning students find it useful (or required!) to write a complete description of an experiment, with sections describing the purpose, methods, results, and conclusions. Arranging your notebook to accept numerical data prior to coming to the lab is an excellent way to prepare for an experiment.

It is good practice to write a balanced chemical equation for every reaction you use. This helps you understand what you are doing and may point out what you do not understand.

Record in your notebook the names of computer disks and files where programs and data are stored. Printed copies of important data collected on a computer should be pasted into your notebook. The lifetime of a printed page is 10 to 100 times greater than that of a disk.

Ask Yourself

2-A. What are the three essential attributes of a lab notebook?

2-3 *The Analytical Balance*

Figure 2-2 shows a typical analytical **electronic balance** with a capacity of 100–200 g and a sensitivity of 0.01–0.1 mg. *Sensitivity* refers to the smallest increment of mass that can be measured. A *microbalance* weighs milligram quantities with a sensitivity of 0.1 μg. An electronic balance works by generating an electromagnetic force to balance the gravitational force acting on the object being weighed. The current required by the electromagnet is proportional to the mass being weighed.

FIGURE 2-2 Analytical electronic balance.

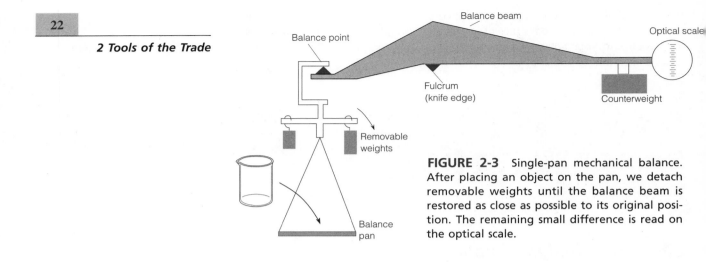

FIGURE 2-3 Single-pan mechanical balance. After placing an object on the pan, we detach removable weights until the balance beam is restored as close as possible to its original position. The remaining small difference is read on the optical scale.

Balances and other lab equipment are delicate and expensive. Be gentle when you place objects on the pan and when you adjust the knobs.

Figure 2-3 shows the principle of operation of a single-pan **mechanical balance.** The balance beam is suspended on a sharp fulcrum called a *knife edge.* The mass of the pan hanging from the balance point (another knife edge) at the left is balanced by a counterweight at the right. After placing the object to be weighed on the pan, knobs are adjusted to remove standard weights from the bar above the pan. The balance beam is restored close to its original position when the weights removed from the bar are nearly equal to the mass on the pan. The slight difference from the original position is shown on an optical scale, whose reading is added to that of the knobs.

Using a Balance

A mechanical balance should be in its arrested position when you load or unload the pan and in the half-arrested position when you are dialing weights. This practice protects against the application of abrupt forces that wear down the knife edges and eventually decrease the sensitivity of the balance.

To weigh a chemical, place a clean receiving vessel on the balance pan. The mass of the empty vessel is called the **tare.** On most electronic balances, the tare can be set to zero by pressing a button. Add the chemical to the vessel and read the new mass. If there is not an automatic tare operation, the mass of the empty vessel should be subtracted from that of the filled vessel. Chemicals should never be placed directly on the weighing pan. This precaution protects the balance from corrosion and allows you to recover all the chemical being weighed.

An alternate procedure, called "weighing by difference," is necessary for **hygroscopic** reagents, which rapidly absorb moisture from the air. First weigh a capped bottle containing dry reagent. Then quickly pour some reagent from the weighing bottle into a receiver. Cap the weighing bottle and weigh it again. The difference is the mass of reagent.

Clean up spills on the balance and do not allow chemicals to get into the mechanism below the pan of an electronic balance. Use a paper towel or tissue to handle the vessel you are weighing, because fingerprints will change its mass. Samples should be at *ambient temperature* (the temperature of the surroundings) when weighed to avoid errors due to convective air currents. The doors of the balance in Figure 2-2 must be closed during weighing to prevent air currents from affecting the reading. A top-loading balance without sliding doors has a fence around the pan to deflect air currents. Sensitive balances should be placed on a heavy table, such as a marble slab, to minimize the effect of building vibrations on the reading. Adjustable feet and a bubble meter allow you to maintain the balance in a level position.

Buoyancy

The actual mass of an object (that measured in vacuum) is usually greater than the apparent mass measured in air. This effect arises from an object's **buoyancy** in air: A sample displacing a particular volume of air appears lighter than its actual mass by an amount equal to the mass of the displaced air. The same effect applies to weights used to calibrate an electronic balance or removable weights in a mechanical balance. A buoyancy error occurs whenever the density of the object being weighed is not equal to the density of the standard weights.

If mass m' is read from a balance, the true mass m is

Buoyancy equation:

$$m = \frac{m'\left(1 - \dfrac{d_a}{d_w}\right)}{\left(1 - \dfrac{d_a}{d}\right)} \qquad (2\text{-}1)$$

where d_a is the density of air (0.001 2 g/mL near 1 atm and 25°C); d_w is the density of balance weights (8.0 g/mL); and d is the density of the object being weighed.

Buoyancy The *Earthwinds* balloon has made numerous attempts at a 'round-the-world voyage. A hot-air balloon cannot carry enough fuel for such a trip. A helium balloon cannot carry enough ballast to cope with the daily cycle of warming and cooling that requires helium to be vented in the day and ballast to be released at night to maintain altitude. The upper helium balloon of *Earthwinds* provides lift. Air is pumped into the lower balloon during the day and pumped out at night to provide the correct buoyancy.

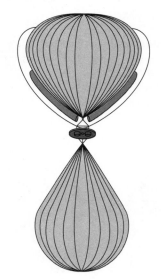

EXAMPLE **Buoyancy Correction**

Find the true mass of water (density = 1.00 g/mL) if the apparent mass is 100.00 g.

SOLUTION Equation 2-1 gives the true mass:

$$m = \frac{100.00 \text{ g}\left(1 - \dfrac{0.001\ 2 \text{ g/mL}}{8.0 \text{ g/mL}}\right)}{\left(1 - \dfrac{0.001\ 2 \text{ g/mL}}{1.00 \text{ g/mL}}\right)} = 100.11 \text{ g}$$

The buoyancy error for water is 0.11%, which is significant for many purposes. For NaCl with a density of 2.16 g/mL, the error is only 0.04%.

Ask Yourself

2-B. (a) Buoyancy corrections are most critical when you calibrate glassware such as a volumetric flask to see how much volume it actually holds. Suppose that you fill a 25-mL volumetric flask with distilled water and find that the mass of water in the flask measured in air is 24.913 g. What is the true mass of the water?

(b) You made the measurement when the lab temperature was 21°C, at which temperature the density of water is 0.998 00 g/mL. What is the true volume of water contained in the volumetric flask?

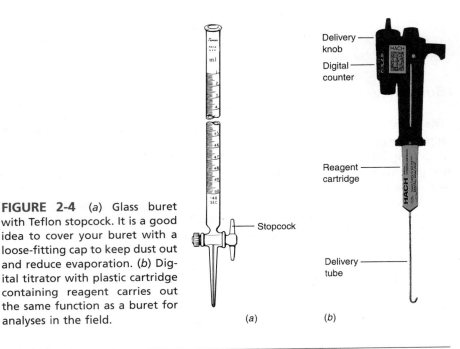

FIGURE 2-4 (*a*) Glass buret with Teflon stopcock. It is a good idea to cover your buret with a loose-fitting cap to keep dust out and reduce evaporation. (*b*) Digital titrator with plastic cartridge containing reagent carries out the same function as a buret for analyses in the field.

(*a*)   (*b*)

2-4 Burets

A **buret** is a precisely manufactured glass tube with graduations enabling you to measure the volume of liquid delivered through the stopcock (the valve) at the bottom (Figure 2-4a). The numbers on the buret increase from top to bottom (with 0 mL near the top). A volume measurement is made by reading the level before and after draining liquid from the buret and subtracting the first reading from the second reading. The graduations of Class A burets (the most accurate grade) are certified to meet the tolerances in Table 2-1. For example, if the reading of a 50-mL buret is 32.50 mL, the true volume can be anywhere in the range 32.45 to 32.55 mL and still be within the manufacturer's stated tolerance of ±0.05 mL.

When reading the liquid level in a buret, your eye should be at the same height as the top of the liquid. If your eye is too high, the liquid seems to be higher than it actually is. If your eye is too low, the liquid appears too low. The error that occurs when your eye is not at the same height as the liquid is called **parallax error.**

The surface of most liquids forms a concave **meniscus,** shown in Figure 2-5. It is helpful to use black tape on a white card as a background for locating the precise position of the meniscus. Align the top of the tape with the bottom of the meniscus and read the position on the buret. Highly colored solutions may appear to have two meniscuses. In such cases, either one may be used. Because volumes are determined by subtracting one reading from another, the most important point is to read the position of the meniscus reproducibly. Always estimate the reading to the nearest tenth of a division between marks.

The thickness of a graduation on a 50-mL buret corresponds to about 0.02 mL. To use the buret most accurately, consider the *top* of a graduation to be zero. When the meniscus is at the bottom of the same graduation, the reading is 0.02 mL greater.

Near the end point of a titration, try to deliver less than one drop at a time from the buret. This practice permits a finer location of the end point. (A drop from a 50-mL buret is about 0.05 mL.) To deliver a fraction of a drop, carefully open the stopcock until part of a drop is hanging from the buret tip. Then touch the inside

TABLE 2-1

Tolerances of Class A burets

Buret volume (mL)	Smallest graduation (mL)	Tolerance (mL)
5	0.01	±0.01
10	0.05 *or* 0.02	±0.02
25	0.1	±0.03
50	0.1	±0.05
100	0.2	±0.10

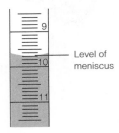

FIGURE 2-5 Buret with the meniscus at 9.68 mL. Estimate the reading of any scale to the nearest tenth of a division. This buret has 0.1-mL divisions, so we estimate the reading to 0.01 mL.

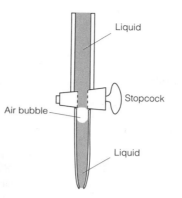

FIGURE 2-6 An air bubble trapped beneath the stopcock of a buret should be expelled before you use the apparatus.

glass wall of the receiving flask to the buret tip to transfer the droplet to the wall of the flask. Carefully tip the flask so that the main body of liquid washes over the newly added droplet. Then swirl the flask to mix the contents. Near the end of a titration, the flask should be tipped and rotated often to ensure that droplets on the wall containing unreacted analyte contact the bulk solution.

Liquid should drain evenly down the wall of a buret. The tendency of liquid to stick to the glass is reduced by draining the buret slowly (<20 mL/min). If many droplets stick to the wall, the buret should be cleaned with detergent and a buret brush. If this cleaning is insufficient, the buret should be soaked in peroxydisulfate-sulfuric acid cleaning solution prepared by your instructor. Cleaning solution eats clothing and people, as well as grease in the buret. Volumetric glassware should not be soaked in alkaline solutions, which attack glass. (A 5 wt % NaOH solution at 95°C dissolves Pyrex glass at a rate of 9 μm/h.)

A common error in using a buret is caused by failure to expel the bubble of air often found directly beneath the stopcock (Figure 2-6). A bubble present at the start of the titration may be filled with liquid during the titration. Therefore some volume that drained out of the graduated portion of the buret did not reach the titration vessel. Usually the bubble can be dislodged by draining the buret for a second or two with the stopcock wide open. A tenacious bubble can be expelled by carefully giving the buret an abrupt shake while draining it into a sink.

When you fill a buret with fresh solution, it is a wonderful idea to rinse the buret several times with small portions of the new solution, discarding each wash. It is not necessary to fill the entire buret with wash solution. Simply tilt the buret to allow the whole surface to be contacted by the wash liquid. This same technique should be used with any vessel (such as a spectrophotometer cuvet or a pipet) that is reused without opportunity for drying.

The *digital titrator* in Figure 2-4b is more convenient and portable, but less accurate, than the glass buret. The digital titrator is useful for conducting titrations in the field where samples are collected. The counter tells how much reagent from the cartridge has been dispensed by rotation of the delivery knob. Its accuracy of 1% is 10 times poorer than the accuracy of a glass buret, but many measurements do not require higher accuracy. The *electronic buret* in Figure 2-7 takes the challenge out of titrations. This battery-operated dispenser fits on a reagent bottle and delivers up to 99.99 mL in 0.01-mL increments. The volume dispensed is displayed on a digital readout.

Buret tips:

- Read bottom of concave meniscus
- Estimate reading to 1/10 of a division
- Avoid parallax
- Account for graduation thickness in readings
- Drain liquid slowly
- Wash buret with new solution
- Deliver fraction of a drop near end point
- Eliminate air bubble before use

FIGURE 2-7 Battery-operated electronic buret with digital readout delivers 0.01-mL increments from a reagent bottle. This device can be used for accurate titrations in the field.

FIGURE 2-8 (*a*) Class A glass volumetric flask meets tolerances in Table 2-2. (*b*) Class B polypropylene plastic flask for trace analysis (ppb concentrations) in which analyte might be lost by *adsorption* (sticking) on glass or contamination with previously adsorbed species. Class B flasks are less accurate than Class A flasks, with twice the tolerances of Table 2-2. (*c*) Short-form volumetric flask with Teflon-lined screw cap fits on analytical balance. Teflon prevents solutions from attacking the inside of the cap.

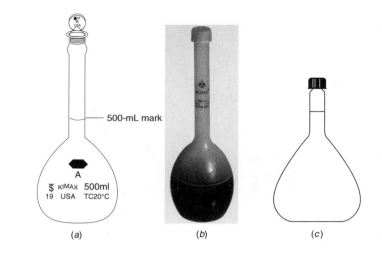

500-mL mark

A
KIMAX   500ml
19 USA   TC20°C

(*a*)   (*b*)   (*c*)

2-5 Volumetric Flasks

A **volumetric flask** (Figure 2-8, Table 2-2) is calibrated to contain a particular volume of solution at 20°C when the bottom of the meniscus is adjusted to the center of the mark on the neck of the flask. Most flasks bear the label "TC 20°C," which means *to contain* at 20°C. (Other types of glassware may be calibrated *to deliver,* "TD," their indicated volume.) The temperature of the container is relevant because both liquid and glass expand when heated.

A volumetric flask is used to prepare a solution of known volume. Typically, reagent is weighed into the flask, dissolved, and diluted to the mark. The mass of reagent and final volume are therefore known. The reagent is first dissolved in *less* than the final volume of liquid. More liquid is added and the solution is mixed again. The final volume adjustment is done with as much well-mixed liquid in the flask as possible. (When two different liquids are mixed, there is generally a small volume change. The total volume is *not* the sum of the two volumes that were mixed. By swirling the liquid in a nearly full volumetric flask before the liquid reaches the thin neck, you minimize the change in volume when the last liquid is added.) For best control, the final drops of liquid are added with a pipet, *not a squirt bottle.* After

TABLE 2-2	Tolerances of Class A volumetric flasks		
Flask capacity (mL)	Tolerance (mL)	Flask capacity (mL)	Tolerance (mL)
1	±0.02	100	±0.08
2	±0.02	200	±0.10
5	±0.02	250	±0.12
10	±0.02	500	±0.20
25	±0.03	1 000	±0.30
50	±0.05	2 000	±0.50

adjusting the liquid to the correct level, the cap should be held firmly in place and the flask inverted 20 times to ensure complete mixing.

Figure 2-9 shows how liquid appears when it is at the *center* of the mark of a volumetric flask or a pipet. Adjustment of the liquid level is done while viewing the flask from above or below the level of the mark. The front and back of the mark describe an ellipse with the meniscus at the center.

Glass is notorious for *adsorbing* traces of chemicals—especially cations. **Adsorption** means to stick to the surface. (In contrast, **absorption** means to take inside, as a sponge takes up water.) For critical work, an **acid wash** is carried out to replace low concentrations of cations on the glass surface with H^+. To do this, already thoroughly cleaned glassware is soaked in 3–6 M HCl (in a fume hood) for >1 h, followed by many rinses with distilled water and a final soak in distilled water. The HCl can be reused many times, as long as it is only used for soaking clean glassware.

Adsorption: to bind a substance on the surface
Absorption: to bind a substance internally

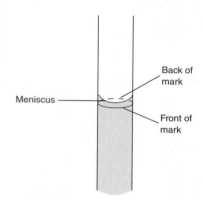

FIGURE 2-9 Proper position of the meniscus: at the center of the ellipse formed by the front and back of the calibration mark when viewed from above or below. Volumetric flasks and transfer pipets are calibrated to this position.

Ask Yourself

2-C. How would you use a volumetric flask to prepare 250.0 mL of 0.150 0 M K_2SO_4?

2-6 Pipets and Syringes

Pipets deliver known volumes of liquid. Four common types are shown in Figure 2-10. The *transfer pipet* is calibrated to deliver one fixed volume. The last drop of liquid does not drain out of the pipet; *it should not be blown out*. The *measuring pipet* is calibrated to deliver a variable volume, which is the difference between the volumes indicated before and after delivery. The measuring pipet in Figure 2-10 could be used to deliver 5.6 mL by starting delivery at the 1-mL mark and terminating at the 6.6-mL mark. The *Ostwald-Folin pipet* is similar to the transfer pipet, except that the last drop *should* be blown out. *Serological pipets* are measuring pipets calibrated all the way to the tip; the last drop *should* be blown out.

A transfer pipet is more accurate than a measuring pipet. The manufacturer's tolerance for a pipet is the allowed uncertainty in the volume that is actually deliv-

Do not blow the last drop out of a transfer pipet.

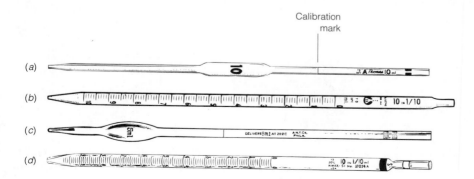

FIGURE 2-10 Common pipets: (a) transfer; (b) measuring (Mohr); (c) Ostwald-Folin (blow out last drop); (d) serological (blow out last drop).

ered. Tolerances for Class A (the most accurate grade) transfer pipets are given in Table 2-3.

Using a Transfer Pipet

Using a rubber bulb, *not your mouth,* suck liquid up past the calibration mark. It is a good idea to discard one or two pipet volumes of liquid to remove traces of previous reagents from the pipet. After taking up a third volume past the calibration mark, quickly replace the bulb with your index finger at the end of the pipet. The liquid should still be above the mark after this maneuver. Pressing the pipet against the bottom of the vessel while removing the rubber bulb helps prevent liquid from draining while you put your finger in place. Wipe the excess liquid off the outside of the pipet with a clean tissue. *Touch the tip of the pipet to the side of a beaker* and drain the liquid until the bottom of the meniscus just reaches the center of the mark, as shown in Figure 2-9. The purpose of touching the beaker wall while draining liquid is to draw liquid out of the pipet without leaving part of a drop hanging from the pipet when the level reaches the calibration mark.

Transfer the pipet to the desired receiving vessel and drain it *while holding the tip against the wall of the vessel.* After the pipet stops draining, hold it against the wall for a few more seconds to complete draining. *Do not blow out the last drop.* The pipet should be nearly vertical at the end of delivery. When you finish with a pipet, it should be rinsed with distilled water or soaked in a pipet container until it is cleaned. Solutions should never be allowed to dry inside a pipet because removing internal deposits is very difficult.

Micropipets

Plastic micropipets, such as that in Figure 2-11a, are used to deliver volumes in the 1 to 1 000 μL range (1 μL = 10^{-6} L). The liquid is contained in the disposable plastic tip. The accuracy is 1–2%, and precision may be as good as 0.5%.

To use a micropipet, place a fresh tip tightly on the barrel. Tips are contained in a package or dispenser so that you do not handle (and contaminate) the points with your fingers. Set the desired volume with the knob at the top of the pipet.

TABLE 2-3

Tolerances of Class A transfer pipets

Volume (mL)	Tolerance (mL)
0.5	±0.006
1	±0.006
2	±0.006
3	±0.01
4	±0.01
5	±0.01
10	±0.02
15	±0.03
20	±0.03
25	±0.03
50	±0.05
100	±0.08

Accuracy refers to the difference between the delivered volume and the desired volume. *Precision* refers to the reproducibility of replicate deliveries.

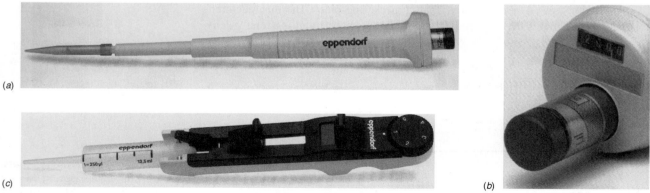

FIGURE 2-11 (*a*) Microliter pipet with disposable plastic tip. (*b*) Volume selection dial of microliter pipet. (*c*) Repeater pipet delivers preset volumes between 10 μL and 5 mL up to 48 times at 1-s intervals without refilling.

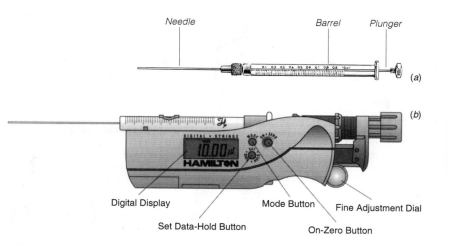

FIGURE 2-12 (*a*) Hamilton syringe with a volume of 1 μL and divisions of 0.01 μL on the glass barrel. (*b*) Digital dispenser for syringes with volumes of 0.5 to 500 μL provides accuracy and precision of 0.5%.

Depress the plunger to the first stop, which corresponds to the selected volume. Hold the pipet *vertically,* dip it 3–5 mm into the reagent solution, and *slowly* release the plunger to suck up liquid. Withdraw the tip from the liquid by sliding it along the wall of the vessel to remove liquid from the outside of the tip. To dispense liquid, touch the micropipet tip to the wall of the receiver and gently depress the plunger to the first stop. After a few seconds to allow liquid to drain down the wall of the pipet tip, depress the plunger further to squirt out the last liquid. It is a good idea to clean and wet a fresh tip by taking up and discarding two or three squirts of reagent first. The tip can be discarded or rinsed well with a squirt bottle and reused.

The volume of liquid taken into the tip depends on the angle at which the pipet is held and how far beneath the surface of reagent the tip is held during uptake. Each person will attain slightly different precision and accuracy with a micropipet.

A microliter *syringe* dispenses tiny volumes. Syringes such as that in Figure 2-12a come in sizes from 1 to 500 μL and have an accuracy and precision near 1%. The digital dispenser in Figure 2-12b improves the accuracy and precision to 0.5%. When using a syringe, take up and discard several volumes of liquid to wash the glass walls free of contaminants and to remove air bubbles from the barrel. The steel needle is attacked by strong acid and will contaminate strongly acidic solutions with iron.

Ask Yourself

2-D. Which is more accurate, a transfer pipet or a measuring pipet? What would you do differently in delivering 1.00 mL of liquid from a 1-mL serological pipet instead of a 1-mL measuring pipet?

2-7 Filtration

In **gravimetric analysis,** the mass of product from a reaction is measured to determine how much unknown was present. Precipitates from gravimetric analyses are collected by filtration, washed, and then dried. Most precipitates are collected in a

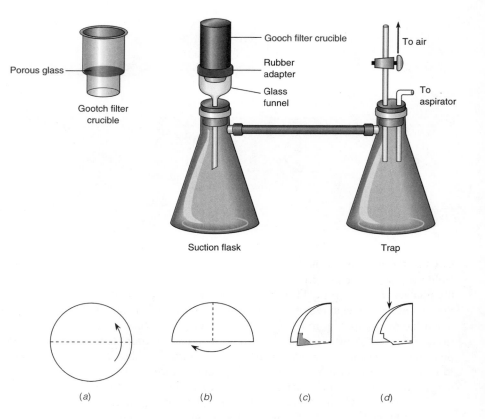

FIGURE 2-13 Filtration with a Gooch filter crucible that has a porous (*fritted*) glass disk through which liquid can pass. Suction is provided by an *aspirator* that uses flowing water from a tap to create a vacuum. The trap prevents possible backup of tap water from the aspirator into the suction flask.

FIGURE 2-14 Folding filter paper for a conical funnel. (*a*) Fold the paper in half and (*b*) in half again. (*c*) Tear off a corner, to better seat the paper in the funnel. (*d*) Open the side that was not torn to fit the paper in the funnel.

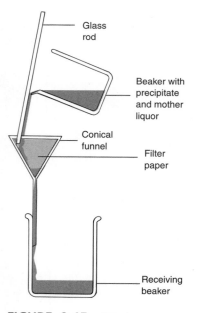

FIGURE 2-15 Filtering a precipitate.

fritted-glass funnel; suction is used to speed filtration (Figure 2-13). The porous glass plate in the funnel allows liquid to pass but retains solids. The empty crucible is first dried at 110°C and weighed. After collecting solid and drying again, the crucible and its contents are weighed again to determine the mass of solid.

Liquid from which a substance precipitates or crystallizes is called the **mother liquor.** Liquid that passes through the filter is called **filtrate.**

In some gravimetric procedures, **ignition** (heating at high temperature over a burner or in a furnace) is used to convert a precipitate to a known, constant composition. For example, Fe^{3+} precipitates as an ill-defined hydrated form of $Fe(OH)_3$ with variable composition. Ignition converts it to Fe_2O_3 prior to weighing. When a gravimetric precipitate is to be ignited, it is collected in **ashless filter paper,** which leaves little residue when burned.

To use filter paper with a conical glass funnel, fold the paper into quarters, tear off one corner (to allow a firm fit into the funnel), and place the paper in the funnel (Figure 2-14). The filter paper should fit snugly and be seated with some distilled water. When liquid is poured in, an unbroken stream of liquid should fill the stem of the funnel. The weight of liquid in the stem helps speed filtration.

For filtration, the slurry of precipitate in the mother liquor is poured down a glass rod to prevent splattering (Figure 2-15). (A *slurry* is a suspension of solid in liquid.) Particles adhering to the beaker or rod are dislodged with a **rubber policeman,** which is a flattened piece of rubber at the end of a glass rod. Use a jet of appropriate wash liquid from a squirt bottle to transfer particles from the rubber and glassware to the filter. If the precipitate is going to be ignited, particles remaining in the beaker should be wiped onto a small piece of moist filter paper; that paper is then added to the filter to be ignited.

2-8 Drying

Reagents, precipitates, and glassware are conveniently dried in an oven at 110°C. (Some chemicals require other temperatures.) Anything that you put in the oven should be labeled. A beaker and watchglass (Figure 2-16) minimize contamination by dust during drying. It is good practice to cover all vessels on the benchtop to prevent dust contamination.

The mass of a gravimetric precipitate is measured by weighing a dry, empty filter crucible before the procedure and weighing the same crucible filled with dry product after the procedure. To weigh the empty crucible, first bring it to "constant mass" by drying it in the oven for 1 h or longer and then cooling it for 30 min in a desiccator. Weigh the crucible and then heat it again for about 30 min. Cool it and reweigh it. When successive weighings agree to ±0.3 mg, the filter has reached "constant mass." A kitchen microwave oven can be used instead of an electric oven for drying reagents and crucibles. Try an initial heating time of 4 min, with subsequent 2-min heatings.

A **desiccator** (Figure 2-17) is a closed chamber containing a drying agent called a **desiccant.** The lid is greased to make an airtight seal. Desiccant is placed beneath the perforated disk at the bottom. Common desiccants in approximate order of decreasing efficiency are magnesium perchlorate $(Mg(ClO_4)_2)$ > barium oxide (BaO) ≈ alumina (Al_2O_3) ≈ phosphorus pentoxide (P_4O_{10}) >> calcium chloride $(CaCl_2)$ ≈ calcium sulfate $(CaSO_4$, called Drierite) ≈ silica gel (SiO_2). After placing a hot object in the desiccator, leave the lid cracked open for a minute until the object has cooled slightly. This practice prevents the lid from popping open when the air inside warms up. To open a desiccator, slide the lid sideways, rather than trying to pull it straight up.

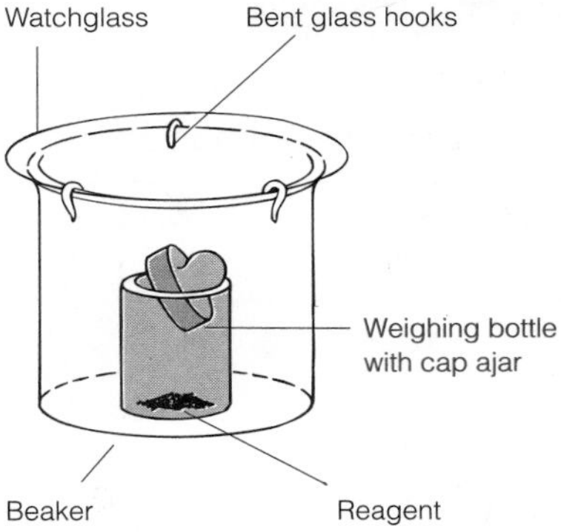

Watchglass   Bent glass hooks

Weighing bottle with cap ajar

Beaker   Reagent

FIGURE 2-16 Use of watchglass as a dustcover while drying reagent or crucible in the oven.

2-9 Calibration of Volumetric Glassware

Volumetric glassware can be calibrated to measure the volume that is actually contained in or delivered by a particular piece of equipment. Calibration is done by measuring the mass of water contained or delivered and using Table 2-4 to convert mass to volume:

$$\text{true volume} = (\text{mass of water}) \times (\text{correction factor in Table 2-4}) \qquad (2\text{-}2)$$

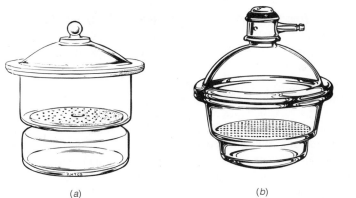

(a)   (b)

FIGURE 2-17 (a) Ordinary desiccator. (b) Vacuum desiccator, which can be evacuated through the sidearm and then sealed by rotating the joint containing the sidearm. Drying is more efficient at low pressure. Drying agents (*desiccants*) are placed at the bottom of each dessicator below the porous porcelain plate.

To calibrate a 25-mL transfer pipet, first weigh an empty weighing bottle like the one in Figure 2-16. Then fill the pipet to the mark with distilled water, drain it into the weighing bottle, and put on the lid to prevent evaporation. Weigh the bottle again to find the mass of water delivered from the pipet. Use Equation 2-2 to convert mass to volume.

TABLE 2-4

Correction factors for volumetric calibration

Temperature (°C)	Correction factor (mL/g)[a]
15	1.002 0
16	1.002 1
17	1.002 3
18	1.002 5
19	1.002 7
20	1.002 9
21	1.003 1
22	1.003 3
23	1.003 5
24	1.003 8
25	1.004 0
26	1.004 3
27	1.004 6
28	1.004 8
29	1.005 1
30	1.005 4

a. Factors are based on the density of water and are corrected for buoyancy with Equation 2-1.

EXAMPLE Calibration of a Pipet

An empty weighing bottle had a mass of 10.283 g. After adding water from a 25-mL pipet, the mass was 35.225 g. If the lab temperature was 23°C, find the volume of water delivered by the pipet.

SOLUTION The mass of water is $35.225 - 10.283 = 24.942$ g. From Equation 2-2 and Table 2-4, the volume of water is $(24.942 \text{ g})(1.003\ 5 \text{ mL/g}) = 25.029$ mL.

Ask Yourself

2-E. A 10-mL pipet delivered 10.000 0 g of water at 15°C to a weighing bottle. What is the true volume of the pipet?

2-10 Methods of Sample Preparation

The analytical process flow chart in Figure 1-1 showed that after a representative bulk sample is selected, a homogeneous laboratory sample must be prepared. Usually, materials are homogenized by grinding to a fine powder or by dissolving the entire sample. Solids can be ground with a **mortar and pestle** like the one in Figure 2-18.

Dissolving Inorganic Materials with Strong Acids

HCl	hydrochloric acid
HBr	hydrobromic acid
HF	hydrofluoric acid
H_3PO_4	phosphoric acid
H_2SO_4	sulfuric acid
HNO_3	nitric acid

HF is extremely harmful to touch or breathe. Flood the affected area with water, coat the skin with calcium gluconate (or another calcium salt), and seek medical help.

The acids HCl, HBr, HF, H_3PO_4, and dilute H_2SO_4 dissolve most metals (M) with heating by the reaction

$$\text{M}(s) + n\text{H}^+(aq) \longrightarrow \text{M}^{n+}(aq) + \tfrac{n}{2}\text{H}_2(g)$$

(In this equation, *s* stands for solid, *aq* stands for aqueous, and *g* stands for gas.) Many other inorganic substances can also be dissolved. Some anions react with H^+ to form **volatile** products (species that evaporate easily), which are lost from hot solutions in open vessels. Examples include carbonate ($CO_3^{2-} + 2H^+ \rightarrow H_2CO_3 \rightarrow CO_2 + H_2O$) and sulfide ($S^{2-} + 2H^+ \rightarrow H_2S$). Hot hydrofluoric acid dissolves silicates found in most rocks. HF also attacks glass, so it is used in Teflon, polyethylene, silver, or platinum vessels. Teflon is inert to attack by most chemicals and can be used up to 260°C.

Substances that do not dissolve in the acids above may dissolve as a result of oxidation by HNO_3 or concentrated H_2SO_4. Nitric acid attacks most metals, but not Au and Pt, which dissolve in the 3:1 (vol/vol) mixture of $HCl:HNO_3$ called *aqua regia*.

Acid dissolution is conveniently carried out with a Teflon-lined **bomb** (a sealed vessel) (Figure 2-19) in a microwave oven, which heats the contents to 200°C in a minute. The bomb cannot be made of metal, which absorbs microwaves. The bomb is cooled prior to opening to prevent loss of volatile products.

Fusion

Inorganic substances that do not dissolve in acid can usually be dissolved by a hot, molten inorganic **flux,** examples of which are lithium tetraborate ($Li_2B_4O_7$) and sodium hydroxide (NaOH). The finely powdered unknown is mixed with 2 to 20 times its mass of solid flux, and **fusion** (melting) is carried out in a platinum-gold alloy crucible at 300° to 1 200°C in a furnace or over a burner. When the sample is homogeneous, the molten flux is poured into a beaker containing 10 wt % aqueous HNO_3 to dissolve the product.

Digestion of Organic Substances

To analyze elements such as N, P, halogens (F, Cl, Br, I), and metals in an organic compound, the compound is first decomposed by combustion (described in Chapter 19) or by *digestion.* **Digestion** is the process in which a substance is decomposed by a reactive liquid and dissolved. For this purpose, sulfuric acid or a mixture of

2-10 Methods of Sample Preparation

Teflon is a *polymer* (a chain of repeating units) with the structure

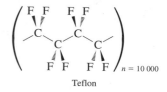

Teflon

Carbon atoms are in the plane of the page. A solid wedge is a bond coming out of the page toward you, and a dashed wedge is a bond going behind the page.

FIGURE 2-18 Agate mortar and pestle. The mortar is the base and the pestle is the grinding tool. Agate is very hard and expensive. Less expensive porcelain mortars are widely used, but they are somewhat porous and easily scratched. These properties can lead to contamination of the sample by porcelain particles or by traces of previous samples embedded in the porcelain.

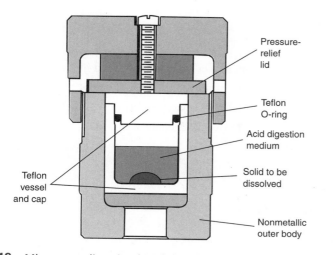

Pressure-relief lid

Teflon O-ring

Acid digestion medium

Solid to be dissolved

Teflon vessel and cap

Nonmetallic outer body

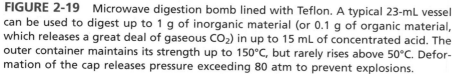

FIGURE 2-19 Microwave digestion bomb lined with Teflon. A typical 23-mL vessel can be used to digest up to 1 g of inorganic material (or 0.1 g of organic material, which releases a great deal of gaseous CO_2) in up to 15 mL of concentrated acid. The outer container maintains its strength up to 150°C, but rarely rises above 50°C. Deformation of the cap releases pressure exceeding 80 atm to prevent explosions.

H_2SO_4 and HNO_3 is added to an organic substance and the mixture is gently boiled (or heated in a microwave bomb) for 10 to 20 min until all particles have dissolved and the solution has a uniform black appearance. After cooling, the dark color is discharged by adding hydrogen peroxide (H_2O_2) or HNO_3, and heat is applied again. Constituents of the decomposed sample are analyzed after it is digested.

Extraction

In **extraction,** analyte is dissolved in a solvent that does not necessarily dissolve the entire sample and does not decompose the analyte. In a typical extraction of pesticides from soil, a mixture containing soil and the solvents acetone and hexane is placed in a Teflon-lined bomb and heated by microwaves to 150°C. This temperature is 50° to 100° higher than the boiling points of the individual solvents in an open vessel at atmospheric pressure. Soluble pesticides dissolve, but most of the soil remains behind. To complete the analysis, the solution is analyzed by chromatography, which is described in Chapters 16–18.

Ask Yourself

2-F. Lead sulfide (PbS) is a black solid that is very sparingly soluble in water but dissolves in concentrated HCl. If such a solution is boiled to dryness, white, crystalline lead chloride ($PbCl_2$) remains. What happened to the sulfide?

Key Equation

Buoyancy
$$m = m'\left(1 - \frac{d_a}{d_w}\right)\Big/\left(1 - \frac{d_a}{d}\right)$$

m = true mass; m' = mass measured in air

d_a = density of air (0.001 2 g/mL near 1 atm and 25°C)

d_w = density of balance weights (8.0 g/mL)

d = density of object being weighed

Important Terms

absorption	buoyancy	electronic balance
acid wash	buret	extraction
adsorption	desiccant	filtrate
ashless filter paper	desiccator	flux
bomb	digestion	fusion

gravimetric analysis mortar and pestle rubber policeman
hygroscopic mother liquor tare
ignition parallax error volatile
mechanical balance pipet volumetric flask
meniscus

Problems

2-1. What do the symbols TD and TC mean on volumetric glassware?

2-2. When would it be preferable to use a plastic volumetric flask instead of a glass flask?

2-3. What is the purpose of the trap in Figure 2-13? What does the watchglass do in Figure 2-16?

2-4. Distinguish absorption from adsorption. When you heat glassware in a drying oven, are you removing absorbed or adsorbed water?

2-5. What is the difference between digestion and extraction?

2-6. What is the true mass of water if the mass measured in air is 5.397 4 g?

2-7. Pentane (C_5H_{12}) is a liquid with a density of 0.626 g/mL. Find the true mass of pentane when the mass weighed in air is 14.82 g.

2-8. Ferric oxide (Fe_2O_3, density = 5.24 g/mL) obtained from ignition of a gravimetric precipitate weighed 0.296 1 g in the atmosphere. What is the true mass in vacuum?

2-9. Your professor has recruited you to work in her lab to help her win the Nobel Prize. It is therefore critical that your work be as accurate as possible. Rather than using the stated volumes of glassware in the lab, you decide to calibrate each piece. An empty 10-mL volumetric flask weighed 10.263 4 g. When filled to the mark with distilled water at 20°C, it weighed 20.214 4 g. What is the true volume of the flask?

2-10. Water from a 5-mL pipet was drained into a weighing bottle whose empty mass was 9.974 g to give a new mass of 14.974 g at 26°C. Find the volume of the pipet.

2-11. Water was drained from a buret between the 0.12- and 15.78-mL marks. The apparent volume was 15.78 − 0.12 = 15.66 mL. Measured in air at 25°C, the mass of water delivered was 15.569 g. What was the true volume?

Experimental Error

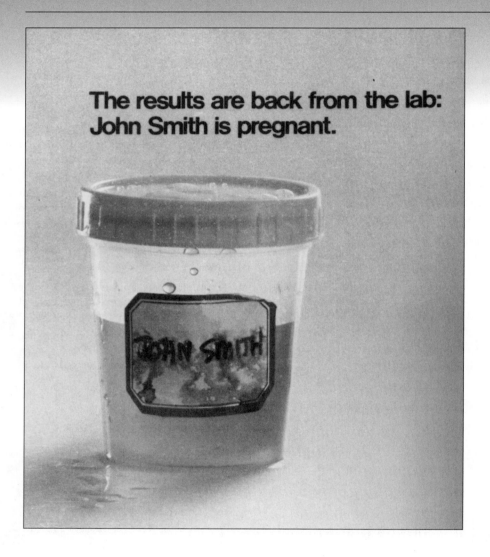

The results are back from the lab:
John Smith is pregnant.

Some laboratory errors are more obvious than others, but there is error associated with every measurement. There is no way to measure the "true" value of anything. The best we can do is to carefully apply a technique that experience tells us is reliable. Repetition of a measurement several times tells us about its reproducibility (*precision*). Measuring the same quantity by different methods gives us confidence of nearness to the "truth" (*accuracy*), if the results agree with one another.

Chapter 3

Math Toolkit

Suppose that you determine the density of a mineral by measuring its mass (4.635 ±0.002 g) and its volume (1.13 ±0.05 mL). Density is mass per unit volume: 4.635/1.13 = 4.101 8 g/mL. The uncertainties in measured mass and volume are ±0.002 g and ±0.05 mL, but what is the uncertainty in the computed density? And how many significant figures should be used for the density? This chapter answers these questions and introduces you to spreadsheets—a powerful numerical tool that will be invaluable to you in and out of this course.

3-1 Significant Figures

The number of **significant figures** is the minimum number of digits needed to write a given value in scientific notation without loss of accuracy. The number 142.7 has four significant figures, because it can be written 1.427×10^2. If you write $1.427\ 0 \times 10^2$, you imply that you know the value of the digit after 7, which is not the case for the number 142.7. The number $1.427\ 0 \times 10^2$ has five significant figures.

The number 6.302×10^{-6} has four significant figures, because all four digits are necessary. You could write the same number as 0.000 006 302, which also has just *four* significant figures. The zeros to the left of the 6 are merely holding decimal places. The number 92 500 is ambiguous. It could mean any of the following:

Significant figures: minimum number of digits required to express a value in scientific notation without loss of accuracy

9.25×10^4	3 significant figures
9.250×10^4	4 significant figures
$9.250\ 0 \times 10^4$	5 significant figures

You should write one of the three numbers above, instead of 92 500, to indicate how many figures are actually known.

Zeros are significant when they occur (1) in the middle of a number or (2) at the end of a number on the right-hand side of a decimal point.

Significant zeros below are **bold:**

10**6**   0.010 **6**   0.**10**6   0.106 **0**

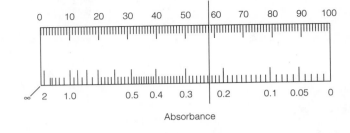

FIGURE 3-1 Scale of a Bausch and Lomb Spectronic 20 spectrophotometer. Percent transmittance is a linear scale and absorbance is a logarithmic scale.

The last (farthest to the right) significant figure in a measured quantity always has some associated uncertainty. The minimum uncertainty is ±1 in the last digit. The scale of a Spectronic 20 spectrophotometer is drawn in Figure 3-1. The needle in the figure appears to be at an absorbance value of 0.234. We say that there are three significant figures because the numbers 2 and 3 are completely certain and the number 4 is an estimate. The value might be read 0.233 or 0.235 by other people. The percent transmittance is near 58.3. Because the transmittance scale is smaller than the absorbance scale at this point, there is more uncertainty in the last digit of transmittance. A reasonable estimate of the uncertainty might be 58.3 ±0.2. There are three significant figures in the number 58.3.

Interpolation: Estimate all readings to the nearest tenth of the distance between scale divisions.

When reading the scale of any apparatus, interpolate between the markings. Try to estimate to the nearest tenth of the distance between two marks. Thus on a 50-mL buret, which is graduated to 0.1 mL, read the level to the nearest 0.01 mL. When using a ruler calibrated in millimeters, estimate distances to the nearest tenth of a millimeter.

Ask Yourself

3-A. How many significant figures are there in each number below?
   (a) 1.903 0   **(b)** 0.039 10   **(c)** 1.40×10^4

3-2 Significant Figures in Arithmetic

We now address the question of how many digits to retain in the answer after you have performed arithmetic operations with your data. Rounding should only be done on the *final answer* (not intermediate results), to avoid accumulating round-off errors.

Addition and Subtraction

If the numbers to be added or subtracted have equal numbers of digits, the answer goes to the *same decimal place* as in any of the individual numbers:

$$
\begin{array}{r}
1.362 \times 10^{-4} \\
+ \; 3.111 \times 10^{-4} \\
\hline
4.473 \times 10^{-4}
\end{array}
$$

The number of significant figures in the answer may exceed or be less than that in the original data.

$$\begin{array}{r} 5.345 \\ + 6.728 \\ \hline 12.073 \end{array} \qquad \begin{array}{r} 7.26 \times 10^{14} \\ - 6.69 \times 10^{14} \\ \hline 0.57 \times 10^{14} \end{array}$$

If the numbers being added do not have the same number of significant figures, we are limited by the least certain one. For example, in a calculation of the molecular weight of KrF_2, the answer is known only to the second decimal place, because we are limited by our knowledge of the atomic weight of Kr.

$$\begin{array}{rl} 18.998\ 403\ 2 & \text{(F)} \\ + 18.998\ 403\ 2 & \text{(F)} \\ + 83.80 & \text{(Kr)} \\ \hline 121.796\ 806\ 4 \end{array}$$

Not significant

The number 121.796 806 4 should be rounded to 121.80 as the final answer.

When rounding off, look at *all* the digits *beyond* the last place desired. In the example above, the digits 6 806 4 lie beyond the last significant decimal place. Because this number is more than halfway to the next higher digit, we round the 9 up to 10 (i.e., we round up to 121.80 instead of down to 121.79). If the insignificant figures were less than halfway, we would round down. For example, 121.794 8 is correctly rounded to 121.79.

In the special case where the number is exactly halfway, round to the nearest *even* digit. Thus, 43.550 00 is rounded to 43.6, if we can only have three significant figures. If we are retaining only three figures, 1.425×10^{-9} becomes 1.42×10^{-9}. The number $1.425\ 01 \times 10^{-9}$ would become 1.43×10^{-9}, because 501 is more than halfway to the next digit. The rationale for rounding to an even digit is to avoid systematically increasing or decreasing results through successive round-off errors. Half the round-offs will be up and half down.

In adding or subtracting numbers expressed in scientific notation, all numbers should first be expressed with the same exponent:

$$\begin{array}{r} 1.632 \times 10^5 \\ + 4.107 \times 10^3 \\ + 0.984 \times 10^6 \end{array} \Rightarrow \begin{array}{r} 1.632\quad \times 10^5 \\ + 0.041\ 07 \times 10^5 \\ + 9.84\quad\ \times 10^5 \\ \hline 11.51\quad \times 10^5 \end{array}$$

The sum $11.513\ 07 \times 10^5$ is rounded to 11.51×10^5 because the number 9.84×10^5 limits us to two decimal places when all numbers are expressed as multiples of 10^5.

Multiplication and Division

In multiplication and division, we are normally limited to the number of digits contained in the number with the fewest significant figures:

$$\begin{array}{r} 3.26 \times 10^{-5} \\ \times 1.78 \\ \hline 5.80 \times 10^{-5} \end{array} \qquad \begin{array}{r} 4.317\ 9 \times 10^{12} \\ \times 3.6\quad\ \times 10^{-19} \\ \hline 1.6\quad\ \times 10^{-6} \end{array} \qquad \begin{array}{r} 34.60 \\ \div\ 2.462\ 87 \\ \hline 14.05 \end{array}$$

The power of 10 has no influence on the number of figures that should be retained.

Rules for rounding off numbers

Addition and subtraction: Express all numbers with the same exponent and align all numbers with respect to the decimal point. Round off the answer according to the number of decimal places in the number with the fewest decimal places.

Challenge Show that the answer has four significant figures even if all numbers are expressed as multiples of 10^4 instead of 10^5.

Logarithms and Antilogarithms

The base 10 **logarithm** of n is the number a, whose value is such that $n = 10^a$:

Logarithm of n:

$$n = 10^a \text{ means that } \log n = a \qquad (3\text{-}1)$$

$10^{-3} = \frac{1}{10^3} = \frac{1}{1\,000} = 0.001$

For example, 2 is the logarithm of 100 because $100 = 10^2$. The logarithm of 0.001 is -3 because $0.001 = 10^{-3}$. To find the logarithm of a number with your calculator, enter the number and press the *log* function.

In Equation 3-1, the number n is said to be the **antilogarithm** of a. That is, the antilogarithm of 2 is 100 because $10^2 = 100$ and the antilogarithm of -3 is 0.001 because $10^{-3} = 0.001$. Your calculator has either a *10^x* key or an *antilog* key. To find the antilogarithm of a number, enter the number in your calculator and press *10^x* (or *antilog*).

A logarithm is composed of a **character** and a **mantissa.** The character is the integer part and the mantissa is the decimal part:

$$\log 339 = \underbrace{2}_{}.\underbrace{530}_{} \qquad \log 3.39 \times 10^{-5} = -\underbrace{4}_{}.\underbrace{470}_{}$$

Character Mantissa   Character Mantissa
$= 2$ $= 0.530$ $= -4$ $= 0.470$

Number of digits in **mantissa** of $\log x$ = number of significant figures in x:

$$\log (5.403 \times 10^{-8}) = -7.\underbrace{267\,4}_{}$$
4 digits (on 5.403) ... 4 digits

The number 339 can be written 3.39×10^2. *The number of digits in the mantissa of log 339 should equal the number of significant figures in 339.* The logarithm of 339 is properly expressed as 2.530. The *character*, 2, corresponds to the exponent in 3.39×10^2.

To see that the third decimal place is the last significant place, consider the following results:

$$10^{2.531} = 340 \ (339.6)$$
$$10^{2.530} = 339 \ (338.8)$$
$$10^{2.529} = 338 \ (338.1)$$

The numbers in parentheses are the results prior to rounding to three figures. Changing the exponent by one digit in the third decimal place changes the answer by one digit in the last (third) place of 339.

In converting a logarithm to its antilogarithm, *the number of significant figures in the antilogarithm should equal the number of digits in the mantissa.* Thus

$$\text{antilog} (-3.\underbrace{42}_{}) = 10^{-3.42} = \underbrace{3.8}_{} \times 10^{-4}$$
2 digits ... 2 digits 2 digits

Number of digits in antilog x $(= 10^x)$ = number of significant figures in **mantissa** of x:

$$10^{6.142} = 1.39 \times 10^6$$
3 digits ... 3 digits

Here are some examples showing the proper use of significant figures:

$$\log 0.001\,237 = -2.907\,6 \qquad \text{antilog } 4.37 = 2.3 \times 10^4$$
$$\log 1\,237 = 3.092\,4 \qquad 10^{4.37} = 2.3 \times 10^4$$
$$\log 3.2 = 0.51 \qquad 10^{-2.600} = 2.51 \times 10^{-3}$$

Ask Yourself

3-B. How would you express each answer with the correct number of digits?
  (a) $1.021 + 2.69 = 3.711$
  (b) $12.3 - 1.63 = 10.67$
  (c) $4.34 \times 9.2 = 39.928$
  (d) $0.060\ 2 \div (2.113 \times 10^4) = 2.849\ 03 \times 10^{-6}$
  (e) $\log (4.218 \times 10^{12}) = ?$
  (f) antilog $(-3.22) = ?$
  (g) $10^{2.384} = ?$

3-3 Types of Error

Every measurement has some uncertainty, which is called *experimental error.* Scientific conclusions can be expressed with a high or low degree of confidence, but never with complete certainty. Experimental error is classified as either *systematic* or *random.*

Systematic error is a consistent error that can be detected and corrected. Box 3-1 describes Standard Reference Materials designed to reduce systematic errors.

Systematic Error

A **systematic error,** also called a **determinate error,** can in principle be discovered and corrected. For example, using a pH meter that has been standardized incorrectly produces a systematic error in your results. Suppose you think that the pH of the buffer used to standardize the meter is 7.00, but it is really 7.08. If the meter is otherwise working properly, all of your pH readings will be 0.08 pH unit too low. When you read a pH of 5.60, the actual pH of the sample is 5.68. This systematic error could be discovered by using another buffer of known pH to test the meter.

Another systematic error arises when you use an uncalibrated buret. The manufacturer's tolerance for a Class A 50-mL buret is ± 0.05 mL. When you think you have delivered 29.43 mL, the real volume could be 29.40 mL and still be within tolerance. One way to correct for an error of this type is by constructing an experimental calibration curve, such as Figure 3-2. To do this, distilled water is delivered

Ways to detect systematic error:

1. Analyze samples of known composition, such as a Standard Reference Material. Your method should reproduce the known answer. (See Box 14-1 for an example.)
2. Analyze "blank" samples containing none of the analyte being sought. If you observe a nonzero result, your method responds to more than you intend.
3. Use different analytical methods to measure the same quantity. If the results do not agree, there is error in one (or more) of the methods.
4. *Round robin* experiment: Identical samples are analyzed in several laboratories by other persons using the same or different methods. Disagreement beyond the expected random error is systematic error.

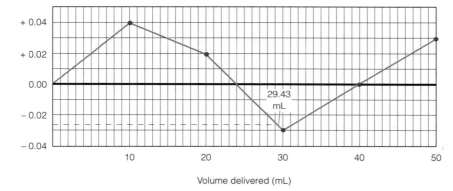

FIGURE 3-2 Calibration curve for a 50-mL buret.

What Are Standard Reference Materials?

Inaccurate laboratory measurements can mean wrong medical diagnosis and treatment, lost production time, wasted energy and materials, manufacturing rejects, and product liability problems. To minimize errors in laboratory measurements, the U.S. National Institute of Standards and Technology distributes more than 1 000 standard reference materials, such as metals, chemicals, rubber, plastics, engineering materials, radioactive substances, and environmental and clinical standards that can be used to test the accuracy of analytical procedures used in different laboratories.

For example, in treating patients with epilepsy, physicians depend on laboratory tests to measure blood serum concentrations of anticonvulsant drugs. Drug levels that are too low lead to seizures; high levels are toxic. Tests of identical serum specimens at different laboratories gave an unacceptably wide range of results. Therefore, the National Institute of Standards and Technology developed a standard reference material containing known levels of antiepilepsy drugs in serum. The reference material allows different laboratories to detect and correct errors in their assay procedures.

Before introduction of this reference material, five laboratories analyzing identical samples reported a range of results with relative errors of 40 to 110% of the expected value. After distribution of the reference material, the error was reduced to 20 to 40%.

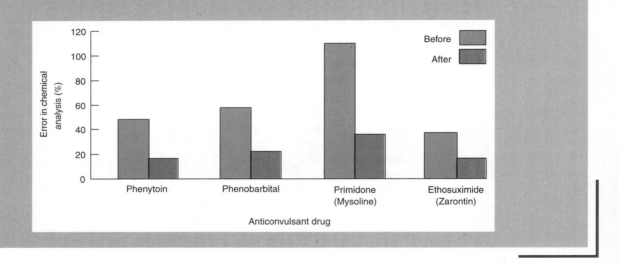

from the buret into a flask and weighed. You can determine the volume of water from its mass by using Table 2-4. The graph tells us to apply a correction factor of −0.03 mL to the measured value of 29.43 mL to reach the correct value of 29.40 mL.

Systematic error may be positive in some regions and negative in others. The key feature of systematic error is that, with care and cleverness, you can detect and correct it.

Random Error

Random error, also called **indeterminate error,** arises from limitations on our ability to make physical measurements. Random error has an equal chance of being positive or negative. It is always present and cannot be corrected. One random error is that associated with reading a scale. Different people reading the scale in

Figure 3-1 report a range of values representing their subjective interpolation between the markings. One person reading the same instrument several times might report several different readings. Another indeterminate error results from random electrical noise in an instrument. Positive and negative fluctuations occur with approximately equal frequency and cannot be completely eliminated.

Random error cannot be eliminated, but it might be reduced by a better experiment.

Precision and Accuracy

Precision is a measure of the reproducibility of a result. **Accuracy** refers to how close a measured value is to the "true" value.

Precision: reproducibility
Accuracy: nearness to the "truth"

A measurement might be reproducible, but wrong. For example, if you made a mistake preparing a solution for a titration, the solution would not have the desired concentration. You might then do a series of reproducible titrations but report an incorrect result because the concentration of the titrating solution was not what you intended. In this case, the precision is good but the accuracy is poor. Conversely, it is possible to make poorly reproducible measurements clustered around the correct value. In this case, the precision is poor but the accuracy is good. An ideal procedure is both precise and accurate.

Accuracy is defined as nearness to the "true" value. The word *true* is in quotes because somebody must *measure* the "true" value, and there is error associated with *every* measurement. The "true" value is best obtained by an experienced person using a well-tested procedure. It is desirable to test the result by using different procedures, because, even though each method might be precise, systematic error could lead to poor agreement between methods. Good agreement among several methods affords us confidence, but never proof, that the results are "true."

Absolute and Relative Uncertainty

Absolute uncertainty expresses the margin of uncertainty associated with a measurement. If the estimated uncertainty in reading a calibrated buret is ± 0.02 mL, we say that ± 0.02 mL is the absolute uncertainty associated with the reading.

Relative uncertainty compares the size of the absolute uncertainty to the size of its associated measurement. The relative uncertainty of a buret reading of 12.35 ± 0.02 mL is a dimensionless quotient:

Relative uncertainty:
$$\text{relative uncertainty} = \frac{\text{absolute uncertainty}}{\text{magnitude of measurement}} \qquad (3\text{-}2)$$

$$= \frac{0.02 \text{ mL}}{12.35 \text{ mL}} = 0.002$$

The percent relative uncertainty is simply

Percent relative uncertainty:
$$\text{percent relative uncertainty} = 100 \times \text{relative uncertainty} \qquad (3\text{-}3)$$

$$= 100 \times 0.002 = 0.2\%$$

If the absolute uncertainty in reading a buret is constant at ± 0.02 mL, the percent relative uncertainty is 0.2% for a volume of 10 mL and 0.1% for a volume of 20 mL.

Ask Yourself

3-C. Cheryl, Cynthia, Carmen, and Chastity shot these targets at Girl Scout camp. Match each target with the proper description.

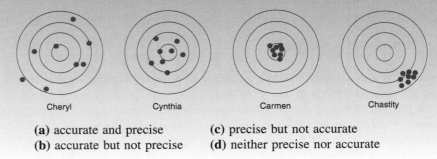

Cheryl   Cynthia   Carmen   Chastity

(a) accurate and precise   **(c)** precise but not accurate
(b) accurate but not precise   **(d)** neither precise nor accurate

3-4   *Propagation of Uncertainty*

We can usually estimate or measure the random error associated with a measurement, such as the length of an object or the temperature of a solution. The uncertainty might be based on how well we can read an instrument or on our experience with a particular method. If possible, uncertainty is expressed as the *standard deviation* or as a *confidence interval;* these parameters are based on a series of replicate measurements. The following discussion applies only to random error. We assume that systematic error has been detected and corrected.

In most experiments it is necessary to perform arithmetic operations on several numbers, each of which has an associated random error. The most likely uncertainty in the result is not simply the sum of the individual errors, because some of these are likely to be positive and some negative. We expect some cancellation of errors.

Standard deviation and confidence interval are discussed in the next chapter.

Addition and Subtraction

Suppose you wish to perform the following arithmetic, in which the experimental uncertainties, designated e_1, e_2, and e_3, are given in parentheses.

$$
\begin{aligned}
&1.76 \ (\pm 0.03) \leftarrow e_1 \\
+ \ &1.89 \ (\pm 0.02) \leftarrow e_2 \\
- \ &0.59 \ (\pm 0.02) \leftarrow e_3 \\
\hline
&3.06 \ (\pm e_4)
\end{aligned}
\tag{3-4}
$$

The arithmetic answer is 3.06; but what is the uncertainty associated with this result?

For addition and subtraction, the uncertainty in the answer is obtained from the *absolute uncertainties* of the individual terms as follows:

For addition and subtraction, use absolute *uncertainty.*

Uncertainty in addition and subtraction:

$$
e_4 = \sqrt{e_1^2 + e_2^2 + e_3^2}
\tag{3-5}
$$

For the sum in Equation 3-4, we can write

$$
e_4 = \sqrt{(0.03)^2 + (0.02)^2 + (0.02)^2} = 0.04_1
$$

The absolute uncertainty e_4 is ± 0.04, and we can write the answer as 3.06 ± 0.04. Although there is only one significant figure in the uncertainty, we wrote it initially as 0.04_1, with the first insignificant figure subscripted. We retain one or more insignificant figures to avoid introducing round-off errors into later calculations through the number 0.04_1. The insignificant figure was subscripted to remind us where the last significant figure should be at the conclusion of the calculations.

To find the percent relative uncertainty in the sum of Equation 3-4, we write

$$\text{percent relative uncertainty} = \frac{0.04_1}{3.06} \times 100 = 1._3\%$$

The uncertainty, 0.04_1, is $1._3\%$ of the result, 3.06. The subscript 3 in $1._3\%$ is not significant. It is sensible to drop the insignificant figures now and express the final result as

$$3.06 \ (\pm 0.04) \qquad \text{(absolute uncertainty)}$$
$$3.06 \ (\pm 1\%) \qquad \text{(relative uncertainty)}$$

For addition and subtraction, use absolute uncertainty. Relative uncertainty can be found at the end of the calculation.

EXAMPLE   Uncertainty in a Buret Reading

The volume delivered by a buret is the difference between the final reading and the initial reading. If the uncertainty in each reading is ± 0.02 mL, what is the uncertainty in the volume delivered?

SOLUTION Suppose that the initial reading is $0.05 \ (\pm 0.02)$ mL and the final reading is $17.88 \ (\pm 0.02)$ mL. The volume delivered is the difference:

$$
\begin{array}{r}
17.88 \ (\pm 0.02) \\
- \ \ 0.05 \ (\pm 0.02) \\
\hline
17.83 \ (\pm e)
\end{array}
\qquad e = \sqrt{0.02^2 + 0.02^2} = 0.03
$$

Regardless of the initial and final readings, if the uncertainty in each one is ± 0.02 mL, the uncertainty in volume delivered is ± 0.03 mL.

Multiplication and Division

For multiplication and division, first convert all uncertainties to percent relative uncertainties. Then calculate the error of the product or quotient as follows:

For multiplication and division, use percent relative uncertainty.

Uncertainty in multiplication and division:
$$e_4 = \sqrt{(\%e_1)^2 + (\%e_2)^2 + (\%e_3)^2} \qquad (3\text{-}6)$$

For example, consider the following operations:

$$\frac{1.76 \ (\pm 0.03) \times 1.89 \ (\pm 0.02)}{0.59 \ (\pm 0.02)} = 5.64 \pm e_4$$

First convert absolute uncertainties to percent relative uncertainties:

$$\frac{1.76 \ (\pm 1._7\%) \times 1.89 \ (\pm 1._1\%)}{0.59 \ (\pm 3._4\%)} = 5.64 \pm e_4$$

Advice Retain one or more extra insignificant figures until you have finished your entire calculation. Then round to the correct number of digits. When storing intermediate results in a calculator, keep all digits without rounding.

Then find the percent relative uncertainty of the answer by using Equation 3-6.

$$\%e_4 = \sqrt{(1._7)^2 + (1._1)^2 + (3._4)^2} = 4._0\%$$

The answer is $5.6_4\ (\pm4._0\%)$.

To convert relative uncertainty to absolute uncertainty, find $4._0\%$ of the answer:

$$4._0\% \times 5.6_4 = 0.04_0 \times 5.6_4 = 0.2_3$$

The answer is $5.6_4\ (\pm0.2_3)$. Finally, drop the insignificant digits:

$$5.6\ (\pm0.2) \qquad \text{(absolute uncertainty)}$$
$$5.6\ (\pm4\%) \qquad \text{(relative uncertainty)}$$

For multiplication and division, use percent relative uncertainty. Absolute uncertainty can be found at the end of the calculation.

The denominator of the original problem, 0.59, limits the answer to two digits.

Mixed Operations

Now consider an operation containing subtraction and division:

$$\frac{[1.76\ (\pm0.03) - 0.59\ (\pm0.02)]}{1.89\ (\pm0.02)} = 0.619_0 \pm\ ?$$

First work out the difference in the numerator, using absolute uncertainties.

$$1.76\ (\pm0.03) - 0.59\ (\pm0.02) = 1.17\ (\pm0.03_6)$$

because $\sqrt{(0.03)^2 + (0.02)^2} = 0.03_6$.

Then convert to percent relative uncertainties:

$$\frac{1.17\ (\pm0.03_6)}{1.89\ (\pm0.02)} = \frac{1.17\ (\pm3._1\%)}{1.89\ (\pm1._1\%)} = 0.619_0\ (\pm3._3\%)$$

because $\sqrt{(3._1\%)^2 + (1._1\%)^2} = 3._3\%$.

The percent relative uncertainty is $3._3\%$, so the absolute uncertainty is $0.03_3 \times 0.619_0 = 0.02_0$. The final answer can be written as follows:

$$0.619\ (\pm0.02_0) \qquad \text{(absolute uncertainty)}$$
$$0.619\ (\pm3._3\%) \qquad \text{(relative uncertainty)}$$

The result of a calculation ought to be written in a manner consistent with the uncertainty in the result.

Because the uncertainty begins in the 0.01 decimal place, it is reasonable to round the result to the 0.01 decimal place:

$$0.62\ (\pm0.02) \qquad \text{(absolute uncertainty)}$$
$$0.62\ (\pm3\%) \qquad \text{(relative uncertainty)}$$

The Real Rule for Significant Figures

The real rule: The first uncertain figure is the last significant figure.

The first uncertain figure of the answer is the last significant figure. For example, in the quotient

$$\frac{0.002\ 364\ (\pm 0.000\ 003)}{0.025\ 00\ (\pm 0.000\ 05)} = 0.094\ 6\ (\pm 0.000\ 2)$$

the uncertainty ($\pm 0.000\ 2$) occurs in the fourth decimal place. Therefore the answer is properly expressed with *three* significant figures, even though the original data have four figures. The first uncertain figure of the answer is the last significant figure. The quotient

$$\frac{0.002\ 664\ (\pm 0.000\ 003)}{0.025\ 00\ (\pm 0.000\ 05)} = 0.106\ 6\ (\pm 0.000\ 2)$$

is expressed with *four* significant figures because the uncertainty occurs in the fourth place. The quotient

$$\frac{0.821\ (\pm 0.002)}{0.803\ (\pm 0.002)} = 1.022\ (\pm 0.004)$$

is expressed with *four* figures even though the dividend and divisor each have *three* figures.

EXAMPLE | **Significant Figures in Laboratory Work**

You prepared a 0.250 M NH_3 solution by diluting 8.45 (± 0.04) mL of 28.0 (± 0.5) wt % NH_3 (density = 0.899 (± 0.003) g/mL) up to 500.0 (± 0.2) mL. Find the uncertainty in 0.250 M. Consider the molecular weight of NH_3, 17.031 g/mol, to have negligible uncertainty.

SOLUTION To find the uncertainty in molarity, you need to find the uncertainty in moles delivered to the 500-mL flask. The concentrated reagent contains 0.899 (± 0.003) g of solution per milliliter. The weight percent tells you that the reagent contains 0.280 (± 0.005) g of NH_3 per gram of solution. In the following calculations, you should retain extra insignificant digits and round off only at the end.

grams of NH_3 per mL in concentrated reagent

$$= 0.899\ (\pm 0.003)\ \frac{g\ solution}{mL} \times 0.280\ (\pm 0.005)\ \frac{g\ NH_3}{g\ solution}$$

$$= 0.899\ (\pm 0.334\%)\ \frac{g\ solution}{mL} \times 0.280\ (\pm 1.79\%)\ \frac{g\ NH_3}{g\ solution}$$

$$= 0.251\ 7\ (\pm 1.82\%)\ \frac{g\ NH_3}{mL}$$

For multiplication and division, convert absolute uncertainty to percent relative uncertainty.

because $\sqrt{(0.334\%)^2 + (1.79\%)^2} = 1.82\%$.

Next, find the moles of ammonia contained in 8.45 (± 0.04) mL of concentrated reagent. The relative uncertainty in volume is $\pm 0.04/8.45 = \pm 0.473\%$.

$$mol\ NH_3 = \frac{0.251\ 7\ (\pm 1.82\%)\ \frac{g\ NH_3}{mL} \times 8.45\ (\pm 0.473\%)\ mL}{17.031\ \frac{g\ NH_3}{mol}}$$

$$= 0.124\ 9\ (\pm 1.88\%)\ mol$$

because $\sqrt{(1.82\%)^2 + (0.473\%)^2 + (0\%)^2} = 1.88\%$.

This much ammonia was diluted to 0.500 0 ($\pm$0.000 2) L. The relative uncertainty in final volume is $\pm$0.000 2/0.500 0 = $\pm$0.04%. The diluted molarity is

$$\frac{\text{mol NH}_3}{\text{L}} = \frac{0.124\ 9\ (\pm 1.88\%)\ \text{mol}}{0.500\ 0\ (\pm 0.04\%)\ \text{L}}$$

$$= 0.249\ 8\ (\pm 1.88\%)\ \text{M}$$

because $\sqrt{(1.88\%)^2 + (0.04\%)^2} = 1.88\%$. The absolute uncertainty is 1.88% of 0.249 8 M = 0.018 8 $\times$ 0.249 8 M = 0.004 7 M. The uncertainty in molarity is in the third decimal place, so your final, rounded answer is

$$[\text{NH}_3] = 0.250\ (\pm 0.005)\ \text{M}$$

Ask Yourself

3-D. To help identify an unknown mineral in your geology class, you measured its mass and volume and found them to be 4.635 $\pm$0.002 g and 1.13 $\pm$0.05 mL.
   (a) Find the percent relative uncertainty in the mass and in the volume.
   (b) Write the density (= mass/volume) and its uncertainty with the correct number of digits.

3-5 Introducing Spreadsheets

Spreadsheets are simple, powerful tools for manipulating quantitative information with a computer. Spreadsheets allow us to conduct "what if" experiments in which we investigate effects such as changing acid strength or concentration on the shape of a titration curve. Any spreadsheet program is suitable for the exercises in this book. Our specific instructions apply to Microsoft Excel operating on an Apple Macintosh computer. You will need directions for your particular computer and software. Although this book can be used with no loss of continuity if you skip the spreadsheet exercises, they will enrich your understanding and give you a tool that is valuable outside of this course.

A Spreadsheet for Temperature Conversions

Let's prepare a spreadsheet to convert temperature from degrees Celsius to kelvins and degrees Fahrenheit by using formulas derived from Table 1-4:

$$\text{K} = °\text{C} + C_0 \tag{3-7a}$$
$$°\text{F} = \left(\tfrac{9}{5}\right) * °\text{C} + 32 \tag{3-7b}$$

where C_0 is the constant 273.15.

Figure 3-3a shows a blank spreadsheet as it would appear on your computer screen. Rows are numbered 1, 2, 3, . . . and columns are lettered A, B, C, ... Each rectangular box is called a *cell*. The fourth cell down in the second column, for example, is designated cell B4.

We adopt a standard format in this book in which constants are collected in column A. Select cell A1 and type "Constant:" as a column heading. Select cell A2 and type "C0 =" to indicate that the constant C_0 will be written in the next cell down. Now select cell A3 and type the number 273.15. Your spreadsheet should now look like the one in Figure 3-3b.

In cell B1 type the label "°C" (or "Celsius" or whatever you like). For illustration, we will enter the numbers -200, -100, 0, 100, and 200 in cells B2 through B6. This is our *input* to the spreadsheet. The *output* will be computed values of kelvins and °F in columns C and D. (If you want to enter very large or very small numbers, you can write, for example, 6.02e23 for 6.02×10^{23} and 2e-8 for 2×10^{-8}.)

Label column C "kelvin" in cell C1. In cell C2 we enter our first *formula*—an entry beginning with an equals sign. Select cell C2 and type "=B2+$A$3". This tells the computer to calculate the contents of cell C2 by taking the contents of cell B2 and adding the contents of cell A3 (which contains the constant, 273.15). We will explain the dollar signs shortly. When this formula is entered, the computer responds by calculating the number 73.15 in cell C2. This is the kelvin equivalent of -200°C.

Now comes the beauty of a spreadsheet. Instead of typing many similar formulas, select cells C2, C3, C4, C5, and C6 all together and give a FILL DOWN command. (Other software uses the command COPY, instead of FILL DOWN.) This command tells the computer to do the same thing in cells C3 through C6 that was done in cell C2. The numbers 173.15, 273.15, 373.15, and 473.15 will appear in cells C3 through C6.

The formula "=B2+$A$3" in cell C2 is equivalent to writing $K = °C + C_0$.

Columns

(a)

	A	B	C	D
1				
2				
3				
4		cell B4		
5				
6				
7				
8				
9				
10				

Rows

(b)

	A	B	C	D
1	Constant:			
2	C0 =			
3	273.15			
4				
5				
6				
7				
8				
9				
10				

(c)

	A	B	C	D
1	Constant:	°C	kelvin	°F
2	C0 =	−200	73.15	−328
3	273.15	−100	173.15	−148
4		0	273.15	32
5		100	373.15	212
6		200	473.15	392
7				
8				
9				
10				

(d)

	A	B	C	D
1	Constant:	°C	kelvin	°F
2	C0 =	−200	73.15	−328
3	273.15	−100	173.15	−148
4		0	273.15	32
5		100	373.15	212
6		200	473.15	392
7				
8	Formulas:			
9	C2 = B2+$A$3			
10	D2 = (9/5)*B2+32			

FIGURE 3-3 Constructing a spreadsheet for temperature conversions.

Absolute reference: $A$3
Relative reference: B2

When computing the output in cell C3, the computer automatically uses input from cell B3 instead of cell B2. The reason for the dollar signs in $A$3 is that we do not want the computer to go down to cell A4 to find input for cell C3. $A$3 is called an *absolute reference* to cell A3. No matter what cell uses the constant C_0, we want it to come from cell A3. The reference to cell B2 is a *relative reference*. Cell C2 will use the contents of cell B2. Cell C6 will use the contents of cell B6. In general, references to constants in column A will be absolute (with dollar signs). References to numbers in the remainder of a spreadsheet will usually be relative (without dollar signs).

In cell D1 enter the label "°F". In cell D2, type the formula "=(9/5)*B2+32". This is equivalent to writing $°F = (\frac{9}{5}) * °C + 32$. The slash (/) is a division sign and the asterisk (*) is a multiplication sign. Parentheses are used to make the computer do what we intend. Operations inside parentheses are carried out before operations outside the parentheses. The computer responds to this formula by writing -328 in cell D2. This is the Fahrenheit equivalent to $-200°C$. Select cells D2 through D6 all together and use FILL DOWN to complete the table shown in Figure 3-3c.

Order of Operations

The arithmetic operations in a spreadsheet are addition, subtraction, multiplication, division, and exponentiation (which uses the symbol ^). The order of operations in formulas is ^ first, followed by * and / (evaluated in order from left to right as they appear), finally followed by + and - (also evaluated from left to right). Make liberal use of parentheses to be sure that the computer does what you intend. The contents of parentheses are evaluated first, before carrying out operations outside the parentheses. Here are some examples:

$$9/5*100+32 = (9/5)*100+32 = (1.8)*100+32 = (1.8*100)+32 = (180)+32 = 212$$
$$9/5*(100+32) = 9/5*(132) = (1.8)*(132) = 237.6$$
$$9+5*100/32 = 9+(5*100)/32 = 9+(500)/32 = 9+(500/32) = 9+(15.625) = 24.625$$
$$9/5\text{^}2+32 = 9/(5\text{^}2)+32 = (9/25)+32 = (0.36)+32 = 32.36$$

When in doubt about how an expression will be evaluated by the computer, use parentheses to force it to do what you intend.

Documentation and Readability

If your spreadsheet cannot be read by another person without your help, it needs better documentation. (The same is true of your lab notebook!)

If you look at your spreadsheet next week, you will probably not know what formulas were used. Therefore, we add the text (labels) in cells A8, A9, and A10 in Figure 3-3d. In cell A8 write "Formulas". In cell A9 write "C2=B2+$A$3" and in cell A10 write "D2=(9/5)*B2+32". Such *documentation* is an excellent practice for every spreadsheet. As you learn to use your spreadsheet, you should use CUT and PASTE commands to copy the formulas used in cells C2 and D2 into the labels in cells A9 and A10. This saves time and reduces transcription errors.

For additional readability, you should learn to control how many decimal places are displayed in a given cell or column, even though the computer retains more digits for calculations. It does not throw away the digits that are not displayed. You can also control whether numbers are displayed in decimal or exponential notation.

	A	B	C
1	Constants:	Temp (°C)	Density (g/mL)
2	a0 =	5	0.99997
3	0.99989	10	0.99970
4	a1 =	15	0.99911
5	5.3322E-05	20	0.99821
6	a2 =	25	0.99705
7	−7.5899E-06	30	0.99565
8	a3 =	35	0.99403
9	3.6719E-08	40	0.99223
10			
11	Formula:		
12	C2 = $A$3+($A$5*B2)+($A$7*B2^2)+($A$9*B2^3)		

FIGURE 3-4 Spreadsheet for computing the density of water as a function of temperature.

Ask Yourself

3-E. Can you reproduce the spreadsheet in Figure 3-3 on your computer? The boiling point of nitrogen (N_2) at 1 atm pressure is −196°C. Use your spreadsheet to find the kelvin and degrees Fahrenheit equivalents of −196°C. Check your answers with your calculator.

3-6   *Spreadsheet + Graph = Power*

Although a spreadsheet has great power for crunching numbers, humans require a visual display to appreciate the relationship between two columns of numbers. Therefore, results of a spreadsheet calculation are plotted directly by the spreadsheet program or can be pasted into a graphing program for display.

Figure 3-4 shows a spreadsheet for computing the density of water as a function of temperature (°C) with the equation

$$\text{density (g/mL)} = a_0 + a_1 * T + a_2 * T^2 + a_3 * T^3 \tag{3-8}$$

where $a_0 = 0.999\ 89$, $a_1 = 5.332\ 2 \times 10^{-5}$, $a_2 = -7.589\ 9 \times 10^{-6}$, and $a_3 = 3.671\ 9 \times 10^{-8}$. The constants a_0 to a_3 are entered in column A. Column B is labeled "Temp (°C)" and column C is labeled "Density (g/mL)". Enter values of temperature in column B. In column C2, type the formula "=$A$3+($A$5*B2)+($A$7*B2^2)+($A$9*B2^3)", which uses the exponent symbol ^ to compute T^2 and T^3. When you enter the formula, the number 0.999 97 is computed in cell C2. The remainder of column C is completed with a FILL DOWN (or COPY) command. The spreadsheet is not finished until it is documented by entering text in cells A11 and A12 to show what formula was used in column C.

To see the relationship between the input in column B and the output in column C we need a graph. At this point, you will need instruction for using your spreadsheet software or a graphing program to plot column C on the y-axis (the ordinate) and column B on the x-axis (the abscissa). Results displayed in Figure 3-5 show that the change in density from 5° to 15°C is small relative to the change in density between 30° and 40°C.

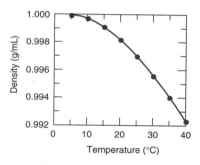

FIGURE 3-5 Density of water computed with the spreadsheet in Figure 3-4.

Equation 3-8 is accurate to five decimal places over the range 4° to 40°C.

A spreadsheet is not complete until it is documented.

Ask Yourself

3-F. Can you reproduce the spreadsheet in Figure 3-4 on your computer? Can you use a graphing program to reproduce the graph in Figure 3-5?

Key Equations

Definition of logarithm	If $n = 10^a$, then a is the logarithm of n.
Definition of antilogarithm	If $n = 10^a$, then n is the antilogarithm of a.
Relative uncertainty	$\text{relative uncertainty} = \dfrac{\text{absolute uncertainty}}{\text{magnitude of measurement}}$
Percent relative uncertainty	$\text{percent relative uncertainty} = 100 \times \text{relative uncertainty}$
Uncertainty in addition and subtraction	$e_4 = \sqrt{e_1^2 + e_2^2 + e_3^2}$ (use absolute uncertainties) $e_4 = $ uncertainty in final answer $e_1, e_2, e_3 = $ uncertainty in individual terms
Uncertainty in multiplication and division	$\%e_4 = \sqrt{\%e_1^2 + \%e_2^2 + \%e_3^2}$ (use percent relative uncertainties)

Important Terms

absolute uncertainty	determinate error	mantissa	relative uncertainty
accuracy	indeterminate error	precision	significant figure
antilogarithm	logarithm	random error	systematic error
character			

Problems

3-1. Round each number as indicated:
 (a) 1.236 7 to 4 significant figures
 (b) 1.238 4 to 4 significant figures
 (c) 0.135 2 to 3 significant figures
 (d) 2.051 to 2 significant figures
 (e) 2.005 0 to 3 significant figures

3-2. Round each number to three significant figures:
 (a) 0.216 74   **(b)** 0.216 5   **(c)** 0.216 500 3

3-3. Indicate how many significant figures there are in
 (a) 0.305 0   **(b)** 0.003 050   **(c)** 1.003×10^4

3-4. Write each answer with the correct number of digits:
 (a) $1.0 + 2.1 + 3.4 + 5.8 = 12.300\ 0$
 (b) $106.9 - 31.4 = 75.500\ 0$
 (c) $107.868 - (2.113 \times 10^2) + (5.623 \times 10^3) = $
   $\qquad\qquad\qquad\qquad\qquad\qquad\qquad 5\ 519.568$
 (d) $(26.14/37.62) \times 4.38 = 3.043\ 413$

(e) $(26.14/37.62 \times 10^8) \times (4.38 \times 10^{-2}) =$
$$3.043\ 413 \times 10^{-10}$$

(f) $(26.14/3.38) + 4.2 = 11.933\ 7$

(g) $\log (3.98 \times 10^4) = 4.599\ 9$

(h) $10^{-6.31} = 4.897\ 79 \times 10^{-7}$

3-5. Write each answer with the correct number of digits:

(a) $3.021 + 8.99 = 12.011$

(b) $12.7 - 1.83 = 10.87$

(c) $6.345 \times 2.2 = 13.959\ 0$

(d) $0.030\ 2 \div (2.114\ 3 \times 10^{-3}) = 14.283\ 69$

(e) $\log (2.2 \times 10^{-18}) = ?$

(f) antilog $(-2.224) = ?$

(g) $10^{-4.555} = ?$

3-6. Find the formula weights of **(a)** $BaCl_2$ and **(b)** $C_{31}H_{32}O_8N_2$ with the correct number of significant figures.

3-7. Why do we use quotation marks around the word *true* in the statement that accuracy refers to how close a measured value is to the "true" value?

3-8. Explain the difference between systematic and random error.

3-9. Rewrite the number $3.123\ 56$ ($\pm 0.167\ 89\%$) in the forms **(a)** number ($\pm$absolute uncertainty) and **(b)** number ($\pm$percent relative uncertainty) with an appropriate number of digits.

3-10. Write each answer with the correct number of digits. Find the absolute uncertainty and percent relative uncertainty for each answer.

(a) $6.2\ (\pm 0.2) - 4.1\ (\pm 0.1) = ?$

(b) $9.43\ (\pm 0.05) \times 0.016\ (\pm 0.001) = ?$

(c) $[6.2\ (\pm 0.2) - 4.1\ (\pm 0.1)] \div 9.43\ (\pm 0.05) = ?$

(d) $9.43\ (\pm 0.05) \times \{[6.2\ (\pm 0.2) \times 10^{-3}] + [4.1\ (\pm 0.1) \times 10^{-3}]\} = ?$

3-11. Write each answer with a reasonable number of figures. Find the absolute uncertainty and percent relative uncertainty for each answer.

(a) $[12.41\ (\pm 0.09) \div 4.16\ (\pm 0.01)] \times$
$$7.068\ 2\ (\pm 0.000\ 4) = ?$$

(b) $[3.26\ (\pm 0.10) \times 8.47\ (\pm 0.05)] - 0.18\ (\pm 0.06) = ?$

(c) $6.843\ (\pm 0.008) \times 10^4 \div [2.09\ (\pm 0.04) - 1.63\ (\pm 0.01)] = ?$

3-12. Write each answer with the correct number of digits. Find the absolute uncertainty and percent relative uncertainty for each answer.

(a) $9.23\ (\pm 0.03) + 4.21\ (\pm 0.02) - 3.26\ (\pm 0.06) = ?$

(b) $91.3\ (\pm 1.0) \times 40.3\ (\pm 0.2) / 21.2\ (\pm 0.2) = ?$

(c) $[4.97\ (\pm 0.05) - 1.86\ (\pm 0.01)] / 21.2\ (\pm 0.2) = ?$

(d) $2.016\ 4\ (\pm 0.000\ 8) + 1.233\ (\pm 0.002) + 4.61\ (\pm 0.01) = ?$

(e) $2.016\ 4\ (\pm 0.000\ 8) \times 10^3 + 1.233\ (\pm 0.002) \times 10^2 + 4.61\ (\pm 0.01) \times 10^1 = ?$

3-13. Find the absolute and percent relative uncertainty and express each answer with a reasonable number of significant figures.

(a) $3.4\ (\pm 0.2) + 2.6\ (\pm 0.1) + ?$

(b) $3.4\ (\pm 0.2) \div 2.6\ (\pm 0.1) = ?$

(c) $[3.4\ (\pm 0.2) \times 10^{-8}] \div [2.6\ (\pm 0.1) \times 10^3] = ?$

(d) $[3.4\ (\pm 0.2) - 2.6\ (\pm 0.1)] \times 3.4\ (\pm 0.2) = ?$

3-14. Express the molecular weight ($\pm$uncertainty) of benzene, C_6H_6, with the correct number of significant figures. The periodic table inside the cover of this book has a note in the legend about uncertainties in atomic weights.

3-15. Express the molecular weight of $C_6H_{13}B$ with the correct number of significant figures and find its uncertainty.

3-16. (a) Show that the formula weight of NaCl is $58.442\ 5$ ($\pm 0.000\ 9$) g/mol.

(b) To prepare a solution of NaCl, you weigh out $2.634\ (\pm 0.002)$ g and dissolve it in a volumetric flask with a volume of $100.00\ (\pm 0.08)$ mL. Express the molarity of the resulting solution, along with its uncertainty, with an appropriate number of digits.

3-17. (a) For use in an iodine titration, you prepare a solution from $0.222\ 2\ (\pm 0.000\ 2)$ g of KIO_3 [FW $214.001\ 0\ (\pm 0.000\ 9)$] in $50.00\ (\pm 0.05)$ mL. Find the molarity and its uncertainty with an appropriate number of significant figures.

(b) Would your answer be affected significantly if the reagent were only 99.9% pure?

3-18. Your instructor has asked you to prepare 2.00 L of 0.169 M NaOH from a stock solution of $53.4\ (\pm 0.4)$ wt % NaOH with a density of $1.52\ (\pm 0.01)$ g/mL.

(a) How many milliliters of stock solution will you need?

(b) If the uncertainty in delivering the NaOH is ± 0.10 mL, calculate the absolute uncertainty in the molarity (0.169 M). Assume negligible uncertainty in the molecular weight of NaOH and in the final volume, 2.00 L.

3-19. Create a spreadsheet to convert energy from joules (in column B) into calories (in column C), British thermal units (in column D), and electron volts (in column E). In column A, write constants from Table 1-4 to convert one set of units to another. For input in column B, use 2×10^{-20}, 1, 100, 1 000, and 2×10^{20} J. Document your spreadsheet with readable formulas at the bottom.

Is My Red Blood Cell Count High Today?

Red blood cells (erythrocytes, Er) tangled in fibrin threads (Fi) in a blood clot. Stacks of erythrocytes in a clot are called a rouleaux formation (Ro).

All measurements contain experimental error, so it is impossible to be completely certain of a result. Nevertheless, we seek to answer questions such as "Is my red blood cell count today higher than usual?" If today's count is twice as high as usual, it is probably truly higher than normal. But what if the "high" count is not excessively above "normal" counts?

Count on "normal" days	Today's count
5.1 5.3 4.8 $\times 10^6$ cells/μL 5.4 5.2	5.6×10^6 cells/μL

The number 5.6 is higher than the five normal values, but the random variation in normal values might lead us to expect that 5.6 will be observed on some "normal" days.

The study of statistics allows us to say that over a long period, today's value will be observed on 1 out of 20 normal days. It is still up to you to decide what to do with this information.

Chapter 4

Statistics

Experimental measurements always have some random error, so no conclusion can be drawn with complete certainty. However, statistics gives us tools to accept conclusions that have a high probability of being correct and to reject conclusions that do not. This chapter describes basic statistical tests and introduces the method of least squares for creating calibration curves.

4-1 *The Gaussian Distribution*

Nerve cells communicate with muscle cells by releasing neurotransmitter molecules adjacent to the muscle. As shown in Figure 4-1, neurotransmitters bind to membrane proteins of the muscle cell and open up channels that permit cations to diffuse into the muscle cell. Ions entering the cell trigger contraction of the muscle. The study of signal transmission at the neuromuscular junction led to the Nobel Prize in Medicine and Physiology in 1991 for Bert Sakmann and Erwin Neher.

Channels are all the same size, so each should allow a similar rate of ion passage across the membrane. Because ions are charged particles, the flow of ions is equiv-

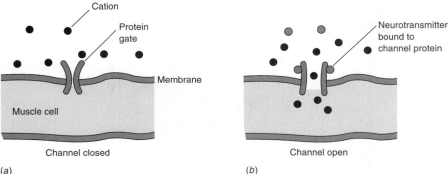

(a)   (b)

FIGURE 4-1 (a) In the absence of neurotransmitter, the ion channel is closed and cations cannot enter the muscle cell. (b) In the presence of neurotransmitter, the channel opens, cations enter the cell, and muscle action is initiated.

55

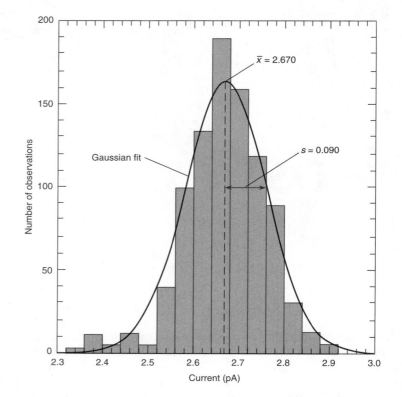

FIGURE 4-2 Bar chart showing observed cation current passing through individual channels of a frog muscle cell. The smooth line is the Gaussian curve that has the same mean and standard deviation as the measured data.

alent to a flow of electricity across the membrane. Of the 922 ion channel responses recorded for Figure 4-2, 190 are in the narrow range 2.64 to 2.68 pA (picoamperes, 10^{-12} amperes), represented by the tallest bar at the center of the chart. The next most probable responses fall in the range just to the right of the tallest bar, and the third most probable responses fall in the range just to the left of the tallest bar.

The bar graph in Figure 4-2 is typical of the vast majority of laboratory measurements: The most probable response is at the center, and the probability of observing other responses decreases as the distance from the center increases. The smooth, bell-shaped curve superimposed on the data in Figure 4-2 is called a **Gaussian distribution.** The more measurements made on any physical system, the closer will the bar chart approach the smooth curve.

Mean and Standard Deviation

Mean locates center of distribution. *Standard deviation* measures width of distribution.

A Gaussian distribution is characterized by a *mean* and a *standard deviation.* The mean is the *center* of the distribution and the standard deviation is a measure of the *width* of the distribution.

The arithmetic **mean,** $\bar{x}$, also called the **average,** is the sum of the measured values divided by the number of measurements.

Mean:
$$\bar{x} = \frac{\sum_i x_i}{n} = \frac{1}{n}(x_1 + x_2 + x_3 + \ldots + x_n)$$
(4-1)

where each x_i is a measured value. Capital Greek sigma, Σ, is the symbol for a sum. In Figure 4-2 the mean value is indicated by the arrow at 2.670 pA.

The **standard deviation,** *s*, is a measure of the width of the distribution. *The smaller the standard deviation, the narrower the distribution.*

Standard deviation:

$$s = \sqrt{\frac{\sum_i (x_i - \bar{x})^2}{n - 1}}$$

(4-2)

In Figure 4-2, $s = 0.090$ pA. Figure 4-3 shows that if the standard deviation were doubled, the Gaussian curve for the same number of observations would be shorter and broader.

The *relative standard deviation* is the standard deviation divided by the average. It is usually expressed as a percentage. For $s = 0.090$ pA and $\bar{x} = 2.670$ pA, the relative standard deviation is $(0.090/2.670) \times 100 = 3.4\%$.

The quantity $n - 1$ in the denominator of Equation 4-2 is called the *degrees of freedom*. The problem begins with n independent data points. After computing the average, there are only $n - 1$ independent pieces of information left, because it is possible to calculate the nth data point if you know $n - 1$ data points and the average.

The symbols $\bar{x}$ and s apply to a finite set of measurements. For the Gaussian curve describing an infinite set of data, the true mean (called the population mean) is designated μ (Greek mu) and the true standard deviation is denoted by σ (lowercase Greek sigma).

The smaller the standard deviation, the more *precise* (reproducible) the results. Greater precision does not necessarily imply greater *accuracy,* which means nearness to the "truth."

EXAMPLE   Mean and Standard Deviation

Find the mean, standard deviation, and relative standard deviation for the set of measurements (7, 18, 10, 15).

SOLUTION The mean is

$$\bar{x} = \frac{7 + 18 + 10 + 15}{4} = 12._5$$

To avoid accumulating round-off errors, retain one more digit for the mean and the standard deviation than was present in the original data. The standard deviation is

$$s = \sqrt{\frac{(7 - 12._5)^2 + (18 - 12._5)^2 + (10 - 12._5)^2 + (15 - 12._5)^2}{4 - 1}} = 4._9$$

The mean and the standard deviation should both end at the *same decimal place.* For $\bar{x} = 12._5$, we write $s = 4._9$. The relative standard deviation is $(4._9/12._5) \times 100 = 39\%$.

Your calculator may have mean and standard deviation functions. Now is the time to learn to use them. Check that your calculator reproduces the results in this example.

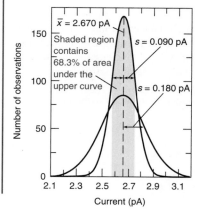

FIGURE 4-3 Gaussian curves showing the effect of doubling the standard deviation. The number of observations described by each curve is the same.

Other terms you should know are the median and the range. The *median* is the middle number in a series of measurements. When the measurements (8, 17, 11, 14, 12) are ordered from lowest to highest to give (8, 11, 12, 14, 17), the middle number (12) is the median. For an even number of measurements, the median is the average of the two middle numbers. For (8, 11, 12, 14), the median is 11.5. Some people prefer to report the median instead of the average, because the median is less influenced by outlying data. The *range* is the difference between the highest and lowest values. The range of (8, 17, 11, 14, 12) is $17 - 8 = 9$.

Standard Deviation and Probability

For an ideal Gaussian distribution, about two-thirds of the measurements (68.3%) lie within one standard deviation on either side of the mean (i.e., in the interval $\mu \pm \sigma$). That is, 68.3% of the area beneath a Gaussian curve lies in the interval $\mu \pm \sigma$, as shown in Figure 4-3. The percentage of measurements lying in the interval $\mu \pm 2\sigma$ is 95.5% and the percentage in the interval $\mu \pm 3\sigma$ is 99.7%. On the average, in real data with a standard deviation s, about 1 in 20 measurements (4.5%) will lie outside the range $\bar{x} \pm 2s$ and only 3 in 1 000 measurements (0.3%) lie outside the range $\bar{x} \pm 3s$. Table 4-1 shows the good correspondence between ideal Gaussian behavior and the observations in Figure 4-2.

The Gaussian distribution is symmetric. If 4.5% of measurements lie outside the range $\mu \pm 2\sigma$, 2.25% of measurements are above $\mu + 2\sigma$ and 2.25% are below $\mu - 2\sigma$.

TABLE 4-1

Percentage of observations in Gaussian distribution

Range	Gaussian distribution	Observed in Figure 4-2
$\mu \pm 1\sigma$	68.3%	71.0%
$\mu \pm 2\sigma$	95.5	95.6
$\mu \pm 3\sigma$	99.7	98.5

Question What fraction of observations in a Gaussian distribution is expected to be below $\mu - 3\sigma$?

Ask Yourself

4-A. What are the mean, standard deviation, relative standard deviation, median, and range for the numbers 821, 783, 834, and 855? Express each to the nearest tenth.

4-2 Student's t

"Student" was the pseudonym of W. S. Gossett, whose employer, the Guinness Breweries of Ireland, restricted publications for proprietary reasons. Because of the importance of his work, Gossett published it under an assumed name in 1908.

Student's t is the statistical tool used to express confidence intervals and to compare results from different experiments. You can use it to evaluate the probability that your red blood cell count will be found in a certain range on "normal" days.

Confidence Intervals

From a limited number of measurements, it is impossible to find the true mean, μ, or the true standard deviation, σ. What we can determine are $\bar{x}$ and s, the sample mean and the sample standard deviation. The **confidence interval** is an expression stating that the true mean, μ, is likely to lie within a certain distance from the measured mean, $\bar{x}$. The confidence interval of μ is given by

Confidence interval:

$$\mu = \bar{x} \pm \frac{ts}{\sqrt{n}}$$

(4-3)

where s is the measured standard deviation, n is the number of observations, and t is Student's t, taken from Table 4-2. Remember that in this table the *degrees of freedom* are equal to $n - 1$. If there are five data points, there are four degrees of freedom.

TABLE 4-2 **Values of Student's *t***

Degrees of freedom	Confidence level (%)						
	50	90	95	98	99	99.5	99.9
1	1.000	6.314	12.706	31.821	63.657	127.32	636.619
2	0.816	2.920	4.303	6.965	9.925	14.089	31.598
3	0.765	2.353	3.182	4.541	5.841	7.453	12.924
4	0.741	2.132	2.776	3.747	4.604	5.598	8.610
5	0.727	2.015	2.571	3.365	4.032	4.773	6.869
6	0.718	1.943	2.447	3.143	3.707	4.317	5.959
7	0.711	1.895	2.365	2.998	3.500	4.029	5.408
8	0.706	1.860	2.306	2.896	3.355	3.832	5.041
9	0.703	1.833	2.262	2.821	3.250	3.690	4.781
10	0.700	1.812	2.228	2.764	3.169	3.581	4.587
15	0.691	1.753	2.131	2.602	2.947	3.252	4.073
20	0.687	1.725	2.086	2.528	2.845	3.153	3.850
25	0.684	1.708	2.068	2.485	2.787	3.078	3.725
30	0.683	1.697	2.042	2.457	2.750	3.030	3.646
40	0.681	1.684	2.021	2.423	2.704	2.971	3.551
60	0.679	1.671	2.000	2.390	2.660	2.915	3.460
120	0.677	1.658	1.980	2.358	2.617	2.860	3.373
∞	0.674	1.645	1.960	2.326	2.576	2.807	3.291

Note: In calculating confidence intervals, σ may be substituted for s in Equation 4-3 if you have a great deal of experience with a particular method and have therefore determined its "true" population standard deviation. If σ is used instead of s, the value of t to use in Equation 4-3 comes from the bottom row of Table 4-2.

EXAMPLE Calculating Confidence Intervals

In replicate analyses, the carbohydrate content of a glycoprotein (a protein with sugars attached to it) is found to be 12.6, 11.9, 13.0, 12.7, and 12.5 g of carbohydrate per 100 g of protein. Find the 50 and 90% confidence intervals for the carbohydrate content.

SOLUTION First we calculate $\bar{x} = 12.5_4$ and $s = 0.4_0$ for the five measurements. To find the 50% confidence interval, look up t in Table 4-2 under 50 and across from *four* degrees of freedom (degrees of freedom $= n - 1$). The value of t is 0.741, so the confidence interval is

$$\mu(50\%) = \bar{x} \pm \frac{ts}{\sqrt{n}} = 12.5_4 \pm \frac{(0.741)(0.4_0)}{\sqrt{5}} = 12.5_4 \pm 0.1_3$$

The 90% confidence interval is

$$\mu(90\%) = \bar{x} \pm \frac{ts}{\sqrt{n}} = 12.5_4 \pm \frac{(2.132)(0.4_0)}{\sqrt{5}} = 12.5_4 \pm 0.3_8$$

These calculations mean that there is a 50% chance that the true mean, μ, lies in the range $12.5_4 \pm 0.1_3$ (12.4_1 to 12.6_7). There is a 90% chance that μ lies in the range $12.5_4 \pm 0.3_8$ (12.1_6 to 12.9_2).

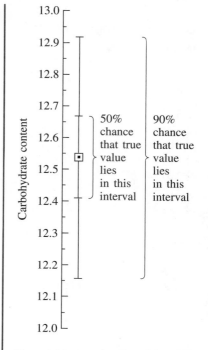

You might appreciate Box 4-1 at this time.

Box 4-1 *Informed Citizen*

Analytical Chemistry and the Law

As a person who will either produce or use analytical results, you should be aware of this warning published in a report entitled "Principles of Environmental Analysis":[†]

Analytical chemists must always emphasize to the public that *the single most important characteristic of any result obtained from one or more analytical measurements is an adequate statement of its uncertainty interval.* Lawyers usually attempt to dispense with uncertainty and try to obtain unequivocal statements; therefore, an uncertainty interval must be clearly defined in cases involving litigation and/or enforcement proceedings. Otherwise, a value of 1.001 without a specified uncertainty, for example, may be viewed as legally exceeding a permissible level of 1.

[†]L. H. Keith, W. Crummett, J. Deegan, Jr., R. A. Libby, J. K. Taylor, and G. Wentler, *Anal. Chem.* **1983,** *55,* 2210.

Some legal limits make no scientific sense. The Delaney Amendment to the U.S. Federal Food, Drug, and Cosmetic Act of 1958 states that "no additive [in processed food] shall be deemed to be safe if it is found to induce cancer when ingested by man or animal . . . ." This means that no detectable level of any carcinogenic (cancer-causing) pesticide may remain in processed foods, even if the level is far below that which can be shown to cause cancer. (Interestingly, the law does not apply to fresh food, which may contain pesticides.) The law was passed at a time when the sensitivity of analytical procedures was relatively poor, so the detection limit was relatively high. Today we can detect concentrations that are 10^3 to 10^6 times lower. A concentration that may have been acceptable 30 years ago is now 10^3 to 10^6 times above the legal limit, regardless of whether there is any evidence that such a low level is harmful. The U.S. Supreme Court upheld this law as recently as 1993.

Comparison of Means with Student's t

Student's t is also used to compare two sets of measurements to decide whether they are "the same" or "different." We adopt the following standard: If there is less than 1 chance in 20 that the difference between the two measurements arises from random variation in the data, then the difference is significant. This criterion gives us 95% confidence in concluding that two measurements are the same or different. There is a 5% probability that our conclusion is wrong.

An example comes from the work of Lord Rayleigh (John W. Strutt), who received the Nobel Prize in 1904 for discovering the inert gas argon—a discovery that came about when he noticed a discrepancy between two sets of measurements of the density of nitrogen. In Rayleigh's time, it was known that dry air is composed of about one-fifth oxygen and four-fifths nitrogen. Rayleigh removed O_2 from air by reaction with red-hot copper ($Cu(s) + \frac{1}{2}O_2(g) \rightarrow CuO(s)$) and measured the density of the remaining gas by collecting it in a fixed volume at a constant temperature and pressure. He then prepared the same volume of nitrogen by decomposition of nitrous oxide (N_2O), nitric oxide (NO), or ammonium nitrite ($NH_4^+NO_2^-$). Table 4-3 and Figure 4-4 show the mass of gas collected in each experiment. The average mass from air was 0.46% greater than the average mass of the same volume of gas from chemical sources.

If Rayleigh's measurements had not been performed with care, a 0.46% difference might have been attributed to experimental error. Instead, Rayleigh understood that the discrepancy was outside his margin of error, and he postulated that nitrogen from the air was mixed with a heavier gas, which turned out to be argon.

TABLE 4-3

Grams of nitrogen-rich gas isolated by Lord Rayleigh

From air	From chemical decomposition
2.310 17	2.301 43
2.309 86	2.298 90
2.310 10	2.298 16
2.310 01	2.301 82
2.310 24	2.298 69
2.310 10	2.299 40
2.310 28	2.298 49
—	2.298 89
Average	
2.310 10$_9$	2.299 47$_2$
Standard deviation	
0.000 14$_3$	0.001 37$_9$

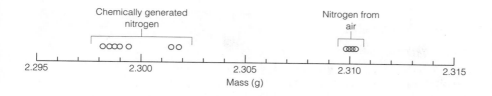

Let's see how to use the *t* **test** to decide whether nitrogen isolated from air is "significantly" heavier than nitrogen isolated from chemical sources. For two sets of data consisting of n_1 and n_2 measurements (with averages $\bar{x}_1$ and $\bar{x}_2$), we calculate a value of t from the formula

t Test for comparison of means:

$$t = \frac{|\bar{x}_1 - \bar{x}_2|}{s_{pooled}} \sqrt{\frac{n_1 n_2}{n_1 + n_2}} \qquad (4\text{-}4)$$

where

$$s_{pooled} = \sqrt{\frac{s_1^2(n_1 - 1) + s_2^2(n_2 - 1)}{n_1 + n_2 - 2}} \qquad (4\text{-}5)$$

Here s_{pooled} is a *pooled* standard deviation making use of both sets of data. The absolute value of $\bar{x}_1 - \bar{x}_2$ is used in Equation 4-4 so that t is always positive. The value of t from Equation 4-4 is to be compared with the value of t in Table 4-2 for $(n_1 + n_2 - 2)$ degrees of freedom. *If the calculated t is greater than the tabulated t at the 95% confidence level, the two results are considered to be different.*

If $t_{calculated} > t_{table}$ (95%), the difference is significant.

EXAMPLE Is Rayleigh's N_2 from Air Denser than N_2 from Chemicals?

The average mass of nitrogen from air in Table 4-3 is $\bar{x}_1 = 2.310\ 10_9$ g, with a standard deviation of $s_1 = 0.000\ 14_3$ (for $n_1 = 7$ measurements). The average mass from chemical sources is $\bar{x}_2 = 2.299\ 47_2$ g, with a standard deviation of $s_2 = 0.001\ 37_9$ (for $n_2 = 8$ measurements). Are the two masses significantly different?

SOLUTION To answer this question, we calculate s_{pooled} with Equation 4-5,

$$s_{pooled} = \sqrt{\frac{0.000\ 14_3{}^2(7 - 1) + 0.001\ 37_9{}^2(8 - 1)}{7 + 8 - 2}} = 0.001\ 01_7$$

and t with Equation 4-4:

$$t = \frac{2.310\ 10_9 - 2.299\ 47_2}{0.001\ 01_7} \sqrt{\frac{7 \cdot 8}{7 + 8}} = 20.2$$

For $7 + 8 - 2 = 13$ degrees of freedom in Table 4-2, t lies between 2.228 and 2.131 at the 95% confidence level. The observed value ($t = 20.2$) is greater than the tabulated t, so the difference is significant. In fact, the tabulated value of t for 99.9% confidence is about 4.3. The difference is significant beyond the 99.9% confidence level. Our eyes do not lie to us in Figure 4-4: N_2 from the air is undoubtedly denser than N_2 from chemical sources. This observation led Rayleigh to discover argon as a heavy constituent of air.

Statistical tests only give us probabilities. They do not relieve us of the ultimate subjective decision to accept or reject a conclusion.

Ask Yourself

4-B. A reliable assay of ATP (adenosine triphosphate) in a certain type of cell gives a value of $111._0$ μmol/100 mL, with a standard deviation of $2._8$ in four replicate measurements. You have developed a new assay, which gave the following values in replicate analyses: 117, 119, 111, 115, 120 μmol/100 mL.

(a) Find the mean and standard deviation for your new analysis.

(b) Can you be 95% confident that your method produces a result different from the "reliable" value?

4-3   Q Test for Bad Data

There was always somebody in my lab section with four thumbs. Freshmen at Phillips University perform an experiment in which they dissolve the zinc from a galvanized nail and measure the mass lost by the nail to tell how much of the nail was zinc. Several students performed the experiment in triplicate and pooled their results:

mass loss (%):   10.2, 10.8, 11.6   9.9, 9.4, 7.8   10.0, 9.2, 11.3   9.5, 10.6, 11.6

Sidney   Cheryl   Tien   Dick

It appears that Cheryl might be the person with four thumbs, because her value 7.8 looks out of line with the other data. Should the group reject the value 7.8 before averaging the rest of the data or should 7.8 be retained?

We answer this question with the **Q test.** To apply the Q test, arrange the data in order of increasing value and calculate Q, defined as

Q test for discarding data:
$$Q = \frac{\text{gap}}{\text{range}}$$
(4-6)

Gap = 1.4

(7.8)   9.2   9.4   9.5   9.9   10.0   10.2   10.6   10.8   11.3   11.6   11.6

Questionable value
(too low?)   Range = 3.8

The *range* is the total spread of the data. The *gap* is the difference between the questionable point and the nearest value. *If Q calculated from Equation 4-6 is greater than Q in Table 4-4, the questionable point should be discarded.*

For the numbers above, $Q = 1.4/3.8 = 0.37$. In Table 4-4, the critical value of Q at the 90% confidence limit is 0.38 for 12 observations. Because the observed Q is smaller than the tabulated Q, the questionable point should be retained. There is more than a 10% chance that the value 7.8 is a member of the same population as the other measurements.

Common sense must always prevail. If Cheryl knew that her measurement was low because she spilled some of her unknown, then the probability that the result is wrong is 100% and the datum should be discarded. Any datum based on a faulty procedure should be discarded, no matter how well it fits the rest of the data.

TABLE 4-4

Values of Q for rejection of data

Q (90% confidence)[a]	Number of observations
0.76	4
0.64	5
0.56	6
0.51	7
0.47	8
0.44	9
0.41	10
0.39	11
0.38	12
0.34	15
0.30	20

a. Q = gap/range. If Q (observed) > Q (tabulated), the value in question can be rejected with 90% confidence.

If $Q_{observed} > Q_{tabulated}$, reject the questionable point.

4-C. Would you reject the value 216 from the set of results 192, 216, 202, 195, and 204?

4-4 Finding the "Best" Straight Line

The *method of least squares* finds the "best" straight line through experimental data points. We will apply this procedure to analytical chemistry calibration curves in the next section.

The equation of a straight line is

Equation of straight line: $$y = mx + b \qquad (4\text{-}7)$$

in which m is the **slope** and b is the **y-intercept** (Figure 4-5). If we measure between two points that lie on the line, the slope is $\Delta y / \Delta x$, which is constant for any pair of points on the line. The y-intercept is the point at which the line crosses the y-axis.

Method of Least Squares

The **method of least squares** finds the "best" line by adjusting the line to minimize the vertical deviations between the points and the line (Figure 4-6). The reasons for minimizing only the vertical deviations are that (1) experimental uncertainties in y values are often greater than uncertainties in x values and (2) the calculation for minimizing the vertical deviations is relatively simple.

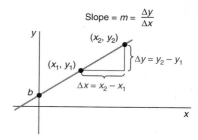

FIGURE 4-5 Parameters of a straight line:

Equation	$y = mx + b$
Slope	$m = \dfrac{\Delta y}{\Delta x} = \dfrac{y_2 - y_1}{x_2 - x_1}$
y-Intercept	b = crossing point on y-axis

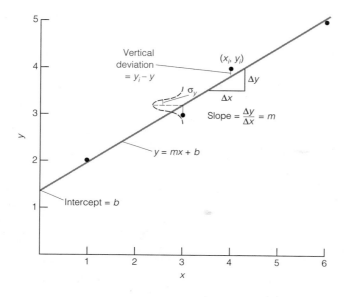

FIGURE 4-6 Least-squares curve fitting minimizes the sum of the squares of the vertical deviations of the measured points from the line. The Gaussian curve drawn over the point (3, 3) is a schematic indication of the distribution of measured y values about the straight line. The most probable value of y falls on the line, but there is a finite probability of measuring y some distance from the line.

The *ordinate* is the value of y that really lies on the line; y_i is the measured value that does not lie exactly on the line.

In Figure 4-6, the vertical deviation for the point (x_i, y_i) is $y_i - y$, where y is the ordinate of the straight line when $x = x_i$.

$$\text{vertical deviation} = d_i = y_i - y = y_i - (mx_i + b) \tag{4-8}$$

Some of the deviations are positive and some are negative. To minimize the magnitude of the deviations irrespective of their signs, we square the deviations to create positive numbers:

$$d_i^2 = (y_i - y)^2 = (y_i - mx_i - b)^2$$

Because we minimize the squares of the deviations, this is called the *method of least squares*.

When we use such a procedure to minimize the sum of squares of the vertical deviations, the slope and the intercept of the "best" straight line fitted to n points turn out to be

Remember that Σ means summation: $\Sigma x_i = x_1 + x_2 + x_3 + \ldots$

Least-squares slope:

$$m = \frac{n\sum(x_i y_i) - \sum x_i \sum y_i}{D} \tag{4-9}$$

Least-squares intercept:

$$b = \frac{\sum(x_i^2)\sum y_i - \sum(x_i y_i)\sum x_i}{D} \tag{4-10}$$

where the denominator, D, is given by

$$D = n\sum(x_i^2) - \left(\sum x_i\right)^2 \tag{4-11}$$

These equations are not as terrible as they appear. Table 4-5 sets out an example in which the four points ($n = 4$) in Figure 4-6 are treated. The first two columns list x_i and y_i for each point. The third column gives the product $x_i y_i$ and the fourth column lists the square x_i^2. At the bottom of each column is the sum for that column. That is, beneath column 1 is Σx_i and beneath column 3 is $\Sigma(x_i y_i)$. The last two columns at the right will be used later.

With the sums from Table 4-5, we compute the slope and intercept by substituting into Equations 4-11, 4-9, and 4-10:

$$D = n\sum(x_i^2) - \left(\sum x_i\right)^2 = 4 \cdot 62 - 14^2 = 52$$

$$m = \frac{n\sum(x_i y_i) - \sum x_i \sum y_i}{D} = \frac{4 \cdot 57 - 14 \cdot 14}{52} = 0.615\ 38$$

$$b = \frac{\sum(x_i^2)\sum y_i - \sum(x_i y_i)\sum x_i}{D} = \frac{62 \cdot 14 - 57 \cdot 14}{52} = 1.346\ 15$$

TABLE 4-5 Calculations for least-squares analysis

x_i	y_i	$x_i y_i$	x_i^2	$d_i(= y_i - mx_i - b)$	d_i^2
1	2	2	1	0.038 47	0.001 479 9
3	3	9	9	-0.192 29	0.036 975
4	4	16	16	0.192 33	0.036 991
6	5	30	36	-0.038 43	0.001 476 9
$\sum x_i = 14$	$\sum y_i = 14$	$\sum(x_i y_i) = 57$	$\sum(x_i^2) = 62$		$\sum(d_i^2) = 0.076\ 923$

The equation of the best straight line through the points in Figure 4-6 is therefore

$$y = 0.615\ 38x + 1.346\ 15$$

We address significant figures for m and b in the next section.

How Reliable Are Least-Squares Parameters?

The uncertainties in m and b are related to the uncertainty in measuring each value of y. Therefore, we first estimate the standard deviation describing the population of y values. This standard deviation, σ_y, characterizes the little Gaussian curve inscribed in Figure 4-6. The deviation of each y_i from the center of its Gaussian curve is $d_i = y_i - y = y_i - (mx_i + b)$ (Equation 4-8). The standard deviation of these vertical deviations turns out to be

$$\sigma_y \approx \sqrt{\frac{\sum(d_i^2)}{n-2}} \tag{4-12}$$

Analysis of uncertainty for Equations 4-10 and 4-11 leads to the following results:

standard deviation of slope: $\quad \sigma_m = \sigma_y\sqrt{\dfrac{n}{D}} \tag{4-13}$

standard deviation of intercept: $\quad \sigma_b = \sigma_y\sqrt{\dfrac{\sum(x_i^2)}{D}} \tag{4-14}$

where σ_y is given by Equation 4-12 and D is given by Equation 4-11.

At last, we can address significant figures for the slope and the intercept of the line in Figure 4-6. In Table 4-5 we see that $\sum(d_i^2) = 0.076\ 923$. Inserting this value in Equation 4-12 gives

$$\sigma_y = \sqrt{\frac{0.076\ 923}{4-2}} = 0.196\ 12$$

Now, we can plug numbers into Equations 4-13 and 4-14 to find

$$\sigma_m = \sigma_y\sqrt{\frac{n}{D}} = (0.196\ 12)\sqrt{\frac{4}{52}} = 0.054\ 394$$

$$\sigma_b = \sigma_y\sqrt{\frac{\sum(x_i^2)}{D}} = (0.196\ 12)\sqrt{\frac{62}{52}} = 0.214\ 15$$

Combining the results for m, σ_m, b, and σ_b, we write

slope: $\quad \dfrac{0.615\ 38}{\pm 0.054\ 39} = 0.62 \pm 0.05 \text{ or } 0.61_5 \pm 0.05_4$

intercept: $\quad \dfrac{1.346\ 15}{\pm 0.214\ 15} = 1.3 \pm 0.2 \text{ or } 1.3_5 \pm 0.2_1$

where the uncertainties represent one standard deviation. *The first decimal place of the standard deviation is the last significant figure of the slope or intercept.*

The first digit of the uncertainty is the last significant figure.

Ask Yourself

4-D. Construct a table analogous to Table 4-5 to calculate the equation of the best straight line going through the points (1,3), (3,2), and (5,0). Express your answer in the form $y(\pm\sigma_y) = [m(\pm\sigma_m)]x + [b(\pm\sigma_b)]$, with a reasonable number of significant figures.

4-5 Constructing a Calibration Curve

Real data from a spectrophotometric analysis are given in Table 4-6. In this procedure, a color develops in proportion to the amount of protein in the sample. Color is measured by the absorbance of light recorded on a spectrophotometer. The first row in Table 4-6 shows readings obtained when no protein was present. The nonzero values arise from color in the reagents themselves. A result obtained with zero analyte is called a **blank** (or *reagent blank*), because it measures effects due to the analytical reagents. The second row shows three readings obtained with 5 μg of protein. Successive rows give results for 10, 15, 20, and 25 μg of protein. A solution containing a known quantity of analyte (or other reagent) is called a **standard solution.**

A **calibration curve** is a graph showing how the experimentally measured property (absorbance) depends on the known concentrations of the standards. To construct the calibration curve in Figure 4-7, we first subtract the average absorbance of the blanks (0.099_3) from those of the standards to obtain *corrected absorbance.* When all points are plotted in Figure 4-7 and a rough straight line is drawn through them, two features stand out:

Inspect your data and use judgment before mindlessly using a computer to draw a calibration curve!

1. One data point for 15 μg of protein (shown by the square in Figure 4-7) is ridiculously high. When we inspect the range of values for each set of three measurements in Table 4-6, we discover that the range for 15 μg is almost four times greater than the next greatest range. The absorbance value 0.392 is clearly out of line, so we discard it as "bad data." Perhaps the glassware was contaminated with protein from a previous experiment?

2. All three points at 25 μg lie slightly below the straight line through the remaining data. The only way to know whether or not this effect is

TABLE 4-6 Spectrophotometer readings for protein analysis by the Lowry method

Sample (μg)	Absorbance of three independent samples			Range	Corrected absorbance (after subtracting average blank)		
0	0.099	0.099	0.100	0.001	-0.000_3	-0.000_3	0.000_7
5	0.185	0.187	0.188	0.003	0.085_7	0.087_7	0.088_7
10	0.282	0.272	0.272	0.010	0.182_7	0.172_7	0.172_7
15	0.392	0.345	0.347	0.047	—	0.245_7	0.247_7
20	0.425	0.425	0.430	0.005	0.325_7	0.325_7	0.330_7
25	0.483	0.488	0.496	0.013	0.383_7	0.388_7	0.396_7

Data used for calibration curve

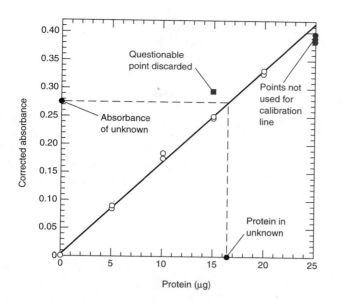

FIGURE 4-7 *Calibration curve* showing the average absorbance values in Table 4-6 versus micrograms of protein analyzed. The average *blank value* for 0 μg of protein has been subtracted from each point.

real is to repeat the experiment. Many repetitions show that these points are consistently below the straight line. Therefore the *linear range* for this determination extends from 0 to 20 μg, but not to 25 μg.

In view of these observations, we discard the 0.392 absorbance value and we do not use the three points at 25 μg for the least-squares straight line.

To construct the calibration curve (the straight line) in Figure 4-7, we use the method of least squares with $n = 14$ data points from Table 4-6 (including three blank values) covering the range 0 to 20 μg of protein. The results of applying Equations 4-9 through 4-14 are

$$m = 0.016\,3_0 \qquad \sigma_m = 0.000\,2_2$$
$$b = 0.004_7 \qquad \sigma_b = 0.002_6$$
$$\qquad\qquad \sigma_y = 0.005_9$$

Equation of calibration line:

$$y\,(\pm\sigma_y) = [m\,(\pm\sigma_m)]x + [b\,(\pm\sigma_b)]$$
$$y = [0.016\,3_0\,(\pm0.000\,2_2)]x + [0.004_7\,(\pm0.002_6)]$$

Finding the Protein in an Unknown

Suppose that the measured absorbance of an unknown sample is 0.373. How many micrograms of protein does it contain, and what uncertainty is associated with the answer?

The first question is easy. The equation of the calibration line is

$$y = mx + b = (0.016\,3_0)x + 0.004_7$$

where y is corrected absorbance (= observed absorbance − blank absorbance) and x is micrograms of protein. If the absorbance of unknown is 0.373, its corrected absorbance is $0.373 - 0.099_3 = 0.273_7$. Plugging this value in for y in the equation above permits us to solve for x:

$$0.273_7 = (0.016\,3_0)x + (0.004_7) \qquad (4\text{-}15a)$$

$$x = \frac{0.273_7 - 0.004_7}{0.016\,3_0} = 16.50 \text{ μg of protein} \qquad (4\text{-}15b)$$

Example: If four replicate samples of an unknown have an average absorbance of 0.376, use the value $0.376 - 0.099_3 = 0.276_7$ on the left side of Equation 4.15a. Use the value 1/4 in place of 1 for the first term of the square root in Equation 4-16 because you measured $k = 4$ replicate unknowns. The value of n in Equation 4-16 is 14 because there are 14 points on the calibration curve (Table 4-6).

But what is the uncertainty in 16.50 μg?

The uncertainty in x in Equation 4-15 turns out to be:

$$\text{uncertainty in } x = \frac{\sigma_y}{|m|}\sqrt{1 + \frac{x^2 n}{D} + \frac{\sum(x_i^2)}{D} - \frac{2x\sum x_i}{D}} \tag{4-16}$$

where $|m|$ is the absolute value of the slope and D is given by Equation 4-11. If you measure k values of y, the uncertainty in x will be reduced. In this case, use the average value of y and change the first term from 1 to $1/k$ in the square root.

Applying Equation 4-16 to the problem at hand gives an uncertainty of ± 0.38 for x. The final answer is therefore expressed with a reasonable number of significant figures as

$$x = 16.5\ (\pm 0.4)\ \mu\text{g of protein}$$

Ask Yourself

4-E. Using your results from Problem 4-D, find the value of x (and its uncertainty) corresponding to $y = 1.00$.

4-6 A Spreadsheet for Least Squares

Figure 4-8 translates the least-squares computations of Table 4-5 into a spreadsheet. Values of x and y are entered in columns B and C, and the total number of points ($n = 4$) is entered in cell A8. The products xy and x^2 are computed in columns D and E. The sums of columns B through G appear on line 7. For example, the sum of xy values in cells D2 through D5 is computed with the formula "= Sum(D2:D5)" in cell D7. The least-squares parameters D, m, and b are computed in cells A10, A12, and A14 from Equations 4-9 through 4-11. The vertical deviations, d, in column F make use of the slope and intercept. Column G contains the squares of the deviations. Standard deviations σ_y, σ_m, and σ_b are then computed in cells B10, B12, and B14, using Equations 4-12 through 4-14. As should always be your practice, formulas used in the spreadsheet are documented in Figure 4-8.

At the bottom of the spreadsheet, we use Equation 4-16 to evaluate uncertainty in a derived value of x. A measured value of y is entered in cell A20 and derived values of x and its uncertainty are computed in cells C20 and F20. This spreadsheet tells us that for a measured value of $y = 2.72$ in Figure 4-6, the value of x is $2.2_3 \pm 0.3_7$.

Ask Yourself

4-F. Reproduce the spreadsheet in Figure 4-8 to solve linear least-squares problems. Use a graphics program to plot the observed data points and the least-squares straight line. (*Note:* Most graphics packages can draw the least-squares line automatically. Use this feature routinely to examine the quality of your data.)

	A	B	C	D	E	F	G
1	n	x	y	xy	x^2	d	d^2
2	1	1	2	2	1	0.0385	0.0015
3	2	3	3	9	9	−0.1923	0.0370
4	3	4	4	16	16	0.1923	0.0370
5	4	6	5	30	36	−0.0385	0.0015
6	---------------------Column Sums [Example: D7 = Sum(D2:D5)]---------------------						
7	n=	14	14	57	62	0.0000	0.0770
8	4						
9	D=	sigma(y)=		A10: D = A8*E7−B7*B7			
10	52.0000	0.1962		A12: m = (D7*A8−B7*C7)/A10			
11	m=	sigma(m)=		A14: b = (E7*C7−D7*B7)/A10			
12	0.6154	0.0544		B10: sigma(y) = Sqrt(G7/(A8−2))			
13	b=	sigma(b)=		B12: sigma(m) = B10*Sqrt(A8/A10)			
14	1.3462	0.2143		B14: sigma(b) = B10*Sqrt(E7/A10)			
15				F2: d = C2−$A$12*B2−$A$14			
16							
17	Propagation of uncertainty calculation with Equation 4-16:						
18							
19	Measured y =		Derived x =		Uncertainty in x =		
20	2.72		2.2325			0.37368	
21							
22	Derived x in cell C20 = (A20−A14)/A12						
23	Uncertainty in x in cell F20 =						
24	(B10/A12)*Sqrt (1+ (C20*C20*A8/A10) + (E7/A10) − (2*C20*B7/A10))						

FIGURE 4-8 Spreadsheet for least-squares calculations.

Key Equations

Mean

$$\bar{x} = \frac{1}{n} \sum_i x_i = \frac{1}{n}(x_1 + x_2 + x_3 + \ldots + x_n)$$

x_i = individual observation, n = number of observations

Standard deviation

$$s = \sqrt{\sum_i (x_i - \bar{x})^2/(n-1)} \qquad (\bar{x} = \text{average})$$

Confidence interval

$$\mu = \bar{x} \pm \frac{ts}{\sqrt{n}} \qquad (\mu = \text{true mean})$$

Student's t comes from Table 4-2 for $n-1$ degrees of freedom at the selected confidence level.

t test

$$t = \frac{\bar{x}_1 - \bar{x}_2}{s_{\text{pooled}}} \sqrt{\frac{n_1 n_2}{n_1 + n_2}} \qquad s_{\text{pooled}} = \sqrt{\frac{s_1^2(n_1 - 1) + s_2^2(n_2 - 1)}{n_1 + n_2 - 2}}$$

If $t_{\text{calculated}} > t_{\text{table}}$ (for 95% confidence and $n_1 + n_2 - 2$ degrees of freedom), the difference is significant.

Straight line

$$y = mx + b \qquad m = \text{slope} = \Delta y/\Delta x$$
$$b = \text{y-intercept}$$

Least-squares equations

You should know how to use Equations 4-9 through 4-14 to derive least-squares slope and intercept and uncertainties.

Calibration curve   You should be able to use Equation 4-16 to find the uncertainty in a result derived from a calibration curve.

Important Terms

average	Gaussian distribution	slope	*t* test
blank	mean	standard deviation	*y*-intercept
calibration curve	method of least squares	standard solution	
confidence interval	*Q* test	Student's *t*	

Problems

4-1. What is the relation between the standard deviation and the precision of a procedure? What is the relation between standard deviation and accuracy?

4-2. What fraction of observations in an ideal Gaussian distribution lies within $\mu \pm \sigma$? Within $\mu \pm 2\sigma$? Within $\mu \pm 3\sigma$?

4-3. The ratio of the number of atoms of the isotopes ^{69}Ga and ^{71}Ga in samples from different sources is listed in the following table:

Sample	$^{69}Ga/^{71}Ga$	Sample	$^{69}Ga/^{71}Ga$
1	1.526 60	5	1.528 94
2	1.529 74	6	1.528 04
3	1.525 92	7	1.526 85
4	1.527 31	8	1.527 93

(a) Find the mean value of $^{69}Ga/^{71}Ga$.

(b) Find the standard deviation and relative standard deviation.

(c) Sample 8 was analyzed seven times, with $\bar{x} = 1.527\ 93$ and $s = 0.000\ 07$. Find the 99% confidence interval for sample 8.

4-4. What is the meaning of a confidence interval?

4-5. For the numbers 116.0, 97.9, 114.2, 106.8, and 108.3, find the mean, standard deviation, and 90% confidence interval for the mean.

4-6. The calcium content of a mineral was analyzed five times by each of two methods. Are the mean values significantly different at the 95% confidence level?

Method	Ca (percent composition)				
1	0.027 1,	0.028 2,	0.027 9,	0.027 1,	0.027 5
2	0.027 1,	0.026 8,	0.026 3,	0.027 4,	0.026 9

4-7. Find the 95 and 99% confidence intervals for the mean mass of nitrogen from chemical sources, Table 4-3.

4-8. Two methods were used to measure the specific activity (units of enzyme activity per milligram of protein) of an enzyme. One unit of enzyme activity is defined as the amount of enzyme that catalyzes the formation of one micromole of product per minute under specified conditions.

Method	Enzyme activity (five replications)				
1	139	147	160	158	135
2	148	159	156	164	159

Is the mean value of method 1 significantly different from the mean value of method 2 at the 95% confidence level?

4-9. Students measured the concentration of HCl in a solution by various titrations in which different indicators were used to find the end point.

Indicator	Mean HCl concentration (M) ($\pm$ standard deviation)	Number of measurements
1. Bromothymol blue	0.095 65 $\pm$0.002 25	28
2. Methyl red	0.086 86 $\pm$0.000 98	18
3. Bromocresol green	0.086 41 $\pm$0.001 13	29

Is the difference between indicators 1 and 2 significant at the 95% confidence level? Answer the same question for indicators 2 and 3.

4-10. The calcium content of a person's urine was determined on two different days (see table). Are the average values significantly different at the 95% confidence level?

Day	[Ca](mg/L) Average ± standard deviation	Number of measurements
1	238 ± 8	4
2	255 ± 10	5

4-11. Using the Q test, decide whether the value 0.195 should be rejected from the set of results 0.217, 0.224, 0.195, 0.221, 0.221, 0.223.

4-12. Find the values of m and b in the equation $y = mx + b$ for the straight line going through the points $(x_1, y_1) = (6, 3)$ and $(x_2, y_2) = (8, -1)$. You can do this by writing

$$m = \frac{\Delta y}{\Delta x} = \frac{(y_2 - y_1)}{(x_2 - x_1)} = \frac{(y - y_1)}{(x - x_1)}$$

and rearranging to the form $y = mx + b$. Sketch the curve and satisfy yourself that the value of b is a sensible y-intercept.

4-13. A straight line is drawn through the points (3.0, -3.87×10^4), (10.0, -12.99×10^4), (20.0, -25.93×10^4), (30.0, -38.89×10^4), and (40.0, -51.96×10^4) by using the method of least squares. The results are $m = -1.298\ 72 \times 10^4$, $b = 256.695$, $\sigma_m = 13.190$, $\sigma_b = 323.57$, and $\sigma_y = 392.9$. Express the slope and intercept and their uncertainties with the correct significant figures.

4-14. Consider the least-squares problem illustrated in Figure 4-6. Suppose that a single new measurement produces a y value of 2.58.

(a) Calculate the corresponding x value and its uncertainty.

(b) Suppose you measure y four times and the average is 2.58. Calculate the uncertainty in x based on four measurements, not one.

4-15. In a common protein analysis, a dye binds to the protein and the color of the dye changes from brown to blue. The intensity of blue color is proportional to the amount of protein present.

Protein (μg):	0.00	9.36	18.72	28.08	37.44
Absorbance:	0.466	0.676	0.883	1.086	1.280

(a) After subtracting the blank absorbance (0.466) from the remaining absorbances, use the method of least squares to determine the equation of the best straight line through these 5 points ($n = 5$). Use the standard deviation of the slope and intercept to express the equation in the form $y (\pm\sigma_m) = [m (\pm\sigma_m)]x + [b (\pm\sigma_b)]$ with a reasonable number of significant figures.

(b) Make a graph showing the experimental data and the calculated straight line.

(c) An unknown gave an observed absorbance of 0.973. Calculate the number of micrograms of protein in the unknown, and estimate its uncertainty.

4-16. The equation for a Gaussian curve is

$$y = \frac{1}{\sigma \sqrt{2\pi}} e^{-(x-\mu)^2/2\sigma^2}$$

The square root function on most spreadsheets is Sqrt and the exponential function is Exp. To find $e^{-3.4}$, write Exp(−3.4). The exponential function above is written

$$e^{-(x-\mu)^2/2\sigma^2} = \text{Exp(-(x-}\mu\text{)\^{}2/(2*}\sigma\text{\^{}2))}$$

(a) Suppose that $\mu = 10.0$ and $\sigma = 1.0$. Use a spreadsheet to compute values of y for the range $4 \le x \le 16$. For x, use the values 4.0, 4.4, 4.8, . . . , 16.0.

(b) Repeat the same calculation for $\sigma = 2.0$.

(c) Use graphics software to plot the results of (a) and (b) on one graph.

(d) For the curve in (c) with $\sigma = 2.0$, shade in the region that should contain 68.3% of all observations. Use different shading to show the region that should contain 95.5% of observations.

Chemical Equilibrium in the Environment

Paper mill on the Potomac River near Westernport, Maryland, neutralizes acid mine drainage in the water. Upstream of the mill, the river is acidic and lifeless; below the mill the river teems with life.

Part of the North Branch of the Potomac River runs crystal clear through the scenic Appalachian Mountains, but it is lifeless—a victim of acid drainage from abandoned coal mines. As the river passes a paper mill and wastewater treatment plant near Westernport, Maryland, the pH rises from an acidic, lethal value of 4.5 to a neutral value of 7.2, at which fish and plants thrive. This happy accident comes about when the calcium carbonate by-product from paper-making exits the mill and reacts with massive quantities of carbon dioxide from bacterial respiration at the sewage treatment plant. The resulting soluble bicarbonate neutralizes the acidic river and restores life downstream of the plant.

$$\underset{\text{Calcium carbonate}}{CaCO_3(s)} + CO_2(aq) + H_2O(l) \rightleftharpoons \underset{\text{Dissolved calcium bicarbonate}}{Ca^{2+}(aq) + 2HCO_3^-(aq)}$$

$$\underset{\text{Acid from river}}{HCO_3^-(aq) + H^+(aq)} \xrightarrow{\text{Neutralization}} CO_2(g) + H_2O(l)$$

Chapter 5

A Skirmish with Equilibrium

Equilibrium describes the state that a system will reach "if you wait long enough." Most reactions of interest in analytical chemistry reach equilibrium in times ranging from fractions of a second to many minutes. In this chapter we study equilibria involving the solubility of ionic compounds and the reactions of acids and bases. These equilibria are fundamental to understanding analytical measurements with electrodes, the principles of acid-base titrations, and the acid-base behavior of proteins.

5-1 The Equilibrium Constant

If the reactants A and B are converted to products C and D with the stoichiometry

$$aA + bB \rightleftharpoons cC + dD \qquad (5\text{-}1)$$

we write the **equilibrium constant**, K, in the form

Equilibrium constant:
$$K = \frac{[C]^c[D]^d}{[A]^a[B]^b} \qquad (5\text{-}2)$$

The equilibrium constant is more correctly expressed as a ratio of *activities* rather than of concentrations. See Box 5-2.

where the superscript letters denote stoichiometric coefficients and each capital letter stands for a chemical species. The symbol [A] stands for the concentration of A relative to its standard state (defined below). By definition, *a reaction is favored whenever $K > 1$.*

In deriving the equilibrium constant, each quantity in Equation 5-2 is expressed as the *ratio* of the concentration of a species to its concentration in its *standard state*.

Equilibrium constants are dimensionless.

Equilibrium constants are dimensionless; but when specifying concentrations, you must use units of molarity (M) for solutes and atmospheres (atm) for gases.

For solutes, the standard state is 1 M. For gases, the standard state is 1 atm; and for solids and liquids, the standard states are the pure solid or liquid. It is understood (but rarely written) that the term [A] in Equation 5-2 really means [A]/(1 M) if A is a solute. If D is a gas, [D] really means (pressure of D in atmospheres)/(1 atm). To emphasize that [D] means pressure of D, we usually write P_D in place of [D]. If C were a pure liquid or solid, the ratio [C]/(concentration of C in its standard state) would be unity (1) because the standard state *is* the pure liquid or solid. If [C] is a solvent, the concentration is so close to that of pure liquid C that the value of [C] is essentially 1. The terms of Equation 5-2 are each dimensionless; therefore all equilibrium constants are dimensionless.

The take-home lesson is this: In evaluating an equilibrium constant

1. The concentrations of solutes should be expressed as moles per liter.

2. The concentrations of gases should be expressed in atmospheres.

3. The concentrations of pure solids, pure liquids, and solvents are omitted because they are unity.

These conventions are arbitrary, but you must use them if you wish to use tabulated values of equilibrium constants and standard reduction potentials.

EXAMPLE **Writing an Equilibrium Constant**

Write the equilibrium constant for the reaction

$$Zn(s) + NH_4^+(aq) \rightleftharpoons Zn^{2+}(aq) + \tfrac{1}{2}H_2(g) + NH_3(aq)$$

SOLUTION Omit the concentration of pure solid and express the concentration of gas as a pressure in atmospheres:

P_{H_2} stands for the pressure of $H_2(g)$, which is expressed in atmospheres in the equilibrium constant.

$$K = \frac{[Zn^{2+}]P_{H_2}^{1/2}[NH_3]}{[NH_4^+]}$$

Manipulating Equilibrium Constants

Consider the reaction of the compound HA that dissociates into H^+ and A^-:

Throughout this text, you should assume that all species in chemical equations are in aqueous solution, unless otherwise specified.

$$HA \xrightleftharpoons{K} H^+ + A^- \qquad K = \frac{[H^+][A^-]}{[HA]}$$

If the reverse reaction is written, the new K' is the reciprocal of the original K:

Equilibrium constant for reverse reaction:

$$H^+ + A^- \xrightleftharpoons{K'} HA \qquad \boxed{K' = 1/K} = \frac{[HA]}{[H^+][A^-]}$$

If reactions are added, the new K is the product of the original K's. The equilibrium of H^+ between the species HA and CH^+ can be derived by adding two equations:

$$\begin{aligned} HA &\rightleftharpoons H^+ + A^- & K_1 \\ H^+ + C &\rightleftharpoons CH^+ & K_2 \\ \hline HA + C &\rightleftharpoons A^- + CH^+ & K_3 \end{aligned}$$

Equilibrium constant for sum of reactions:

$$K_3 = K_1 K_2 = \frac{[H^+][A^-]}{[HA]} \cdot \frac{[CH^+]}{[H^+][C]} = \frac{[A^-][CH^+]}{[HA][C]}$$

If a reaction is reversed, then $K' = 1/K$. If two reactions are added, then $K_3 = K_1 K_2$.

If n reactions are added, the overall equilibrium constant is the product of all n individual equilibrium constants.

EXAMPLE Combining Equilibrium Constants

From the equilibria below

$$H_2O \rightleftharpoons H^+ + OH^- \qquad K_w = [H^+][OH^-] = 1.0 \times 10^{-14}$$

$$NH_3(aq) + H_2O \rightleftharpoons NH_4^+ + OH^- \qquad K_{NH_3} = \frac{[NH_4^+][OH^-]}{[NH_3(aq)]} = 1.8 \times 10^{-5}$$

H_2O is omitted from K because it is a pure liquid. Its concentration remains nearly constant.

find the equilibrium constant for the reaction

$$NH_4^+ \rightleftharpoons NH_3(aq) + H^+$$

SOLUTION The third reaction is obtained by reversing the second reaction and adding it to the first reaction:

$$\begin{aligned} H_2O &\rightleftharpoons H^+ + OH^- & K_1 = K_w \\ NH_4^+ + OH^- &\rightleftharpoons NH_3(aq) + H_2O & K_2 = 1/K_{NH_3} \\ \hline NH_4^+ &\rightleftharpoons H^+ + NH_3(aq) & K_3 = K_w \cdot \frac{1}{K_{NH_3}} = 5.6 \times 10^{-10} \end{aligned}$$

Le Châtelier's Principle

Le Châtelier's principle states that if a system at equilibrium is disturbed, the direction in which the system proceeds back to equilibrium is such that the disturbance is partially offset. Let's see what happens when we change the concentration of one species in the reaction:

$$\underset{\text{Bromate}}{BrO_3^-} + \underset{\text{Chromium(III)}}{2Cr^{3+}} + 4H_2O \rightleftharpoons \underset{\text{Bromide}}{Br^-} + \underset{\text{Dichromate}}{Cr_2O_7^{2-}} + 8H^+ \qquad (5\text{-}3)$$

for which the equilibrium constant is

$$K = \frac{[Br^-][Cr_2O_7^{2-}][H^+]^8}{[BrO_3^-][Cr^{3+}]^2} = 1 \times 10^{11} \text{ at } 25°C$$

H_2O is omitted from K because it is the solvent. Its concentration remains nearly constant.

In one particular equilibrium state of this system, the following concentrations exist:

$$[H^+] = 5.0 \text{ M} \qquad [Cr_2O_7^{2-}] = 0.10 \text{ M} \qquad [Cr^{3+}] = 0.003\ 0 \text{ M}$$

$$[Br^-] = 1.0 \text{ M} \qquad [BrO_3^-] = 0.043 \text{ M}$$

Suppose that equilibrium is disturbed by increasing the concentration of dichromate from 0.10 to 0.20 M. In what direction will the reaction proceed to reach equilibrium?

According to the principle of Le Châtelier, the reaction should go in the reverse direction to partially offset the increase in dichromate, which is a product in Reaction 5-3. We can verify this algebraically by setting up the *reaction quotient, Q,* which has the same form as the equilibrium constant. The only difference is that Q is evaluated with whatever concentrations happen to exist, even though the solution is not at equilibrium. When the system reaches equilibrium, $Q = K$. For Reaction 5-3,

> *Q has the same form as K, but the concentrations are generally not the equilibrium concentrations.*

$$Q = \frac{(1.0)(0.20)(5.0)^8}{(0.043)(0.003\ 0)^2} = 2 \times 10^{11} > K$$

Because $Q > K$, the reaction must go in reverse to decrease the numerator and increase the denominator, until $Q = K$.

In general,

> *If Q < K, the reaction must proceed to the right to reach equilibrium. If Q > K, the reaction must proceed to the left to reach equilibrium.*

1. If a reaction is at equilibrium and products are added (or reactants are removed), the reaction goes in the reverse direction (to the left).

2. If a reaction is at equilibrium and reactants are added (or products are removed), the reaction goes in the forward direction (to the right).

In equilibrium problems, we predict what must happen for a system to reach equilibrium, but not how long it will take. Some reactions are over in an instant; others do not reach equilibrium in a million years. A stick of dynamite remains unchanged indefinitely, until a spark sets off the spontaneous, explosive decomposition. The size of an equilibrium constant tells us nothing about the rate of reaction. A large equilibrium constant does not imply that a reaction is fast.

Ask Yourself

5-A. (a) Show how the equations below can be rearranged and added to give the reaction $HOBr \rightleftharpoons H^+ + OBr^-$:

$$HOCl \rightleftharpoons H^+ + OCl^- \qquad\qquad K = 3.0 \times 10^{-8}$$
$$HOCl + OBr^- \rightleftharpoons HOBr + OCl^- \qquad K = 15$$

(b) Find the value of K for the reaction $HOBr \rightleftharpoons H^+ + OBr^-$.

(c) If the reaction $HOBr \rightleftharpoons H^+ + OBr^-$ is at equilibrium and a substance is added that consumes H^+, will the reaction proceed in the forward or reverse direction to reestablish equilibrium?

5-2 Solubility Product

The **solubility product** is the equilibrium constant for the reaction in which a solid salt dissolves to give its constituent ions in solution. The concentration of the solid is omitted from the equilibrium constant because the solid is in its standard state.

> We will use solubility products later to understand precipitation titrations and electrode measurements.

Calculating the Solubility of an Ionic Compound

Consider the dissolution of lead(II) iodide in water:

$$\underset{\text{Lead(II) iodide}}{\text{PbI}_2(s)} \rightleftharpoons \underset{\text{Lead(II)}}{\text{Pb}^{2+}} + \underset{\text{Iodide}}{2\text{I}^-} \qquad K_{sp} = [\text{Pb}^{2+}][\text{I}^-]^2 = 7.9 \times 10^{-9} \quad (5\text{-}4)$$

> Pure solid is omitted from the equilibrium constant because its concentration is not significantly changed.

for which the solubility product, K_{sp}, is listed in Appendix A. A solution that contains all the solid capable of being dissolved and exists in the presence of excess, undissolved solid is said to be **saturated.** What is the concentration of Pb^{2+} in a solution saturated with PbI_2?

Reaction 5-4 produces two I^- ions for each Pb^{2+} ion. If the concentration of dissolved Pb^{2+} is x M, the concentration of dissolved I^- must be $2x$ M. We can display this relationship neatly in a little concentration table:

	$\text{PbI}_2(s) \rightleftharpoons \text{Pb}^{2+} + 2\text{I}^-$		
Initial concentration:	solid	0	0
Final concentration:	solid	x	$2x$

Putting these concentrations into the solubility product gives

$$[\text{Pb}^{2+}][\text{I}^-]^2 = (x)(2x)^2 = 7.9 \times 10^{-9}$$
$$4x^3 = 7.9 \times 10^{-9}$$
$$x = \left(\frac{7.9 \times 10^{-9}}{4}\right)^{1/3}$$
$$x = 0.001\ 3 \text{ M}$$

The concentration of Pb^{2+} is 0.001 3 M and the concentration of I^- is $2x = (2)(0.001\ 3) = 0.002\ 6$ M.

The physical meaning of the solubility product is this: If an aqueous solution is left in contact with excess solid PbI_2, solid dissolves until the condition $[\text{Pb}^{2+}][\text{I}^-]^2 = K_{sp}$ is satisfied. Thereafter, no more solid dissolves. Unless excess solid remains, there is no guarantee that $[\text{Pb}^{2+}][\text{I}^-]^2 = K_{sp}$. If Pb^{2+} and I^- are mixed together (with appropriate counterions) such that the product $[\text{Pb}^{2+}][\text{I}^-]^2$ exceeds K_{sp}, then $\text{PbI}_2(s)$ will precipitate (Figure 5-1).

The solubility product does not tell the entire story for some ionic compounds. The concentration of *undissociated* species may be significant. In the case of CaSO_4, for example, about half of the dissolved material dissociates to Ca^{2+} and SO_4^{2-} and half dissolves as undissociated $\text{CaSO}_4(aq)$ (a tightly bound *ion pair*).

FIGURE 5-1 The yellow solid, lead(II) iodide (PbI_2), precipitates when a colorless solution of lead nitrate ($\text{Pb(NO}_3)_2$) is added to a colorless solution of potassium iodide (KI).

The Common Ion Effect

Consider what happens if we add a second source of I^- to a solution saturated with $PbI_2(s)$. Let's add 0.030 M NaI, which dissociates completely to Na^+ and I^-. What is the concentration of Pb^{2+} in this solution?

$$PbI_2(s) \rightleftharpoons Pb^{2+} + 2I^- \qquad (5\text{-}4)$$

Initial concentration:	solid	0	0.030
Final concentration:	solid	x	$2x + 0.030$

The initial concentration of I^- is from dissolved NaI. The final concentration of I^- has contributions from NaI and PbI_2.

The solubility product is

$$[Pb^{2+}][I^-]^2 = (x)(2x + 0.030)^2 = K_{sp} = 7.9 \times 10^{-9} \qquad (5\text{-}5)$$

But think about the size of x. When there was no added I^-, we found $x = 0.001\ 3$ M. In the present case, we anticipate that x will be smaller than $0.001\ 3$ M, because of Le Châtelier's principle. Addition of the product I^- to Reaction 5-4 displaces the reaction in the reverse direction. In the presence of extra I^-, there will be less dissolved Pb^{2+}.

Common ion effect: A salt is less soluble if one of its ions is already present in the solution.

This application of Le Châtelier's principle is called the **common ion effect**. *A salt will be less soluble if one of its constituent ions is already present in the solution.*

Getting back to Equation 5-5, we suspect that $2x$ may be much smaller than 0.030. As an approximation, we will ignore $2x$ in comparison to 0.030. The equation simplifies to

$$(x)(0.030)^2 = K_{sp} = 7.9 \times 10^{-9}$$

$$x = \frac{7.9 \times 10^{-9}}{(0.030)^2} = 8.8 \times 10^{-6}$$

It is important to confirm at the end of the calculation that the approximation $2x \ll 0.030$ is valid. Box 5-1 discusses approximations.

Because $2x = 1.8 \times 10^{-5} \ll 0.030$, it was justifiable to neglect $2x$ to solve the problem. The answer also illustrates the common ion effect. In the absence of added I^-, the solubility of Pb^{2+} was $0.001\ 3$ M. In the presence of 0.030 M I^-, $[Pb^{2+}]$ is reduced to 8.8×10^{-6} M.

What is the maximum I^- concentration at equilibrium in a solution in which the concentration of Pb^{2+} is somehow *fixed* at 1.0×10^{-4} M? Our concentration table looks like this now:

$$PbI_2(s) \rightleftharpoons Pb^{2+} + 2I^-$$

Initial concentration:	solid	1.0×10^{-4}	0
Final concentration:	solid	1.0×10^{-4}	x

$[Pb^{2+}]$ is not x in this example, so there is no reason to set $[I^-] = 2x$. The problem is solved by plugging each concentration into the solubility product:

$$[Pb^{2+}][I^-]^2 = K_{sp}$$
$$(1.0 \times 10^{-4})(x)^2 = 7.9 \times 10^{-9}$$
$$x = [I^-] = 8.9 \times 10^{-3}\ M$$

If I^- is added above a concentration of 8.9×10^{-3} M, then $PbI_2(s)$ will precipitate.

Box 5-1 Explanation

The Logic of Approximations

Many real problems are difficult to solve without judicious approximations. For example, rather than solving the equation

$$(x)(2x + 0.030)^2 = 7.9 \times 10^{-9}$$

we hoped and prayed that $2x \ll 0.030$, and therefore solved the much simpler equation

$$(x)(0.030)^2 = 7.9 \times 10^{-9}$$

But how can we be sure that our solution fits the original problem?

When we use an approximation, we assume it is true. *If the assumption is true, it does not create a contradiction. If the assumption is false, it leads to a contradiction.* You can test an assumption by using it and seeing whether you are right or wrong afterward.

You may object to this reasoning, feeling "How can the truth of an assumption be tested by using the assumption?" Suppose you wish to test the statement "Gail can swim 100 meters." To see whether or not the statement is true, you can assume it is. If Gail can swim 100 meters, then you can dump her in the middle of a lake with a radius of 100 meters and expect her to swim to shore. If she comes ashore alive, then your assumption was correct and no contradiction is created. If she does not make it to shore, then there is a contradiction. Your assumption must have been wrong. There are only two possibilities: Either the assumption is correct and using it is correct, or the assumption is wrong and using it is wrong. (A third possibility in this case is that there are sharks in the lake.)

Example 1. $(x)(2x + 0.030)^2 = 7.9 \times 10^{-9}$

$(x)(0.030)^2 = 7.9 \times 10^{-9}$

(assuming $2x \ll 0.030$)

$x = (7.9 \times 10^{-9})/(0.030)^2 =$

8.8×10^{-6}

No contradiction: $2x = 1.76 \times 10^{-5}$

$\ll 0.030.$

The assumption is true.

Example 2. $(x)(2x + 0.030)^2 = 7.9 \times 10^{-5}$

$(x)(0.030)^2 = 7.9 \times 10^{-5}$

(assuming $2x \ll 0.030$)

$x = (7.9 \times 10^{-5})/(0.030)^2 = 0.088$

A contradiction: $2x = 0.176 > 0.030.$

The assumption is false.

In Example 2 the assumption leads to a contradiction, so the assumption cannot be correct. When this happens, you must solve the cubic equation $(x)(2x + 0.030)^2 = 7.9 \times 10^{-5}$.

A reasonable way to solve this equation is by trial-and-error, as shown below:

Guess	$x(2x + 0.030)^2$	Result is
$x = 0.01$	2.5×10^{-5}	too low
$x = 0.02$	9.8×10^{-5}	too high
$x = 0.015$	5.4×10^{-5}	too low
$x = 0.018$	7.84×10^{-5}	too low
$x = 0.019$	8.79×10^{-5}	too high
$x = 0.018\ 1$	7.93×10^{-5}	too high
$x = 0.018\ 05$	7.89×10^{-5}	too low
$x = 0.018\ 06$	7.90×10^{-5}	not bad!

You can create the table above by hand, and even more easily with a spreadsheet. In cell A1 enter a guess for x. In cell A2 enter the formula "=A1*(2*A1+0.030)^2". When you guess x correctly in cell A1, cell B2 will have the value 7.9×10^{-5}.

Gail's lake

EXAMPLE   **Using the Solubility Product**

The compound silver ferrocyanide dissociates into Ag^+ and $Fe(CN)_6^{4-}$ in solution. Find the solubility of $Ag_4Fe(CN)_6$ in water expressed as **(a)** moles of $Ag_4Fe(CN)_6$ per liter, **(b)** grams of $Ag_4Fe(CN)_6$ per 100 mL, and **(c)** ppb Ag^+ ($\approx$ ng Ag^+/mL).

SOLUTION **(a)** We start off with a concentration table:

$$Ag_4Fe(CN)_6(s) \rightleftharpoons 4Ag^+ + Fe(CN)_6^{4-}$$

	Silver ferrocyanide FW 643.43		Ferrocyanide
Initial concentration:	solid	0	0
Final concentration:	solid	$4x$	x

The solubility product of $Ag_4Fe(CN)_6$ in Appendix A is 8.5×10^{-45}. Therefore,

$$[Ag^+]^4[Fe(CN)_6^{4-}] = (4x)^4(x) = 8.5 \times 10^{-45}$$
$$256x^5 = 8.5 \times 10^{-45}$$
$$x^5 = \frac{8.5 \times 10^{-45}}{256}$$
$$x = \left(\frac{8.5 \times 10^{-45}}{256}\right)^{1/5} = 5.0_6 \times 10^{-10} \text{ M}$$

There are $5.0_6 \times 10^{-10}$ moles of $Ag_4Fe(CN)_6$ dissolved per liter. The concentration of Ag^+ is $4x$ and the concentration of $Fe(CN)_6^{4-}$ is x.
(b) To find g/100 mL, we convert mol/L into g/L and then into g/100 mL:

$$\left(5.0_6 \times 10^{-10} \frac{mol}{L}\right)\left(643.43 \frac{g}{mol}\right)\left(\frac{0.100 \text{ } L}{100 \text{ mL}}\right) = 3.3 \times 10^{-8} \frac{g}{100 \text{ mL}}$$

(c) The concentration of Ag^+ is $4x = 4(5.0_6 \times 10^{-10} \text{ M}) = 2.0_2 \times 10^{-9}$ M. Use the atomic weight of Ag to convert mol/L to g/L:

$$\left(2.0_2 \times 10^{-9} \frac{mol}{L}\right)\left(107.868 \frac{g}{mol}\right) = 2.1_8 \times 10^{-7} \frac{g}{L}$$

Atomic weight of Ag

To find ppb, we need units of ng/mL:

$$\left(2.1_8 \times 10^{-7} \frac{g}{L}\right)\left(\frac{1}{1\,000} \frac{L}{mL}\right)\left(10^9 \frac{ng}{g}\right) = 0.21_8 \frac{ng}{mL} = 0.22 \text{ ppb}$$

Ask Yourself

5-B. How many grams of $PbBr_2$ (FW 367.0) will dissolve in 0.500 L of **(a)** water and **(b)** 0.063 4 M NaBr?

In aqueous chemistry, an **acid** is a substance that increases the concentration of H_3O^+ (**hydronium ion**). Conversely, a **base** decreases the concentration of H_3O^+ in aqueous solution. As we shall see shortly, a decrease in H_3O^+ concentration necessarily requires an increase in OH^- concentration. Therefore, a base is a substance that increases the concentration of OH^- in aqueous solution.

The species H^+ is called a *proton,* because a proton is all that remains when a hydrogen atom loses its electron. Hydronium ion, H_3O^+, is a combination of H^+ with H_2O (Figure 5-2). Although H_3O^+ is a more accurate representation than H^+ for the hydrogen ion in aqueous solution, we will use H_3O^+ and H^+ interchangeably in this book.

A more general definition of acids and bases given by Brønsted and Lowry is that an *acid* is a *proton donor* and a *base* is a *proton acceptor.* This definition includes the one stated above. For example, HCl is an acid because it donates a proton to H_2O to form H_3O^+:

$$HCl + H_2O \rightleftharpoons H_3O^+ + Cl^-$$

The Brønsted-Lowry definition can be extended to nonaqueous solvents and to the gas phase:

$$\underset{\substack{\text{Hydrochloric acid} \\ \text{(acid)}}}{HCl(g)} + \underset{\substack{\text{Ammonia} \\ \text{(base)}}}{NH_3(g)} \rightleftharpoons \underset{\substack{\text{Ammonium chloride} \\ \text{(salt)}}}{NH_4^+Cl^-(s)}$$

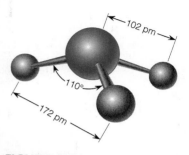

FIGURE 5-2 Structure of hydronium ion, H_3O^+.

Brønsted-Lowry acid: proton donor
Brønsted-Lowry base: proton acceptor

Salts

Any ionic solid, such as ammonium chloride, is called a **salt.** In a formal sense, a salt can be thought of as the product of an acid-base reaction. When an acid and a base react stoichiometrically, they are said to **neutralize** each other. Most salts are *strong electrolytes,* meaning that they dissociate completely into their component ions when dissolved in water. Thus, ammonium chloride gives NH_4^+ and Cl^- in aqueous solution:

$$NH_4^+Cl^-(s) \longrightarrow NH_4^+(aq) + Cl^-(aq)$$

Conjugate Acids and Bases

The products of a reaction between an acid and a base are also acids and bases:

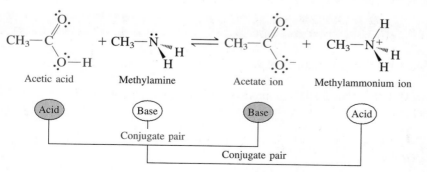

A solid wedge is a bond coming out of the page toward you. A dashed wedge is a bond going behind the page.

Conjugate acids and bases are related by the gain or loss of one proton.

Acetate is a base because it can accept a proton to make acetic acid. The methylammonium ion is an acid because it can donate a proton and become methylamine. Acetic acid and the acetate ion are said to be a **conjugate acid-base pair.** Methylamine and the methylammonium ion are likewise conjugate. *Conjugate acids and bases are related to each other by the gain or loss of one H+.*

Ask Yourself

5-C. When an acid and base react, they are said to _____ each other. Acids and bases related by the gain or loss of one proton are said to be _____.

5-4 *The Relation Between [H+], [OH−], and pH*

In **autoprotolysis,** one substance acts as both an acid and a base:

Autoprotolysis of water:
$$H_2O + H_2O \xrightleftharpoons{K_w} H_3O^+ + OH^- \qquad (5\text{-}6a)$$
Hydronium ion   Hydroxide ion

We abbreviate Reaction 5-6a in the following manner:

$$H_2O \xrightleftharpoons{K_w} H^+ + OH^- \qquad (5\text{-}6b)$$

and we designate its equilibrium constant (the autoprotolysis constant) as K_w.

Autoprotolysis constant for water:
$$K_w = [H^+][OH^-] = 1.0 \times 10^{-14} \text{ at } 25°C \qquad (5\text{-}7)$$

Equation 5-7 provides a tool with which we can find the concentration of H^+ and OH^- in pure water. Also, given that the product $[H^+][OH^-]$ is constant, we can always find the concentration of either species if the concentration of the other is known. Because the product is a constant, *as the concentration of H+ increases, the concentration of OH− necessarily decreases, and vice versa.*

EXAMPLE   Concentration of H+ and OH− in Pure Water at 25°C

Calculate the concentrations of H^+ and OH^- in pure water at 25°C.

SOLUTION H^+ and OH^- are produced in a 1:1 mole ratio in Reaction 5-6b. Calling each concentration x, we write

$$K_w = 1.0 \times 10^{-14} = [H^+][OH^-] = [x][x] \Rightarrow x = 1.0 \times 10^{-7} \text{ M}$$

The concentrations of H^+ and OH^- are both 1.0×10^{-7} M.

What is the concentration of OH^- if $[H^+] = 1.0 \times 10^{-3}$ M at 25°C?

SOLUTION Setting $[H^+] = 1.0 \times 10^{-3}$ M in Equation 5-7 gives

$$K_w = [H^+][OH^-] \Rightarrow [OH^-] = \frac{K_w}{[H^+]} = \frac{1.0 \times 10^{-14}}{1.0 \times 10^{-3}} = 1.0 \times 10^{-11} \text{ M}$$

When $[H^+] = 1.0 \times 10^{-3}$ M, $[OH^-] = 1.0 \times 10^{-11}$ M. A concentration of $[OH^-] = 1.0 \times 10^{-3}$ M would give $[H^+] = 1.0 \times 10^{-11}$ M. As one concentration increases, the other decreases.

To simplify the writing of H^+ concentration, we define **pH** as

Approximate definition of pH:

$$pH = -\log [H^+]$$

(5-8)   Box 5-2 tells you that Equation 5-8 is only approximate.

Box 5-2 *Explanation*

Activity

The equilibrium constant in Equation 5-2 and the definition of pH in Equation 5-8 are not strictly correct. Consider the equilibrium constant for the dissociation of acetic acid:

$$CH_3CO_2H \rightleftharpoons CH_3CO_2^- + H^+$$

$$K = \frac{[CH_3CO_2^-][H^+]}{[CH_3CO_2H]} \qquad (A)$$

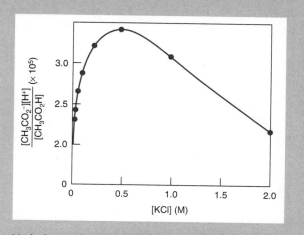

Variation of the quotient $[CH_3CO_2^-][H^+]/[CH_3CO_2H]$ when KCl is added to a solution of acetic acid.

If this quotient were a true equilibrium "constant," it would be constant under all conditions. In fact, if KCl is added to a solution of acetic acid, the quotient in Equation A varies over a range of about 50%, as shown in the illustration.

To account for this behavior, concentrations in Equation A are replaced by *activities* ($\mathcal{A}$), which are proportional to concentration but also depend on environmental factors such as the ionic strength (concentration of other ions) and solvent composition of the solution.

True equilibrium constant:

$$K = \frac{\mathcal{A}_{CH_3CO_2^-}\mathcal{A}_{H^+}}{\mathcal{A}_{CH_3CO_2H}} \qquad (B)$$

Expression B is truly constant.

In a similar vein, the definition of pH in Equation 5-8 should really be written in terms of the activity of H^+:

Correct definition of pH:

$$pH = -\log \mathcal{A}_{H^+}$$

A pH electrode measures $-\log \mathcal{A}_{H^+}$, not $-\log [H^+]$.

It is beyond the scope of this book to use activities. We will write equilibrium constants and pH in terms of concentrations, recognizing that the expressions are only approximate.

Here are some examples:

$$[H^+] = 10^{-3} \text{ M} \quad \Rightarrow \text{pH} = -\log(10^{-3}) = 3$$
$$[H^+] = 10^{-10} \text{ M} \quad \Rightarrow \text{pH} = -\log(10^{-10}) = 10$$
$$[H^+] = 3.8 \times 10^{-8} \text{ M} \Rightarrow \text{pH} = -\log(3.8 \times 10^{-8}) = 7.42$$

A solution is **acidic** if $[H^+] > [OH^-]$. A solution is **basic** if $[H^+] < [OH^-]$. The previous example demonstrated that in pure water (which is neither acidic nor basic and is said to be *neutral*), $[H^+] = [OH^-] = 10^{-7}$ M, so the pH is $-\log(10^{-7}) = 7$. *An acidic solution has a pH below 7, and a basic solution has a pH above 7.*

Example: An acidic solution has pH = 4. This means $[H^+] = 10^{-4}$ M and $[OH^-] = K_w/[H^+] = 10^{-10}$ M. Therefore, $[H^+] > [OH^-]$.

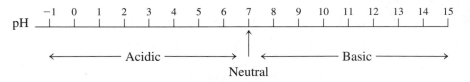

pH values of various substances are shown in Figure 5-3.

Although pH generally falls in the range 0 to 14, these are not limits. A pH of -1, for example, means $-\log[H^+] = -1$, or $[H^+] = 10^{+1} = 10$ M. This pH is attained in a concentrated solution of a strong acid such as HCl.

Ask Yourself

5-D. A solution of 0.050 M Mg^{2+} is treated with NaOH until $Mg(OH)_2$ precipitates.

    (a) At what concentration of OH^- does this occur?
    (b) At what pH does this occur?

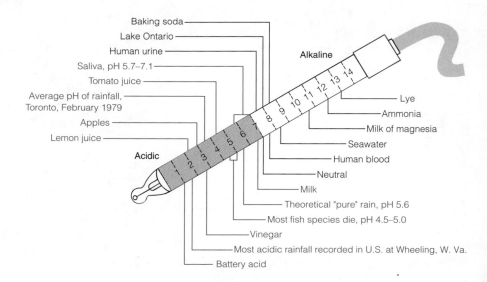

FIGURE 5-3 pH values of various substances. The most acidic rainfall in the United States is more acidic than lemon juice. See the opening highlight of Chapter 9 for more on acid rain.

5-5 Strengths of Acids and Bases

Acids and bases are classified as strong or weak, depending on whether they react "completely" or only "partly" to produce H^+ or OH^-. Because there is a continuous range for a "partial" reaction, there is no sharp distinction between weak and strong. However, some compounds react so completely that they are unquestionably strong acids or bases—and everything else is defined as weak.

Strong Acids and Bases

Common strong acids and bases are listed in Table 5-1, which you must memorize. Note that even though HCl, HBr, and HI are strong acids, HF is *not*. A strong acid or base is completely dissociated in aqueous solution. That is, the equilibrium constants for the following reactions are very large:

$$HCl(aq) \longrightarrow H^+ + Cl^-$$
$$KOH(aq) \longrightarrow K^+ + OH^-$$

Virtually no undissociated HCl or KOH exists in aqueous solution. To indicate that the equilibrium constant is very large, we write $\longrightarrow$ instead of $\rightleftharpoons$. Demonstration 5-1 shows one consequence of the strong acid behavior of HCl.

Weak Acids and Bases

All weak acids, HA, react with water by donating a proton to H_2O:

$$HA + H_2O \xrightarrow{K_a} H_3O^+ + A^-$$

which means the same as

Dissociation of weak acid:
$$HA \xrightarrow{K_a} H^+ + A^- \qquad K_a = \frac{[H^+][A^-]}{[HA]} \qquad (5\text{-}9)$$

The equilibrium constant, K_a, is called the **acid dissociation constant.** A weak acid is only partially dissociated in water. This definition means that, for a weak acid, K_a is "small."

Weak bases, B, react with water by abstracting (grabbing) a proton from H_2O:

Base hydrolysis:
$$B + H_2O \xrightarrow{K_b} BH^+ + OH^- \qquad K_b = \frac{[BH^+][OH^-]}{[B]} \qquad (5\text{-}10)$$

The equilibrium constant K_b is called the **base hydrolysis constant.** A weak base is one for which K_b is "small."

TABLE 5-1

Common strong acids and bases

Formula	Name
Acids	
HCl	Hydrochloric acid (hydrogen chloride)
HBr	Hydrogen bromide
HI	Hydrogen iodide
H_2SO_4[a]	Sulfuric acid
HNO_3	Nitric acid
$HClO_4$	Perchloric acid
Bases	
LiOH	Lithium hydroxide
NaOH	Sodium hydroxide
KOH	Potassium hydroxide
RbOH	Rubidium hydroxide
CsOH	Cesium hydroxide
R_4NOH[b]	Quaternary ammonium hydroxide

a. For H_2SO_4, only the first proton ionization is complete. Dissociation of the second proton has an equilibrium constant of 1.0×10^{-2}.
b. This is a general formula for any hydroxide salt of an ammonium cation containing four organic groups. An example is tetrabutylammonium hydroxide: $(CH_3CH_2CH_2CH_2)_4N^+OH^-$.

HCl Fountain

The complete dissociation of HCl into H^+ and Cl^- makes HCl(g) extremely soluble in water.

$$HCl(g) \rightleftharpoons HCl(aq) \qquad \text{(A)}$$

$$HCl(aq) \longrightarrow H^+(aq) + Cl^-(aq) \qquad \text{(B)}$$

Reaction B consumes the product of Reaction A, thereby pulling Reaction A to the right.

An HCl fountain is assembled as shown below. In panel a, an inverted 250-mL round-bottom flask containing air is set up with its inlet tube leading to a source of HCl(g) and its outlet tube directed into an inverted bottle of water. As HCl is admitted to the flask, air is displaced. When the bottle is filled with air, the flask is filled mostly with HCl(g).

The hoses are disconnected and replaced with a beaker of indicator and a rubber bulb (panel b). For an indicator, we use slightly alkaline methyl purple, which is green above pH 5.4 and purple below pH 4.8. When 1 mL of water is squirted from the rubber bulb into the flask, a vacuum is created and indicator solution is drawn up into the flask, creating a colorful fountain (Color Plate 1).

Questions Why is vacuum created when water is squirted into the flask? Why does the indicator change color when it enters the flask?

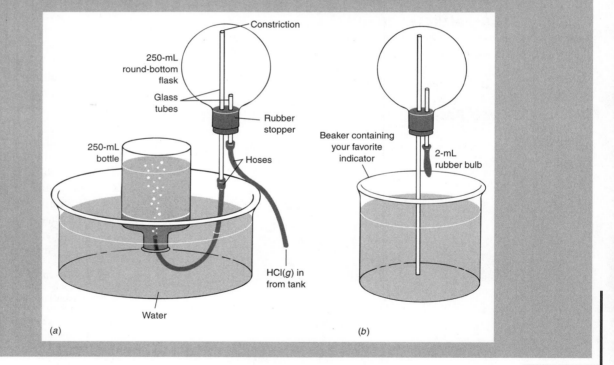

(a)

(b)

Carboxylic Acids Are Weak Acids and Amines Are Weak Bases

As we begin to study acids and bases, the shorthand for drawing chemical structures will be helpful. At this point you should study Box 5-3 and work the exercise at the end of the box.

Acetic acid is a typical weak acid:

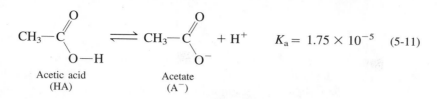

Acetic acid
(HA)

Acetate
(A⁻)

$K_a = 1.75 \times 10^{-5}$ (5-11)

Acetic acid is representative of carboxylic acids, which have the general structure below, where R is an organic substituent. *Most* **carboxylic acids** *are weak acids, and most* **carboxylate anions** *are weak bases.*

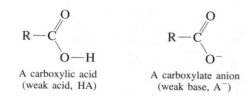

A carboxylic acid
(weak acid, HA)

A carboxylate anion
(weak base, A⁻)

Methylamine is a typical weak base. It forms a bond to H^+ by sharing the lone pair of electrons from the nitrogen atom of the amine:

Carboxylic acids (RCO_2H) and ammonium ions (R_3NH^+) are weak acids. Carboxylate anions (RCO_2^-) and amines (R_3N) are weak bases.

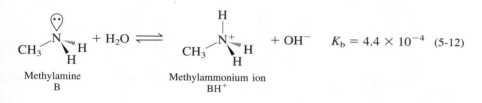

Methylamine
B

Methylammonium ion
BH^+

$K_b = 4.4 \times 10^{-4}$ (5-12)

Methylamine is a representative amine, a nitrogen-containing compound:

$R\ddot{N}H_2$ a primary amine RNH_3^+

$R_2\ddot{N}H$ a secondary amine $R_2NH_2^+$ ⎫ ammonium ions

$R_3\ddot{N}$ a tertiary amine R_3NH^+ ⎭

Amines *are weak bases, and* **ammonium ions** *are weak acids.* The "parent" of all amines is ammonia, NH_3. When methylamine reacts with water, the product is the conjugate acid. That is, the methylammonium ion produced in Reaction 5-12 is a weak acid:

Weak acids: **HA** and **BH⁺**
Weak bases: **A⁻** and **B**

$$CH_3\overset{+}{N}H_3 \underset{}{\overset{K_a}{\rightleftharpoons}} CH_3\ddot{N}H_2 + H^+ \qquad K_a = 2.3 \times 10^{-11} \qquad (5\text{-}13)$$

BH^+ B

The methylammonium ion is the conjugate acid of methylamine.

You should learn to recognize whether a compound is acidic or basic. For example, the salt methylammonium chloride dissociates completely in water to give methylammonium cation and chloride anion:

$$CH_3\overset{+}{N}H_3Cl^-(s) \longrightarrow CH_3\overset{+}{N}H_3(aq) + Cl^-(aq)$$

Methylammonium
chloride

Methylammonium chloride is a weak acid because

1. It dissociates into $CH_3NH_3^+$ and Cl^-.
2. $CH_3NH_3^+$ is a weak acid, being conjugate to CH_3NH_2, a weak base.
3. Cl^- has no basic properties. It is conjugate to HCl, a strong acid. That is, HCl dissociates completely.

Box 5-3 *Explanation*

Shorthand for Organic Structures

Chemists and biochemists use simple conventions for drawing structures of carbon-containing compounds to avoid drawing every atom. Each vertex of a structure is understood to be a carbon atom, unless otherwise labeled. In the shorthand, we usually omit bonds from carbon to hydrogen. Carbon forms four chemical bonds. If you see carbon forming fewer than four bonds, the remaining bonds are assumed to go to hydrogen atoms that are not written. Here are some examples:

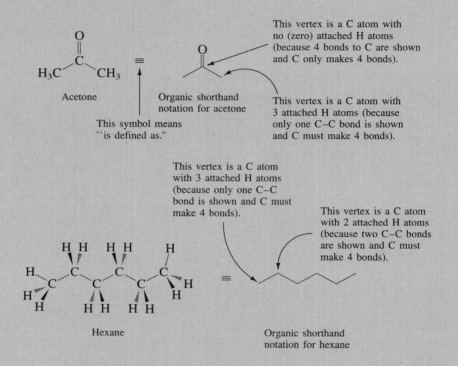

Atoms other than carbon and hydrogen are always shown. Hydrogen atoms attached to atoms other than carbon are always shown. Oxygen and sulfur normally make two bonds. Nitrogen makes three bonds if it is neutral and four bonds if it is a cation. Here are some examples:

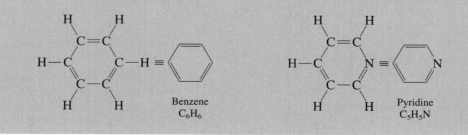

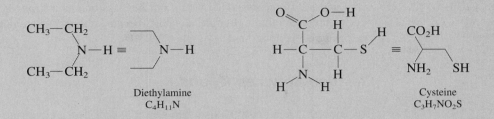

Diethylamine
$C_4H_{11}N$

Cysteine
$C_3H_7NO_2S$

Because of the two equivalent resonance structures of a benzene ring, the alternating single and double bonds are often replaced by a circle:

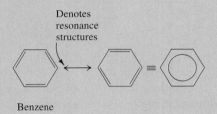

Denotes resonance structures

Benzene

Exercise: Write the chemical formula (e.g., C_4H_8O) for each structure below.

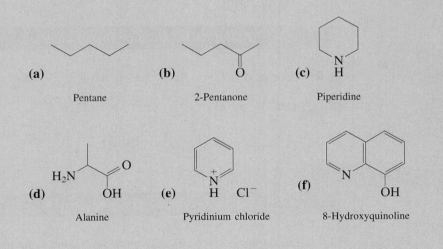

(a)

Pentane

(b)

2-Pentanone

(c)

Piperidine

(d)

Alanine

(e)

Pyridinium chloride

(f)

8-Hydroxyquinoline

89

Challenge Phenol (C_6H_5OH) is a weak acid. Explain why a solution of the ionic compound potassium phenolate ($C_6H_5O^-K^+$) will be basic.

The methylammonium ion, being the conjugate acid of methylamine, is a weak acid (Reaction 5-13). The chloride ion is neither an acid nor a base. It is the conjugate base of HCl, a strong acid. In other words, Cl^- *has virtually no tendency to associate with* H^+; otherwise HCl would not be classified as a strong acid. We predict that methylammonium chloride is acidic, because the methylammonium ion is an acid and Cl^- is not a base.

Relation Between K_a and K_b

An important relationship exists between K_a and K_b of a conjugate acid-base pair in aqueous solution. We can derive this result with the acid HA and its conjugate base A^-.

$$\text{\sout{HA}} \rightleftharpoons H^+ + \text{\sout{A}}^- \qquad K_a = \frac{[H^+][A^-]}{[HA]}$$

$$\text{\sout{A}}^- + H_2O \rightleftharpoons \text{\sout{HA}} + OH^- \qquad K_b = \frac{[HA][OH^-]}{[A^-]}$$

$$H_2O \rightleftharpoons H^+ + OH^- \qquad K_w = K_a \cdot K_b$$

$$= \frac{[H^+][\text{\sout{A}}^-]}{[\text{\sout{HA}}]} \cdot \frac{[\text{\sout{HA}}][OH^-]}{[\text{\sout{A}}^-]}$$

When the reactions above are added, their equilibrium constants must be multiplied, thereby obtaining a most useful result:

$K_a \cdot K_b = K_w$ for a conjugate acid-base pair in aqueous solution.

Relation between K_a and K_b for a conjugate pair:

$$K_a \cdot K_b = K_w \qquad (5\text{-}14)$$

Equation 5-14 applies to any acid and its conjugate base in aqueous solution.

EXAMPLE **Finding K_b for the Conjugate Base**

K_a for acetic acid is 1.75×10^{-5} (Reaction 5-11). Find K_b for the acetate ion.

SOLUTION

$$K_b = \frac{K_w}{K_a} = \frac{1.0 \times 10^{-14}}{1.75 \times 10^{-5}} = 5.7 \times 10^{-10}$$

EXAMPLE **Finding K_a for the Conjugate Acid**

K_b for methylamine is 4.4×10^{-4} (Reaction 5-12). Find K_a for methylammonium ion.

SOLUTION

$$K_a = \frac{K_w}{K_b} = 2.3 \times 10^{-11}$$

Ask Yourself

5-E. Which is a stronger acid, A or B? Write the K_a reaction for each.

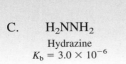

A. Cl_2HCCOH B. ClH_2CCOH

  Dichloroacetic acid   Chloroacetic acid
  $K_a = 5.0 \times 10^{-2}$ $K_a = 1.36 \times 10^{-3}$

Which is a stronger base, C or D? Write the K_b reaction for each.

C. H_2NNH_2 D. H_2NCNH_2

  Hydrazine   Urea
  $K_b = 3.0 \times 10^{-6}$ $K_b = 1.5 \times 10^{-14}$

Key Equations

Equilibrium constant	$aA + bB \overset{K}{\rightleftharpoons} cC + dD \qquad K = \dfrac{[C]^c[D]^d}{[A]^a[B]^b}$
	Concentrations of solutes are in M and gases are in atmospheres. Omit solvents and pure solids and liquids.
Reversed reaction	$K' = 1/K$
Add two reactions	$K_3 = K_1K_2$
Le Châtelier's principle	1. Adding product (or removing reactant) drives reaction in reverse.
	2. Adding reactant (or removing product) drives reaction forward.
Solubility product	$PbI_2(s) \overset{K_{sp}}{\rightleftharpoons} Pb^{2+} + 2I^- \qquad K_{sp} = [Pb^{2+}][I^-]^2$
	Common ion effect: A salt is less soluble in the presence of one of its constituent ions.
Conjugate acids and bases	$HA + B \rightleftharpoons A^- + BH^+$
	Acid Base Base Acid
Autoprotolysis of water	$H_2O \overset{K_w}{\rightleftharpoons} H^+ + OH^- \quad K_w = [H^+][OH^-] = 1.0 \times 10^{-14}$ at 25°C
pH	$pH = -\log[H^+]$
Acid dissociation constant	$HA \overset{K_a}{\rightleftharpoons} H^+ + A^- \qquad K_a = \dfrac{[H^+][A^-]}{[HA]}$
Base hydrolysis constant	$B + H_2O \overset{K_b}{\rightleftharpoons} BH^+ + OH^- \qquad K_b = \dfrac{[BH^+][OH^-]}{[B]}$
Relation between K_a and K_b for conjugate acid-base pair	$K_a \cdot K_b = K_w$

Common weak acids RCO_2H R_3NH^+

                                 Carboxylic acid   Ammonium ion

Common weak bases RCO_2^- R_3N

                                 Carboxylate anion Amine

Important Terms

acid

acid dissociation constant

acidic solution

amine

ammonium ion

autoprotolysis

base

base hydrolysis constant

basic solution

carboxylate anion

carboxylic acid

common ion effect

conjugate acid-base pair

equilibrium constant

hydronium ion

Le Châtelier's principle

neutralization

pH

salt

saturated solution

solubility product

Problems

5-1. Even though you need to express concentrations of solutes in mol/L and the concentrations of gases in atm, why do we say that equilibrium constants are dimensionless?

5-2. Write the expression for the equilibrium constant for each of the following reactions. Write the pressure of a gaseous molecule, X, as P_X.

  (a) $3Ag^+(aq) + PO_4^{3-}(aq) \rightleftharpoons Ag_3PO_4(s)$
  (b) $C_6H_6(l) + \frac{15}{2}O_2(g) \rightleftharpoons 3H_2O(l) + 6CO_2(g)$

5-3. For the reaction $2A(g) + B(aq) + 3C(l) \rightleftharpoons D(s) + 3E(g)$, the concentrations at equilibrium are $P_A = 2.8 \times 10^3$ Pa, $[B] = 1.2 \times 10^{-2}$ M, $[C] = 12.8$ M, $[D] = 16.5$ M, and $P_E = 3.6 \times 10^4$ torr.

  (a) Gas pressure is conventionally expressed in atmospheres when writing the equilibrium constant. Express the pressures of A and E in atm.

  (b) Find the numerical value of the equilibrium constant that would appear in a table of equilibrium constants.

5-4. Suppose that the reaction $Br_2(l) + I_2(s) + 4Cl^-(aq) \rightleftharpoons 2Br^-(aq) + 2ICl_2^-(aq)$ has come to equilibrium. If more $I_2(s)$ is added, will the concentration of ICl_2^- in the aqueous phase increase, decrease, or remain unchanged?

5-5. From the reactions below,

$$CuN_3(s) \rightleftharpoons Cu^+ + N_3^- \qquad K = 4.9 \times 10^{-9}$$
$$HN_3 \rightleftharpoons H^+ + N_3^- \qquad K = 2.2 \times 10^{-5}$$

find the value of K for the reaction $Cu^+ + HN_3 \rightleftharpoons CuN_3(s) + H^+$.

5-6. Consider the following equilibria in aqueous solution:

  (1) $Ag^+ + Cl^- \rightleftharpoons AgCl(aq)$ $\qquad K = 2.0 \times 10^3$
  (2) $AgCl(aq) + Cl^- \rightleftharpoons AgCl_2^-$ $\qquad K = 93$
  (3) $AgCl(s) \rightleftharpoons Ag^+ + Cl^-$ $\qquad K = 1.8 \times 10^{-10}$

  (a) Find the value of K for the reaction $AgCl(s) \rightleftharpoons AgCl(aq)$.

  (b) Find $[AgCl(aq)]$ in equilibrium with excess $AgCl(s)$.

  (c) Find the value of K for the reaction $AgCl_2^- \rightleftharpoons AgCl(s) + Cl^-$.

5-7. Calculate the solubility of CuBr (FW 143.45) in water expressed as (a) moles per liter and (b) grams per 100 mL.

5-8. Find the solubility of silver chromate in water:

$$Ag_2CrO_4 \rightleftharpoons 2Ag^+ + CrO_4^{2-}$$

FW 331.73

Express your answer as **(a)** moles per liter; **(b)** g/100 mL; and **(c)** ppm Ag^+ ($\approx \mu g$ Ag^+/mL).

5-9. Ag^+ at 10–100 ppb (ng/mL) disinfects swimming pools. One way to maintain an appropriate concentration of Ag^+ is to add a slightly soluble silver salt to the pool. Calculate the ppb of Ag^+ at equilibrium in saturated solutions of AgCl, AgBr, and AgI.

5-10. The mercury(I) ion (Hg_2^{2+}, called mercurous ion) is a diatomic ion with a charge of +2. Mercury(I) iodate dissociates into three ions:

$$Hg_2(IO_3)_2 \rightleftharpoons Hg_2^{2+} + 2IO_3^- \qquad K_{sp} = [Hg_2^{2+}][IO_3^-]^2$$
FW 750.99

Find the concentrations of Hg_2^{2+} and IO_3^- in **(a)** a saturated solution of $Hg_2(IO_3)_2(s)$ and **(b)** a 0.010 M solution of KIO_3 saturated with $Hg_2(IO_3)_2(s)$.

5-11. Look up the solubility products of AgCl, AgBr, AgI, and $AgCrO_4$. If a solution containing 0.10 M Cl^-, Br^-, I^-, and CrO_4^{2-} is treated with Ag^+, in what order will the anions precipitate?

5-12. Identify the conjugate acid-base pairs in the following reactions:

(a) $CN^- + HCO_2H \rightleftharpoons HCN + HCO_2^-$
(b) $PO_4^{3-} + H_2O \rightleftharpoons HPO_4^{2-} + OH^-$
(c) $HSO_3^- + OH^- \rightleftharpoons SO_3^{2-} + H_2O$

5-13. A solution is *acidic* if _____. A solution is *basic* if _____.

5-14. Find the pH of a solution containing
(a) 10^{-4} M H^+ (c) 5.8×10^{-4} M H^+
(b) 10^{-5} M OH^- (d) 5.8×10^{-5} M OH^-

5-15. The concentration of H^+ in your blood is 3.5×10^{-8} M.
(a) What is the pH of blood?
(b) Find the concentration of OH^- in blood.

5-16. Sulfuric acid is the principal acidic component of acid rain. The mean pH of rainfall in southern Norway in the 1970s was 4.3. What concentration of H_2SO_4 will produce this pH by the reaction $H_2SO_4 \rightleftharpoons 2H^+ + SO_4^{2-}$?

5-17. An acidic solution containing 0.010 M La^{3+} is treated with NaOH until $La(OH)_3$ precipitates. Use the solubility product for $La(OH)_3$ to find the concentration of OH^- when La^{3+} first precipitates. At what pH does this occur?

5-18. Make a list of the common strong acids and strong bases. Memorize this list.

5-19. Write the structures and names for two classes of weak acids and two classes of weak bases.

5-20. Calculate the concentration of H^+ and the pH of the following solutions:
(a) 0.010 M HNO_3 (d) 3.0 M HCl
(b) 0.035 M KOH   (e) 0.010 M $[(CH_3)_4N^+]OH^-$
(c) 0.030 M HCl   Tetramethylammonium hydroxide

5-21. **(a)** Write the K_a reaction for trichloroacetic acid, Cl_3CCO_2H ($K_a = 0.22$), and for the anilinium ion,

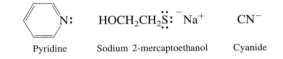

(b) Which is the stronger acid?

5-22. **(a)** Write the K_b reactions for pyridine, sodium 2-mercaptoethanol, and cyanide.

Pyridine   Sodium 2-mercaptoethanol   Cyanide

H^+ binds to S in sodium 2-mercaptoethanol and to C in cyanide.

(b) Values of K_a for the conjugate acids are shown below. Which base in part **(a)** is strongest?

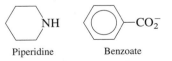

Pyridinium ion	2-Mercaptoethanol	Hydrogen cyanide
$K_a = 5.90 \times 10^{-6}$	$K_a = 1.91 \times 10^{-10}$	$K_a = 6.2 \times 10^{-10}$

5-23. Write the autoprotolysis reaction of H_2SO_4, whose structure is

$$HO-\overset{\overset{\displaystyle O}{\|}}{\underset{\underset{\displaystyle O}{\|}}{S}}-OH$$

5-24. Write the K_b reactions for piperidine and benzoate.

Piperidine   Benzoate

5-25. Hypochlorous acid has the structure H–O–Cl. Write the base hydrolysis reaction of hypochlorite, OCl^-. Given that K_a for HOCl is 3.0×10^{-8}, find K_b for hypochlorite.

The Earliest Known Buret

Good Titrations
—Ode to a Lab Partner
Kurt Wood and Jeff Lederman
(University of California, Davis, 1977)

(Sung to the tune of *Good Vibrations* by the Beach Boys)

Ah! I love the color of pink you get,
And the way the acid drips from your buret.
All the painful things in life seem alien
As I mix in several drops of phenolphthalein.
>> I'm pickin' up good titrations
>> She's givin' me neutralizations
>>> Dew drop drop, good titrations . . .

I look at you and drift away;
The ruby red turns slowly to rosé.
You gaze at me and light my fire
As drop on drop falls to the Erlenmeyer . . .
>> I'm pickin' up good titrations
>> She's givin' me neutralizations
>>> Dew drop drop, good titrations . . .

I look longingly to your eyes,
But you stare down at the lab bench in surprise.
Gone our love before it grew much sweeter,
'Cause we passed the end point by a milliliter.
>> I'm pickin' up back titrations
>> She's prayin' for neutralizations
>>> Dew drop drop, good titrations . . .

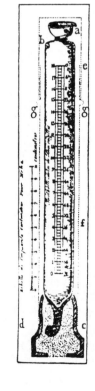

The buret, invented by F. Descroizilles in the early 1800s, was used in the same manner as a graduated cylinder is used today. The stopcock was introduced in 1846. This buret and its progeny have terrorized generations of analytical chemistry students.

Chapter 6

Good Titrations

I n **volumetric analysis,** the volume of a known reagent needed to react with analyte is measured. From this volume, we calculate how much analyte is in the unknown. In this chapter we discuss general principles that apply to any volumetric procedure, and then we illustrate a few particular analyses.

6-1 Principles of Volumetric Analysis

A **titration** is a procedure in which increments of the known reagent solution—the **titrant**—are added to analyte until the reaction is complete. Titrant is usually delivered from a buret, as shown in Figure 6-1. Each increment of titrant should be completely and quickly consumed by reaction with analyte until the analyte is used up. Common titrations are based on acid-base, oxidation-reduction, complex formation, or precipitation reactions.

Methods of determining when analyte has been consumed include (1) detecting a sudden change in voltage or current between a pair of electrodes, (2) observing an *indicator* color change (Color Plate 2), and (3) monitoring the absorbance of light by species in the reaction. An **indicator** is a compound with a physical property (usually color) that changes abruptly when the titration is complete. The change is caused by the disappearance of analyte or appearance of excess titrant.

We will study end-point detection methods later:
  electrodes: Chapters 11, 13, and 14
  indicators: Chapters 6, 10, 11, and 15
  absorbance: Chapter 8

The **equivalence point** occurs when the quantity of titrant added is the exact amount necessary for stoichiometric reaction with the analyte. For example, five moles of oxalic acid react with two moles of permanganate in hot acidic solution:

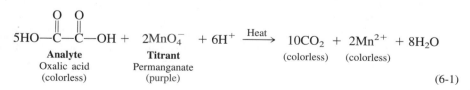

$$5HO-\underset{\substack{\text{Analyte} \\ \text{Oxalic acid} \\ \text{(colorless)}}}{\overset{\overset{\displaystyle O}{\|}}{C}}-\underset{}{\overset{\overset{\displaystyle O}{\|}}{C}}-OH + \underset{\substack{\text{Titrant} \\ \text{Permanganate} \\ \text{(purple)}}}{2MnO_4^-} + 6H^+ \xrightarrow{\text{Heat}} \underset{\substack{\text{(colorless)}}}{10CO_2} + \underset{\substack{\text{(colorless)}}}{2Mn^{2+}} + 8H_2O$$

$$(6\text{-}1)$$

If the unknown contains 5.00 mmol of oxalic acid, the equivalence point is reached when 2.00 mmol of MnO_4^- has been added.

The equivalence point is the ideal result we seek in a titration. What we actually measure is the **end point,** which is marked by a sudden change in a physical property of the solution. For Reaction 6-1, a convenient end point is the abrupt appearance of the purple color of permanganate in the flask. Up to the equivalence point, all of the added permanganate is consumed by oxalic acid, and the titration solution remains colorless. After the equivalence point, unreacted MnO_4^- ion builds up until there is enough to see. The *first trace* of purple color marks the end point. The better your eyes, the closer the measured end point will be to the true equivalence point. The end point cannot exactly equal the equivalence point because extra MnO_4^-—more than that needed to react with oxalic acid—is required to create perceptible purple color.

The difference between the end point and the equivalence point is an inescapable **titration error.** By choosing an appropriate physical property, in which a change is easily observed (such as indicator color, optical absorbance of a reactant or product, or pH), it is possible to have an end point that is very close to the equivalence point. It is also possible to estimate the titration error with a **blank titration,** in which the same procedure is carried out without analyte. For example, a solution containing no oxalic acid could be titrated with MnO_4^- to see how much is needed to create observable purple color. This volume of MnO_4^- is then subtracted from the volume observed in the titration of unknown.

The validity of an analytical result depends on knowing the amount of one of the reactants used. A **primary standard** is a reagent that is pure enough to be weighed out and used directly to provide a known number of moles. For example, if you want to titrate unknown hydrochloric acid with base, you could weigh out primary standard grade sodium carbonate and dissolve it in water to make titrant:

$$2HCl \; + \; \underset{\substack{\text{Sodium carbonate} \\ \text{FW 105.99}}}{Na_2CO_3} \longrightarrow H_2CO_3 + 2NaCl \qquad (6\text{-}2)$$

$$\underset{\text{(unknown)}}{} \qquad \underset{\text{(primary standard)}}{}$$

Two moles of HCl react with one mole of Na_2CO_3, which has a mass of exactly 105.99 g. You could not carry out the same procedure beginning with solid NaOH

$$HCl \; + \; \underset{\substack{\text{Sodium hydroxide} \\ \text{FW 40.00}}}{NaOH} \longrightarrow H_2O + NaCl$$

because the solid is not pure. NaOH is normally contaminated with some Na_2CO_3 (from reaction with CO_2 in the air) and H_2O (also from the air). If you weighed out 40.00 g of sodium hydroxide, it would not contain exactly one mole.

A primary standard should be 99.9% pure, or better. It should not decompose under ordinary storage, and it should be stable when dried by heating or vacuum, because drying is required to remove traces of adsorbed water from the atmosphere.

In most cases, titrant is not available as a primary standard. Instead, a solution having approximately the desired concentration is used to titrate a weighed, primary standard. From the volume of titrant required to react with the primary standard, we calculate the concentration of titrant. The process of titrating a standard to determine the concentration of titrant is called **standardization.** We say that a solution whose concentration is known is a **standard solution.** The validity of the analytical result ultimately depends on knowing the composition of some primary standard.

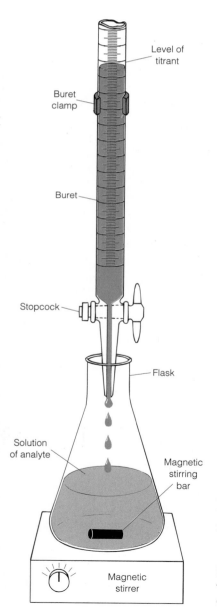

Level of titrant

Buret clamp

Buret

Stopcock

Flask

Solution of analyte

Magnetic stirring bar

Magnetic stirrer

FIGURE 6-1   Typical setup for a titration. The analyte is contained in the flask, and the titrant in the buret. The stirring bar is a magnet coated with Teflon, which is inert to almost all solutions. The bar is spun by a rotating magnet inside the stirrer.

In a **direct titration,** titrant is added to analyte until the end point is observed.

Direct titration: analyte + titrant $\longrightarrow$ product
                        _{Unknown}   _{Known}

In a **back titration,** a known *excess* of a standard reagent is added to the analyte. Then a second standard reagent is used to titrate the excess of the first reagent.

analyte + reagent 1 $\longrightarrow$ product + excess reagent 1   (6-3a)
_{Unknown}   _{Known}   _{Unknown quantity}

*Back
titration:*   excess reagent 1 + reagent 2 $\longrightarrow$ product   (6-3b)
                 _{Unknown}   _{Known}

Back titrations are useful when the end point of the back titration is clearer than the end point of the direct titration or when an excess of the first reagent is required for complete reaction with the analyte.

As examples of direct and back titrations, consider first the addition of permanganate titrant to oxalic acid analyte in Reaction 6-1; this reaction is a direct titration. For a back titration, an *excess* (but known amount) of permanganate is added to consume the oxalic acid. Then the excess permanganate is back-titrated with standard Fe^{2+} to measure how much permanganate was left unreacted. The difference between the initial quantity of permanganate and the amount left unreacted tells us how much was consumed by the oxalic acid.

Ask Yourself

6-A. (a) Why does the validity of an analytical result ultimately depend on knowing the composition of a primary standard?
   (b) How does a blank titration reduce titration error?
   (c) What is the difference between a direct titration and a back titration?

Box 3-1 describes Standard Reference Materials that allow different laboratories to test the accuracy of their procedures.

6-2 *Titration Calculations*

To interpret the results of a direct titration, the key steps are

1. From the volume of titrant, calculate the number of moles of titrant consumed.

2. From the stoichiometry of the titration reaction, relate the unknown moles of analyte to the known moles of titrant.

1:1 Stoichiometry

Consider the titration of an unknown chloride solution with standard Ag^+:

$$Ag^+ + Cl^- \overset{K = 1/K_{sp}}{\rightleftharpoons} AgCl(s)$$

When K is large, we sometimes write $\rightarrow$ instead of $\rightleftharpoons$:

$$Ag^+ + Cl^- \longrightarrow AgCl(s)$$

The reaction is rapid and the equilibrium constant is large ($1/K_{sp}$[for AgCl] = 5.6×10^9), so the reaction essentially goes to completion with each addition of titrant. White AgCl precipitate forms as soon as the two reagents are mixed.

Suppose that 10.00 mL of unknown chloride solution (measured with a transfer pipet) requires 22.97 mL of 0.052 74 M $AgNO_3$ solution (delivered from a buret) for complete reaction. What is the concentration of Cl^- in the unknown? Following the two-step recipe, we first find the moles of Ag^+:

Retain an extra, insignificant digit until the end of the calculations to avoid round-off error.

$$\text{mol } Ag^+ = \text{volume} \times \text{molarity} = (0.022\ 97\ \text{L})\left(0.052\ 74\ \frac{\text{mol}}{\text{L}}\right) = 0.001\ 211_4\ \text{mol}$$

Next, relate the unknown moles of Cl^- to the known moles of Ag^+. We know that one mole of Cl^- reacts with one mole of Ag^+. If $0.001\ 211_4$ mol of Ag^+ is required, then $0.001\ 211_4$ mol of Cl^- must have been in 10.00 mL of unknown. Therefore

$$[Cl^-] \text{ in unknown} = \frac{\text{mol } Cl^-}{\text{L of unknown}} = \frac{0.001\ 211_4\ \text{mol}}{0.010\ 00\ \text{L}} = 0.121\ 1_4\ \text{M}$$

The solution we just titrated was made by dissolving 1.004 g of an unknown solid in a total volume of 100.0 mL. What is the weight percent of chloride in the solid?

We know that 10.00 mL of unknown solution contain $0.001\ 211_4$ mol of Cl^-. Therefore 100.0 mL must contain 10 times as much, or $0.012\ 11_4$ mol of Cl^-. This much Cl^- weighs $(0.012\ 11_4\ \text{mol } Cl^-)(35.453\ \text{g/mol } Cl^-) = 0.429\ 4_8\ \text{g } Cl^-$. The weight percent of Cl^- in the unknown is

$$\text{wt \% } [Cl^-] = \frac{\text{g } Cl^-}{\text{g unknown}} \times 100 = \frac{0.429\ 4_8\ \text{g } Cl^-}{1.004\ \text{g unknown}} \times 100 = 42.78\ \text{wt \%}$$

EXAMPLE **A Slightly More Complicated Example**

(a) Standard Ag^+ solution was prepared by dissolving 1.224 3 g of dry $AgNO_3$ (FW 169.87) in water in a 500.0-mL volumetric flask. A dilution was made by delivering 25.00 mL of solution with a pipet to a second 500.0-mL volumetric flask and diluting to the mark. Find the concentration of Ag^+ in the dilute standard solution. **(b)** A 25.00-mL aliquot of unknown solution containing Cl^- was titrated with the dilute Ag^+ solution and the equivalence point was reached when 37.38 mL of Ag^+ solution had been delivered. Find the concentration of Cl^- in the unknown.

SOLUTION **(a)** The concentration of the initial $AgNO_3$ solution is

$$[Ag^+] = \frac{(1.224\ 3\ \text{g})/(169.87\ \text{g/mol})}{0.500\ 0\ \text{L}} = 0.014\ 41_6\ \text{M}$$

To find the concentration of the dilute solution, use the dilution formula 1-5:

Notice that the general form of all dilution problems is

$$[X]_{final} = \frac{V_{initial}}{V_{final}} \cdot [X]_{initial}$$

$$[Ag^+]_{conc} \cdot V_{conc} = [Ag^+]_{dil} \cdot V_{dil} \tag{1-5}$$
$$(0.014\ 41_6\ \text{M}) \cdot (25.00\ \text{mL}) = [Ag^+]_{dil} \cdot (500.0\ \text{mL})$$

$$[Ag^+]_{dil} = \left(\frac{25.00\ \text{mL}}{500.0\ \text{mL}}\right)(0.014\ 41_6\ \text{M}) = 7.207_3 \times 10^{-4}\ \text{M}$$

(b) One mole of Cl^- requires one mole of Ag^+. The number of moles of Ag^+ required to reach the equivalence point is

$$mol\ Ag^+ = (7.207_3 \times 10^{-4}\ M)\ (0.037\ 38\ L) = 2.694_1 \times 10^{-5}\ mol$$

The concentration of Cl^- in 25.00 mL of unknown is therefore

$$[Cl^-] = \frac{2.694_1 \times 10^{-5}\ mol}{0.025\ 00\ L} = 1.078 \times 10^{-3}\ M$$

Silver ion titrations are especially nice because $AgNO_3$ is a primary standard. After drying at 110°C for 1 h to remove residual moisture, the solid has the exact composition $AgNO_3$. Methods for finding the end point in silver titrations are described at the end of this chapter. Silver compounds and solutions should be stored in the dark to protect against photodecomposition and never exposed to direct sunlight.

$AgNO_3$ solution is an antiseptic. If you spill $AgNO_3$ solution on yourself, your skin will turn black for a few days until the affected skin is shed.

x:y Stoichiometry

Consider Reaction 6-1 in which 5 moles of oxalic acid ($H_2C_2O_4$) react with 2 moles of permanganate (MnO_4^-).

$$\text{stoichiometry relationship for Reaction 6-1: } \frac{5\ mol\ H_2C_2O_4}{2\ mol\ MnO_4^-}$$

Photodecomposition of white $AgCl(s)$:

$$AgCl(s) \xrightarrow{\text{light}} Ag(s) + \tfrac{1}{2}Cl_2(g)$$

Finely divided $Ag(s)$ imparts a faint violet color to the solid.

In permanganate titrations, we customarily run a *blank titration* without analyte and subtract the blank volume from the volume required to titrate unknown. In principle this procedure eliminates the titration error associated with locating the purple end point.

EXAMPLE **Titration of $H_2C_2O_4$ with MnO_4^-**

A 10.00-mL aliquot of unknown oxalic acid solution required 15.44 mL of 0.011 17 M $KMnO_4$ solution to reach the purple end point. A blank titration of 10 mL of similar solution containing no oxalic acid required 0.04 mL to exhibit detectable color. Find the concentration of oxalic acid in the unknown.

SOLUTION Subtract the blank volume from the end-point volume to find the equivalence volume: $15.44 - 0.04 = 15.40$ mL. This much titrant solution contains

$$mol\ MnO_4^- = (0.011\ 17\ M)(0.015\ 40\ L) = 1.720_2 \times 10^{-4}\ mol$$

The moles of oxalic acid in the unknown must be

$$mol\ H_2C_2O_4 = (1.720_2 \times 10^{-4}\ mol\ MnO_4^-)\left(\frac{5\ mol\ H_2C_2O_4}{2\ mol\ MnO_4^-}\right) = 4.300_4 \times 10^{-4}\ mol$$

Keep in mind the ratio $\dfrac{5\ mol\ H_2C_2O_4}{2\ mol\ MnO_4^-}$

The concentration of oxalic acid in the unknown is

$$[H_2C_2O_4] = \frac{4.300_4 \times 10^{-4}\ mol}{0.010\ 00\ L} = 4.300 \times 10^{-2}\ M$$

Standardization of Titrant Followed by Analysis of Unknown

The calcium content of urine can be determined by the scheme in Figure 6-2:

1. Ca^{2+} is precipitated with excess oxalate in basic solution:

$$Ca^{2+} + C_2O_4^{2-} \longrightarrow CaC_2O_4 \cdot H_2O(s)$$

Oxalate   Calcium oxalate

2. The precipitate is washed with ice-cold water to remove excess oxalate.

3. The remaining solid is dissolved in hot acid, giving Ca^{2+} and $H_2C_2O_4(aq)$.

4. $H_2C_2O_4(aq)$ is titrated with standard $KMnO_4$ until the purple end point is observed (Reaction 6-1). For best results, the oxalic acid solution is treated at 25°C with 90–95% of the expected volume of $KMnO_4$. The solution is then heated to 60°C and the titration is completed.

$KMnO_4$ is not pure enough to be a primary standard. Also, traces of organic impurities in distilled water consume some of the freshly dissolved MnO_4^- to produce $MnO_2(s)$. To prepare a permanganate solution for titrations, $KMnO_4$ is dissolved in distilled water to give the approximately desired concentration and the solution is boiled for 1 h to complete the reaction between MnO_4^- and organic impurities. The mixture is filtered through a sintered glass filter (not a paper filter, which is organic) to remove MnO_2, and the solution is then cooled and standardized against primary standard sodium oxalate.

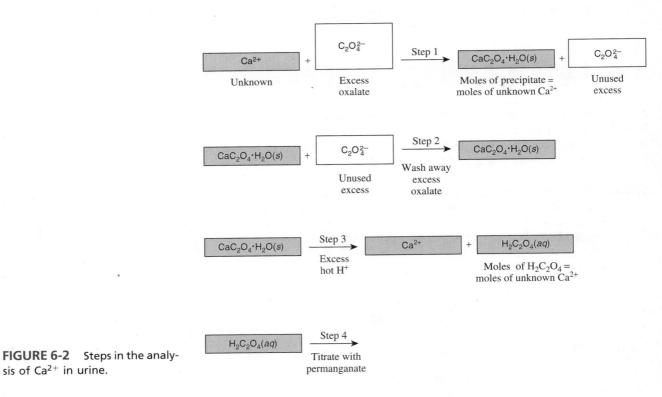

FIGURE 6-2 Steps in the analysis of Ca^{2+} in urine.

EXAMPLE Standardizing KMnO₄ with Oxalate

Standard oxalate solution was prepared by dissolving 0.356 2 g of $Na_2C_2O_4$ (FW 134.00) in 1.0 M H_2SO_4 in a 250.0-mL volumetric flask. A 10.00-mL aliquot of this solution required 48.39 mL of $KMnO_4$ solution for titration, and a blank titration required 0.03 mL of $KMnO_4$. Find the molarity of $KMnO_4$.

SOLUTION   The number of moles of $Na_2C_2O_4$ dissolved in 250.0 mL is 0.356 2 g/(134.00 g/mol) = 0.002 658$_2$ mol. The concentration of oxalate solution is

$$[Na_2C_2O_4] = \frac{0.002\ 658_2\ \text{mol}}{0.250\ 0\ \text{L}} = 0.010\ 63_3\ \text{M}$$

The $C_2O_4^{2-}$ in 10.00 mL is $(0.010\ 63_3\ \text{M})(0.010\ 00\ \text{L}) = 1.063_3 \times 10^{-4}$ mol. Reaction 6-1 requires 2 mol of permanganate for 5 mol of oxalate. Therefore

$$\text{mol MnO}_4^- = (\text{mol C}_2\text{O}_4^{2-})\left(\frac{2\ \text{mol MnO}_4^-}{5\ \text{mol H}_2\text{C}_2\text{O}_4}\right) = 4.253_1 \times 10^{-5}\ \text{mol}$$

The equivalence volume of $KMnO_4$ is 48.39 − 0.03 = 48.36 mL. The concentration of MnO_4^- titrant is

You can use either ratio, $\dfrac{5\ \text{mol H}_2\text{C}_2\text{O}_4}{2\ \text{mol MnO}_4^-}$ or $\dfrac{2\ \text{mol MnO}_4^-}{5\ \text{mol H}_2\text{C}_2\text{O}_4}$, whenever you please, as long as the units work out.

$$[MnO_4^-] = \frac{4.253_1 \times 10^{-5}\ \text{mol}}{0.048\ 36\ \text{L}} = 8.794_7 \times 10^{-4}\ \text{M}$$

EXAMPLE Analysis of Calcium in Urine

The calcium in a 5.00-mL urine sample was precipitated by the procedure in Figure 6-2, redissolved, and titrated. The titration required 16.21 mL of standard MnO_4^- solution. A blank titration required 0.04 mL to produce perceptible color. Find the concentration of Ca^{2+} in the urine.

SOLUTION The equivalence volume is 16.21 − 0.04 = 16.17 mL of MnO_4^-, which contains $(0.016\ 17\ \text{L})(8.794_7 \times 10^{-4}\ \text{M}) = 1.422_1 \times 10^{-5}$ mol of MnO_4^-. This quantity reacts with

$$\text{mol of C}_2\text{O}_4^{2-} = (\text{mol of MnO}_4^-)\left(\frac{5\ \text{mol H}_2\text{C}_2\text{O}_4}{2\ \text{mol MnO}_4^-}\right) = 3.555_3 \times 10^{-5}\ \text{mol}$$

Because there is one oxalate ion for each calcium ion in $CaC_2O_4 \cdot H_2O$, there must have been $3.555_3 \times 10^{-5}$ mol of Ca^{2+} in 5.00 mL of urine:

$$[Ca^{2+}] = \frac{3.555_3 \times 10^{-5}\ \text{mol}}{0.005\ 00\ \text{L}} = 0.007\ 11\ \text{M}$$

A Back Titration

Reactions 6-3a and 6-3b gave the general sequence for a back titration:

$$\text{analyte} + \text{reagent 1} \longrightarrow \text{product} + \text{excess reagent 1} \qquad \text{(6-3a)}$$
$$\underset{\text{Unknown}}{} \quad \underset{\text{Known}}{} \qquad \qquad \underset{\text{Unknown quantity}}{}$$

Back titration:
$$\text{excess reagent 1} + \text{reagent 2} \longrightarrow \text{product} \qquad \text{(6-3b)}$$
$$\underset{\text{Unknown}}{} \qquad \underset{\text{Known}}{}$$

To interpret the results of a back titration, the key steps are

1. From the volume of reagent 2, calculate the number of moles of excess reagent 1 in Reaction 6-3b. This answer is equal to the excess of reagent 1 left after Reaction 6-3a.

2. Subtract the moles of excess reagent 1 from the known initial moles of reagent 1 in Reaction 6-3a. This answer tells how much reagent 1 reacted with analyte.

3. From the stoichiometry of Reaction 6-3a, relate the unknown moles of analyte to the known moles of reagent 1 in Reaction 6-3a.

EXAMPLE **Calculations for a Back Titration**

A cyanide solution with a volume of 25.00 mL was treated with 20.00 mL of 0.038 56 M Ni^{2+} solution (containing excess Ni^{2+}) to convert CN^- to $Ni(CN)_4^{2-}$:

$$\underset{\text{Cyanide}}{4CN^-} + Ni^{2+} \longrightarrow \underset{\text{Tetracyanonickelate(II)}}{Ni(CN)_4^{2-}} \qquad \text{(A)}$$

The excess Ni^{2+} was then back-titrated with 11.77 mL of 0.023 07 M ethylenediaminetetraacetic acid (EDTA). One mole of EDTA reacts with one mole of Ni^{2+}:

$$Ni^{2+} + EDTA^{4-} \longrightarrow Ni(EDTA)^{2-} \qquad \text{(B)}$$

Calculate the molarity of CN^- in the 25.00-mL cyanide sample.

SOLUTION The EDTA required to react with excess Ni^{2+} is (0.011 77 L)(0.023 07 M) = $2.715_3 \times 10^{-4}$ mol EDTA. In Reaction B, the moles of EDTA are equal to the moles of Ni^{2+}. Therefore the excess Ni^{2+} left from Reaction A must also have been $2.715_3 \times 10^{-4}$ mol. The number of initial moles of Ni^{2+} in Reaction A was (0.020 00 L)(0.038 56 M) = $7.712_0 \times 10^{-4}$ mol. The number of moles of Ni^{2+} consumed in Reaction A is the difference between the initial moles and the moles left over for Reaction B:

mol of Ni^{2+} consumed in Reaction A
$$= \text{initial mol } Ni^{2+} - \text{mol } Ni^{2+} \text{ left for Reaction B}$$
$$= 7.712_0 \times 10^{-4} \text{ mol} - 2.715_3 \times 10^{-4} \text{ mol} = 4.996_7 \times 10^{-4} \text{ mol } Ni^{2+}$$

In Reaction A, 1 mol Ni^{2+} reacts with 4 mol CN^-. Therefore the CN^- in Reaction A must have been

$$mol\ CN^- = (4.996_7 \times 10^{-4}\ mol\ Ni^{2+}) \left(\frac{4\ mol\ CN^-}{1\ mol\ Ni^{2+}}\right) = 1.998_7 \times 10^{-3}\ mol$$

The concentration of cyanide in 25.00 mL of unknown is

$$[CN^-] = \frac{1.998_7 \times 10^{-3}\ mol}{0.025\ 00\ L} = 0.079\ 95\ M$$

Ask Yourself

6-B. Vitamin C (ascorbic acid) from foods can be measured by titration with I_3^-:

$$\underset{\substack{\text{Ascorbic acid} \\ \text{FW 176.126}}}{C_6H_8O_6} + \underset{\text{Triiodide}}{I_3^-} + H_2O \longrightarrow C_6H_8O_7 + 3I^- + 2H^+$$

Starch is used as an indicator in the reaction. The end point is marked by the appearance of a deep blue starch-iodine complex when unreacted I_3^- is present.
   (a) If 29.41 mL of I_3^- solution is required to react with 0.197 0 g of pure ascorbic acid, what is the molarity of the I_3^- solution?
   (b) A vitamin C tablet containing ascorbic acid plus inert binder was ground to a powder, and 0.424 2 g was titrated by 31.63 mL of I_3^-. How many moles of vitamin C are present in the 0.424 2-g sample?
   (c) Find the weight percent of ascorbic acid in the tablet.

6-3 Titration of a Mixture

When Ag^+ is added to a solution containing Cl^- and I^-, the less soluble $AgI(s)$ precipitates first:

$$\left. \begin{array}{l} Ag^+ + I^- \longrightarrow AgI(s) \\ Ag^+ + Cl^- \longrightarrow AgCl(s) \end{array} \right\} \begin{array}{l} AgI\ (K_{sp} = 8.3 \times 10^{-17})\ precipitates \\ before\ AgCl\ (K_{sp} = 1.8 \times 10^{-10}) \end{array} \quad \begin{array}{l} (6\text{-}4a) \\ (6\text{-}4b) \end{array}$$

Because the two solubility products are sufficiently different, the first precipitation is nearly complete before the second commences.
   Figure 6-3 shows how the reaction is monitored with a silver electrode to find both end points. We will learn how this electrode responds to silver ion concentration in Chapter 14. Figure 6-4 shows experimental curves for the titration of I^- or a mixture of I^- plus Cl^- by Ag^+. The voltage in Figure 6-4 decreases as the concentration of Ag^+ increases.
   In the titration of I^- (curve b in Figure 6-4), the voltage difference between the two electrodes remains almost constant near 650 mV until the equivalence point

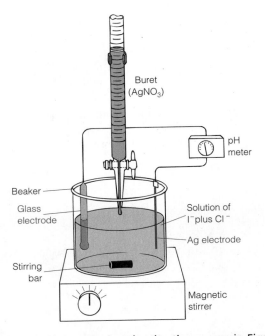

FIGURE 6-3 Apparatus for measuring the titration curves in Figure 6-4. The silver electrode responds to changes in the Ag^+ concentration, and the glass electrode provides a constant reference potential in this experiment. The measured voltage changes by approximately 59 mV for each factor-of-10 change in $[Ag^+]$. All solutions, including $AgNO_3$, were maintained at pH 2.0 by using 0.010 M sulfate buffer prepared from H_2SO_4 and KOH.

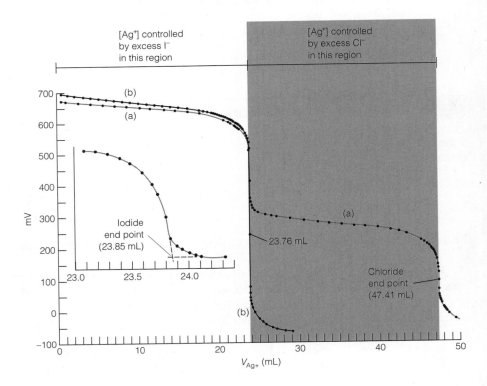

FIGURE 6-4 Experimental titration curves. The ordinate (*y*-axis) is the electric potential difference (in millivolts) between the two electrodes in Figure 6-3. (*a*) Titration of KI plus KCl with 0.084 5 M $AgNO_3$. The inset is an expanded view of the region near the first equivalence point. (*b*) Titration of I^- with 0.084 5 M Ag^+.

(23.76 mL) when I^- is used up. At this point, there is an abrupt decrease in the electrode potential. The reason for the abrupt change is that the electrode is responding to the concentration of Ag^+ in the solution. Prior to the equivalence point, virtually all of the added Ag^+ reacts with I^- to precipitate $AgI(s)$. The concentration of Ag^+ in solution is very low and nearly constant. When I^- has been consumed, the concentration of Ag^+ suddenly increases because Ag^+ is being added from the buret and is no longer consumed by I^-. This change gives rise to the abrupt decrease in electrode potential.

In the titration of $I^- + Cl^-$ (curve a in Figure 6-4), there are two abrupt changes in electrode potential. The first occurs when I^- is used up, and the second comes when Cl^- is used up. Prior to the first equivalence point, the very low concentration of Ag^+ is governed by the solubility of AgI. Between the first and second equivalence points, essentially all I^- has precipitated and Cl^- is in the process of being consumed. The concentration of Ag^+ is still small but governed by the solubility of AgCl, which is greater than that of AgI. After the second equivalence point, the concentration of Ag^+ shoots upward as Ag^+ is added from the buret. Therefore we observe two abrupt changes of electric potential in this experiment.

The I^- end point is taken as the intersection of the steep and nearly horizontal curves shown at 23.85 mL in the inset of Figure 6-4. The reason for using the intersection is that the precipitation of I^- is not quite complete when Cl^- begins to precipitate. Therefore the end of the steep portion (the intersection) is a better approximation of the equivalence point than is the middle of the steep section. The Cl^- end point is taken as the midpoint of the second steep section, at 47.41 mL. The moles of Cl^- in the sample correspond to the moles of Ag^+ delivered between the first and second end points. That is, it requires 23.85 mL of Ag^+ to precipitate I^-, and $(47.41 - 23.85) = 23.56$ mL of Ag^+ to precipitate Cl^-.

EXAMPLE Extracting Results from Figure 6-4

In curve a of Figure 6-4, 40.00 mL of unknown solution containing both I^- and Cl^- was titrated with 0.084 5 M Ag^+. Find the concentrations of each halide ion in the unknown.

The elements F, Cl, Br, I, and At are called *halogens*. Their anions are called *halides*.

SOLUTION The inset shows the first end point at 23.85 mL. Reaction 6-4a tells us that one mole of I^- consumes one mole of Ag^+. The moles of Ag^+ delivered at this point are $(0.084\ 5\ M)\ (0.023\ 85\ L) = 2.015 \times 10^{-3}$ mol. The molarity of iodide in the unknown is therefore

$$[I^-] = \frac{2.015 \times 10^{-3}\ \text{mol}}{0.040\ 00\ \text{L}} = 0.050\ 3_8\ M$$

The second end point is at 47.41 mL. The quantity of Ag^+ titrant required to react with Cl^- is the difference between the two end points: $(47.41 - 23.85) = 23.56$ mL. The moles of Ag^+ required to react with Cl^- are $(0.084\ 5\ M)(0.023\ 56\ L) = 1.991 \times 10^{-3}$ mol. The molarity of chloride in the unknown is

$$[Cl^-] = \frac{1.991 \times 10^{-3}\ \text{mol}}{0.040\ 00\ \text{L}} = 0.049\ 7_7\ M$$

Linda A. Hughes

NH_3 (FW 17.031) contains 82.24 wt % N. Therefore, a solution containing 1.216 g NH_3/mL contains

$$\left(1.216 \; \frac{g \; NH_3}{mL}\right)\left(0.822 \; 4 \; \frac{g \; N}{g \; NH_3}\right)$$
$$= 1.000 \; \mu g \; N/mL = 1 \; ppm \; N$$

Ask Yourself

6-C. A 25.00-mL solution containing Br^- and Cl^- was titrated with 0.033 33 M $AgNO_3$.

(a) Write the two titration reactions and use solubility products to find which occurs first.

(b) In an experiment analogous to that in Figures 6-3 and 6-4, the first end point was observed at 15.55 mL. Find the concentration of the first halide that precipitated. Is it Br^- or Cl^-?

(c) The second end point was observed at 42.23 mL. Find the concentrations of the other halide.

6-4 | *Chemistry in a Fishtank*

Students at the Georgia Institute of Technology have studied analytical chemistry by measuring chemical changes in a saltwater aquarium in their laboratory. One of the chemicals measured is nitrite, NO_2^-, which is a key species in the natural cycle of nitrogen (Figure 6-5). Box 6-1 shows concentrations of ammonia (NH_3), nitrite, and nitrate (NO_3^-) measured in the aquarium. Concentrations are expressed in parts per million of nitrogen, which means μg of nitrogen per gram of seawater. Because 1 g of water $\approx$ 1 mL, we will consider 1 ppm to be 1 μg/mL.

The nitrite measurement was done by a spectrophotometric procedure described in Chapter 8. Because there is no convenient primary standard for nitrite, a titration is used to standardize a $NaNO_2$ solution that serves as the standard for the spectrophotometric procedure. As always, the validity of any analytical procedure ultimately depends on knowing the composition of a primary standard, which is sodium oxalate in this case.

FIGURE 6-5 Nitrogen is exchanged among different forms of life through the *nitrogen cycle.* Only a few organisms, such as blue-green algae, have the ability to use N_2 directly from the air. Our existence depends on the health of all organisms in the nitrogen cycle.

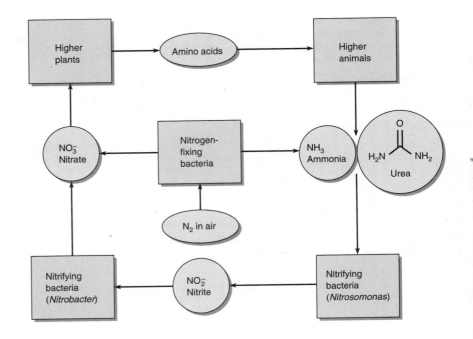

Box 6-1 *Informed Citizen*

Studying a Marine Ecosystem

A saltwater aquarium was set up at Georgia Tech to study the chemistry of a marine ecosystem. When fish and food were introduced into the aquarium, bacteria began to grow and to metabolize organic compounds into ammonia (NH_3). Ammonia is toxic to marine animals when the level exceeds 1 ppm; but, fortunately, it is removed by *Nitrosomonas* bacteria, which oxidize ammonia to nitrite (NO_2^-). Alas, nitrite is also toxic at levels above 1 ppm, but it is further oxidized to nitrate (NO_3^-) by *Nitrobacter* bacteria.

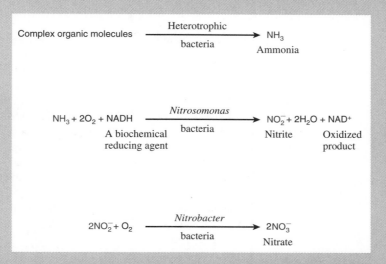

Some actions of nitrogen-metabolizing bacteria. *Heterotrophic* bacteria require complex organic molecules from the breakdown of other organisms for nourishment. By contrast, *autotrophic* bacteria can utilize CO_2 as their carbon source for biosynthesis.

The following graph shows ammonia and nitrite concentrations observed by students at Georgia Tech. About 18 days after introducing fish, significant levels of ammonia were observed. The first dip in NH_3 concentration occurred during a 48-h period when no food was added to the tank. The third peak in NH_3 concentration arose when changing flow patterns through the aquarium filter exposed fresh surfaces devoid of the bacteria that remove ammonia. When the population of *Nitrosomonas* bacteria was sufficient, ammonia levels decreased but nitrite became perilously high. After 60 days, the population of *Nitrobacter* bacteria was great enough to convert most nitrite to nitrate.

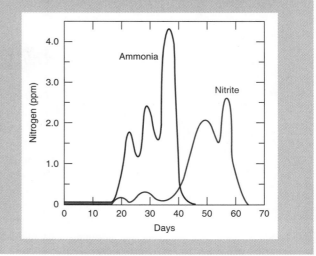

Ammonia and nitrite concentrations in a saltwater aquarium at Georgia Tech after fish and food were introduced into the "sterile" tank on day 0. Concentrations are expressed in parts per million of nitrogen (i.e., μg of N per mL of solution).

The concentrations of $NaNO_2$ and $KMnO_4$ are only approximate. The solutions will be standardized in subsequent titrations.

Three solutions are required for the measurement of nitrite:

a. Prepare ~0.018 M $NaNO_2$ (FW 68.995) by dissolving 1.25 g of $NaNO_2$ in 1.00 L of distilled water.

b. Prepare 0.025 00 M $Na_2C_2O_4$ (FW 134.00) by dissolving 3.350 g of primary standard grade $Na_2C_2O_4$ in water and diluting to 1.000 L.

c. Prepare ~0.010 M $KMnO_4$ (FW 158.03) by dissolving 1.6 g of $KMnO_4$ in 1.00 L of distilled water. (As mentioned on page 100, this solution must be heated and filtered prior to storage in a dark bottle.) Use the procedure in Section 6-2 to standardize the $KMnO_4$ solution by titrating standard $Na_2C_2O_4$ (solution b above):

$$5H_2C_2O_4 + 2MnO_4^- + 6H^+ \longrightarrow 10CO_2 + 2Mn^{2+} + 8H_2O \qquad (6\text{-}1)$$

Here is the procedure for standardizing a solution of $NaNO_2$:

1. Pipet 50.00 mL of standard $KMnO_4$, 5 mL of concentrated (18 M) H_2SO_4, and 50.00 mL of stock $NaNO_2$ solution into a glass-stoppered flask. Mix gently and warm to 70°–80°C on a hot plate. At this point, the nitrite is consumed and there is excess permanganate in the solution:

$$5NO_2^- + 2MnO_4^- + 6H^+ \longrightarrow 5NO_3^- + 2Mn^{2+} + 3H_2O \qquad (6\text{-}5)$$

2. Add standard $Na_2C_2O_4$ solution in 10.00-mL aliquots until the purple color of permanganate disappears. Now MnO_4^- has been consumed and there is excess $H_2C_2O_4$.

3. *Back-titrate* the excess $H_2C_2O_4$ in the warm solution with standard $KMnO_4$.

4. Perform a *blank titration* with water in place of the 50 mL of $KMnO_4$, the 50 mL of $NaNO_2$, and the $Na_2C_2O_4$. The purpose is to see how much $KMnO_4$ must be added to observe purple color.

Knowing how much $Na_2C_2O_4$ and how much $KMnO_4$ were used allows us to find the concentration of $NaNO_2$.

EXAMPLE   Using a Back Titration to Standardize $NaNO_2$

In step 1, 50.00 mL of 0.010 54 M $KMnO_4$, 5 mL of H_2SO_4, and 50.00 mL of $NaNO_2$ solution were mixed and warmed to take Reaction 6-5 to completion. Addition of 10.00 mL of 0.025 00 M $Na_2C_2O_4$ in step 2 did not decolorize the permanganate after a few minutes of heating. Addition of a second 10.00-mL aliquot of $Na_2C_2O_4$ left the solution colorless, meaning that permanganate was used up and there was excess oxalate. In step 3 the excess oxalate was back-titrated, which required 2.11 mL of 0.010 54 M $KMnO_4$ to give perceptible purple color. The blank titration in step 4 required 0.06 mL of $KMnO_4$. This volume should be subtracted from the volume required in step 3, which is therefore $2.11 - 0.06 = 2.05$ mL. Find the concentration of $NaNO_2$ solution.

SOLUTION This problem is not nearly as complicated as it seems. Our strategy is to compute the total moles of $KMnO_4$ used and subtract the moles that reacted with $Na_2C_2O_4$. The remainder must have reacted with $NaNO_2$. So here goes:

total volume of $KMnO_4$ = 50.00 (step 1) + 2.05 (steps 3 and 4) = 52.05 mL

total moles of $KMnO_4$ = (0.010 54 M)(0.052 05 L) = $5.486_0 \times 10^{-4}$ mol

volume of $Na_2C_2O_4$ = 20.00 mL

moles of $Na_2C_2O_4$ = (0.025 00 M)(0.020 00 L) = $5.000_0 \times 10^{-4}$ mol

From the stoichiometry of Reaction 6-1, we know that $5.000_0 \times 10^{-4}$ mol $Na_2C_2O_4$ reacts with

mol $KMnO_4$

$$= (5.000_0 \times 10^{-4} \text{ mol } Na_2C_2O_4) \left(\frac{2 \text{ mol } KMnO_4}{5 \text{ mol } H_2C_2O_4} \right) = 2.000_0 \times 10^{-4} \text{ mol}$$

Therefore the mol of $KMnO_4$ that reacted with $NaNO_2$ was

$KMnO_4$ reacting with $NaNO_2$ = total $KMnO_4$ − $KMnO_4$ reacting with $Na_2C_2O_4$
$$= 5.486_0 \times 10^{-4} \text{ mol} - 2.000_0 \times 10^{-4} \text{ mol} = 3.486_0 \times 10^{-4} \text{ mol}$$

From Reaction 6-5, we can say that $3.486_0 \times 10^{-4}$ mol of $KMnO_4$ reacts with

mol $NaNO_2$ reacting with $KMnO_4$

$$= (3.486_0 \times 10^{-4} \text{ mol } KMnO_4) \left(\frac{5 \text{ mol } NaNO_2}{2 \text{ mol } KMnO_4} \right) = 8.715_0 \times 10^{-4} \text{ mol}$$

The volume of $NaNO_2$ containing this many moles is 50.00 mL, so the concentration is

$$[NaNO_2] = \frac{8.715_0 \times 10^{-4} \text{ mol}}{0.050 00 \text{ L}} = 0.017 \ 43 \text{ M}$$

Ask Yourself

6-D. The procedure above used 0.009 22 M $KMnO_4$. Step 2 required 30.00 mL of $Na_2C_2O_4$, step 3 required 9.68 mL of $KMnO_4$, and step 4 required 0.07 mL of $KMnO_4$.

(a) Find the total moles of $KMnO_4$ in the reaction. Remember to subtract the blank volume of step 4 before computing the total moles of $KMnO_4$.

(b) Find the total moles of $Na_2C_2O_4$ in the reaction.

(c) How many moles of $KMnO_4$ reacted with $Na_2C_2O_4$?

(d) How many moles of $KMnO_4$ reacted with $NaNO_2$?

(e) How many moles of $NaNO_2$ were delivered to the reaction?

(f) What was the molarity of the $NaNO_2$ stock solution?

The Latin word for silver is *argentum,* from which the symbol Ag is derived.

6-5 Titrations Involving Silver Ion

We now introduce three indicator methods that apply to the titration of Cl^- with Ag^+. Titrations with Ag^+ are called *argentometric titrations.* The three indicator methods are

1. **Mohr titration:** formation of a colored precipitate at the end point
2. **Volhard titration:** formation of a soluble, colored complex at the end point
3. **Fajans titration:** adsorption of a colored indicator on the precipitate at the end point

Mohr Titration

In the Mohr titration, Cl^- is titrated with Ag^+ in the presence of chromate (CrO_4^{2-}):

$$\text{titration reaction:} \quad Ag^+ + Cl^- \longrightarrow \underset{\text{White}}{AgCl(s)}$$

$$\text{end-point reaction:} \quad 2Ag^+ + CrO_4^{2-} \longrightarrow \underset{\text{Red}}{Ag_2CrO_4(s)}$$

If the pH is too low, the concentration of CrO_4^{2-} is reduced by the equilibrium $2CrO_4^{2-} + 2H^+ \rightleftharpoons Cr_2O_7^{2-} + H_2O$. If the pH is too high, $AgOH(s)$ may precipitate.

The AgCl precipitates before Ag_2CrO_4. The color of AgCl is white; dissolved CrO_4^{2-} is yellow; and Ag_2CrO_4 is red. The end point is marked by the first appearance of red Ag_2CrO_4. The pH should be in the range 4 to 10.5.

Because excess Ag_2CrO_4 is necessary for visual detection, the color is not seen until after the true equivalence point. One way to correct for the titration error is by means of a blank titration with no chloride present. The volume of Ag^+ needed to form detectable color is subtracted from the volume in the Cl^- titration. Alternatively, we can standardize $AgNO_3$ by the Mohr method, using standard NaCl solution and titrant volumes similar to those for the unknown. The Mohr method is useful for Cl^- and Br^-, but not for I^- or SCN^- (thiocyanate).

Volhard Titration

The Volhard titration is actually a procedure for the titration of Ag^+. To determine Cl^-, a back titration is necessary. First, the Cl^- is precipitated by a known, excess quantity of standard $AgNO_3$:

$$Ag^+ + Cl^- \longrightarrow AgCl(s)$$

Because the Volhard method is a titration of Ag^+, it can be adapted for the determination of any anion that forms an insoluble silver salt.

The AgCl is isolated, and excess Ag^+ is titrated with standard KSCN in the presence of Fe^{3+}:

$$Ag^+ + SCN^- \longrightarrow AgSCN(s)$$

When Ag^+ has been consumed, SCN^- reacts with Fe^{3+} to form a red complex:

$$Fe^{3+} + SCN^- \longrightarrow \underset{\text{Red}}{FeSCN^{2+}}$$

The appearance of red color signals the end point. Knowing how much SCN^- was required for the back titration tells us how much Ag^+ was left over from the reaction with Cl^-. Because the total amount of Ag^+ is known, the amount consumed by Cl^- can then be calculated.

In the analysis of Cl^- by the Volhard method, the end point slowly fades because AgCl is more soluble than AgSCN. The AgCl slowly dissolves and is replaced by AgSCN. To prevent this secondary reaction from happening, two techniques are commonly used. One is to filter off the AgCl and titrate only the Ag^+ in the filtrate. An easier procedure is to shake a few milliliters of nitrobenzene, $C_6H_5NO_2$, with the precipitated AgCl prior to the back titration. Nitrobenzene coats the AgCl and effectively isolates it from attack by SCN^-. Br^- and I^-, whose silver salts are *less* soluble than AgSCN, can be titrated by the Volhard method without isolating the silver halide precipitate.

Fajans Titration

The Fajans titration uses an **adsorption indicator.** To see how this works, consider the electric charge of a precipitate. When Ag^+ is added to Cl^-, there is excess Cl^- in solution prior to the equivalence point. Some Cl^- is selectively adsorbed on the AgCl surface, imparting a negative charge to the crystal surface (Figure 6-6a). After the equivalence point, there is excess Ag^+ in solution. Adsorption of Ag^+

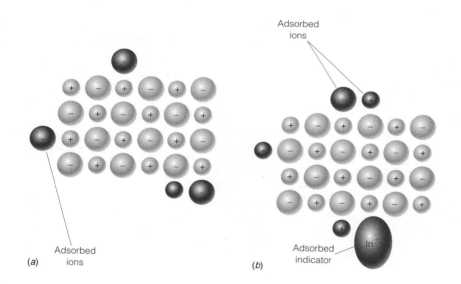

(a)

(b)

FIGURE 6-6 Ions from a solution are adsorbed on the surface of a growing crystallite. (*a*) A crystal growing in the presence of excess lattice anions (anions that belong in the crystal) will have a slight negative charge because anions are predominantly adsorbed. (*b*) A crystal growing in the presence of excess lattice cations will have a slight positive charge on its surface and can therefore adsorb a negative indicator ion. Anions and cations in the solution that do not belong in the crystal lattice are less likely to be adsorbed than are ions belonging to the lattice. These diagrams omit other ions in solution. Overall, each solution plus its growing crystallites must have zero total charge.

Fajans Titration

The Fajans titration of Cl^- with Ag^+ convincingly demonstrates the utility of indicator end points in precipitation titrations. Dissolve 0.5 g of NaCl plus 0.15 g of dextrin in 400 mL of water. The purpose of the dextrin is to retard coagulation of the AgCl precipitate. Add 1 mL of dichlorofluorescein indicator solution containing 1 mg/mL of dichlorofluorescein in 95 wt % aqueous ethanol or 1 mg/mL of the sodium salt in water. Titrate the NaCl solution with a solution containing 2 g of AgNO$_3$ in 30 mL. About 20 mL are required to reach the end point.

Color Plate 2a shows the yellow color of the indicator in the NaCl solution prior to the titration. Color Plate 2b shows the milky white appearance of the AgCl suspension during titration, before the end point is reached. The pink suspension in Color Plate 2c appears at the end point, when the anionic indicator becomes adsorbed to the cationic particles of precipitate.

TABLE 6-1 **Applications of precipitation titrations**

Species analyzed	Notes
	MOHR METHOD
Cl^-, Br^-	Ag_2CrO_4 end point is used.
	VOLHARD METHOD
Br^-, I^-, SCN^-, CNO^-, AsO_4^{3-}	Precipitate removal is unnecessary.
Cl^-, PO_4^{3-}, CN^-, $C_2O_4^{2-}$, CO_3^{2-}, S^{2-}, CrO_4^{2-}	Precipitate removal required.
	FAJANS METHOD
Cl^-, Br^-, I^-, SCN^-, $Fe(CN)_6^{4-}$	Titration with Ag^+. Detection with such dyes as fluorescein, dichlorofluorescein, eosin, bromophenol blue.
F^-	Titration with $Th(NO_3)_4$ to produce ThF_4. End point detected with alizarin red S.
Zn^{2+}	Titration with $K_4Fe(CN)_6$ to produce $K_2Zn_3[Fe(CN)_6]_2$. End-point detection with diphenylamine.
SO_4^{2-}	Titration with $Ba(OH)_2$ in 50 vol % aqueous methanol using alizarin red S as indicator.
Hg_2^{2+}	Titration with NaCl to produce Hg_2Cl_2. End point detected with bromophenol blue.
PO_4^{3-}, $C_2O_4^{2-}$	Titration with $Pb(CH_3CO_2)_2$ to give $Pb_3(PO_4)_2$ or PbC_2O_4. End point detected with dibromofluorescein (PO_4^{3-}) or fluorescein ($C_2O_4^{2-}$).

cations on the crystal surface creates a positive charge on the particles of precipitate (Figure 6-6b). The abrupt change from negative charge to positive charge occurs at the equivalence point.

Common adsorption indicators are anionic (negatively charged) dyes, which are attracted to the positively charged precipitate produced immediately after the equivalence point. Adsorption of the dye on the surface of the solid precipitate changes the color of the dye by interactions that are not well understood. The color change signals the end point in the titration. Because the indicator reacts with the precipitate surface, it is desirable to have as much surface area as possible. In other words, the titration is performed under conditions that tend to keep the particles as small as possible, because small particles have more surface area than an equal volume of large particles. Low electrolyte concentration helps prevent coagulation of the precipitate and maintain small particle size.

The indicator most commonly used for AgCl is dichlorofluorescein, which has a greenish yellow color in solution but turns pink when adsorbed on AgCl (Demonstration 6-1). Because the indicator is a weak acid and must be present in its anionic form, the pH of the reaction must be controlled. The dye eosin is useful in the titration of Br^-, I^-, and SCN^-. It gives a sharper end point than dichlorofluorescein and is more sensitive (i.e., less halide can be titrated). It cannot be used for AgCl because the eosin anion is more strongly bound to AgCl than is Cl^- ion. Eosin will bind to the AgCl crystallites even before the particles become positively charged.

Applications of precipitation titrations are listed in Table 6-1. Whereas the Mohr and Volhard methods are specifically applicable to argentometric titrations, the Fajans method has wider application. Because the Volhard titration is carried out in acidic solution (typically 0.2 M HNO_3), it avoids certain interferences that would affect other titrations. Silver salts of anions such as CO_3^{2-}, $C_2O_4^{2-}$, and AsO_4^{3-} are soluble in acidic solution, so these anions do not interfere with the analysis.

Key Equation

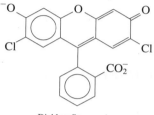

Dichlorofluorescein

Avoid strong light in all silver titrations.

Question Why are salts of the anions CO_3^{2-}, $C_2O_4^{2-}$, and AsO_4^{3-} soluble in acidic solution?

Ask Yourself

6-E. **(a)** Why is a blank titration useful in the Mohr titration?
(b) Why is nitrobenzene used in the Volhard titration of chloride?
(c) Why does the surface charge of a precipitate change sign at the equivalence point?
(d) Examine the procedure in Table 6-1 for the Fajans titration of Zn^{2+}. Do you expect the charge on the precipitate to be positive or negative after the equivalence point?

Key Equation

Stoichiometry   For the reaction $aA + bB \rightarrow$ products, use the ratio $\left(\dfrac{a \text{ mol A}}{b \text{ mol B}}\right)$
for stoichiometry calculations.

Important Terms

adsorption indicator	Fajans titration	titrant
back titration	indicator	titration
blank titration	Mohr titration	titration error
direct titration	primary standard	Volhard titration
end point	standardization	volumetric analysis
equivalence point	standard solution	

Problems

6-1. Distinguish between the terms *end point* and *equivalence point*.

6-2. For Reaction 6-1, how many milliliters of 0.165 0 M $KMnO_4$ are needed to react with 108.0 mL of 0.165 0 M oxalic acid? How many milliliters of 0.165 0 M oxalic acid are required to react with 108.0 mL of 0.165 0 M $KMnO_4$?

6-3. Ammonia reacts with hypobromite, OBr^-, according to

$$2NH_3 + 3OBr^- \longrightarrow N_2 + 3Br^- + 3H_2O$$

Find the molarity of OBr^- if 1.00 mL of OBr^- solution reacts with 1.69 mg of NH_3 (MW 17.03).

6-4. How many milliliters of 0.100 M KI are needed to react with 40.0 mL of 0.040 0 M $Hg_2(NO_3)_2$ if the reaction is $Hg_2^{2+} + 2I^- \rightarrow Hg_2I_2(s)$?

6-5. Cl^- in blood serum, cerebrospinal fluid, or urine can be measured by titration with mercuric ion: $Hg^{2+} + 2Cl^- \rightarrow HgCl_2(aq)$. When the reaction is complete, excess Hg^{2+} reacts with the indicator, diphenylcarbazone, which forms a violet-blue color.

    (a) Mercuric nitrate was standardized by titrating a solution containing 147.6 mg of NaCl (FW 58.442), which required 28.06 mL of $Hg(NO_3)_2$ solution. Find the molarity of the $Hg(NO_3)_2$.

    (b) When this same $Hg(NO_3)_2$ solution was used to titrate 2.000 mL of urine, 22.83 mL was required. Find the concentration of Cl^- (mg/mL) in the urine.

6-6. *Volhard titration:* A 30.00-mL solution containing an unknown amount of I^- was treated with 50.00 mL of 0.365 0 M $AgNO_3$. The precipitated AgI was filtered off, and the filtrate (plus Fe^{3+}) was titrated with 0.287 0 M KSCN. When

37.60 mL had been added, the solution turned red. How many milligrams of I^- were present in the original solution?

6-7. How many milligrams of oxalic acid dihydrate, $H_2C_2O_4 \cdot 2H_2O$ (FW 126.07), will react with 1.00 mL of 0.027 3 M ceric sulfate $(Ce(SO_4)_2)$ if the reaction is $H_2C_2O_4 + 2Ce^{4+} \rightarrow 2CO_2 + 2Ce^{3+} + 2H^+$?

6-8. Arsenic(III) oxide (As_2O_3) is available in pure form and is a useful (and poisonous) primary standard for standardizing many oxidizing agents, such as MnO_4^-. The As_2O_3 is first dissolved in base and then titrated with MnO_4^- in acidic solution. A small amount of iodide (I^-) or iodate (IO_3^-) is used to catalyze the reaction between H_3AsO_3 and MnO_4^-. The reactions are

$$As_2O_3 + 4OH^- \rightleftharpoons 2HAsO_3^{2-} + H_2O$$
$$HAsO_3^{2-} + 2H^+ \rightleftharpoons H_3AsO_3$$
$$5H_3AsO_3 + 2MnO_4^- + 6H^+ \longrightarrow$$
$$5H_3AsO_4 + 2Mn^{2+} + 3H_2O$$

    (a) A 3.214-g aliquot of $KMnO_4$ (FW 158.034) was dissolved in 1.000 L of water, heated to cause any reactions with impurities to occur, cooled, and filtered. What is the theoretical molarity of this solution if no MnO_4^- was consumed by impurities?

    (b) What mass of As_2O_3 (FW 197.84) would be just sufficient to react with 25.00 mL of the $KMnO_4$ solution in part **(a)**?

    (c) It was found that 0.146 8 g of As_2O_3 required 29.98 mL of $KMnO_4$ solution for the faint color of unreacted MnO_4^- to appear. In a blank titration, 0.03 mL of MnO_4^- was required to produce enough color to be seen. Calculate the molarity of the permanganate solution.

6-9. A cyanide solution with a volume of 12.73 mL was treated with 25.00 mL of Ni^{2+} (see example at the end of Section 6-2). The excess Ni^{2+} was then back-titrated with 10.15 mL of 0.013 07 M ethylenediaminetetraacetic acid (EDTA). If 39.35 mL of EDTA was required to react with 30.10 mL of the original Ni^{2+} solution, calculate the molarity of CN^- in the 12.73-mL cyanide sample.

6-10. This problem describes a gravimetric titration in which the *mass* of titrant is measured instead of the *volume*. A solution of NaOH was standardized by titration of a known quantity of the primary standard, potassium hydrogen phthalate:

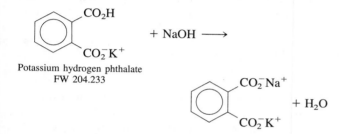

Potassium hydrogen phthalate
FW 204.233

The NaOH was then used to find the concentration of an unknown solution of H_2SO_4:

$$H_2SO_4 + 2NaOH \longrightarrow Na_2SO_4 + 2H_2O$$

(a) Titration of 0.824 g of potassium hydrogen phthalate required 38.314 g of NaOH solution to reach the end point detected by phenolphthalein indicator. Find the concentration of NaOH expressed as mol NaOH/kg solution.

(b) A 10.00-mL aliquot of H_2SO_4 solution required 57.911 g of NaOH solution to reach the phenolphthalein end point. Find the molarity of H_2SO_4.

6-11. An unknown molybdate (MoO_4^{2-}) solution (50.00 mL) was passed through a column containing $Zn(s)$ to convert molybdate to Mo^{3+}. One mole of MoO_4^{2-} gives one mole of Mo^{3+}. The resulting sample required 22.11 mL of 0.012 34 M $KMnO_4$ to reach a purple end point from the reaction

$$3MnO_4^- + 5Mo^{3+} + 4H^+ \longrightarrow 3Mn^{2+} + 5MoO_2^{2+} + 2H_2O$$

A blank required 0.07 mL. Find the molarity of molybdate in the unknown.

6-12. A 25.00-mL sample of La^{3+} was treated with excess $Na_2C_2O_4$ to precipitate $La_2(C_2O_4)_3$, which was washed, dissolved in acid, and titrated with 12.34 mL of 0.004 321 M $KMnO_4$ to reach a purple end point. Find the molarity of La^{3+} in the unknown.

6-13. A glycerol solution weighing 153.2 mg was treated with 50.0 mL of 0.089 9 M Ce^{4+} in 4 M $HClO_4$ at 60°C for 15 min to convert glycerol to formic acid:

$$C_3H_8O_3 + 8Ce^{4+} + 3H_2O \longrightarrow 3HCO_2H + 8Ce^{3+} + 8H^+$$

Glycerol
MW 92.095

Formic acid

The excess Ce^{4+} required 10.05 mL of 0.043 7 M Fe^{2+} for a back titration by the reaction $Ce^{4+} + Fe^{2+} \rightarrow Ce^{3+} + Fe^{3+}$. What was the weight percent of glycerol in the unknown?

6-14. *Propagation of uncertainty.* Consider the titration of 50.00 (± 0.05) mL of a mixture of I^- and SCN^- with 0.068 3 (± 0.000 1) M Ag^+. Look up the solubility products of AgI and AgSCN to decide which precipitate is formed first. The first equivalence point is observed at 12.6 (± 0.4) mL, and the second occurs at 27.7 (± 0.3) mL. Find the molarity and the uncertainty in molarity of thiocyanate in the original mixture.

The Ozone Hole

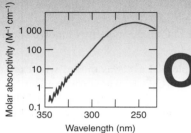

Molar absorptivity (M^{-1} cm^{-1})

Wavelength (nm)

Spectrum of ozone showing maximum absorption of ultraviolet radiation at a wavelength near 260 nm.

Ozone, formed at altitudes of 20 to 40 km by the action of solar ultraviolet radiation ($h\nu$) on O_2, absorbs ultraviolet radiation that causes sunburns and skin cancer:

$$O_2 \xrightarrow{h\nu} 2O \qquad O + O_2 \longrightarrow O_3$$
$$\text{Ozone}$$

In 1985 the British Antarctic Survey reported that ozone in the Antarctic stratosphere had decreased by 50% in early spring, relative to levels observed in the preceding 20 years. Observations have since shown that this "ozone hole" occurs only in early spring and is gradually becoming worse.

The current explanation of this phenomenon begins with chlorofluorocarbons such as Freon-12 (CCl_2F_2) from refrigerants. These long-lived compounds diffuse to the stratosphere, where they catalyze ozone decomposition:

(1) $CCl_2F_2 \xrightarrow{h\nu} CClF_2 + Cl$ Photochemical Cl formation

(2) $Cl + O_3 \longrightarrow ClO + O_2$

(3) $O_3 \xrightarrow{h\nu} O + O_2$ Net reaction of (2)–(4): Catalytic O_3 destruction

(4) $O + ClO \longrightarrow Cl + O_2$ $2O_3 \longrightarrow 3O_2$

Cl produced in step 4 goes back to destroy another ozone molecule in step 1.

A single Cl atom in this chain reaction can destroy $>10^5$ molecules of O_3. The chain is terminated when Cl or ClO reacts with hydrocarbons or NO_2 to form HCl or $ClONO_2$.

Stratospheric clouds formed during the Antarctic winter catalyze the reaction of HCl with $ClONO_2$ to form Cl_2, which is subsequently split by sunlight into Cl atoms to initiate O_3 destruction:

$$HCl + ClONO_2 \xrightarrow[\text{polar clouds}]{\text{Surface of}} Cl_2 + HNO_3 \qquad Cl_2 \xrightarrow{h\nu} 2Cl$$

Formation of polar stratospheric clouds requires winter cold. It is only when the sun is rising in September and October, and the clouds are still present, that conditions are right for ozone destruction.

To protect life from ultraviolet radiation, international treaties now ban or phase out chlorofluorocarbons. However, so much of these compounds has already been released, and so much more remains in use in your house and mine, that ozone depletion is expected to become more severe for the next decade. Ozone levels may not return to historic values until late in the next century. An important pending issue is the widespread use of the agricultural fumigant bromomethane (CH_3Br), which is also a potent ozone destroyer.

Chapter 7

Let There Be Light

M easuring absorption and emission of *electromagnetic radiation* (a fancy term for light) is a key tool in analytical chemistry. We use light in quantitative experiments to determine chemical concentrations, and we use light in a qualitative way to detect compounds as they exit a chromatography column. In this chapter we discuss basic aspects of the absorption of light by molecules. The next chapter applies what we learn to quantitative analysis.

7-1 Properties of Light

Light can be described as both waves and particles. Light waves consist of perpendicular, oscillating electric and magnetic fields (Figure 7-1). The **wavelength,** λ, is the crest-to-crest distance between waves. The **frequency,** ν, is the number of complete oscillations that the wave makes each second. The unit of frequency is reciprocal seconds, s^{-1}. One oscillation per second is also called one **hertz** (Hz). A frequency of $10^6\ s^{-1}$ is therefore said to be 10^6 Hz, or one *megahertz* (MHz). The product of frequency times wavelength is c, the speed of light (2.998×10^8 m/s in vacuum):

Relation between frequency and wavelength:

$$\nu\lambda = c \qquad (7\text{-}1)$$

EXAMPLE   Relating Wavelength and Frequency

What is the wavelength of radiation in your microwave oven, whose frequency is 2.45 GHz?

Following the discovery of the Antarctic ozone "hole" in 1985, atmospheric chemist Susan Solomon led the first expedition in 1986 specifically intended to make chemical measurements of the Antarctic atmosphere by using high-altitude balloons and ground-based spectroscopy. The expedition discovered that ozone depletion occurred after polar sunrise and that the concentration of chemically active chlorine in the stratosphere was 100 times greater than had been predicted from gasphase chemistry. Solomon's group identified chlorine as the culprit in ozone destruction and polar stratospheric clouds as the catalytic surface for the release of so much chlorine.

117

FIGURE 7-1 *Plane-polarized electromagnetic radiation of wavelength λ, propagating along the x-axis. The electric field oscillates in the xy-plane and the magnetic field oscillates in the xz-plane. Ordinary, unpolarized light has electric field components in all planes.*

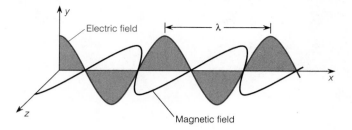

SOLUTION First recognize that 2.45 GHz means 2.45×10^9 Hz $= 2.45 \times 10^9$ s^{-1}. From Equation 7-1, we write

$$\lambda = \frac{c}{\nu} = \frac{2.998 \times 10^8 \text{ m/s}}{2.45 \times 10^9 \text{ s}^{-1}} = 0.122 \text{ m}$$

Light can also be thought of as particles called **photons.** The energy, E, of a photon is proportional to its frequency:

Relation between energy and frequency: $$E = h\nu \qquad (7\text{-}2)$$

Physical constants are listed on the inside cover.

where h is *Planck's constant* ($= 6.626 \times 10^{-34}$ J·s).

Combining Equations 7-1 and 7-2, we can write

$$E = h\frac{c}{\lambda} = hc\frac{1}{\lambda} = hc\tilde{\nu} \qquad (7\text{-}3)$$

Energy increases if

- frequency (ν) increases
- wavelength (λ) decreases
- wavenumber ($\tilde{\nu}$) increases

where $\tilde{\nu}(= 1/\lambda)$ is called the **wavenumber.** Energy is inversely proportional to wavelength and directly proportional to wavenumber. Red light, with a wavelength longer than that of blue light, is less energetic than blue light. The SI unit for wavenumber is m^{-1}. However, the most common unit of wavenumber is cm^{-1}, read "reciprocal centimeters" or "wavenumbers." Wavenumber units are used extensively in infrared spectroscopy.

Regions of the **electromagnetic spectrum** are shown in Figure 7-2. Visible light—the kind our eyes detect—represents only a small fraction of the electromagnetic spectrum.

The lowest energy state of a molecule is called the **ground state.** When the molecule absorbs a photon, the energy of the molecule increases and we say that the molecule is promoted to an **excited state** (Figure 7-3). If the molecule emits a photon, the energy of the molecule decreases. Figure 7-2 indicates that microwave radiation cause molecules to rotate faster. A microwave oven heats food by increasing the rotational energy of water in the food. Infrared radiation stimulates vibrations of molecules. Visible light and ultraviolet radiation promote electrons from lower energy states to higher energy states. (A molecule does not absorb visible light unless the molecule is colored.) X-rays and short-wavelength ultraviolet radiation are harmful because they break chemical bonds and ionize molecules (which is why you should minimize your exposure to medical X-rays).

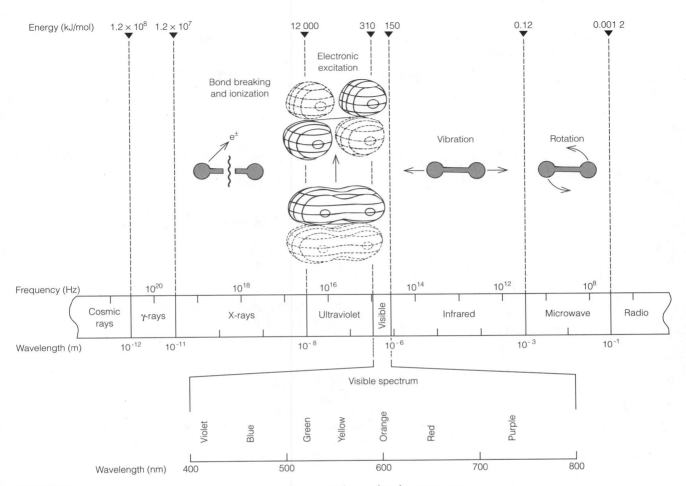

FIGURE 7-2 Electromagnetic spectrum showing representative molecular processes that occur when light in each region is absorbed. The visible spectrum spans the wavelength range 380 to 780 nanometers (1 nm = 10^{-9} m).

By how many joules is the energy of a molecule increased when it absorbs (a) visible light with a wavelength of 500 nm or (b) infrared radiation with a wavenumber of 1 251 cm^{-1}?

SOLUTION (a) The visible wavelength is 500 nm = 500×10^{-9} m.

$$E = h\nu = h\frac{c}{\lambda}$$

$$= (6.626 \times 10^{-34} \text{ J·s}) \left(\frac{2.998 \times 10^8 \text{ m/s}}{500 \times 10^{-9} \text{ m}} \right) = 3.97 \times 10^{-19} \text{ J}$$

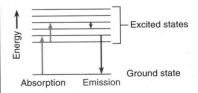

FIGURE 7-3 Absorption of light increases the energy of a molecule. Emission of light decreases its energy.

119

This is the energy of one photon absorbed by one molecule. If a mole of molecules absorbed a mole of photons, the energy increase is

$$E = \left(3.97 \times 10^{-19} \frac{J}{molecule}\right)\left(6.022 \times 10^{23} \frac{molecules}{mol}\right) = 2.39 \times 10^5 \frac{J}{mol}$$

$$= \left(2.39 \times 10^5 \frac{J}{mol}\right)\left(\frac{1\ kJ}{1\ 000\ J}\right) = 239 \frac{kJ}{mol}$$

(b) Given the wavenumber, we use Equation 7-3. However, we must remember to convert the wavenumber unit cm^{-1} to the SI unit m^{-1} with the conversion factor 100 cm/m. The energy of one photon is

$$E = hc\tilde{\nu} = (6.626 \times 10^{-34}\ J \cdot s)\left(2.998 \times 10^8\ \frac{m}{s}\right)\underbrace{(1\ 251\ cm^{-1})\left(100\ \frac{cm}{m}\right)}_{\text{Conversion of } cm^{-1} \text{ to } m^{-1}}$$

$$= 2.485 \times 10^{-20}\ J$$

Multiplying by Avogadro's number, we find that this photon energy corresponds to 14.97 kJ/mol, which falls in the infrared region and excites molecular vibrations.

Ask Yourself

7-A. What is the frequency (Hz), wavenumber (cm^{-1}), and energy (kJ/mol) of light with a wavelength of **(a)** 100 nm; **(b)** 500 nm; **(c)** 10 μm; and **(d)** 1 cm? In which spectral region does each kind of radiation lie and what molecular process occurs when the radiation is absorbed?

7-2 *Absorption of Light*

A **spectrophotometer** is an instrument that measures transmission of light through a substance. When light is absorbed by a sample, the *radiant power* of the light beam decreases. Radiant power, P, refers to the energy per second per unit area of the beam. In Figure 7-4, light passes through a *monochromator* that selects one wavelength. Light of this wavelength, with radiant power P_0, strikes a sample of length b. The radiant power of the beam emerging from the other side of the sample is P. Some of the light may be absorbed by the sample, so $P \leq P_0$.

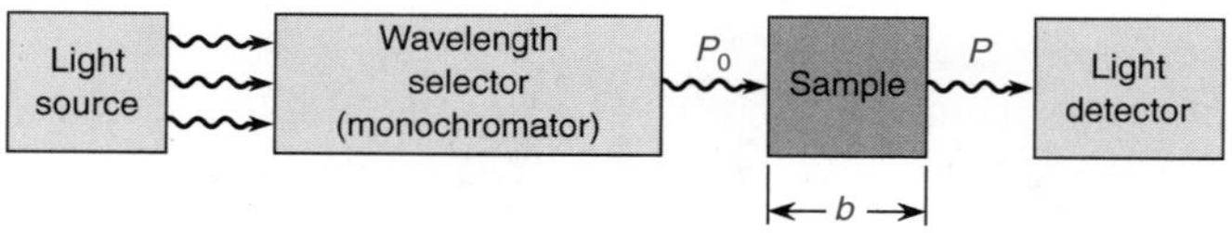

Monochromatic light consists of a single color (wavelength).

FIGURE 7-4 Schematic diagram of a single-beam spectrophotometric experiment.

Transmittance, Absorbance, and Beer's Law

Transmittance, T, is the fraction of incident light that passes through a sample.

Transmittance:
$$T = \frac{P}{P_0} \qquad (7\text{-}4)$$

Transmittance has the range 0 to 1. If no light is absorbed by the sample, the transmittance is 1. If all light is absorbed, the transmittance is 0. *Percent transmittance* (100T) ranges from 0% to 100%. A transmittance of 30% means that 70% of the light does not pass through the sample.

The most useful quantity for chemical analysis is **absorbance,** A, defined as

Absorbance:
$$A = \log \frac{P_0}{P} = -\log \frac{P}{P_0} = -\log T \qquad (7\text{-}5)$$

When no light is absorbed, $P = P_0$ and $A = 0$. If 90% of the light is absorbed, 10% is transmitted and $P = P_0/10$. This ratio gives $A = 1$. If 1% of the light is transmitted, $A = 2$.

Of course you remember that
$$\log \frac{1}{x} = -\log x.$$

P/P_0	% T	A
1	100	0
0.1	10	1
0.01	1	2

EXAMPLE Absorbance and Transmittance

What absorbance corresponds to 99% transmittance? To 0.10% transmittance?

SOLUTION Use the definition of absorbance in Equation 7-5:

99% T: $\qquad A = -\log T = -\log 0.99 = 0.004\ 4$

0.10% T: $\qquad A = -\log T = -\log 0.001\ 0 = 3.0$

The higher the absorbance, the less light is transmitted through a sample.

The reason why absorbance is so important is that *absorbance is proportional to the concentration of light-absorbing molecules in the sample,* as given by **Beer's law:**

Beer's law:
$$A = \epsilon b c \qquad (7\text{-}6)$$

Absorbance (A) is dimensionless. Concentration (c) has units of moles per liter (M), and pathlength (b) is commonly expressed in centimeters. The quantity ϵ (epsilon) is a constant called the **molar absorptivity.** It has the units $M^{-1}\ cm^{-1}$ because the product $\epsilon b c$ must be dimensionless. Molar absorptivity tells how much light is absorbed at a particular wavelength.

Color Plate 3 shows the colors of standard solutions of an iron compound. You can see that the color intensity increases as the concentration increases. Absorbance is a measure of color intensity: The more intense the color, the greater the absorbance.

Box 7-1 gives a physical picture of Beer's law that could be the basis for a classroom exercise.

EXAMPLE Absorbance, Transmittance, and Beer's Law

Find the absorbance and transmittance of a 0.002 40 M solution of a substance with a molar absorptivity of (a) 1.00×10^2 or (b) 2.00×10^2 M^{-1} cm^{-1} in a cell with a 2.00-cm pathlength.

SOLUTION Beer's law tells us the absorbance:

(a) $A = \epsilon bc = (100 \text{ M}^{-1} \text{ cm}^{-1})(2.00 \text{ cm})(0.002 \text{ 40 M}) = 0.480$

(b) $A = (200 \text{ M}^{-1} \text{ cm}^{-1})(2.00 \text{ cm})(0.002 \text{ 40 M}) = 0.960$

Doubling the molar absorptivity doubles the absorbance. Doubling the concentration would also double the absorbance.

Box 7-1 *Explanation*

Discovering Beer's Law

Each photon passing through a solution has a certain probability of striking a light-absorbing molecule and being absorbed. Let's model this process by thinking of an inclined plane with holes representing absorbing molecules. The number of molecules is equal to the number of holes, and the pathlength is equal to the length of the plane. Suppose that 1 000 small balls, representing 1 000 photons, are rolled down the incline. Whenever a ball drops through a hole, we consider it to have been "absorbed" by a molecule.

Let the plane be divided into 10 equal intervals and let the probability that a ball will fall through a hole in the first interval be 1/10. Of the 1 000 balls entering the first interval, one-tenth—100 balls—are absorbed (dropping through the holes) and 900 pass into the second interval. Of the 900 balls entering the second interval, one-tenth—90 balls—are absorbed and 810 proceed to the third interval. Of these 810 balls, 81 are absorbed and 729 proceed to the fourth interval. The table summarizes the action.

Transmittance is defined as

$$T = \frac{\text{number of surviving balls}}{\text{initial number of balls} (= 1\ 000)}$$

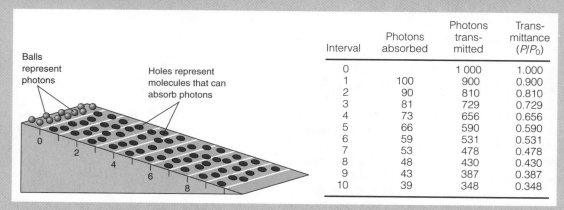

Interval	Photons absorbed	Photons trans-mitted	Trans-mittance (P/P_0)
0		1 000	1.000
1	100	900	0.900
2	90	810	0.810
3	81	729	0.729
4	73	656	0.656
5	66	590	0.590
6	59	531	0.531
7	53	478	0.478
8	48	430	0.430
9	43	387	0.387
10	39	348	0.348

Balls represent photons

Holes represent molecules that can absorb photons

Inclined plane model for photon absorption.

Transmittance is obtained from Equation 7-5 by raising 10 to the power on each side of the equation:

$$\log T = -A$$
$$\underbrace{10^{\log T}}_{10^{\log T} \text{ is the same as } T} = 10^{-A}$$

(a) $T = 10^{-0.480} = 0.331 = 33.1\%$

(b) $T = 10^{-0.960} = 0.110 = 11.0\%$

Transmittance is not linearly related to molar absorptivity (or concentration). Doubling the molar absorptivity or the concentration does not decrease transmittance by a factor of 2.

To evaluate $10^{-0.480}$ on your calculator, use y^x or *antilog*. If you use y^x, $y = 10$ and $x = -0.480$. If you use *antilog*, find the antilog of -0.480. Be sure you can show that $10^{-0.480} = 0.331$.

Graph a shows that a plot of transmittance versus interval number (which is analogous to plotting transmittance versus pathlength in a spectrophotometric experiment) is not linear. However, the plot of $-\log(\text{transmittance})$ versus interval number in graph b is linear and passes through the origin. Graph b shows that $-\log T$ is proportional to pathlength.

To investigate how transmittance depends on the concentration of absorbing molecules, you could do the same mental experiment with a different number of holes in the inclined plane. For example, try setting up a table to show what happens if the probability of absorption in each interval is 1/20 instead of 1/10. This change corresponds to decreasing the concentration of absorbing molecules to half of its initial value. You will discover that a graph of $-\log T$ versus interval number has a slope equal to one-half that in graph b. That is, $-\log T$ is proportional to concentration as well as to pathlength. So, we have just shown that $-\log T$ is proportional to both concentration and pathlength. Defining absorbance as $-\log T$ gives us the essential terms in Beer's law:

$$A \equiv -\log T \propto \text{concentration} \times \text{pathlength}$$

The symbol $\propto$ means "is proportional to."

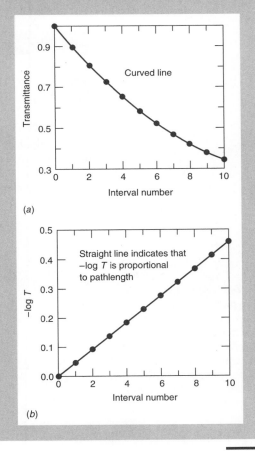

(a)

(b)

EXAMPLE Finding Concentration from the Absorbance

Gaseous ozone has a molar absorptivity of 2 700 M^{-1} cm^{-1} at the absorption peak near 260 nm in the spectrum at the beginning of this chapter. Find the concentration of ozone (mol/L) in air if a sample has an absorbance of 0.23 in a 10.0-cm cell. Air has negligible absorbance at 260 nm.

SOLUTION We rearrange Beer's law to solve for concentration:

$$c = \frac{A}{\epsilon b} = \frac{0.23}{(2\ 700\ M^{-1}\ cm^{-1})(10.0\ cm)} = 8.5 \times 10^{-6}\ M$$

Absorption Spectra and Color

The plural of "spectrum" is "spectra."

An **absorption spectrum** is a graph showing how A (or ϵ) varies with wavelength (or frequency or wavenumber). The opening highlight of this chapter shows the ultraviolet absorption spectrum of ozone, which peaks near 260 nm. Figure 7-5 shows the absorption spectrum of a typical sunscreen lotion, which absorbs harmful solar radiation below about 350 nm. Demonstration 7-1 illustrates the meaning of an absorption spectrum.

EXAMPLE How Effective Is Sunscreen?

What fraction of ultraviolet radiation is transmitted through the sunscreen in Figure 7-5 at the peak absorbance near 300 nm?

SOLUTION From the spectrum in Figure 7-5, the absorbance at 300 nm is approximately 0.35. Therefore the transmittance is $T = 10^{-A} = 10^{-0.35} = 0.45 = 45\%$. Just over half the ultraviolet radiation (55%) is absorbed by the sunscreen and does not reach your skin.

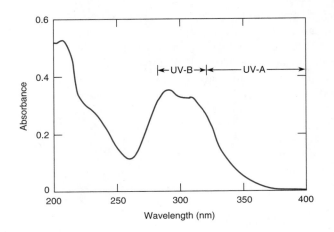

FIGURE 7-5 Absorption spectrum of typical sunscreen lotion shows absorbance versus wavelength in the ultraviolet region. Sunscreen was thinly coated onto a transparent window to make this measurement. Sunscreen makers refer to the region 400–320 nm as UV-A and 320–280 nm as UV-B.

Absorption Spectra

The spectrum of visible light can be projected on a screen in a darkened room in the following manner: Four layers of plastic diffraction grating[†] are mounted on a cardboard frame that has a square hole large enough to cover the lens of an overhead projector. This assembly is taped over the projector lens facing the screen. An opaque cardboard surface with two 1×3 cm slits is placed on the working surface of the projector.

When the lamp is turned on, the white image of each slit is projected on the center of the screen. A visible spectrum appears on either side of each image. When a beaker of colored solution is placed over one slit, you can see color projected on the screen where the white image previously appeared. The spectrum beside the colored image loses its intensity in regions where the colored solution absorbs light.

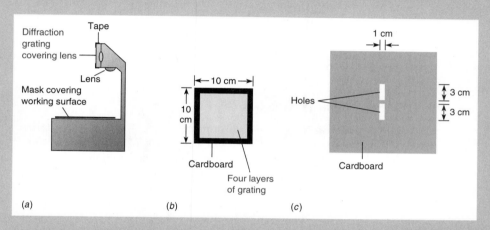

(a) Overhead projector. (b) Diffraction grating mounted on cardboard. (c) Mask for working surface.

Color Plate 4a shows the spectrum of white light and the absorption spectra of three different colored solutions. We see that potassium dichromate, which appears orange or yellow, absorbs blue wavelengths. Bromophenol blue absorbs yellow and orange wavelengths and appears blue to our eyes. The absorption of phenolphthalein is located near the center of the visible spectrum. For comparison, the spectra of these three solutions recorded with a spectrophotometer are shown in Color Plate 4b.

[†]Diffraction grating is available from Edmund Scientific Co., 5975 Edscorp Building, Barrington, NJ 08007, catalog no. 40,267.

White light contains all of the colors of the rainbow. Any substance that absorbs visible light appears colored when white light is transmitted through it or reflected from it. The substance absorbs certain wavelengths of the white light, and our eyes detect the wavelengths that are not absorbed. A rough guide to colors is given in Table 7-1. The observed color is called the *complement* of the absorbed color. As an example, bromophenol blue in Color Plate 4 has a visible absorbance maximum at 591 nm, and its observed color is blue.

The color of a substance is the complement of the color of the light that it absorbs.

125

TABLE 7-1	Colors of visible light	
Wavelength of maximum absorption (nm)	*Color absorbed*	*Color observed*
380–420	Violet	Green-yellow
420–440	Violet-blue	Yellow
440–470	Blue	Orange
470–500	Blue-green	Red
500–520	Green	Purple
520–550	Yellow-green	Violet
550–580	Yellow	Violet-blue
580–620	Orange	Blue
620–680	Red	Blue-green
680–780	Purple	Green

Ask Yourself

7-B. (a) What is the absorbance of a 2.33×10^{-4} M solution of a compound with a molar absorptivity of 1.05×10^3 M^{-1} cm^{-1} in a 1.00-cm cell?

(b) What is the transmittance of the solution in **(a)**?

(c) Find A and % T when the pathlength is doubled to 2.00 cm.

(d) Find A and % T when the pathlength is 1.00 cm but the concentration is doubled.

(e) What would be the absorbance in **(a)** for a different compound with twice as great a molar absorptivity ($\epsilon = 2.10 \times 10^3$ M^{-1} cm^{-1})? The concentration and pathlength are unchanged from **(a).**

7-3 The Spectrophotometer

The minimum requirements for a spectrophotometer were shown in Figure 7-4. The source of visible light is simply a tungsten light bulb. For ultraviolet radiation, a deuterium arc lamp is usually employed. In this lamp, an electric spark dissociates D_2 gas, creating ultraviolet emission in the range 200–400 nm. A hot silicon carbide rod called a *globar* is used as a source of infrared radiation.

The wavelength selector, called a **monochromator,** is a prism, grating, or filter that selects a narrow band of wavelengths from the light source. A reflection *grating,* which is most common, consists of closely ruled grooves on a reflective metal surface. Different wavelengths of light striking this surface are reflected at different angles. Therefore, white light is spread out into its component colors, as shown in Color Plate 5.

Monochromatic light from the wavelength selector then travels through a sample of pathlength b. The sample is usually contained in a cell, called a **cuvet** (Figure 7-6), that has flat quartz faces. Quartz transmits visible and ultraviolet light.

"Monochromatic" means one wavelength. Although it is impossible to produce truly monochromatic light, the better the monochromator, the narrower is the range of wavelengths in the emerging beam.

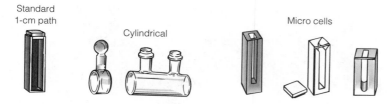

FIGURE 7-6 Common cuvets for ultraviolet and visible measurements.

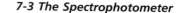

Glass and plastic absorb ultraviolet radiation, so glass or clear plastic can be used only for measurements at visible wavelengths. Infrared cells are typically made of sodium chloride or potassium bromide crystals.

The radiant power of light emerging from the sample cell is measured by a detector, which may be a *photomultiplier tube* in visible and ultraviolet spectrophotometers. The photomultiplier is a vacuum tube that creates an electric current proportional to the number of photons striking a photosensitive metal surface.

Single-Beam and Double-Beam Instruments

The instrument represented in Figure 7-4 is called a *single-beam spectrophotometer* because it has only one beam of light. We do not measure the incident radiant power, P_0, directly. Rather, the radiant power of light passing through a reference cuvet containing pure solvent is *defined* as P_0. This cuvet is then removed and replaced by an identical one containing sample. The radiant power of light striking the detector is then taken as P, permitting T or A to be determined. The reference cuvet, containing pure solvent, compensates for reflection, scattering, or absorption of light by the cuvet and solvent. The radiant power of light striking the detector would not be the same if the reference cuvet were removed from the beam. A single-beam spectrophotometer is inconvenient because two different samples must be placed alternately in the beam.

In a *double-beam spectrophotometer* (Figure 7-7), light is passed alternately through the sample and reference cuvets by a rotating mirror called a *beam chopper*. When light passes through the sample cuvet, the detector measures the radiant power that we call P in Equation 7-4. When the chopper diverts the beam to the reference cuvet, the detector measures P_0. The beam is chopped several times per second, and the circuitry computes P/P_0 to obtain transmittance and absorbance. Most research-quality spectrophotometers provide for automatic wavelength scanning and continuous recording of the absorbance.

The double-beam instrument makes essentially continuous measurements of the light emerging from the sample and the reference cells.

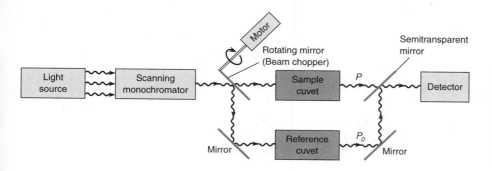

FIGURE 7-7 Schematic diagram of a double-beam scanning spectrophotometer. The incident beam is passed alternately through the sample and reference cuvets by the rotating beam chopper.

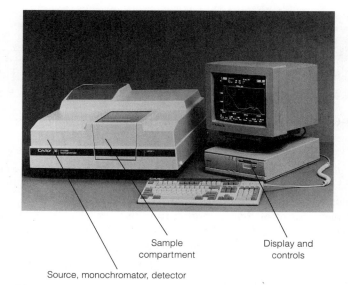

Sample
compartment

Display and
controls

Source, monochromator, detector

(a)

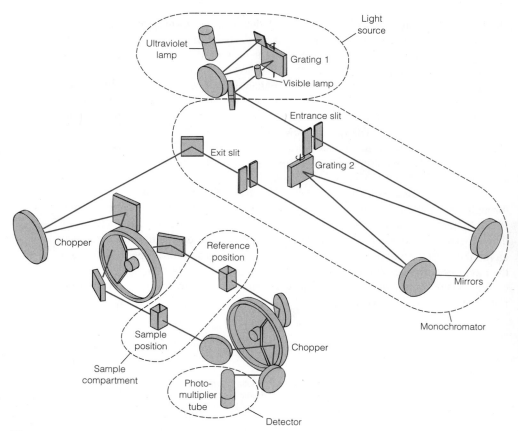

(b)

FIGURE 7-8 (a) Varian Cary 3E Ultraviolet-Visible Spectrophotometer. (b) Schematic
diagram of the Varian's optical train.

A double-beam ultraviolet-visible spectrophotometer is shown in Figure 7-8. The source of visible light is a quartz-halogen lamp (like an automobile headlight) and the ultraviolet source is a deuterium arc lamp. Only one lamp is used at a time. Grating 1 selects some of the source wavelengths to enter the monochromator, which, in turn, directs an even narrower band to the exit slit. After being chopped and passing through the sample and reference cells, the signal is detected by a photomultiplier tube. Results are displayed on the computer screen.

Good Operating Techniques

Spectrophotometry is most accurate at intermediate levels of absorbance ($A \approx$ 0.4–0.9). If too little light gets through the sample (high absorbance), the intensity is hard to measure. If too much light gets through (low absorbance), it is hard to distinguish the transmittance of the sample from that of the reference. It is therefore desirable to adjust the concentration of the sample so that its absorbance falls in the range 0.4–0.9.

For spectrophotometric analysis, measurements are made at a wavelength (λ_{max}) corresponding to a peak in the absorbance spectrum. This wavelength gives the greatest sensitivity—maximum response for a given concentration of analyte. Errors due to wavelength drift and the finite bandwidth of wavelengths selected by the monochromator are minimized because the spectrum varies least with wavelength at the absorbance maximum.

In measuring a spectrum, it is routine to first record a baseline with pure solvent or a reagent blank in *both* cuvets. In principle, the baseline absorbance should be zero. However, small mismatches between the two cuvets and instrumental imperfections lead to small positive or negative baseline absorbance. The absorbance of the sample is then recorded and the absorbance of the baseline is subtracted from that of the sample to obtain true absorbance.

All vessels should be covered to protect them from dust, which scatters light and therefore makes it look like the absorbance of the sample has increased. Cuvets should be handled with a tissue to avoid putting fingerprints on the faces and must be kept scrupulously clean to avoid surface contamination, which also leads to scattering.

Do not touch the clear faces of a cuvet. Fingerprints scatter and absorb light.

Slight mismatch between sample and reference cuvets, over which you have little control, leads to systematic errors in spectrophotometry. It is important to place a cuvet in the spectrophotometer as reproducibly as possible. Slight misplacement of the cuvet in its holder, or turning a flat cuvet around by 180°, or rotation of a circular cuvet, all lead to random errors in absorbance measurements.

Ask Yourself

7-C. (a) What is the difference between a single-beam and a double-beam spectrophotometer?

(b) Why is it most accurate to measure absorbances in the range $A =$ 0.4–0.9?

(c) Why should you not touch the optical surfaces of a cuvet with your fingers?

Key Equations

Frequency-wavelength relation $\nu\lambda = c$

ν = frequency λ = wavelength c = speed of light

Wavenumber $\tilde{\nu} = 1/\lambda$

Photon energy $E = h\nu = hc/\lambda = hc\tilde{\nu}$

h = Planck's constant

Transmittance $T = P/P_0$

P_0 = radiant intensity of light incident on sample

P = radiant intensity of light emerging from the sample

Absorbance $A = -\log T$

Beer's law $A = \epsilon bc$

ϵ = molar absorptivity of the absorbing species

b = pathlength

c = concentration of absorbing species

Important Terms

absorbance	frequency	photon
absorption spectrum	ground state	spectrophotometer
Beer's law	hertz	transmittance
cuvet	molar absorptivity	wavelength
electromagnetic spectrum	monochromator	wavenumber
excited state		

Problems

7-1. (a) When you double the frequency of electromagnetic radiation, you _____ the energy.

(b) When you double the wavelength, you _____ the energy.

(c) When you double the wavenumber, you _____ the energy.

7-2. How much energy (J) is carried by one photon of **(a)** red light with $\lambda = 650$ nm? **(b)** violet light with $\lambda = 400$ nm?

After finding the energy of one photon of each wavelength, express the energy of a mole of each type of photon in kJ/mol.

7-3. What color would you expect for light transmitted through a solution with an absorption maximum at **(a)** 450; **(b)** 550; and **(c)** 650 nm?

7-4. State the difference between transmittance, absorbance, and molar absorptivity. Which one is proportional to concentration?

7-5. An absorption spectrum is a graph of _____ or _____ versus _____.

7-6. Why does a compound whose visible absorption maximum is at 480 nm (blue-green) appear to be red?

7-7. What color would you expect to observe for a solution with a visible absorbance maximum at 562 nm?

7-8. Calculate the frequency (Hz), wavenumber (cm^{-1}), and energy (J/photon and kJ per mole of photons) of **(a)** ultraviolet light with a wavelength of 250 nm and **(b)** infrared light with a wavelength of 2.50 μm.

7-9. Convert transmittance to absorbance:

Transmittance:
0.99   0.90   0.50   0.10   0.010   0.001 0   0.000 10
Absorbance:
1.0

7-10. The absorbance of a 2.31×10^{-5} M solution is 0.822 at a wavelength of 266 nm in a 1.00-cm cell. Calculate the molar absorptivity at 266 nm.

7-11. The iron-transport protein in your blood is called transferrin. When its two iron-binding sites do not contain metal ions, the protein is called apotransferrin.

(a) Apotransferrin has a molar absorptivity of 8.83×10^4 M^{-1} cm^{-1} at 280 nm. Find the concentration of apotransferrin in water if the absorbance is 0.244 in a 0.100-cm cell.

(b) The molecular weight of apotransferrin is 81 000. Express the concentration from **(a)** in g/L.

7-12. A 15.0-mg sample of a compound with a molecular weight of 384.63 was dissolved in a 5-mL volumetric flask. A 1.00-mL aliquot was withdrawn, placed in a 10-mL volumetric flask, and diluted to the mark.

(a) Find the concentration of sample in the 5-mL flask.

(b) Find the concentration in the 10-mL flask.

(c) The 10-mL sample was placed in a 0.500-cm cuvet and gave an absorbance of 0.634 at 495 nm. Find the molar absorptivity at 495 nm.

7-13. (a) From Figure 7-5, measure the peak absorbance of sunscreen near 215 nm.

(b) What fraction of ultraviolet radiation is transmitted through the sunscreen near 215 nm?

7-14. (a) What value of absorbance corresponds to 45.0% T?

(b) When the concentration of a solution is doubled, the *absorbance* is doubled. If a 0.010 0 M solution exhibits 45.0% T at some wavelength, what will be the percent transmittance for a 0.020 0 M solution of the same substance?

7-15. A 0.267-g quantity of a compound with a molecular weight of 337.69 was dissolved in 100.0 mL of ethanol. Then 2.000 mL was withdrawn and diluted to 100.0 mL. The spectrum of this solution exhibited a maximum absorbance of 0.728 at 438 nm in a 2.000-cm cell. Find the molar absorptivity of the compound.

7-16. Four sections of the optical train of the spectrophotometer in Figure 7-8 are enclosed by dotted lines. Describe the purpose and essential components of each section.

Fiber-Optic Glucose Sensor

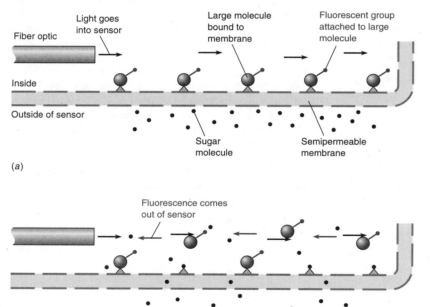

Light goes into sensor

Fiber optic

Large molecule bound to membrane

Fluorescent group attached to large molecule

Inside

Outside of sensor

Sugar molecule

Semipermeable membrane

(a)

Fluorescence comes out of sensor

(b)

The biosensor shown above, which fits inside a hypodermic needle, measures blood sugar concentration in the vein of a patient. The end of the sensor near the needle is covered with a *semipermeable membrane* through which small molecules such as sugar can diffuse but large molecules cannot pass (diagram a). Sugar diffusing into the needle displaces large molecules that were bound to the inner side of the membrane (diagram b). Laser light from the optical fiber causes the liberated molecules to fluoresce. The fluorescence is carried back through the optical fiber to a detector whose response is proportional to the blood sugar concentration.

Chapter **8**

Spectrophotometry

Spectrophotometry is any procedure that uses electromagnetic radiation to
measure chemical concentrations. We now discuss how absorption and emission of
radiation are used in quantitative analysis. This foundation should allow you to use
spectrophotometry in the lab and to appreciate its application to other topics in this
book.

8-1 Using Beer's Law

For a compound to be analyzed by spectrophotometry, it must absorb electromagnetic radiation, and this absorption should be distinguishable from that of other
species in the sample. Biochemists assay proteins in the ultraviolet region at 280 nm
because benzene rings present in virtually every protein have an absorbance maximum at 280 nm. Other common solutes such as salts, buffers, and carbohydrates
have little or no absorbance at this wavelength. In this section, we use Beer's law
for a simple analysis and then discuss the measurement of nitrite in an aquarium.

Spectrophotometric analyses employing visible radiation are called
colorimetric analyses.

EXAMPLE **Measuring Benzene in Hexane**

(a) The solvent hexane has negligible ultraviolet absorbance above a wavelength of
200 nm. A solution prepared by dissolving 25.8 mg of benzene (C_6H_6, MW 78.114)
in hexane and diluting to 250.0 mL has an absorption peak at 256 nm with an absorbance of 0.266 in a 1.000-cm cell. Find the molar absorptivity of benzene at this
wavelength.

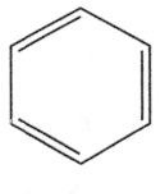

Benzene
C_6H_6

133

Beer's law: $\boxed{A = \epsilon bc}$

A = absorbance (dimensionless)
ϵ = molar absorptivity (M^{-1} cm^{-1})
b = pathlength (cm)
c = concentration (M)
ϵ has funny units so that the product ϵbc will be dimensionless.

Linda A. Hughes

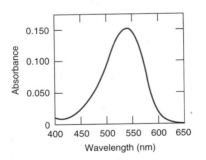

FIGURE 8-1 Spectrum of the red-purple product of Reaction 8-1 beginning with a standard nitrite solution containing 0.915 ppm nitrogen.

Straight line: $y = \underset{\underset{\text{Slope}}{\uparrow}}{m} \cdot x + \underset{\underset{\text{Intercept}}{\uparrow}}{b}$

Beer's law: $A = \underset{\underset{\text{Slope}}{\uparrow}}{\epsilon b} \cdot c + \underset{\underset{\text{Intercept}}{\uparrow}}{0}$

SOLUTION The concentration of benzene is

$$[C_6H_6] = \frac{(0.025\ 8\ \text{g})/(78.114\ \text{g/mol})}{0.250\ 0\ \text{L}} = 1.32_1 \times 10^{-3}\ \text{M}$$

We find the molar absorptivity from Beer's law:

$$\text{molar absorptivity} = \epsilon = \frac{A}{bc} = \frac{0.266}{(1.000\ \text{cm})(1.32_1 \times 10^{-3}\ \text{M})} = 201._3\ \text{M}^{-1}\ \text{cm}^{-1}$$

(b) A sample of hexane contaminated with benzene has an absorbance of 0.070 at 256 nm in a cell with a 5.000-cm pathlength. Find the concentration of benzene.

SOLUTION Use the molar absorptivity from **(a)** in Beer's law:

$$[C_6H_6] = \frac{A}{\epsilon b} = \frac{0.070}{(201._3\ \text{M}^{-1}\ \text{cm}^{-1})(5.000\ \text{cm})} = 6.95_3 \times 10^{-5}\ \text{M}$$

Using a Standard Curve to Measure Nitrite

Box 6-1 showed that nitrogen compounds derived from animals and plants are broken down to ammonia by heterotrophic bacteria. Ammonia is oxidized first to nitrite (NO_2^-) and then to nitrate (NO_3^-) by nitrifying bacteria. In Section 6-4 we saw how a permanganate titration was used to standardize a nitrite stock solution. The nitrite solution is used here to prepare standards for a spectrophotometric analysis of nitrite in aquarium water.

The aquarium nitrite analysis is based on a reaction whose colored product has an absorbance maximum at 543 nm (Figure 8-1):

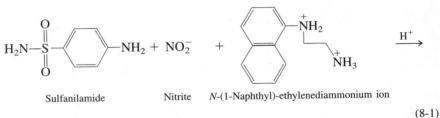

Sulfanilamide   Nitrite   N-(1-Naphthyl)-ethylenediammonium ion

(8-1)

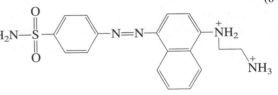

Red-purple product (λ_{max} = 543 nm)

For quantitative analysis, a **standard curve** (called a *calibration curve* in Section 4-5) is prepared in which absorbance at 543 nm is plotted against nitrite concentration in a series of standards (Figure 8-2). Beer's law (Equation 7-6) says that absorbance is proportional to concentration. Therefore the standard curve should be a straight line passing through the origin.

The general procedure for measuring nitrite is to add color-forming reagent to an unknown or standard, wait 10 min for the reaction to be completed, and measure the absorbance. A *reagent blank* is prepared with nitrite-free, artificial seawater in place of unknown or standards. *The absorbance of the blank is subtracted from the absorbance of all other samples prior to any calculations.* The purpose of the blank is to subtract absorbance at 543 nm arising from starting materials or impurities. Here are the details:

Reagents:

1. *Color-forming reagent* is prepared by mixing 1.0 g of sulfanilamide, 0.10 g of *N*-(1-naphthyl)-ethylenediamine dihydrochloride, and 10 mL of 85 wt % phosphoric acid and diluting to 100 mL. Store the solution in a dark bottle in the refrigerator to prevent thermal and photochemical degradation.

2. *Standard nitrite* (~0.02 M) is prepared by dissolving $NaNO_2$ in water and standardizing (measuring the concentration) by the titration described in Section 6-4. Dilute the concentrated standard with artificial seawater (containing no nitrite) to prepare standards containing 0.5–3 ppm nitrite nitrogen.

General procedure:
For each analysis, dilute 10.00 mL of standard or unknown up to 100.0 mL with water. Then place 50.00 mL of the diluted solution in a flask and add 2.00 mL of color-forming reagent. After 10 min, measure the absorbance in a cuvet with a 1.000-cm pathlength.

a. Construct a *standard curve* from known nitrite solutions. Prepare a *reagent blank* by carrying artificial seawater through the same steps as a standard.

b. Analyze duplicate samples of *unknown* aquarium water that has been filtered prior to dilution to remove suspended solids. Several trial dilutions may be required before the aquarium water is dilute enough to have a nitrite concentration that falls within the calibration range.

Table 8-1 and Figure 8-2 show typical results. The equation of the calibration line in Figure 8-2, determined by the method of least squares, is

$$\text{absorbance} = 0.176\,9\,[\text{ppm}] + 0.001\,5 \qquad (8\text{-}2)$$

The saltwater aquarium is filled with artificial seawater made by adding water to a mixture of salts.

The unknown should always be adjusted to fall within the calibration range, because you have not verified that the response remains linear outside the calibration range.

TABLE 8-1   Aquarium nitrite analysis

Sample	Absorbance at 543 nm in 1.000-cm cuvet	Corrected absorbance (blank subtracted)
Blank	0.003	—
Standards		
0.457 5 ppm	0.085	0.082
0.915 0 ppm	0.167	0.164
1.830 ppm	0.328	0.325
Unknown	0.281	0.278
Unknown	0.277	0.274

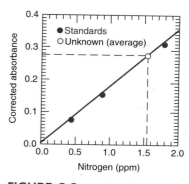

FIGURE 8-2 Calibration curve for nitrite analysis using corrected absorbance values from Table 8-1.

where [ppm] represents micrograms of nitrite nitrogen per milliliter. In principle, the intercept should be 0, but we will use the observed intercept (0.001 5) in our calculations. By plugging the average absorbance of unknown into Equation 8-2, we can solve for the concentration of nitrite (ppm) in the unknown.

EXAMPLE | Using the Standard Curve

From the data in Table 8-1, find the molarity of nitrite in the aquarium.

SOLUTION The average corrected absorbance of unknowns in Table 8-1 is 0.276. Substituting this value into Equation 8-2 gives ppm of nitrite nitrogen in the aquarium:

$$0.276 = 0.176\ 9\ [\text{ppm}] + 0.001\ 5$$

$$[\text{ppm}] = \frac{0.276 - 0.001\ 5}{0.176\ 9} = 1.55\ \text{ppm} = 1.55\ \frac{\mu g\ N}{mL}$$

To find the molarity of nitrite nitrogen, we first find the mass of nitrogen in a liter, which is

$$1.55 \times 10^{-6}\ \frac{g\ N}{mL} \times 1\ 000\ \frac{mL}{L} = 1.55 \times 10^{-3}\ \frac{g\ N}{L}$$

Then we convert mass of nitrogen into moles of nitrogen:

$$[\text{nitrite nitrogen}] = \frac{1.55 \times 10^{-3}\ g\ N/L}{14.007\ g\ N/mol} = 1.11 \times 10^{-4}\ M$$

Because one mole of nitrite (NO_2^-) contains one mole of nitrogen, the concentration of nitrite is also 1.11×10^{-4} M.

EXAMPLE | Preparing Nitrite Standards

How would you prepare a nitrite standard containing approximately 2 ppm nitrite nitrogen from a concentrated standard containing 0.018 74 M $NaNO_2$?

SOLUTION First let's find out how many ppm of nitrogen are in 0.018 74 M $NaNO_2$. Because one mole of nitrite contains one mole of nitrogen, the concentration of nitrogen in the concentrated standard is 0.018 74 M. The mass of nitrogen in 1 mL is

$$\frac{g\ N}{mL} = \left(0.018\ 74\ \frac{mol}{L}\right)\left(14.007\ \frac{g}{mol}\right)\left(0.001\ \frac{L}{mL}\right) = 2.625 \times 10^{-4}\ \frac{g}{mL}$$

Assuming that 1.00 mL of solution has a mass of 1.00 g, we use the definition of parts per million to convert the mass of nitrogen to ppm:

$$\text{ppm} = \frac{g\ N}{g\ \text{solution}} \times 10^6 = \frac{2.625 \times 10^{-4}\ g\ N}{1.00\ g\ \text{solution}} \times 10^6 = 262.5\ \text{ppm}$$

Dilution formula 1-5:

$$M_{conc} \cdot V_{conc} = M_{dil} \cdot V_{dil}$$

Use any units you like for M and V as long as you use the same units on both sides of the equation. M could be ppm and V could be mL.

To prepare a standard containing ~2 ppm N, you could dilute the concentrated standard by a factor of 100 to give 2.625 ppm N. This dilution could be done by pipeting 10.00 mL of concentrated standard into a 1-L volumetric flask and diluting to the mark.

Ask Yourself

8-A. You have been sent to India to investigate the occurrence of goiter disease attributed to iodine deficiency. As part of your investigation, you must make field measurements of traces of iodide (I^-) in groundwater. The procedure is to oxidize I^- to I_2 and convert the product into an intensely colored complex with the dye brilliant green in the organic solvent toluene.

(a) A 3.15×10^{-6} M solution of the colored complex exhibited an absorbance of 0.267 at 635 nm in a 1.000-cm cuvet. A blank solution made from distilled water in place of groundwater had an absorbance of 0.019. Find the molar absorptivity of the colored complex.

(b) The absorbance of an unknown solution prepared from groundwater was 0.175. Subtract the blank absorbance from the unknown absorbance and use Beer's law to find the concentration of the unknown.

8-2 *Spectrophotometric Titrations*

Iron for biosynthesis is transported through the bloodstream, not as a free ion, but attached to the protein *transferrin* (Figure 8-3). A solution of transferrin can be titrated with iron to measure the transferrin content. Transferrin without iron, called *apotransferrin,* is colorless. Each protein molecule with a molecular weight of 81 000 binds two Fe^{3+} ions. When the iron binds to the protein, a red color with an absorbance maximum at 465 nm develops. The appearance of the red color is used to follow

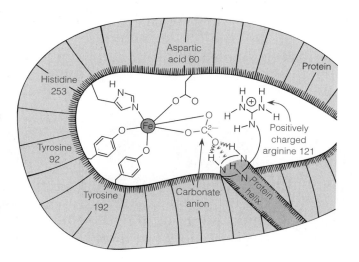

FIGURE 8-3 Each of the two iron-binding sites of transferrin is located in a cleft of the protein. The Fe^{3+} ion binds to one nitrogen atom from the amino acid histidine and three oxygen atoms from tyrosine and aspartic acid. Two more binding sites of the metal are occupied by oxygen atoms from a carbonate anion (CO_3^{2-}), which is anchored in place by electrostatic interaction with the positively charged amino acid arginine and by hydrogen bonding to part of the protein helix. When transferrin is taken up by a cell, it is brought to a compartment whose pH is lowered to 5.5. H^+ then reacts with the carbonate ligand to make HCO_3^- and H_2CO_3, thereby releasing Fe^{3+} from the protein.

the course of the titration of an unknown amount of apotransferrin with a standard solution of Fe^{3+}.

$$\text{apotransferrin} + 2Fe^{3+} \longrightarrow (Fe^{3+})_2\text{transferrin} \qquad (8\text{-}3)$$

Colorless   Red

A **spectrophotometric titration** is one in which absorption or emission of electromagnetic radiation is used to detect the end point.

Figure 8-4 shows the results of a titration of 2.000 mL of a purified solution of apotransferrin with 1.79×10^{-3} M ferric nitrilotriacetate. As iron is added to the protein, red color develops and the absorbance increases. When the protein is saturated with iron, no further color forms, and the curve levels off. The end point is the extrapolated intersection of the two straight portions of the titration curve at 203 μL in Figure 8-4. Absorbance rises slowly after the equivalence point because ferric nitrilotriacetate has some absorbance at 465 nm.

In constructing the graph in Figure 8-4, the effect of dilution should be considered, because the volume is different at each point. Each point plotted on the graph represents the absorbance that would be observed *if the solution had not been diluted from its original volume of 2.000 mL.*

$$\text{corrected absorbance} = \left(\frac{\text{total volume}}{\text{initial volume}} \right) (\text{observed absorbance}) \qquad (8\text{-}4)$$

Ferric nitrilotriacetate is used because Fe^{3+} reacts with H_2O to give $Fe(OH)_3$, which precipitates in neutral solution. Nitrilotriacetate binds Fe^{3+} through four atoms shown in **bold** type:

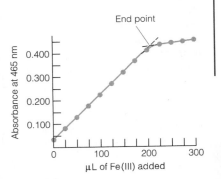

Nitrilotriacetate anion

EXAMPLE   **Correcting Absorbance for the Effect of Dilution**

The absorbance measured after adding 125 μL ($=0.125$ mL) of ferric nitrilotriacetate to 2.000 mL of apotransferrin was 0.260. Calculate the corrected absorbance that should be plotted in Figure 8-4.

SOLUTION The total volume was $2.000 + 0.125 = 2.125$ mL. If the volume had been 2.000 mL, the absorbance would have been greater than 0.260 by a factor of 2.125/2.000.

$$\text{corrected absorbance} = \left(\frac{2.125 \text{ mL}}{2.000 \text{ mL}} \right) (0.260) = 0.276$$

The absorbance plotted in Figure 8-4 is 0.276.

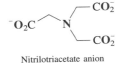

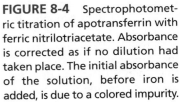

FIGURE 8-4 Spectrophotometric titration of apotransferrin with ferric nitrilotriacetate. Absorbance is corrected as if no dilution had taken place. The initial absorbance of the solution, before iron is added, is due to a colored impurity.

Ask Yourself

8-B. A 2.00-mL solution of apotransferrin, titrated as in Figure 8-4, required 163 μL of 1.43 mM ferric nitrilotriacetate to reach the end point.

   (a) How many moles of Fe^{3+} were required to reach the end point?

   (b) Each apotransferrin molecule binds two Fe^{3+} ions. Find the concentration of apotransferrin in the 2.00-mL solution.

   (c) Why does the slope in Figure 8-4 change abruptly at the equivalence point?

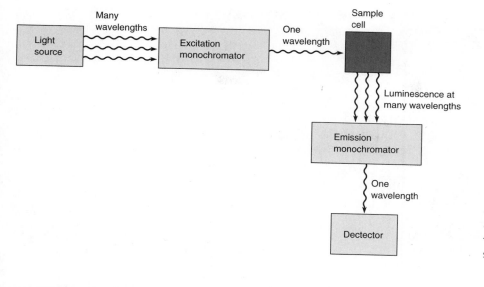

FIGURE 8-5 Schematic diagram of luminescence experiment. The sample is irradiated at one wavelength and emission is observed over a range of wavelengths. The emission monochromator selects one emission wavelength at a time to measure the emission spectrum.

After absorbing a photon and being promoted to an excited state, a molecule can return to its ground state by (1) emitting a photon (Figure 7-3), (2) dissipating heat, or (3) breaking a chemical bond. Emission of electromagnetic radiation, which is called **luminescence,** is illustrated in Color Plates 6 and 7. Chemistry initiated by the action of radiation in breaking chemical bonds (as in the reaction $O_2 \xrightarrow{h\nu} 2O$ in the upper atmosphere) is called *photochemistry.* Some chemical reactions (not initiated by light) release energy in the form of light, which is called *chemiluminescence.* The light from a firefly is chemiluminescence.

Luminescence is measured by exciting a sample at one wavelength in an absorption band and observing emission perpendicular to the incident beam over a range of wavelengths (Figure 8-5). Short-lived luminescence that occurs on a time scale of 10^{-8} to 10^{-4} s after absorption is called **fluorescence** (Demonstration 8-1). Long-lived emission occurring in a time of 10^{-4} to 10^2 s after absorption is called **phosphorescence.** Figure 8-6 shows that fluorescence comes at lower energy (longer wavelength) than absorption and is nearly the mirror image of absorption.

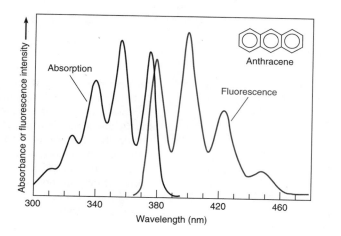

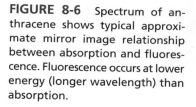

FIGURE 8-6 Spectrum of anthracene shows typical approximate mirror image relationship between absorption and fluorescence. Fluorescence occurs at lower energy (longer wavelength) than absorption.

In Which Your Class Really Shines

A fluorescent lamp is a glass tube filled with mercury vapor; the inner walls are coated with a *phosphor* (luminescent substance) consisting of a calcium halophosphate ($Ca_5(PO_4)_3F_{1-x}Cl_x$) doped with Mn^{2+} and Sb^{3+}. (*Doping* means adding an intentional impurity, called a *dopant*.) The mercury atoms, promoted to an excited state by electric current passing through the lamp, emit mostly ultraviolet radiation at 254 and 185 nm. This radiation is absorbed by the Sb^{3+}, and some of the energy is passed on to Mn^{2+}. Sb^{3+} emits blue light and Mn^{2+} emits yellow light, with the combined emission appearing white. The emission spectrum is shown in the next column. Fluorescent lamps are important energy-saving devices because they are more efficient than incandescent lamps in the conversion of electricity to light.

White fabrics are sometimes made "whiter" by treatment with a fluorescent dye. Turn on an ultraviolet lamp in a darkened classroom and illuminate some

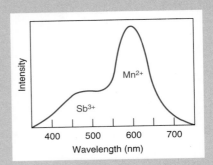

people standing at the front of the room. (*The victims should not look directly at the lamp,* because ultraviolet light is harmful to eyes.) You will discover a surprising amount of emission from white fabrics, including shirts, pants, shoelaces, and unmentionables. You may also be surprised to see fluorescence from teeth and from recently bruised areas of skin that show no surface damage.

Luminescence is used in analytical chemistry because over some concentration range the intensity of luminescence (I) is proportional to the concentration of the emitting species (c):

Equation 8-5 tells us that the two ways to increase luminescence are to increase the concentration of the emitting compound and to increase the intensity of incident light.

Relation of emission intensity to concentration:

$$I = kP_0c \qquad (8\text{-}5)$$

where P_0 is the intensity of the incident radiation and k is a constant. Figure 8-7 is a calibration curve illustrating Equation 8-5. Fluorescence is used in the same manner as absorbance (Figure 8-2) for chemical analysis. The biosensor at the beginning of this chapter measures glucose in blood from the fluorescence intensity of molecules displaced from a membrane by glucose.

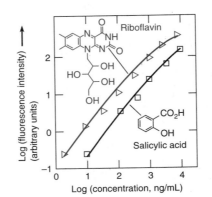

FIGURE 8-7 Calibration curves demonstrating nearly linear relation between fluorescence intensity (*I*) and concentration (*c*) for riboflavin (vitamin B_2) and salicyclic acid. A log-log scale is used because the data span several orders of magnitude. If the slope of log *I* versus log *c* is unity, *I* is proportional to *c*.

Luminescence measurements are inherently more sensitive than absorption measurements. Imagine yourself in a stadium at night with the lights off, but each of the 50 000 raving fans is holding a lighted candle. If 500 people blow out their candles, you will hardly notice the difference. Now imagine that the stadium is completely dark, but then 500 people light their candles. The change would be dramatic. The first example is analogous to changing the transmittance from 100% to 99%. It is hard to measure such a small change because the background is so bright. The second example is analogous to observing fluorescence from 1% of the molecules in a sample. Against the black background, this fluorescence is easy to detect.

Luminescence is generally more sensitive than absorption. Luminescence can detect lower concentrations of analyte.

Ask Yourself

8-C. Red tide is a phenomenon in which the sea becomes red and mass fish kills may occur. Chemiluminescence can be used to detect the early stages of red tide caused by the plankton *Chattonella marina,* which secretes superoxide ion (O_2^-). The procedure uses a flow cell with a spiral mixing chamber to combine unknown seawater with a chemiluminescent probe designated MCLA at pH 9.7. When O_2^- reacts with MCLA, the product emits light that is detected with a photomultiplier tube. The table below shows the detector signal as a function of known cell concentrations of *Chattonella marina* in seawater. Prepare a calibration curve and find the concentration of *Chattonella marina* in a seawater sample that produces a detector signal of 15.9 units.

Cells/mL in seawater	Detector signal (arbitrary units)	Cells/mL in seawater	Detector signal (arbitrary units)
188	2.2	2 336	16.5
376	4.6	3 087	20.6
738	6.0	3 973	26.6
1 154	9.1	4 913	31.5
1 544	11.6		

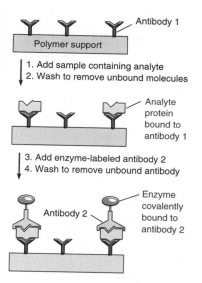

FIGURE 8-8 Enzyme-linked immunosorbent assay. Antibody 1, which is specific for the analyte of interest, is bound to a polymer support and treated with unknown. After washing away excess, unbound molecules, the analyte remains bound to antibody 1. The bound analyte is then treated with antibody 2, which recognizes a different site on the analyte and to which an enzyme is covalently attached. After washing away unbound material, each molecule of analyte is coupled to an enzyme, which will be used in Figure 8-9.

8-4 Immunoassays

An important application of luminescence is in **immunoassays,** which employ antibodies to detect analyte. An **antibody** is a protein produced by the immune system of an animal in response to a foreign molecule, which is called an **antigen.** An antibody specifically recognizes and binds to the antigen that stimulated its synthesis.

Figure 8-8 illustrates the principle of an *enzyme-linked immunosorbent assay,* abbreviated ELISA in biochemical literature. Antibody 1, which is specific for the analyte of interest (the antigen), is bound to a polymer support. In steps 1 and 2, analyte is incubated with the polymer-bound antibody to form the antibody-antigen complex. The fraction of antibody sites that bind analyte is proportional to the concentration of analyte in the unknown. The surface is then washed to remove unbound substances. In steps 3 and 4, the antibody-antigen complex is treated with antibody 2, which recognizes a different region of the analyte. An enzyme that will be used later was covalently attached to antibody 2 (prior to step 3). Again, excess unbound substances are washed away.

Box 8-1 *Informed Citizen*

Immunoassays in Environmental Analysis

Immunoassays are becoming available for screening and analysis of environmental samples in the field. An advantage of screening in the field is that uncontaminated regions that require no further attention are readily identified. Assays have been developed to monitor pesticides, industrial chemicals, and microbial toxins at the parts-per-trillion to parts-per-million levels in groundwater, soil, and food. In some cases, the immunoassay field test is 20–40 times less expensive than a chromatographic analysis in the laboratory. Immunoassays require less than a milliliter of sample and can be completed in 2–3 h in the field. Chromatographic analyses might require up to 2 days, because analyte must first be extracted or concentrated from liter-quantity samples to obtain a sufficient concentration.

The diagram shows how an assay works. In step 1, antibody for the intended analyte is adsorbed to the bottom of a microtiter well, which is a depression in a plate having as many as 96 wells for simultaneous analyses. In step 2, known volumes of sample and standard solution containing enzyme-labeled analyte are added to the well. In step 3, analyte in the sample competes with enzyme-labeled analyte for binding sites on the antibody. This is the key step: *The greater the concentration of analyte in the unknown sample, the more it will bind to antibody and the less enzyme-labeled analyte will bind.* After an incubation period, unbound sample and standard are washed away. In step 4, a *chromogenic* substance is added to the well. This is a colorless substance that reacts in the presence of enzyme to make a colored product. In step 5, colored product is measured visually by comparison to standards or quantitatively with a spectrophotometer. The greater the concentration of analyte in the unknown sample, the less color will be present in step 5.

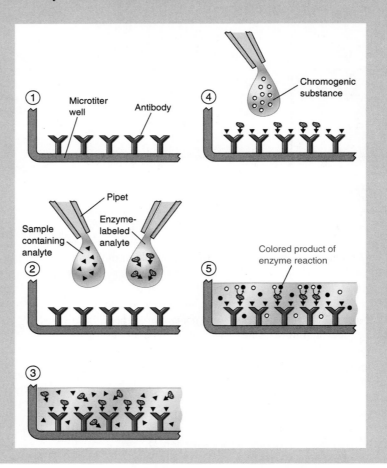

Figure 8-9 shows two ways in which the enzyme attached to antibody 2 is used for quantitative analysis. In Figure 8-9a, the enzyme transforms a colorless reactant into a colored product. Because one enzyme molecule catalyzes the same reaction many times, many molecules of colored product are created for each molecule of antigen. The enzyme thereby *amplifies* the signal in chemical analysis. The higher the concentration of analyte in the unknown, the more enzyme is bound and the greater the extent of the enzyme-catalyzed reaction. In Figure 8-9b, the enzyme converts a nonfluorescent reactant into a fluorescent product. Enzyme-linked immunosorbent assays are sensitive to <1 ng of analyte. Pregnancy tests are based on the immunoassay of a placental protein in urine. Box 8-1 shows how immunoassays are used in field-portable environmental analyses.

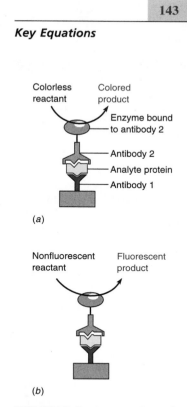

(a)

(b)

FIGURE 8-9 Enzyme bound to antibody 2 can catalyze reactions that produce (a) colored or (b) fluorescent products. Each molecule of analyte bound in the immunoassay leads to many molecules of colored or fluorescent product that are easily measured.

Ask Yourself

8-D. This problem describes an "electronic dog" being developed to sniff trinitrotoluene (TNT) explosives at airports.

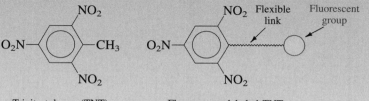

Trinitrotoluene (TNT)   Fluorescence-labeled TNT

1. A column containing covalently bound antibodies against TNT was prepared. Fluorescence-labeled TNT was passed through this column to saturate all of the antibodies with labeled TNT. The column was washed with excess solvent until no fluorescence was detected at the outlet.

2. Air containing TNT vapors was drawn through a sampling device that concentrated traces of the gaseous vapors into distilled water.

3. Aliquots of water from the sampler were injected into the antibody column. The fluorescence intensity of the liquid exiting the column was proportional to the concentration of unlabeled TNT in the water from the sampler over the range of 20 to 1 200 ng/mL.

Draw pictures showing the state of the column in steps 1 and 3 and explain how this detector works.

Key Equations

Absorption of light (Beer's law) $A = \epsilon bc$ A = absorbance (dimensionless)

ϵ = molar absorptivity ($M^{-1}\,cm^{-1}$)

b = pathlength (cm)

c = concentration (M)

Fluorescence intensity $\qquad I = kP_0c \qquad$ I = fluorescence intensity
k = constant
P_0 = intensity of incident radiation
c = concentration of fluorescing species

Important Terms

antibody

antigen

fluorescence

immunoassay

luminescence

phosphorescence

spectrophotometric titration

spectrophotometry

standard curve

Problems

8-1. (a) A 3.96×10^{-4} M solution of compound A exhibited an absorbance of 0.624 at 238 nm in a 1.000-cm cuvet. A blank solution containing only solvent had an absorbance of 0.029 at the same wavelength. Find the molar absorptivity of compound A.

(b) The absorbance of an unknown solution of compound A in the same solvent and cuvet was 0.375 at 238 nm. Subtract the blank absorbance from the unknown absorbance and use Beer's law to find the concentration of the unknown.

(c) A concentrated solution of compound A in the same solvent was diluted from an initial volume of 2.00 mL to a final volume of 25.00 mL and then had an absorbance of 0.733. What is the concentration of A in the 25.00-mL solution?

(d) Considering the dilution from 2.00 mL up to 25.00 mL, what was the concentration of A in the 2.00-mL solution in **(c)**?

8-2. A compound with a molecular weight of 292.16 was dissolved in a 5-mL volumetric flask. A 1.00-mL aliquot was withdrawn, placed in a 10-mL volumetric flask, and diluted to the mark. The absorbance measured at 340 nm was 0.427 in a 1.000-cm cuvet. The molar absorptivity for this compound at 340 nm is $\epsilon_{340} = 6\ 130\ \text{M}^{-1}\ \text{cm}^{-1}$.

(a) Calculate the concentration of compound in the cuvet.

(b) What was the concentration of compound in the 5-mL flask?

(c) How many milligrams of compound were used to make the 5-mL solution?

8-3. A nitrite analysis was conducted according to the procedure in Section 8-1, giving the data in the table below. Fill in the corrected absorbance, which is the measured absorbance minus the average blank absorbance (0.023). Construct a calibration line to find the **(a)** ppm of nitrite nitrogen and **(b)** molar concentration of nitrite in the aquarium. Use the average blank and the average unknown absorbances.

Sample	Absorbance	Corrected absorbance
Blank	0.022	—
Blank	0.024	—
Standards		
0.538 ppm	0.121	0.098
1.076 ppm	0.219	
2.152 ppm	0.413	
3.228 ppm	0.600	
4.034 ppm	0.755	
Unknown	0.333	
Unknown	0.339	
Unknown	0.338	

8-4. Use the method of least squares in Chapter 4 to find the *uncertainty* in nitrite nitrogen concentration (ppm) in the previous problem.

8-5. Starting with 0.015 83 M $NaNO_2$ solution, explain how you would prepare standards containing *approximately* 0.5, 1, 2, and 3 ppm nitrogen (1 ppm = 1 μg/mL). You may use any volumetric flasks and transfer pipets in Tables 2-2 and 2-3. What would be the exact concentrations of the standards prepared by your method?

8-6. Ammonia (NH_3) is determined spectrophotometrically by reaction with phenol in the presence of hypochlorite (OCl^-):

$$\text{phenol} + \text{ammonia} \xrightarrow{OCl^-} \text{blue product}$$

Colorless   Colorless $\lambda_{max} = 625$ nm

1. A 4.37-mg sample of protein was chemically digested to convert its nitrogen to ammonia and then diluted to 100.0 mL.
2. Then 10.0 mL of the solution was placed in a 50-mL volumetric flask and treated with 5 mL of phenol solution plus 2 mL of sodium hypochlorite solution. The sample was diluted to 50.0 mL, and the absorbance at 625 nm was measured in a 1.00-cm cuvet after 30 min.
3. A standard solution was prepared from 0.010 0 g of NH_4Cl (FW 53.49) dissolved in 1.00 L of water. A 10.0-mL aliquot of this standard was placed in a 50-mL volumetric flask and analyzed in the same manner as the unknown.
4. A reagent blank was prepared by using distilled water in place of unknown.

Sample	Absorbance at 625 nm
Blank	0.140
Reference	0.308
Unknown	0.592

(a) From step 3, calculate the molar absorptivity of the blue product.

(b) Using the molar absorptivity, find the concentration of ammonia in step 2.

(c) From your answer to **(b)**, find the concentration of ammonia in the 100-mL solution in step 1.

(d) Find the weight percent of nitrogen in the protein.

8-7. The iron-binding site of transferrin in Figure 8-3 can accommodate certain other metal ions besides Fe^{3+} and certain other anions besides CO_3^{2-}. Data are given below for the titration of transferrin (3.57 mg in 2.00 mL) with 6.64 mM

Ga^{3+} solution in the presence of the anion oxalate, $C_2O_4^{2-}$, and in the absence of a suitable anion. Prepare a graph similar to Figure 8-4, showing both sets of data. Indicate the theoretical equivalence point for the binding of two Ga^{3+} ions per molecule of protein and the observed end point. How many Ga^{3+} ions are bound to transferrin in the presence and in the absence of oxalate?

Titration in presence of $C_2O_4^{2-}$		Titration in absence of anion	
Total μL Ga^{3+} added	Absorbance at 241 nm	Total μL Ga^{3+} added	Absorbance at 241 nm
0.0	0.044	0.0	0.000
2.0	0.143	2.0	0.007
4.0	0.222	6.0	0.012
6.0	0.306	10.0	0.019
8.0	0.381	14.0	0.024
10.0	0.452	18.0	0.030
12.0	0.508	22.0	0.035
14.0	0.541	26.0	0.037
16.0	0.558		
18.0	0.562		
21.0	0.569		
24.0	0.576		

8-8. Cu^+ reacts with neocuproine to form the colored complex (neocuproine)$_2Cu^+$, with an absorption maximum at 454 nm. Neocuproine is particularly useful because it reacts with few other metals. The copper complex is soluble in isoamyl alcohol, an organic solvent that does not dissolve appreciably in water. In other words, when isoamyl alcohol is added to water, a two-layered mixture results, with the denser water layer at the bottom. If (neocuproine)$_2Cu^+$ is present, virtually all of it goes into the organic phase. For the purpose of this problem, assume that no isoamyl alcohol dissolves in water and that all of the colored complex will be in the organic phase. Suppose that the following procedure is carried out (see also illustration on page 146):

1. A rock containing copper is pulverized, and all metals are extracted from it with strong acid. The acidic solution is neutralized with base and made up to 250.0 mL in flask A.
2. Next 10.00 mL of the solution is transferred to flask B and treated with 10.00 mL of a reducing agent to reduce Cu^{2+} to Cu^+. Then 10.00 mL of buffer is added to bring the pH to a value suitable for complex formation with neocuproine.

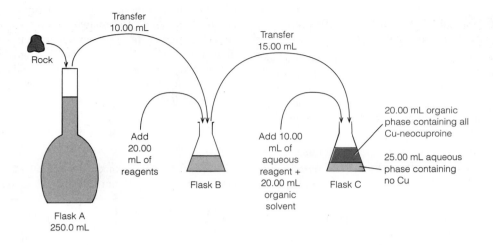

Transfer
10.00 mL

Rock

Transfer
15.00 mL

Add
20.00
mL of
reagents

Flask B

Add 10.00
mL of
aqueous
reagent +
20.00 mL
organic
solvent

Flask C

20.00 mL organic
phase containing all
Cu-neocuproine

25.00 mL aqueous
phase containing
no Cu

Flask A
250.0 mL

3. After that, 15.00 mL of this solution is withdrawn and placed in flask C. To the flask is added 10.00 mL of an aqueous solution containing neocuproine and 20.00 mL of isoamyl alcohol. After shaking well and allowing the phases to separate, all $(neocuproine)_2Cu^+$ is in the organic phase.

4. A few milliliters of the upper layer are withdrawn, and the absorbance at 454 nm is measured in a 1.00-cm cell. A blank carried through the same procedure gave an absorbance of 0.056.

(a) Suppose that the rock contained 1.00 mg Cu. What will the concentration of copper (moles per liter) in the isoamyl alcohol phase be?

(b) If the molar absorptivity of $(neocuproine)_2Cu^+$ is $7.90 \times 10^3 \ M^{-1} \ cm^{-1}$, what will the observed absorbance be? Remember that a blank carried through the same procedure gave an absorbance of 0.056.

(c) A rock is analyzed and found to give a final absorbance of 0.874 (uncorrected for the blank). How many milligrams of copper are in the rock?

8-9. Spectrophotometric analysis of phosphate can be performed by the following procedure:

Standard solutions:

A. KH_2PO_4 (potassium dihydrogen phosphate, FW 136.09): 81.37 mg dissolved in 500.0 mL H_2O

B. $Na_2MoO_4 \cdot 2H_2O$ (sodium molybdate): 1.25 g in 50 mL of 5 M H_2SO_4

C. $H_3NNH_3^{2+}SO_4^{2-}$ (hydrazine sulfate): 0.15 g in 100 mL H_2O

Procedure:

Place the sample (either an unknown or the standard phosphate solution, A) in a 5-mL volumetric flask, and add 0.500 mL of B and 0.200 mL of C. Dilute to almost 5 mL with water, and heat at 100°C for 10 min to form a blue product $(H_3PO_4(MoO_3)_{12}$, 12-molybdophosphoric acid). Cool the flask to room temperature, dilute to the mark with water, mix well, and measure the absorbance at 830 nm in a 1.00-cm cell.

(a) When 0.140 mL of solution A was analyzed, an absorbance of 0.829 was recorded. A blank carried through the same procedure gave an absorbance of 0.017. Find the molar absorptivity of blue product.

(b) A solution of the phosphate-containing iron-storage protein ferritin was analyzed by this procedure. The unknown contained 1.35 mg of ferritin, which was digested in a total volume of 1.00 mL to release phosphate from the protein. Then 0.300 mL of this solution was analyzed by the procedure above and found to give an absorbance of 0.836. A blank carried through this procedure gave an absorbance of 0.038. Find the weight percent of phosphorus in the ferritin.

8-10. Starting with an ammonia solution known to contain 28.6 wt % NH_3, explain how to prepare ammonia standards containing exactly 1.00, 2.00, 4.00, and 8.00 ppm nitrogen (1 ppm = 1 μg/mL) for a spectrophotometric calibration curve. You will need to use a known *mass* of concentrated reagent, and you may use any flasks and pipets from Tables 2-2 and 2-3.

8-11. The metal-binding compound semi-xylenol orange is yellow at pH 5.9 but turns red (λ_{max} = 490 nm) when it reacts with Pb^{2+}. A 2.025-mL sample of semi-xylenol orange was titrated with 7.515×10^{-4} M $Pb(NO_3)_2$, with the following results:

Total μL Pb^{2+} added	Absorbance at 490 nm in 1-cm cell	Total μL Pb^{2+} added	Absorbance at 490 nm in 1-cm cell
0.0	0.227	42.0	0.425
6.0	0.256	48.0	0.445
12.0	0.286	54.0	0.448
18.0	0.316	60.0	0.449
24.0	0.345	70.0	0.450
30.0	0.370	80.0	0.447
36.0	0.399		

Make a graph of corrected absorbance versus microliters of Pb^{2+} added. The corrected absorbance is what would be observed if the volume were not changed from its initial value of 2.025 mL. Assuming that the reaction of semi-xylenol orange with Pb^{2+} has a 1:1 stoichiometry, find the molarity of semi-xylenol orange in the original solution.

8-12. When I was a boy, Uncle Wilbur let me watch as he analyzed the iron content of runoff from his banana ranch. A 25.0-mL sample was acidified with nitric acid and treated with excess KSCN to form a red complex. (KSCN itself is colorless.) The solution was then diluted to 100.0 mL and put in a variable-pathlength cell. For comparison, a 10.0-mL reference sample of 6.80×10^{-4} M Fe^{3+} was treated with HNO$_3$ and KSCN and diluted to 50.0 mL. The reference was placed in a cell with a 1.00-cm light path. The runoff sample exhibited the same absorbance as the reference when the pathlength of the runoff cell was 2.48 cm. What was the concentration of iron in Uncle Wilbur's runoff?

8-13. Explain how signal amplification is achieved in enzyme-linked immunosorbent assays.

Collecting Environmental Samples

Sarah Leen (*Matrix*)

Some people have all the fun! Scientists at the Dorset Research Center in Ontario, Canada, measure the acidity of Lake Muskoka.

Combustion products from automobiles and factories include nitrogen oxides and sulfur dioxide, which react with water and oxidizing agents in the atmosphere to produce acids.

$$\underbrace{NO + NO_2}_{\substack{\text{Nitrogen oxides} \\ \text{designated } NO_x}} \xrightarrow[H_2O]{\text{Oxidation}} \underset{\text{Nitric acid}}{HNO_3} \qquad \underset{\text{Sulfur dioxide}}{SO_2} \xrightarrow[H_2O]{\text{Oxidation}} \underset{\text{Sulfuric acid}}{H_2SO_4}$$

Figure 9-1 shows that *acid rain* in North America is most severe in the eastern half, downwind of many coal-burning power plants and factories. The pH of rain over northern Europe is in the range 4.2 to 4.5. Acid rain is a serious threat to forests and lakes around the world. For example, acid rain has leached half of the essential nutrients Ca^{2+} and Mg^{2+} from the soil of Sweden's forests since 1950. Acid rain increases the solubility of toxic Al^{3+} and other metals in groundwater. The area over which fish populations have declined or disappeared in acidic lakes in Norway has doubled since 1970. Monitoring the acidity of rainwater is a critical component of programs to measure and reduce the production of acid rain by man's activities.

Chapter 9

Monoprotic Acid-Base Equilibria

It is difficult to have a meaningful discussion of many diverse subjects such as protein folding or the weathering of rocks without understanding acids and bases. In this chapter we treat monoprotic systems, which involve one acidic proton. Subsequent chapters will treat buffers, titrations, and polyprotic systems—those with more than one acidic proton. Before beginning, you should already have a foundation in acids and bases from Chapter 5.

9-1 Strong Acids and Bases

Figure 9-1 shows the average pH of rainfall in the United States. The principal components that make rainfall acidic are nitric and sulfuric acids, which are strong acids. Each mole of **strong acid** or **strong base** in aqueous solution dissociates completely to provide one mole of H^+ or OH^-. Nitric acid is a strong acid, so the reaction

$$\underset{\text{Nitric acid}}{HNO_3} \longrightarrow H^+ + \underset{\text{Nitrate}}{NO_3^-}$$

goes to completion. In the case of sulfuric acid, one proton is completely dissociated, but the second is only partially dissociated (depending on conditions):

$$\underset{\text{Sulfuric acid}}{H_2SO_4} \longrightarrow H^+ + \underset{\substack{\text{Hydrogen sulfate} \\ \text{(also called bisulfate)}}}{HSO_4^-} \rightleftharpoons H^+ + \underset{\text{Sulfate}}{SO_4^{2-}}$$

Common strong acids	Common strong bases
HCl	LiOH
HBr	NaOH
HI	KOH
$H_2SO_4{}^a$	RbOH
HNO_3	CsOH
$HClO_4$	R_4NOH^b

See Table 5-1 for more information.
a. Only one proton is strong acid.
b. Tetraalkylammonium hydroxide.

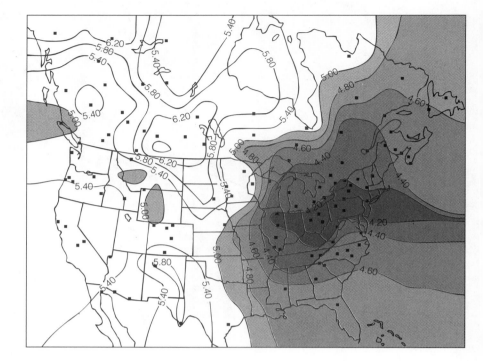

FIGURE 9-1 Average pH of precipitation over North America (from a 1981 study). Any pH below 7 is acidic; the lower the pH, the more acidic the water.

pH of a Strong Acid

Because HBr is completely dissociated, the pH of 0.010 M HBr is

$$pH = -\log[H^+] = -\log(0.010) = 2.00$$

Acid: pH < 7
Base: pH > 7

Is pH 2 sensible? (Always ask yourself that question at the end of a calculation.) Yes—because pH values below 7 are acidic and pH values above 7 are basic.

EXAMPLE pH of a Strong Acid

Find the pH of 4.2×10^{-3} M $HClO_4$.

SOLUTION An easy one! The pH is simply

$$pH = -\log[H^+] = -\log(\underbrace{4.2 \times 10^{-3}}_{\substack{\text{2 significant} \\ \text{figures}}}) = \underbrace{2.38}_{\substack{\text{2 digits} \\ \text{in mantissa}}}$$

How about significant figures? The two significant figures in the mantissa of the logarithm correspond to the two significant figures in the number 4.2×10^{-3}.

For consistency in working problems in this book, *we are generally going to express pH values to the 0.01 decimal place regardless of what is justified by significant figures.* Real pH measurements are rarely more accurate than ±0.02, although differences in pH between two solutions can be accurate to ±0.002 pH units.

pH of a Strong Base

Equation 5-7 told us that the product $[H^+][OH^-]$ is a constant called K_w:

Ion product
of water:
$$K_w = [H^+][OH^-] = 1.0 \times 10^{-14} \quad \text{(at 25°C)} \tag{9-1}$$

Therefore we can always find $[OH^-]$ if we know $[H^+]$, and vice versa.

Now we ask, "What is the pH of 4.2×10^{-3} M KOH?"

$$[OH^-] = 4.2 \times 10^{-3} \text{ M}$$

$$[H^+] = \frac{K_w}{[OH^-]} = \frac{1.0 \times 10^{-14}}{4.2 \times 10^{-3}} = 2.3_8 \times 10^{-12} \text{ M} \tag{9-2}$$

$$pH = -\log[H^+] = -\log(2.3_8 \times 10^{-12}) = 11.62$$

Let's try a tricky one. What is the pH of 4.2×10^{-9} M KOH? By our previous reasoning, we might first say

$$[H^+] = \frac{K_w}{[OH^-]} = \frac{1.0 \times 10^{-14}}{4.2 \times 10^{-9}} = 2.3_8 \times 10^{-6} \text{ M} \Rightarrow pH = 5.62$$

Is this reasonable? Can we dissolve base in water and obtain an acidic pH (<7)? No way!

The fallacy is that we neglected the contribution of the reaction $H_2O \rightleftharpoons H^+ + OH^-$ to the concentration of OH^-. Pure water creates 10^{-7} M OH^-, which is more than the KOH that we added. The pH of water plus added KOH cannot fall below 7. The pH of 4.2×10^{-9} M KOH is very close to 7. Similarly, the pH of 10^{-10} M HNO_3 is very close to 7, not 10. Figure 9-2 shows how pH depends on concentration for a strong acid and a strong base. In a very dilute solution of acid or base, the chemistry of dissolved carbon dioxide ($CO_2 + H_2O \rightleftharpoons HCO_3^- + H^+$) overwhelms the effect of the added acid or base.

Water Almost Never Produces 10^{-7} M H^+ and 10^{-7} M OH^-

There is 10^{-7} M H^+ and 10^{-7} M OH^- *only* in extremely pure water with no added acid or base. In a 10^{-4} solution of HBr, for example, the pH is 4. The concentration of OH^- is $K_w/[H^+] = 10^{-10}$ M. But the source of $[OH^-]$ is dissociation of water. If water produces only 10^{-10} M OH^-, it must also produce only 10^{-10} M H^+, because it makes one H^+ for every OH^-. In a 10^{-4} M HBr solution, water dissociation produces only 10^{-10} M OH^- and 10^{-10} M H^+.

$$[OH^-] = \frac{K_w}{[H^+]} \qquad [H^+] = \frac{K_w}{[OH^-]}$$

Keep at least one extra insignificant figure (or all the digits in your calculator) in the middle of a calculation to avoid round-off errors in the final answer.

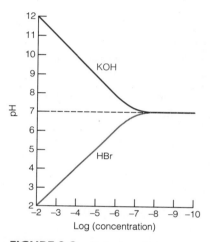

FIGURE 9-2 Calculated pH as a function of the concentration of a strong acid or strong base dissolved in water.

Acids and bases suppress water ionization, as predicted by Le Châtelier's principle.

Question What concentrations of H^+ and OH^- are produced by H_2O dissociation in 10^{-2} M NaOH?

Ask Yourself

9-A. **(a)** What is the pH of **(i)** 1.0×10^{-3} M HBr and **(ii)** 1.0×10^{-2} M KOH?
(b) Calculate the pH of **(i)** 3.2×10^{-5} M HI and **(ii)** 7.7 mM LiOH.
(c) Find the concentration of H^+ in a solution whose pH is 4.44.
(d) Find $[H^+]$ in 7.7 mM LiOH solution. What is the source of this H^+?
(e) Find the pH of 3.2×10^{-9} M tetramethylammonium hydroxide, $(CH_3)_4N^+OH^-$.

9-2 Weak Acids and Bases

A **weak acid** is not completely dissociated. That is, Reaction 9-3 does not go to completion:

Weak-acid
equilibrium:

$$HA \xrightleftharpoons{K_a} H^+ + A^- \qquad K_a = \frac{[H^+][A^-]}{[HA]} \tag{9-3}$$

The **acid dissociation constant, K_a,** is the equilibrium constant for Reaction 9-3. For a base, B, the **base hydrolysis constant, K_b,** is defined by the reaction

Weak-base
equilibrium:

$$B + H_2O \xrightleftharpoons{K_b} BH^+ + OH^- \qquad K_b = \frac{[BH^+][OH^-]}{[B]} \tag{9-4}$$

A **weak base** is one for which Reaction 9-4 does not go to completion. **pK** is the negative logarithm of the equilibrium constant:

$$pK_a = -\log K_a \qquad pK_b = -\log K_b \tag{9-5}$$

The *stronger* an acid, the *smaller* its pK_a.

Stronger acid	Weaker acid
$K_a = 10^{-4}$	$K_a = 10^{-8}$
$pK_a = 4$	$pK_a = 8$

The acid HA and its corresponding base, A^-, are said to be a **conjugate acid-base pair,** because they are related by the gain or loss of one proton. Similarly, B and BH^+ are conjugate. An important relationship derived in Equation 5-14 between K_a and K_b for a conjugate acid-base pair is

Relation between K_a and
K_b for conjugate pair:

$$K_a \cdot K_b = K_w \tag{9-6}$$

Weak Is Conjugate to Weak

The conjugate base of a weak acid is a weak base.
The conjugate acid of a weak base is a weak acid.
Weak is conjugate to weak.

The conjugate base of a weak acid is a weak base. The conjugate acid of a weak base is a weak acid. Consider a weak acid, HA, with $K_a = 10^{-4}$. The conjugate base, A^-, has $K_b = K_w/K_a = 10^{-10}$. That is, if HA is a weak acid, A^- is a weak base. If the K_a value were 10^{-5}, then the K_b value would be 10^{-9}. As HA becomes a weaker acid, A^- becomes a stronger base (but never a strong base). Conversely, the greater the acid strength of HA, the less the base strength of A^-. However, if either A^- or HA is weak, so is its conjugate. If HA is strong (such as HCl), its conjugate base (Cl^-) is *so* weak that it is not a base at all in water.

Using Appendix B

$$CH_3NH_2 \qquad\qquad CH_3NH_3^+$$
Methylamine   Methylammonium
ion

Acid dissociation constants appear in Appendix B. Each compound is shown in its *fully protonated form.* Methylamine, for example, is shown as $CH_3NH_3^+$, which is really the methylammonium ion. The value of K_a (2.3×10^{-11}) given for methylamine is actually K_a for the methylammonium ion. To find K_b for methylamine, we write $K_b = K_w/K_a = (1.0 \times 10^{-14})/(2.3 \times 10^{-11}) = 4.3 \times 10^{-4}$.

For polyprotic acids and bases, several K_a values are given. Pyridoxal phosphate is given in its fully protonated form as follows:

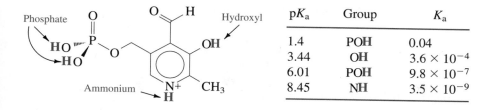

pKₐ	Group	Kₐ
1.4	POH	0.04
3.44	OH	3.6×10^{-4}
6.01	POH	9.8×10^{-7}
8.45	NH	3.5×10^{-9}

pK_1 (1.4) is for dissociation of one of the phosphate protons, and pK_2 (3.44) is for the hydroxyl proton. The third most acidic proton is the other phosphate proton, for which $pK_3 = 6.01$, and the NH^+ group is the least acidic ($pK_4 = 8.45$).

Pyridoxal phosphate is derived from vitamin B_6, which is essential in amino acid metabolism in your body.

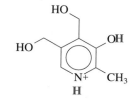

Vitamin B_6

Ask Yourself

9-B. (a) Write the acid dissociation reaction for formic acid, HCO_2H.
(b) What is the conjugate base of formic acid?
(c) Write the K_a equilibrium expression for formic acid and look up its value.
(d) Write the K_b equilibrium expression for formate ion, HCO_2^-.
(e) Find the base hydrolysis constant for formate.

9-3 Weak-Acid Equilibrium

Let's find the pH and composition of a solution containing 0.020 0 mol benzoic acid in 1.00 L of water.

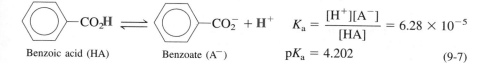

$$K_a = \frac{[H^+][A^-]}{[HA]} = 6.28 \times 10^{-5}$$

Benzoic acid (HA)   Benzoate (A^-) $pK_a = 4.202$ (9-7)

For every mole of A^- formed by dissociation of HA, one mole of H^+ is created. That is, $[A^-] = [H^+]$. (For reasonable concentrations of weak acids of reasonable strength, the contribution of H^+ from Reaction 9-7 is much greater than the contribution from $H_2O \rightleftharpoons H^+ + OH^-$.) Abbreviating the formal concentration of HA as F and using x for the concentration of H^+, we can make a table showing concentrations before and after dissociation of the weak acid:

	HA	$\rightleftharpoons$	A^-	+	H^+
Initial concentration	F		0		0
Final concentration	$F - x$		x		x

Putting these values into the K_a expression in Equation 9-7 gives

$$K_a = \frac{[H^+][A^-]}{[HA]} = \frac{(x)(x)}{F - x}$$

Setting $F = 0.020\ 0$ M and $K_a = 6.28 \times 10^{-5}$ gives

$$\frac{x^2}{0.020\ 0 - x} = 6.28 \times 10^{-5} \qquad (9\text{-}8)$$

The first step in solving Equation 9-8 for x is to cross multiply the numerator on one side by the denominator on the other side:

$$\frac{x^2}{0.020\ 0 - x} = \frac{6.28 \times 10^{-5}}{1}$$

$$(x^2)(1) = (6.28 \times 10^{-5})(0.020\ 0 - x) = (1.25_6 \times 10^{-6}) - (6.28 \times 10^{-5})x$$

Collecting terms gives a quadratic equation (in which the highest power is x^2):

$$x^2 + (6.28 \times 10^{-5})x - (1.25_6 \times 10^{-6}) = 0 \qquad (9\text{-}9)$$

Equation 9-9 has two solutions (called *roots*) described in Box 9-1. One root is positive and one is negative. Because a concentration cannot be negative, we reject the negative solution:

$$x = 1.09 \times 10^{-3} \quad \text{(negative root rejected)}$$

From the value of x, we can find concentrations and pH:

$$[H^+] = [A^-] = x = 1.09 \times 10^{-3}\ \text{M}$$
$$[HA] = F - x = 0.020\ 0 - (1.09 \times 10^{-3}) = 0.018\ 9\ \text{M}$$
$$pH = -\log x = 2.96$$

Was the approximation $[H^+] \approx [A^-]$ justified? The concentration of $[H^+]$ is 1.09×10^{-3} M, which means $[OH^-] = K_w/[H^+] = 9.20 \times 10^{-12}$ M.

$[H^+]$ from HA dissociation $= [A^-]$ from HA dissociation $= 1.09 \times 10^{-3}$ M
$[H^+]$ from H_2O dissociation $= [OH^-]$ from H_2O dissociation $= 9.20 \times 10^{-12}$ M

The assumption that H^+ is derived mainly from HA is excellent because 1.09×10^{-3} M $\gg 9.20 \times 10^{-12}$ M.

Fraction of Dissociation

What fraction of HA is dissociated? If the total concentration of acid ($= [HA] + [A^-]$) is $0.020\ 0$ M and the concentration of A^- is 1.09×10^{-3} M, then the *fraction of dissociation* is

*You should **check your answer** by plugging it back into Equation 9-8 and seeing that the equality is true.*

For uniformity, we are going to express pH to the 0.01 decimal place, even though significant figures in this problem justify another digit.

In a weak-acid solution, H^+ is derived almost entirely from HA, not from H_2O.

Box 9-1 *Explanation*

Quadratic Equations

A quadratic equation of the general form $ax^2 + bx + c = 0$ has two solutions:

$$x = \frac{-b + \sqrt{b^2 - 4ac}}{2a} \qquad x = \frac{-b - \sqrt{b^2 - 4ac}}{2a}$$

The solutions to the equation

$$\underbrace{(1)}_{a = 1}[H^+]^2 + \underbrace{(6.28 \times 10^{-5})}_{b = 6.28 \times 10^{-5}}[H^+] - \underbrace{(1.256 \times 10^{-6})}_{c = -1.256 \times 10^{-6}} = 0$$

are

$$[H^+] = \frac{-(6.28 \times 10^{-5}) + \sqrt{(6.28 \times 10^{-5})^2 - 4(1)(-1.256 \times 10^{-6})}}{2(1)} = 1.09 \times 10^{-3} \text{ M}$$

and

$$[H^+] = \frac{-(6.28 \times 10^{-5}) - \sqrt{(6.28 \times 10^{-5})^2 - 4(1)(-1.256 \times 10^{-6})}}{2(1)} = -1.09 \times 10^{-3} \text{ M}$$

Because the concentration of $[H^+]$ cannot be negative, we reject the negative solution and choose 1.09×10^{-3} M as the correct answer.

When you solve a quadratic equation, retain all of the digits in your calculator at intermediate stages of the computation, or serious round-off errors can occur in some cases. If you have a programmable calculator, we strongly recommend writing a program to solve quadratic equations and using it whenever one arises.

Fraction of dissociation of an acid:
$$\frac{[A^-]}{[A^-] + [HA]} = \frac{1.09 \times 10^{-3}}{0.020\ 0} = 0.054 \qquad (9\text{-}10)$$

The acid is indeed weak. It is only 5.4% dissociated.

Figure 9-3 compares the fraction of dissociation of two weak acids as a function of formal concentration. As the solution becomes more dilute, the fraction of dissociation increases. The stronger acid has a greater fraction of dissociation at any formal concentration. Demonstration 9-1 compares properties of strong and weak acids.

The Essence of a Weak-Acid Problem

When faced with finding the pH of a weak acid, you should immediately realize that $[H^+] = [A^-] = x$ and proceed to set up and solve the equation

Equation for weak acids:
$$\boxed{\frac{[H^+][A^-]}{[HA]} = \frac{x^2}{F - x} = K_a} \qquad (9\text{-}11)$$

where F is the formal concentration of HA. The approximation $[H^+] = [A^-]$ would be poor only if the acid were very dilute or very weak, neither of which constitutes a practical problem.

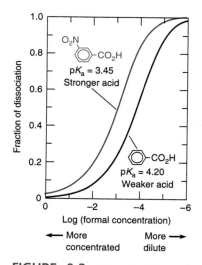

FIGURE 9-3 The fraction of dissociation of a weak acid increases as it is diluted. The stronger of the two acids ($pK_a = 3.45$ is stronger than $pK_a = 4.20$) is more dissociated at any given concentration.

Demonstration 9-1

Conductivity of Weak Electrolytes

The relative conductivity of strong and weak acids is directly related to their different degrees of dissociation in aqueous solution. To demonstrate conductivity, we use a Radio Shack piezo alerting buzzer, but any kind of buzzer or light bulb could be substituted for the electric horn. The voltage required will depend on the buzzer or light chosen.

When a conducting solution is placed in the beaker, the horn sounds. First show that distilled water and sucrose solution are nonconductive. Solutions of the strong electrolytes NaCl or HCl are conductive. Compare strong and weak electrolytes by demonstrating that 1 mM HCl gives a loud sound, whereas 1 mM acetic acid gives little or no sound. With 10 mM acetic acid, the strength of the sound varies noticeably as the electrodes are moved away from each other in the beaker.

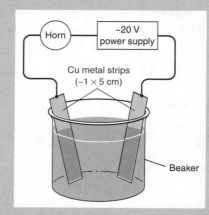

Apparatus for demonstrating conductivity of electrolyte solutions.

EXAMPLE Finding the pH of a Weak Acid

Find the pH of 0.100 M trimethylammonium chloride.

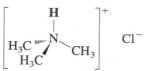

Trimethylammonium chloride

SOLUTION We must first realize that salts of this type are *completely dissociated* to give $(CH_3)_3NH^+$ and Cl^-. We then recognize that trimethylammonium ion is a weak acid, being conjugate to trimethylamine, $(CH_3)_3N$, a weak base. The ion Cl^- has no basic or acidic properties and should be ignored. In Appendix B, we find trimethylammonium ion listed under the name trimethylamine but drawn as the trimethylammonium ion. The value of pK_a is 9.800, so

$$K_a = 10^{-pK_a} = 10^{-9.800} = 1.58 \times 10^{-10}$$

It's all downhill from here:

$$(CH_3)_3NH^+ \overset{K_a}{\rightleftharpoons} (CH_3)_3N + H^+$$
$$F - x \qquad\qquad x \qquad x$$

$$\frac{x^2}{0.100 - x} = 1.58 \times 10^{-10}$$

$$x = 3.97 \times 10^{-6} \text{ M} \Rightarrow pH = -\log(3.97 \times 10^{-6}) = 5.40$$

A 0.100 M solution of hydrazoic acid is found to have pH = 2.83. Find pK_a for this acid.

$$- :\ddot{N}{=}\overset{+}{N}{=}\ddot{N}{-}H \overset{K_a}{\rightleftharpoons} - :\ddot{N}{=}\overset{+}{N}{=}\ddot{N}: {}^- + H^+$$

Hydrazoic acid   Azide (N_3^-)
0.100 − x   x   x

SOLUTION We know that $[H^+] = 10^{-pH} = 10^{-2.83} = 1.4_8 \times 10^{-3}$ M. Because $[N_3^-] = [H^+]$ in this solution, $[N_3^-] = 1.48 \times 10^{-3}$ M and $[HN_3] = 0.100 - (1.4_8 \times 10^{-3}) = 0.098\,5$ M. From these concentrations, we calculate K_a and pK_a:

$$K_a = \frac{[N_3^-][H^+]}{[HN_3]} = \frac{(1.48 \times 10^{-3})^2}{0.098\,5} = 2.2_2 \times 10^{-5}$$

$$\Rightarrow pK_a = -\log(2.2_2 \times 10^{-5}) = 4.65$$

Ask Yourself

9-C. **(a)** What is the pH and fraction of dissociation of a 0.100 M solution of the weak acid HA with $K_a = 1.00 \times 10^{-5}$?

(b) A 0.045 0 M solution of HA has a pH of 2.78. What is pK_a for this acid?

(c) A 0.045 0 M solution of HA is 0.60% dissociated. What is pK_a for this acid?

9-4 *Weak-Base Equilibrium*

The treatment of weak bases is almost the same as that of weak acids.

$$B + H_2O \overset{K_b}{\rightleftharpoons} BH^+ + OH^- \qquad K_b = \frac{[BH^+][OH^-]}{[B]}$$

Nearly all OH^- comes from the reaction of B + H_2O, and little comes from dissociation of H_2O. Setting $[OH^-] = x$, we must also set $[BH^+] = x$, because one BH^+ is produced for each OH^-. Setting $F = [B] + [BH^+]$, we can write

$$[B] = F - [BH^+] = F - x$$

Plugging these values into the K_b equilibrium expression, we get

Equation for weak base:
$$\frac{[BH^+][OH^-]}{[B]} = \frac{x^2}{F - x} = K_b \qquad (9\text{-}12)$$

which looks a lot like a weak-acid problem, except that now $x = [OH^-]$.

A weak-base problem has the same algebra as a weak-acid problem, except $K = K_b$ and $x = [OH^-]$.

EXAMPLE Finding the pH of a Weak Base

Find the pH of a 0.037 2 M solution of the commonly encountered weak base cocaine:

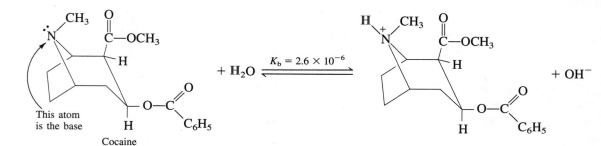

This atom is the base

Cocaine

SOLUTION Designating cocaine as B, we formulate the problem as follows:

$$B + H_2O \rightleftharpoons BH^+ + OH^-$$
$$0.037\ 2 - x \qquad\qquad x \qquad x$$

$$\frac{x^2}{0.037\ 2 - x} = 2.6 \times 10^{-6} \Rightarrow x = 3.1_0 \times 10^{-4}$$

Because $x = [OH^-]$, we can write

$$[H^+] = K_w/[OH^-] = (1.0 \times 10^{-14})/(3.1_0 \times 10^{-4}) = 3.2_2 \times 10^{-11}$$
$$pH = -\log[H^+] = 10.49$$

This is a reasonable pH for a weak base.

Question Were we justified in neglecting water dissociation as a source of OH^-? What concentration of OH^- is produced by H_2O dissociation in this solution?

What fraction of cocaine reacted with water in the previous example?

Fraction of associ-ation of a base: $\qquad \dfrac{[BH^+]}{[BH^+] + [B]} = \dfrac{3.1 \times 10^{-4}}{0.037\ 2} = 0.008\ 3 \qquad$ (9-13)

because $[BH^+] = [OH^-] = 3.1 \times 10^{-4}$ M. Only 0.83% of the base has reacted.

The physical properties of a neutral amine, such as cocaine (B), and its protonated product, cocaine hydrochloride (BH^+Cl^-), are quite different. Neutral cocaine is the extremely dangerous substance called "crack." It is soluble in organic solvents and cell membranes and has a moderate vapor pressure. The ionic hydrochloride is soluble in water, not in cell membranes, and has a much lower vapor pressure. Its physiological action is slower than that of "crack."

Conjugate Acids and Bases—Revisited

HA and A^- are a conjugate acid-base pair. So are BH^+ and B.

The conjugate base of a weak acid is a weak base, and the conjugate acid of a weak base is a weak acid. The exceedingly important relation between the equilibrium constants for a conjugate acid-base pair is

$$K_a \cdot K_b = K_w$$

In Section 9-3 we discussed benzoic acid, designated HA. Now consider 0.050 M sodium benzoate, Na^+A^-, which contains the conjugate base of benzoic acid. When this salt dissolves in water, it dissociates completely into Na^+ and A^-. Na^+ does not react with water, but A^- is a weak base:

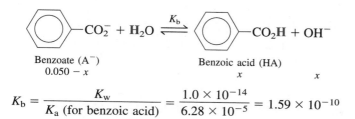

Benzoate (A^-) Benzoic acid (HA)
0.050 − x   x   x

$$K_b = \frac{K_w}{K_a \text{ (for benzoic acid)}} = \frac{1.0 \times 10^{-14}}{6.28 \times 10^{-5}} = 1.59 \times 10^{-10}$$

To find the pH of this solution, we write

$$\frac{[HA][OH^-]}{[A^-]} = \frac{x^2}{0.050 - x} = 1.59 \times 10^{-10} \Rightarrow x = [OH^-] = 2.8 \times 10^{-6} \text{ M}$$

$$[H^+] = K_w/[OH^-] = 3.5 \times 10^{-9} \text{ M} \Rightarrow pH = 8.45$$

This is a reasonable pH for a solution of a weak base.

EXAMPLE   **A Weak-Base Problem**

Find the pH of 0.10 M ammonia.

SOLUTION When ammonia is dissolved in water, its reaction is

$$NH_3 + H_2O \overset{K_b}{\rightleftharpoons} NH_4^+ + OH^-$$

Ammonia   Ammonium
                                        ion

F − x   x   x

In Appendix B we find the ammonium ion, NH_4^+, listed next to ammonia. K_a for ammonium ion is 5.70×10^{-10}. Therefore K_b for NH_3 is

$$K_b = \frac{K_w}{K_a} = \frac{1.0 \times 10^{-14}}{5.70 \times 10^{-10}} = 1.7_5 \times 10^{-5}$$

To find the pH of 0.10 M NH_3, we set up and solve the equation

$$\frac{[NH_4^+][OH^-]}{[NH_3]} = \frac{x^2}{0.10 - x} = K_b = 1.7_5 \times 10^{-5}$$

$$x = [OH^-] = 1.3_1 \times 10^{-3} \text{ M}$$

$$[H^+] = \frac{K_w}{[OH^-]} = 7.6_1 \times 10^{-12} \text{ M}$$

$$pH = -\log [H^+] = 11.12$$

Ask Yourself

9-D. **(a)** What is the pH and fraction of association of a 0.100 M solution of a weak base with $K_b = 1.00 \times 10^{-5}$?

$$\text{fraction of association} = \frac{[BH^+]}{[BH^+] + [B]} = 0.020$$

(b) A 0.10 M solution of a base has pH = 9.28. Find K_b.
(c) A 0.10 M solution of a base is 2.0% associated. Find K_b.

9-5 Fractional Composition Equations

We wish to find expressions for the fraction of an acid in each form (HA and A^-). These equations will be useful in future problems involving acids and bases. Consider an acid with formal concentration F:

$$HA \xrightleftharpoons{K_a} H^+ + A^- \qquad K_a = \frac{[H^+][A^-]}{[HA]}$$

We know that the concentrations of HA and A^- must add up to F because these are the only two species containing the group A. This relationship is called a *mass balance*:

The mass balance would more appropriately be called a mole balance.

Mass balance: $\qquad\qquad F = [HA] + [A^-]$

Rearranging the mass balance gives $[A^-] = F - [HA]$, which can be plugged into the K_a equilibrium to give

$$K_a = \frac{[H^+](F - [HA])}{[HA]}$$

or, with a little algebra,

$$[HA] = \frac{[H^+]F}{[H^+] + K_a} \tag{9-14}$$

The *fraction* of molecules in the form HA is called α_{HA}.

$$\alpha_{HA} = \frac{[HA]}{[HA] + [A^-]} = \frac{[HA]}{F}$$

Dividing Equation 9-14 by F gives

*Fraction in the
form HA:* $\qquad \alpha_{HA} = \frac{[HA]}{F} = \frac{[H^+]}{[H^+] + K_a} \tag{9-15}$

In a similar manner, the fraction in the form A^-, designated α_{A^-}, can be obtained:

α_{HA} = fraction in the form HA
α_{A^-} = fraction in the form A^-
$\alpha_{HA} + \alpha_{A^-} = 1$

*Fraction in the
form A^-:* $\qquad \alpha_{A^-} = \frac{[A^-]}{F} = \frac{K_a}{[H^+] + K_a} \tag{9-16}$

Figure 9-4 shows α_{HA} and α_{A^-} for a system with $pK_a = 5.00$. At low pH, almost all of the acid is in the form HA. At high pH, almost everything is in the form A^-. HA is the predominant species when $pH < pK_a$. A^- is the predominant species when $pH > pK_a$.

If you were dealing with the conjugate pair BH^+ and B instead of HA and A^-, Equation 9-15 gives the fraction in the form BH^+ and Equation 9-16 gives the fraction in the form B. In this case, K_a is the acid of dissociation constant for BH^+ (which is K_w/K_b).

EXAMPLE Fractional Composition for an Acid

pK_a for benzoic acid (HA) is 4.20. **(a)** Find the fraction in each form at pH 4.20. **(b)** Find the concentration of A^- at pH 5.31 if the formal concentration of HA is 0.021 3 M.

SOLUTION (a) At $pH = 4.20$, $[H^+] = 10^{-4.20} = 6.3 \times 10^{-5}$ M. Using $K_a = 10^{-4.20} = 6.3 \times 10^{-5}$ in Equations 9-15 and 9-16, we find

$$\alpha_{HA} = \frac{[H^+]}{[H^+] + K_a} = \frac{6.3 \times 10^{-5}}{(6.3 \times 10^{-5}) + (6.3 \times 10^{-5})} = 0.50$$

$$\alpha_{A^-} = \frac{K_a}{[H^+] + K_a} = \frac{6.3 \times 10^{-5}}{(6.3 \times 10^{-5}) + (6.3 \times 10^{-5})} = 0.50$$

At $pH = pK_a$, we discover that $\alpha_{HA} = \alpha_{A^-} = 0.50$. That is, 50% is in each form. *Whenever $pH = pK_a$ for any acid, $[HA] = [A^-]$.*
   (b) At $pH = 5.31$, $[H^+] = 10^{-5.31} = 4.9 \times 10^{-6}$ M.

$$\alpha_{A^-} = \frac{K_a}{[H^+] + K_a} = \frac{6.3 \times 10^{-5}}{(4.9 \times 10^{-6}) + (6.3 \times 10^{-5})} = 0.92_8$$

From Equation 9-16, we can say

$$[A^-] = \alpha_{A^-}F = (0.92_8)(0.021\ 3) = 0.020\ M$$

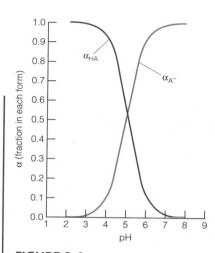

FIGURE 9-4 Fractional composition diagram of a monoprotic system with $pK_a = 5.00$. Below pH 5, HA is the dominant form; above pH 5, A^- dominates.

When $pH = pK_a$, $[HA] = [A^-]$.

EXAMPLE Fractional Composition for a Base

K_a for the ammonium ion, NH_4^+, is 5.70×10^{-10} ($pK_a = 9.244$). Find the fraction in the form BH^+ at pH 10.38.

SOLUTION At $pH = 10.38$, $[H^+] = 10^{-10.38} = 4.1_7 \times 10^{-11}$ M. Using Equation 9-15, with BH^+ in place of HA, we find

$$\alpha_{BH^+} = \frac{[H^+]}{[H^+] + K_a} = \frac{4.1_7 \times 10^{-11}}{4.1_7 \times 10^{-11} + 5.70 \times 10^{-10}} = 0.068$$

Ask Yourself

9-E. The acid HA has $pK_a = 3.00$. Use Equations 9-15 and 9-16 to find the fraction in the form HA and the fraction in the form A^- at pH = **(a)** 2.00; **(b)** 3.00; **(c)** 4.00. Compute the quotient $[HA]/[A^-]$ at each pH.

Key Equations

Definition of pH
$$pH = -\log[H^+]$$
(This relationship is only approximate—but it is the one we use in this book.)

Definitions of pK
$$pK_a = -\log K_a \qquad pK_b = -\log K_b$$

Ion product of water
$$K_w = [H^+][OH^-] = 1.0 \times 10^{-14} \text{ (at 25°C)}$$

Finding $[OH^-]$ from $[H^+]$
$$[OH^-] = K_w/[H^+]$$

Weak-acid equilibrium
$$HA \underset{F-x}{\overset{K_a}{\rightleftharpoons}} \underset{x}{H^+} + \underset{x}{A^-} \qquad K_a = \frac{[H^+][A^-]}{[HA]} = \frac{x^2}{F-x}$$
(F = formal concentration of weak acid or weak base)

Weak-base equilibrium
$$\underset{F-x}{B} + H_2O \overset{K_b}{\rightleftharpoons} \underset{x}{BH^+} + \underset{x}{OH^-} \qquad K_b = \frac{[BH^+][OH^-]}{[B]} = \frac{x^2}{F-x}$$

Conjugate acid-base pair
$$K_a \cdot K_b = K_w$$

Fraction of dissociation of HA
$$\text{fraction of dissociation} = \frac{[A^-]}{[A^-] + [HA]}$$

Fraction of association of B
$$\text{fraction of association} = \frac{[BH^+]}{[BH^+] + [B]}$$

Fraction in acidic form
$$\alpha_{HA} = \frac{[HA]}{F} = \frac{[H^+]}{[H^+] + K_a} = \alpha_{BH^+} = \frac{[BH^+]}{F}$$

Fraction in basic form
$$\alpha_{A^-} = \frac{[A^-]}{F} = \frac{K_a}{[H^+] + K_a} = \alpha_B = \frac{[B]}{F}$$

Important Terms

acid dissociation constant	pK	weak acid
base hydrolysis constant	strong acid	weak base
conjugate acid-base pair	strong base	

Problems

9-1. Calculate the pH of 5.0×10^{-5} M $HClO_4$.

9-2. Calculate the pH of 3.0×10^{-5} M $Mg(OH)_2$, which is completely dissociated to Mg^{2+} and OH^-.

9-3. Find the concentration of H^+ in a solution whose pH is 11.65.

9-4. Find the pH and fraction of dissociation of a 0.010 0 M solution of the weak acid HA with $K_a = 1.00 \times 10^{-4}$.

9-5. Find the pH and fraction of dissociation in a 0.150 M solution of hydroxybenzene (also called phenol).

9-6. Calculate the pH of 0.085 0 M pyridinium bromide, $C_5H_5NH^+Br^-$. Find the concentrations of pyridine (C_5H_5N), pyridinium ion ($C_5H_5NH^+$), and Br^- in the solution.

9-7. A 0.100 M solution of the weak acid HA has a pH of 2.36. Calculate pK_a for HA.

9-8. A 0.022 2 M solution of HA is 0.15% dissociated. Calculate pK_a for this acid.

9-9. Find the pH and concentrations of $(CH_3)_3N$ and $(CH_3)_3NH^+$ in a 0.060 M solution of trimethylammonium chloride.

9-10. Calculate the pH and fraction of dissociation of **(a)** $10^{-2.00}$ M and **(b)** $10^{-10.00}$ M barbituric acid.

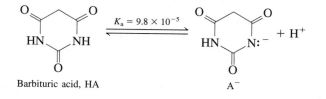

Barbituric acid, HA A^-

9-11. $BH^+ClO_4^-$ is a salt formed from the base B ($K_b = 1.00 \times 10^{-4}$) and perchloric acid. It dissociates into BH^+, a weak acid, and ClO_4^-, which is neither an acid nor a base. Find the pH of 0.100 M $BH^+ClO_4^-$.

9-12. Find K_a for cyclohexylammonium ion and K_b for cyclohexylamine.

Cyclohexylammonium ion   Cyclohexylamine

9-13. Write the chemical reaction whose equilibrium constant is **(a)** K_b for aniline and **(b)** K_a for anilinium ion.

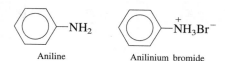

Aniline   Anilinium bromide

9-14. Find the pH and concentrations of $(CH_3)_3N$ and $(CH_3)_3NH^+$ in a 0.060 M solution of trimethylamine.

9-15. Calculate the pH and fraction of association of 1.00×10^{-1}, 1.00×10^{-2}, and 1.00×10^{-12} M sodium acetate.

9-16. Find the pH of 0.050 M NaCN.

9-17. Find the pH and fraction of association of 0.026 M NaOCl.

9-18. If a 0.030 M solution of a base has pH = 10.50, find K_b for the base.

9-19. If a 0.030 M solution of a base is 0.27% hydrolyzed ($\alpha_{BH^+} = 0.002\ 7$), find K_b for the base.

9-20. Find the fraction of pyridine (B) in the form B and the fraction in the form BH^+ at pH = **(a)** 4.00; **(b)** 5.00; **(c)** 6.00.

9-21. Find the fraction of 1-naphthoic acid in the form HA and the fraction in the form A^- at pH = **(a)** 2.00; **(b)** 3.00; **(c)** 3.50.

9-22. The base B has $pK_b = 4.00$. Find the fraction in the form B and the fraction in the form BH^+ at pH = **(a)** 9.00; **(b)** 10.00; **(c)** 10.30.

9-23. The smell (and taste) of fish arise from amine compounds. Suggest a reason why adding lemon juice (which is acidic) reduces the "fishy" smell.

9-24. Cr^{3+} (and most metal ions with charge ≥ 2) is acidic by virtue of the hydrolysis reaction

$$Cr^{3+} + H_2O \rightleftharpoons Cr(OH)^{2+} + H^+ \quad K = 10^{-3.80}$$

[Further reactions that we will neglect produce $Cr(OH)_2^+$, $Cr(OH)_3$, and $Cr(OH)_4^-$.] Find the pH of 0.010 M $Cr(ClO_4)_3$. What fraction of chromium is in the form $Cr(OH)^{2+}$?

9-25. Create a spreadsheet to solve the equation $x^2/(F - x) = K$. The input will be F and K. The output is the positive value of x. Use your spreadsheet to check your answer to Ask Yourself 9-C, part **(a)**.

9-26. Create a spreadsheet using Equations 9-15 and 9-16 to compute the concentrations of HA and A^- in a 0.200 M solution of hydroxybenzene as a function of pH from pH 2 to pH 12. Use graphics software to plot the curves in a beautifully labeled figure.

Measuring pH
Inside Single Cells

R. Kopelman, University of Michigan

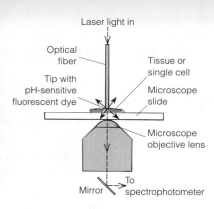

Laser light in

Optical fiber

Tip with pH-sensitive fluorescent dye

Tissue or single cell

Microscope slide

Microscope objective lens

Mirror

To spectrophotometer

Micrograph shows light emitted from the tip of an optical fiber inserted into a rat embryo. The emission spectrum indicates that the pH of fluid surrounding the fiber is 7.50 $\pm$ 0.05 in 10 embryos. Prior to the development of this microscopic method, 1 000 embryos had to be homogenized for a single measurement.

pH controls the rate and thermodynamics of every biological process. To measure pH inside embryos and even single cells, an optical fiber with a pH-sensitive, fluorescent dye bound to its tip can be used. The fiber is inserted into the specimen and laser light is directed down the fiber. Dye molecules at the fiber tip absorb the laser light and then emit fluorescence whose spectrum depends on the pH of the surrounding medium. pH in your blood is closely regulated at 7.40 $\pm$ 0.05 by the buffering action of carbonate, phosphate, and proteins.

Buffers

A buffered solution resists changes in pH when small amounts of acids or bases are added or when dilution occurs. The **buffer** consists of a mixture of an acid and its conjugate base. Biochemists are particularly interested in buffers because the functioning of biological systems is critically dependent on pH. For example, Figure 10-1 shows how the rate of the following enzyme-catalyzed reaction varies with pH:

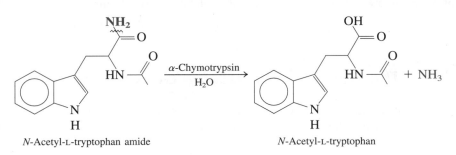

N-Acetyl-L-tryptophan amide   N-Acetyl-L-tryptophan

In the absence of the enzyme α-chymotrypsin, the rate of this reaction is negligible. For any organism to survive, it must control the pH of each subcellular compartment so that each of its enzyme-catalyzed reactions can proceed at the proper rate.

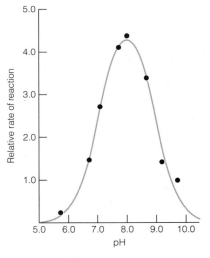

FIGURE 10-1 α-Chymotrypsin catalyzed cleavage of the C—N bond has a maximum rate near pH 8 that is twice as great as the rate near pH 7 or pH 9.

10-1 What You Mix Is What You Get

If you mix A moles of a weak acid with B moles of its conjugate base, the moles of acid remain close to A and the moles of base remain close to B. Little reaction occurs to change either concentration.

    To understand why this should be so, look at the K_a and K_b reactions in terms of Le Châtelier's principle. Consider an acid with $pK_a = 4.00$ and its conjugate base

When you mix a weak acid with its conjugate base, you get what you mix!

165

with $pK_b = 10.00$. We will calculate the fraction of acid that dissociates in a 0.100 M solution of HA.

$$HA \rightleftharpoons H^+ + A^- \qquad pK_a = 4.00$$
$$0.100 - x \qquad x \qquad x$$

$$\frac{x^2}{F - x} = K_a = 1.0 \times 10^{-4} \Rightarrow x = 3.1 \times 10^{-3} \text{ M}$$

$$\text{fraction of dissociation} = \frac{[A^-]}{[A^-] + [HA]} = \frac{x}{F} = 0.031$$

F refers to the formal concentration of HA, which is 0.100 M in this example.

The acid is only 3.1% dissociated under these conditions.

In a solution containing 0.100 mol of A^- dissolved in 1.00 L, the extent of reaction of A^- with water is even smaller:

$$A^- + H_2O \rightleftharpoons HA + OH^- \qquad pK_b = 10.00$$
$$0.100 - x \qquad x \qquad x$$

$$\frac{x^2}{F - x} = K_b = 1.0 \times 10^{-10} \Rightarrow x = 3.2 \times 10^{-6} \text{ M}$$

$$\text{fraction of association} = \frac{[HA]}{[A^-] + [HA]} = \frac{x}{F} = 3.2 \times 10^{-5}$$

The approximation that what you mix is what you get breaks down if the solution is too dilute or the acid or base is too weak. We will not consider these cases.

HA dissociates very little, and Le Châtelier's principle tells us that adding extra A^- to the solution will make the HA dissociate even less. Similarly, A^- does not react very much with water, and adding extra HA makes A^- react even less. If 0.050 mol of A^- plus 0.036 mol of HA are added to water, there will be close to 0.050 mol of A^- and close to 0.036 mol of HA in the solution at equilibrium.

10-2 *The Henderson-Hasselbalch Equation*

The central equation for buffers is the **Henderson-Hasselbalch equation,** which is merely a rearranged form of the K_a equilibrium expression:

$$K_a = \frac{[H^+][A^-]}{[HA]}$$

$$\log K_a = \log\left(\frac{[H^+][A^-]}{[HA]}\right) = \log [H^+] + \log\left(\frac{[A^-]}{[HA]}\right)$$

$$\underbrace{-\log [H^+]}_{pH} = \underbrace{-\log K_a}_{pK_a} + \log\left(\frac{[A^-]}{[HA]}\right)$$

Reminder:

$$\log xy = \log x + \log y$$

Henderson and Hasselbalch recognized that the concentrations $[A^-]$ and $[HA]$ can be set equal to their formal concentrations, and they were among the first people to apply Equation 10-1 to practical problems.

Henderson-Hasselbalch equation for an acid:
$$pH = pK_a + \log\left(\frac{[A^-]}{[HA]}\right) \qquad (10\text{-}1)$$

The Henderson-Hasselbalch equation tells us the pH of a solution, provided we know the ratio of the concentrations of conjugate acid and base, as well as pK_a for the acid.

If a solution is prepared from the weak base B and its conjugate acid, the analogous equation is

Henderson-Hasselbalch equation for a base:

$$pH = pK_a + \log\left(\frac{[B]}{[BH^+]}\right) \quad (10\text{-}2)$$

pK_a applies to *this* acid

where pK_a is the acid dissociation constant of the weak acid BH^+. The important features of Equations 10-1 and 10-2 are (1) that the base (A^- or B) appears in the numerator and (2) pK_a applies to the acid in the denominator.

When [A⁻] = [HA], pH = pK_a

When the concentrations of A^- and HA are equal in Equation 10-1, the log term is 0 because $\log(1) = 0$. Therefore, when $[A^-] = [HA]$, $pH = pK_a$.

$\log 1 = 0$ because $1 = 10^0$

$$pH = pK_a + \log\left(\frac{[A^-]}{[HA]}\right) = pK_a + \log 1 = pK_a$$

Regardless of how complex a solution may be, whenever $pH = pK_a$, $[A^-]$ must equal [HA]. This relation is true because *all equilibria must be satisfied simultaneously in any solution at equilibrium.* If there are 10 different acids and bases in the solution, the 10 forms of Equation 10-1 must all give the same pH, because **there can be only one concentration of H⁺ in a solution.**

When [A⁻]/[HA] Changes by a Factor of 10, the pH Changes by One Unit

Another feature of the Henderson-Hasselbalch equation is that for every power-of-10 change in the ratio $[A^-]/[HA]$, the pH changes by one unit (Table 10-1). As the concentration of base (A^-) increases, the pH goes up. As the concentration of acid (HA) increases, the pH goes down. For any conjugate acid-base pair, you can say, for example, that if $pH = pK_a - 1$, there must be 10 times as much HA as A^-. Therefore ten-elevenths is in the form HA and one-eleventh is in the form A^-.

TABLE 10-1

Change of pH with change of [A⁻]/[HA]

[A⁻]/[HA]	pH
100:1	$pK_a + 2$
10:1	$pK_a + 1$
1:1	pK_a
1:10	$pK_a - 1$
1:100	$pK_a - 2$

EXAMPLE   Using the Henderson-Hasselbalch Equation

Sodium hypochlorite (NaOCl, the active ingredient of bleach) was dissolved in a solution buffered to pH 6.20. Find the ratio $[OCl^-]/[HOCl]$ in this solution.

SOLUTION OCl^- is the conjugate base of hypochlorous acid, HOCl. In Appendix B we find that $pK_a = 7.53$ for HOCl. Knowing the pH, we calculate the ratio $[OCl^-]/[HOCl]$ from the Henderson-Hasselbalch equation:

When we say that the solution is buffered to pH 6.20, we mean that unspecified acids and bases were used to set the pH to 6.20. If buffering is sufficient, then adding a little more acid or base does not change the pH appreciably.

$$HOCl \rightleftharpoons H^+ + OCl^- \qquad pH = pK_a + \log\left(\frac{[OCl^-]}{[HOCl]}\right)$$

$$6.20 = 7.53 + \log\left(\frac{[OCl^-]}{[HOCl]}\right)$$

$$-1.33 = \log\left(\frac{[OCl^-]}{[HOCl]}\right)$$

It doesn't matter how the pH got to 6.20. If we know that pH = 6.20, the Henderson-Hasselbalch equation tells us the ratio [OCl⁻]/[HOCl].

To solve this equation, raise 10 to the power shown on each side:

If $a = b$, then $10^a = 10^b$

$$10^{-1.33} = 10^{\log([OCl^-]/[HOCl])} = \frac{[OCl^-]}{[HOCl]}$$

$$0.047 = \frac{[OCl^-]}{[HOCl]}$$

If your calculator has the *antilog* function instead of the *10^x* function, solve the equation $-1.33 = \log([OCl^-]/[HOCl])$ by computing $[OCl^-]/[HOCl] = \text{antilog}(-1.33)$.

Finding the ratio $[OCl^-]/[HOCl]$ requires only pH and pK_a. We do not care what else is in the solution, how much NaOCl was added, or what the volume of the solution is.

Ask Yourself

10-A. (a) What is the pH of a buffer prepared by dissolving 0.100 mol of the weak acid HA ($K_a = 1.0 \times 10^{-5}$) plus 0.050 mol of its conjugate base Na⁺A⁻ in 1.00 L?

(b) Write the Henderson-Hasselbalch equation for a solution of formic acid, HCO_2H. What is the quotient $[HCO_2^-]/[HCO_2H]$ at pH 3.00, 3.745, and 4.00?

10-3 A Buffer in Action

For illustration, we choose a widely used buffer called "tris."

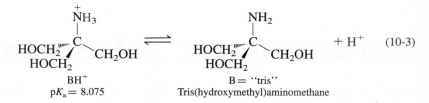

In Appendix B we find pK_a for BH⁺, the conjugate acid of tris, to be 8.075. An example of a salt containing BH⁺ is tris hydrochloride, which is BH⁺Cl⁻. When BH⁺Cl⁻ is dissolved in water, it dissociates completely to BH⁺ and Cl⁻. To find the pH of a known mixture of B and BH⁺, simply plug their concentrations into the Henderson-Hasselbalch equation.

Find the pH of a solution prepared by dissolving 12.43 g of tris (FW 121.136) plus 4.67 g of tris hydrochloride (FW 157.597) in 1.00 L of water.

SOLUTION The concentrations of B and BH^+ added to the solution are

$$[B] = \frac{12.43 \text{ g/L}}{121.136 \text{ g/mol}} = 0.102\ 6 \text{ M} \qquad [BH^+] = \frac{4.67 \text{ g/L}}{157.597 \text{ g/mol}} = 0.029\ 6 \text{ M}$$

Assuming that what we mixed stays in the same form, we insert the concentrations into the Henderson-Hasselbalch equation to find the pH:

$$pH = pK_a + \log\left(\frac{[B]}{[BH^+]}\right) = 8.075 + \log\left(\frac{0.102\ 6}{0.029\ 6}\right) = 8.61$$

> Don't panic over significant figures in pH. For consistency, we are almost always going to express pH to the 0.01 place.

Notice that *the volume of solution is irrelevant to finding the pH,* because volume cancels in the numerator and denominator of the log term:

$$pH = pK_a + \log\left(\frac{\text{moles of B/L of solution}}{\text{moles of } BH^+/\text{L of solution}}\right)$$

$$= pK_a + \log\left(\frac{\text{moles of B}}{\text{moles of } BH^+}\right)$$

> The pH of a buffer is nearly independent of volume.

If strong acid is added to a buffered solution, some of the buffer base will be converted into the conjugate acid and the quotient $[B]/[BH^+]$ changes. If a strong base is added to a buffered solution, some BH^+ is converted into B. Knowing how much strong acid or base is added, we can compute the new quotient $[B]/[BH^+]$ and the new pH.

EXAMPLE   Effect of Adding Acid to a Buffer

If we add 12.0 mL of 1.00 M HCl to the solution in the previous example, what will the new pH be?

SOLUTION The key to this problem is to realize that **when a strong acid is added to a weak base, both react completely to give BH^+** (see Box 10-1). In the present example we are adding 12.0 mL of 1.00 M HCl, which contains $(0.012\ 0 \text{ L})(1.00 \text{ mol/L}) = 0.012\ 0$ mol of H^+. The H^+ consumes 0.012 0 mol of B to create 0.012 0 mol of BH^+:

	B	+	H^+	$\rightarrow$	BH^+
	Tris		From HCl		

	B	H^+	BH^+
Initial moles:	0.102 6	0.012 0	0.029 6
Final moles:	0.090 6	—	0.041 6
	(0.102 6 − 0.012 0)		(0.029 6 + 0.012 0)

The table contains enough information for us to calculate the pH.

Box 10-1 *Explanation*

Strong Plus Weak Reacts Completely

A strong acid reacts with a weak base "completely" because the equilibrium constant is large:

$$\underset{\substack{\text{Weak} \\ \text{base}}}{B} + \underset{\substack{\text{Strong} \\ \text{acid}}}{H^+} \rightleftharpoons BH^+ \qquad K = \frac{1}{K_a} \quad \text{(for } BH^+\text{)}$$

If B is tris, then the equilibrium constant for reaction with HCl is

$$K = \frac{1}{K_a} = \frac{1}{10^{-8.075}} = 1.2 \times 10^8 \leftarrow \text{A big number}$$

A strong base reacts "completely" with a weak acid because the equilibrium constant is, again, very large:

$$\underset{\substack{\text{Strong} \\ \text{base}}}{OH^-} + \underset{\substack{\text{Weak} \\ \text{acid}}}{HA} \rightleftharpoons A^- + H_2O \qquad K = \frac{1}{K_b} \quad \text{(for } A^-\text{)}$$

If HA is acetic acid, then the equilibrium constant for reaction with NaOH is

$$K = \frac{1}{K_b} = \frac{K_a \text{ for HA}}{K_w} = 1.7 \times 10^9 \leftarrow \text{Another big number}$$

The reaction of a strong acid with a strong base is even more complete than a strong plus weak reaction:

$$\underset{\substack{\text{Strong} \\ \text{acid}}}{H^+} + \underset{\substack{\text{Strong} \\ \text{base}}}{OH^-} \rightleftharpoons H_2O \qquad K = \frac{1}{K_w} = 10^{14}$$
$$\underset{\text{Ridiculously big!}}{\uparrow}$$

If you mix a strong acid, a strong base, a weak acid, and a weak base, the strong acid and base will neutralize each other until one is used up. The remaining strong acid or base will then react with the weak base or weak acid.

Question Does the pH change in the right direction when HCl is added?

$$pH = pK_a + \log\left(\frac{\text{moles of B}}{\text{moles of BH}^+}\right) = 8.075 + \log\left(\frac{0.090\ 6}{0.041\ 6}\right) = 8.41$$

The volume of solution is irrelevant.

A buffer resists changes in pH . . .

The preceding example illustrates that *the pH of a buffer does not change very much when a limited amount of a strong acid or base is added.* Addition of 12.0 mL of 1.00 M HCl changed the pH from 8.61 to 8.41. Addition of 12.0 mL of 1.00 M HCl to 1.00 L of unbuffered solution would have lowered the pH to 1.93.

. . . because the buffer "consumes" the added acid or base.

But *why* does a buffer resist changes in pH? **It does so because the strong acid or base is consumed by B or BH⁺.** If you add HCl to tris, B is converted to BH⁺. If you add NaOH, BH⁺ is converted to B. As long as you don't use up the B or BH⁺ by adding too much HCl or NaOH, the log term of the Henderson-Hasselbalch equation does not change very much and the pH does not change very much. Demonstration 10-1 illustrates what happens when the buffer does get used up. The buffer has its maximum capacity to resist changes of pH when pH = pK_a. We will return to this point later.

Ask Yourself

10-B. **(a)** What is the pH of a solution prepared by dissolving 10.0 g tris plus 10.0 g tris hydrochloride in 0.250 L water?

(b) What will the pH be if 10.5 mL of 0.500 M HClO₄ is added to (a)?

(c) What will the pH be if 10.5 mL of 0.500 M NaOH is added to (a)?

Buffers are usually prepared by starting with a measured amount of either a weak acid (HA) or a weak base (B). Then OH^- is added to HA to make a mixture of HA and A^- (a buffer) or H^+ is added to B to make a mixture of B and BH^+ (a buffer).

EXAMPLE   Calculating How to Prepare a Buffer Solution

How many milliliters of 0.500 M NaOH should be added to 10.0 g of tris hydrochloride (BH^+, Equation 10-3) to give a pH of 7.60 in a final volume of 250 mL?

SOLUTION The number of moles of tris hydrochloride in 10.0 g is (10.0 g)/(157.597 g/mol) = 0.063 5. We can make a table to help solve the problem:

Reaction with OH^-	BH^+	+	OH^-	$\rightarrow$	B
Initial moles:	0.063 5		x		—
Final moles:	0.063 5 − x		—		x

The Henderson-Hasselbalch equation allows us to find x, because we know pH and pK_a:

$$pH = pK_a + \log\left(\frac{\text{mol B}}{\text{mol }BH^+}\right)$$

$$7.60 = 8.075 + \log\left(\frac{x}{0.063\ 5 - x}\right)$$

$$-0.475 = \log\left(\frac{x}{0.063\ 5 - x}\right)$$

To solve for x, raise 10 to the power of the terms on both sides, remembering that $10^{\log z} = z$.

$$10^{-0.475} = 10^{\log[x/(0.063\ 5-x)]}$$

$$0.335 = \frac{x}{0.063\ 5 - x} \Rightarrow x = 0.015\ 9 \text{ mol}$$

This many moles of NaOH is contained in

$$\frac{0.015\ 9 \text{ mol}}{0.500 \text{ mol/L}} = 0.031\ 8 \text{ L} = 31.8 \text{ mL}$$

Our calculation tells us to mix 31.8 mL of 0.500 M NaOH with 10.0 g of tris hydrochloride to give a pH of 7.60.

If your calculator has the antilog function, solve the equation $-0.475 = \log z$ by calculating $z = \text{antilog}(-0.475)$.

Preparing a Buffer in Real Life

When you mix calculated quantities of acid and base to make a buffer, the pH *does not* come out exactly as expected. The main reason for the discrepancy is that pH is governed by the *activities* of the conjugate acid-base pair, not by the concentrations. (Activity was discussed in Box 5-2.) If you really want to prepare tris buffer at pH 7.60, you should use a pH electrode to get exactly what you need.

Suppose you wish to prepare 1.00 L of buffer containing 0.100 M tris at pH 7.60. You have solid tris hydrochloride and ~1 M NaOH. Here's how to do it:

1. Weigh out 0.100 mol tris hydrochloride and dissolve it in a beaker containing about 800 mL water.

2. Place a pH electrode in the solution and monitor the pH.

3. Add NaOH solution until the pH is exactly 7.60. The electrode does not respond instantly. Be sure to allow the pH reading to stabilize after each addition of reagent.

4. Transfer the solution to a volumetric flask and wash the beaker a few times. Add the washings to the volumetric flask.

5. Dilute to the mark and mix.

Ionic strength is a measure of the total concentration of ions in a solution. Changing ionic strength changes activities of ionic species such as H^+ and A^-, even though their concentrations are constant. Diluting a buffer with water changes the ionic strength and therefore changes the pH slightly.

The reason for using 800 mL of water in the first step is so that the volume will be close to the final volume during pH adjustment. Otherwise, the pH will change slightly when the sample is diluted to its final volume and the ionic strength changes.

Demonstration 10-1

How Buffers Work

A buffer resists changes in pH because the added acid or base is consumed by the buffer. As the buffer is used up, it becomes less resistant to changes in pH.

In this demonstration, a mixture containing a 10:1 mole ratio of $HSO_3^-:SO_3^{2-}$ is prepared. Because pK_a for HSO_3^- is 7.2, the pH should be approximately

$$pH = pK_a + \log\left(\frac{[SO_3^{2-}]}{[HSO_3^-]}\right) = 7.2 + \log\left(\frac{1}{10}\right) = 6.2$$

We can prepare a table showing how the pH should change as the HSO_3^- reacts:

Percentage of reaction completed	$[SO_3^{2-}]:[HSO_3^-]$	Calculated pH
0	1:10	6.2
90	1:1	7.2
99	1:0.1	8.2
99.9	1:0.01	9.2
99.99	1:0.001	10.2

When formaldehyde is added, the net reaction is the consumption of HSO_3^- but not of SO_3^{2-}.

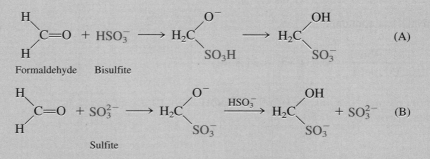

(Sequence A consumes bisulfite. In sequence B, the net reaction is destruction of HSO_3^-, with no change in the SO_3^{2-} concentration.)

(a) (b) (c)

Color Plate 1 HCl Fountain (Demonstration 5-1) (*a*) Basic indicator solution in beaker. (*b*) Indicator is drawn into flask and changes to acidic color. (*c*) Solution levels at end of experiment.

(a) (b) (c)

Color Plate 2 Fajans Titration of Cl^- with $AgNO_3$, Using Dichlorofluorescein (Demonstration 6-1) (*a*) Indicator before beginning titration. (*b*) AgCl precipitate before end point. (*c*) Indicator adsorbed on precipitate after end point.

Color Plate 3   Fe(phenanthroline)$_3^{2+}$ Standards for Spectrophotometric Analysis (Section 7-2; Experiment 21-16) Volumetric flasks containing Fe(phenanthroline)$_3^{2+}$ solutions with iron concentrations ranging from 1 mg/L (left) to 10 mg/L (right).

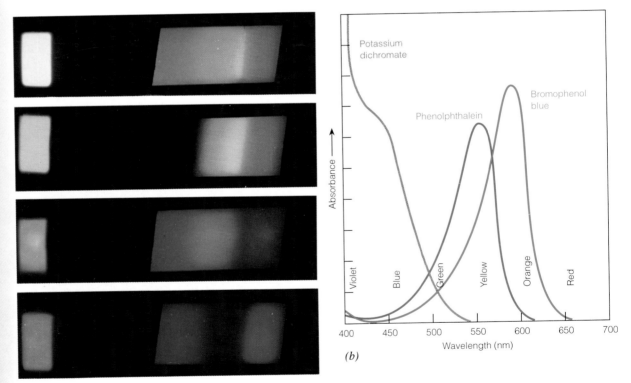

Color Plate 4   Absorption Spectra (Demonstration 7-1) (*a*) Projected visible spectra of (from top to bottom) white light, potassium dichromate, bromophenol blue, and phenolphthalein. (*b*) Absorption spectra recorded with a spectrophotometer.

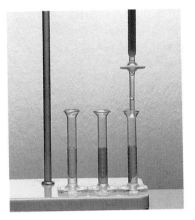

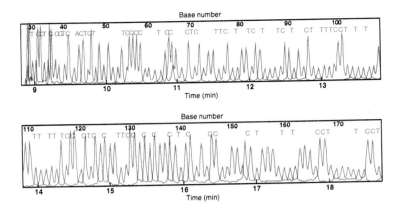

Color Plate 16 **Cation-Exchange Separation of Co(II) and Ni(II) (Section 18-1)** A solution containing Co(II) and Ni(II) in 6 M HCl was applied to the top of the cation-exchange chromatography column. The blue Ni(II) was eluted first and collected in the graduated cylinders. Green Co(II) is retained more strongly and has not yet been eluted from the column.

Color Plate 17 **DNA Sequencing by Capillary Gel Electrophoresis with Fluorescent Labels (Section 18-7)** Tall red peaks correspond to DNA chains terminating in cytosine and short red peaks correspond to thymine. Tall blue peaks arise from adenine and short blue peaks indicate guanine. Two different fluorescent labels and two fluorescence wavelengths were required to generate this information.

(a)   *(b)*   *(c)*

Color Plate 18 **Colloids and Dialysis (Demonstration 19-1)** (*a*) Ordinary aqueous Fe(III) (right) and colloidal Fe(III) (left). (*b*) Dialysis bags containing colloidal Fe(III) (left) and a solution of Cu(II) (right) immediately after placement in flasks of water. (*c*) After 24 h of dialysis, the Cu(II) has diffused out and is dispersed uniformly between the bag and the flask, but the colloidal Fe(III) remains inside the bag.

Color Plate 19 **Titration of VO²⁺ with Potassium Permanganate (Experiment 21-11)** Blue VO^{2+} solution prior to titration (left). Mixture of blue VO^{2+} and yellow VO_2^+ observed during titration (center). Dark color of MnO_4^- at end point (right).

Color Plate 13 Titration of Cu(II) with EDTA, Using Auxiliary Complexing Agent (Section 15-4) 0.02 M $CuSO_4$ before titration (left). Color of Cu(II)-ammonia complex after adding ammonia buffer, pH 10 (center). End-point color when all ammonia ligands have been displaced by EDTA (right). In this case the purpose of the auxiliary complexing agent is not to keep the metal ion in solution but to accentuate the color change at the equivalence point.

Color Plate 14 Iodometric Titration (Section 15-5) (Experiment 21-12) I_3^- solution (left). I_3^- solution before end point in titration with $S_2O_3^{2-}$ (left center). I_3^- solution immediately before end point, with starch indicator present (right center). At the end point (right).

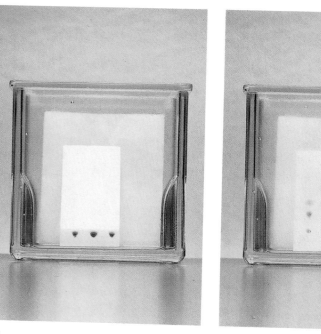

(a)   *(b)*

Color Plate 15 Thin-Layer Chromatography (Section 16-1) *(a)* Solvent ascends past mixture of dyes near bottom of flat plate coated with solid adsorbent. *(b)* Separation achieved after solvent has ascended most of the way up the plate.

Color Plate 5 Grating Dispersion (Section 7-3) Visible spectrum produced by grating inside spectrophotometer.

(a)

(b)

Color Plate 6 Luminescence (Section 8-3) (*a*) Green crystal of yttrium aluminum garnet containing a small amount of Cr^{3+}. (*b*) When irradiated with high-intensity blue light from a laser at the right side, Cr^{3+} absorbs blue light and emits lower energy red light. When the laser is removed, the crystal appears green again.

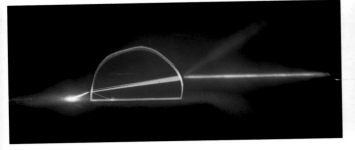

(a)

(b)

Color Plate 7 Transmission, Reflection, Refraction, and Absorption of Light (Section 8-3) (*a*) Blue-green laser is directed into a semicircular crystal of yttrium aluminum garnet containing a small amount of Er^{3+}, which emits yellow light when it absorbs the laser light. Light entering the crystal from the right is refracted (bent) and partially reflected at the right-hand surface of the crystal. The laser beam appears yellow inside the crystal because of luminescence from Er^{3+}. As it exits the crystal at the left side, the laser beam is refracted again, and partially reflected back into the crystal. (*b*) Same experiment, but with blue light instead of blue-green light. The blue light is absorbed by Er^{3+} and does not penetrate very far into the crystal.

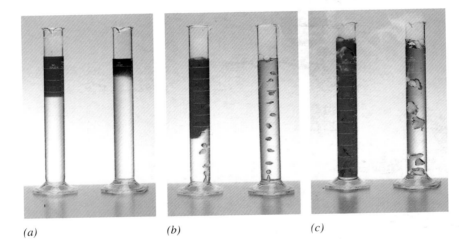

(a)   (b)   (c)

(d)   (e)

Color Plate 8 Indicators and Acidity of CO_2 (Demonstration 10-2) (*a*) Cylinder before adding Dry Ice. Ethanol indicator solutions of phenolphthalein (left) and bromothymol blue (right) have not yet mixed with entire cylinder. (*b*) Adding Dry Ice causes bubbling and mixing. (*c*) Further mixing. (*d*) Phenolphthalein changes to its colorless acidic form. Color of bromothymol blue is due to mixture of acidic and basic forms. (*e*) After addition of HCl and stirring of right-hand cylinder, bubbles of CO_2 can be seen leaving solution, and indicator changes completely to its acidic color.

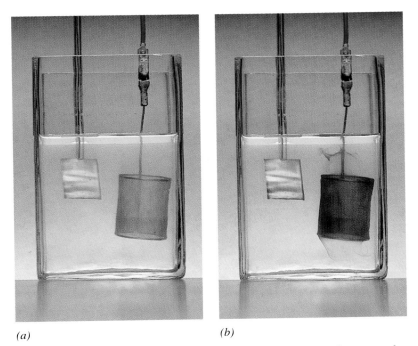

(a)   *(b)*

Color Plate 11   **Electrolysis of I⁻ in Solution to Make I₂ at the Anode (Demonstration 13-1)** *(a)* Cu electrode (flat plate, left) and Pt electrode (mesh basket, right) immersed in solution containing KI and starch, with no electric current. *(b)* I_3^--starch complex forms at surface of Pt anode when current flows.

(a)   *(b)*

Color Plate 12   **Titration of Mg²⁺ by EDTA Using Eriochrome Black T Indicator (Demonstration 15-1)** *(a)* Before (left), near (center), and after (right) equivalence point. *(b)* Same titration with methyl red added as inert dye to alter colors.

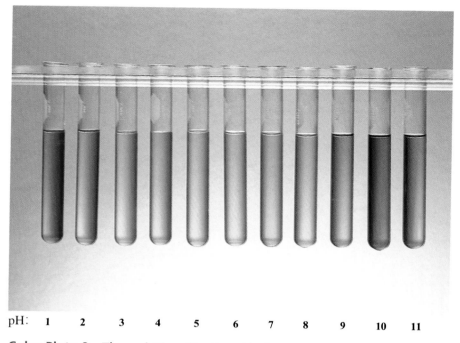

pH: 1 2 3 4 5 6 7 8 9 10 11

Color Plate 9   Thymol Blue (Section 10-6) Acid-base indicator thymol blue between pH 1 (left) and 11 (right). The pK values are 1.7 and 8.9.

(a)

(b)

(c)

Color Plate 10   Electrochemical Writing (Demonstration 13-1) (*a*) Stylus used as cathode. (*b*) Stylus used as anode. (*c*) Foil backing has a polarity opposite that of the stylus and produces reverse color on the bottom sheet of paper.

Ask Yourself

10-C. (a) How many milliliters of 1.20 M HCl should be added to 10.0 g of tris (B, Equation 10-3) to give a pH of 7.60 in a final volume of 250 mL?

   (b) What operations would you perform to prepare exactly 100 mL of 0.200 M acetate buffer, pH 5.00, starting with pure liquid acetic acid and solutions containing ~3 M HCl and ~3 M NaOH?

10-5 *Buffer Capacity*

Buffer capacity measures how well a solution resists changes in pH when acid or base is added. The greater the buffer capacity, the less the pH changes. We will see in this section that *the maximum buffer capacity occurs when pH = pK_a for the buffer.*

The greater the buffer capacity, the less the pH changes when H^+ or OH^- is added. Buffer capacity is maximum when pH = pK_a.

You can see that through 90% completion the pH rises by just one unit. In the next 9% of the reaction, the pH rises by another unit. At the end of the reaction, the change in pH is very abrupt.

In the formaldehyde clock reaction, formaldehyde is added to a solution containing HSO_3^-, SO_3^{2-}, and phenolphthalein indicator. Phenolphthalein is colorless below pH 8.5 and red above this pH. What is observed is that the solution remains colorless for more than a minute. Suddenly the pH shoots up and the liquid turns pink. Monitoring the pH with a glass electrode gave the results shown in the figure.

Procedure: Phenolphthalein solution is prepared by dissolving 50 mg of solid indicator in 50 mL of ethanol and diluting with 50 mL of water. The following solutions should be fresh: Dilute 9 mL of 37 wt % formaldehyde to 100 mL. Dissolve 1.5 g $NaHSO_3$ and 0.18 g Na_2SO_3 in 400 mL of water and add 1 mL of phenolphthalein indicator solution. To initiate the clock reaction, add 23 mL of formaldehyde solution to the well-stirred buffer solution. Reaction time is affected by the temperature, concentrations, and volume.

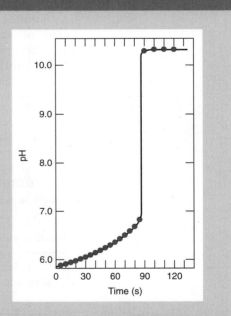

Graph of pH versus time in the formaldehyde clock reaction.

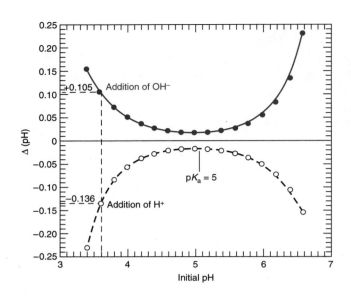

FIGURE 10-2 *Buffer capacity:* Effect of adding 0.01 mol of H^+ or OH^- to a buffer containing HA and A^- (total quantity of HA + $A^- = 1$ mol). The minimum change in pH occurs when the initial pH of the buffer equals pK_a for HA. That is, the maximum buffer capacity occurs when $pH = pK_a$.

Figure 10-2 shows an experiment conducted with a spreadsheet to see how a buffer responds to small additions of H^+ or OH^-. A buffer was made by mixing HA (with acid dissociation constant $K_a = 10^{-5}$) with A^-. The total moles of HA + A^- were set equal to 1. The relative quantities of HA and A^- were varied to give initial pH values from 3.4 to 6.6. Then 0.01 mol of either H^+ or OH^- was added to the solution and the new pH was calculated. Figure 10-2 shows the change in pH, that is, $\Delta(\text{pH})$, as a function of the initial pH of the buffer.

For example, mixing 0.038 3 mol of A^- and 0.961 7 mol of HA gives an initial pH of 3.600:

$$pH = pK_a + \log\left(\frac{\text{mol A}^-}{\text{mol HA}}\right) = 5.000 + \log\left(\frac{0.038\ 3}{0.961\ 7}\right) = 3.600$$

When 0.010 0 mol of OH^- is added to this mixture, the concentrations and pH change in a manner that you are now smart enough to calculate:

Reaction with OH^-:	HA	+	OH^-	→	A^-
Initial moles:	0.961 7		0.010 0		0.038 3
Final moles:	0.961 7 − 0.010 0		—		0.038 3 + 0.010 0
	0.951 7				0.048 3

$$pH = pK_a + \log\left(\frac{\text{mol A}^-}{\text{mol HA}}\right) = 5.000 + \log\left(\frac{0.048\ 3}{0.951\ 7}\right) = 3.705$$

The change in pH is $\Delta(\text{pH}) = 3.705 - 3.600 = +0.105$. This is the value plotted in Figure 10-2 on the upper curve for an initial pH of 3.600. If the same starting mixture had been treated with 0.010 0 mol of H^+, the pH would change by -0.136, which is shown on the lower curve.

We see in Figure 10-2 that the magnitude of the pH change is smallest when the initial pH is equal to pK_a for the buffer. That is, the buffer capacity is greatest when $pH = pK_a$.

In choosing a buffer for an experiment, you should *seek one whose pK_a is as close as possible to the desired pH. The useful pH range of a buffer is usually considered to be $pK_a \pm 1$ pH unit.* Outside this range, there is not enough of either the weak acid or the weak base to react with added base or acid. Clearly, the buffer capacity can be increased by increasing the concentration of the buffer.

Choose a buffer whose pK_a is close to the desired pH.

Table 10-2 lists pK_a values for common buffers. Some of the buffers have more than one acidic proton, so more than one pK is listed. We will consider polyprotic acids and bases in Chapter 12.

Buffer pH Depends on Temperature and Ionic Strength

Most buffers exhibit a noticeable dependence of pK_a on temperature. Tris has an exceptionally large dependence, approximately -0.031 pK_a units per degree, near 25°C. A solution of tris made up to pH 8.08 at 25°C will have $pH \approx 8.7$ at 4°C and $pH \approx 7.7$ at 37°C. When a 0.5 M stock solution of phosphate buffer at pH 6.6 is diluted to 0.05 M, the pH rises to 6.9, because the ionic strength and activities of the buffering species, $H_2PO_4^-$ and HPO_4^{2-}, change.

Summary

A buffer consists of a mixture of a weak acid and its conjugate base. The buffer is most useful when $pH \approx pK_a$. Over a reasonable range of concentrations, the pH of a buffer is nearly independent of concentration. A buffer resists changes in pH because it reacts with added acids or bases. If too much acid or base is added, the buffer will be consumed and no longer resists changes in pH.

Ask Yourself

10-D. (a) Look up pK_a for each of the following acids and decide which would be best for preparing a buffer of pH 3.10:
> **(i)** hydroxybenzene   **(iii)** cyanoacetic acid
> **(ii)** propanoic acid   **(iv)** sulfuric acid

(b) Why does buffer capacity increase as the concentration of buffer increases?

(c) Based on the K_b values listed below, which of the following bases would be best for preparing a buffer of pH 9.00?
> **(i)** NH_3 (ammonia, $K_b = 1.75 \times 10^{-5}$)
> **(ii)** $C_6H_5NH_2$ (aniline, $K_b = 3.99 \times 10^{-10}$)
> **(iii)** H_2NNH_2 (hydrazine, $K_b = 3.0 \times 10^{-6}$)
> **(iv)** C_5H_5N (pyridine, $K_b = 1.69 \times 10^{-9}$)

TABLE 10-2   Structures and pK_a values for common buffers

Name	Structure[a]	pK_a ($\sim$ 25°C)	Formula weight
Phosphoric acid	H_3PO_4	2.15 (pK_1)	97.995
Citric acid	$\begin{matrix} & \text{OH} \\ & | \\ HO_2CCH_2 & CCH_2CO_2H \\ & | \\ & CO_2H \end{matrix}$	3.13 (pK_1)	192.125
Citric acid	$H_2(\text{citrate})^-$	4.76 (pK_2)	192.125
Acetic acid	CH_3CO_2H	4.76	60.053
2-(N-Morpholino)ethanesulfonic acid (MES)	$O\!\!-\!\!\overset{+}{N}HCH_2CH_2SO_3^-$	6.15	195.240
Citric acid	$H(\text{citrate})^{2-}$	6.40 (pK_3)	192.125
Piperazine-N,N'-bis(2-ethanesulfonic acid) (PIPES)	$^-O_3SCH_2CH_2\overset{+}{N}H\!\!-\!\!\overset{+}{H}NCH_2CH_2SO_3^-$	6.80	302.373
3-(N-Morpholino)-2-hydroxypropanesulfonic acid (MOPSO)	$O\!\!-\!\!\overset{+}{N}HCH_2\overset{\text{OH}}{\underset{}{C}}HCH_2SO_3^-$	6.95	225.265
Imidazole hydrochloride	$\overset{+}{H}N\!\!-\!\!\text{(imidazole ring)}\!\!-\!\!N H \quad Cl^-$	6.99	104.539
Phosphoric acid	$H_2PO_4^-$	7.20 (pK_2)	97.995
N-2-Hydroxyethylpiperazine-N'-2-ethanesulfonic acid (HEPES)	$HOCH_2CH_2N\!\!-\!\!\overset{+}{N}HCH_2CH_2SO_3^-$	7.56	238.308
Tris(hydroxymethyl)aminomethane hydrochloride (tris hydrochloride)	$(HOCH_2)_3C\overset{+}{N}H_3 \quad Cl^-$	8.08	157.597
Glycylglycine	$H_3\overset{+}{N}CH_2\overset{O}{\overset{\|}{C}}NHCH_2CO_2^-$	8.40	132.119
Boric acid	$B(OH)_3$	9.24 (pK_1)	61.833
Cyclohexylaminoethanesulfonic acid (CHES)	$\text{(cyclohexyl)}\!\!-\!\!\overset{+}{N}H_2CH_2CH_2SO_3^-$	9.50	207.294
3-(Cyclohexylamino)propanesulfonic acid (CAPS)	$\text{(cyclohexyl)}\!\!-\!\!\overset{+}{N}H_2CH_2CH_2CH_2SO_3^-$	10.40	221.321
Phosphoric acid	HPO_4^{2-}	12.15 (pK_3)	97.995
Boric acid	$OB(OH)_2^-$	12.74 (pK_2)	61.833

a. The protonated form of each molecule is shown. Acidic hydrogen atoms are shown in **bold** type.

10-6 How Indicators Work

An acid-base **indicator** is itself an acid or base whose various protonated species have different colors. In the next chapter we will learn how to choose an indicator to find the end point for a titration. For now, let's try to use the Henderson-Hasselbalch equation to understand the pH range over which color changes are observed.

Consider bromocresol green as an example. We will call pK_a of the indicator pK_{HIn} to distinguish it from pK_a of the acid being titrated in the next chapter.

An indicator is an acid or a base whose various protonated forms have different colors.

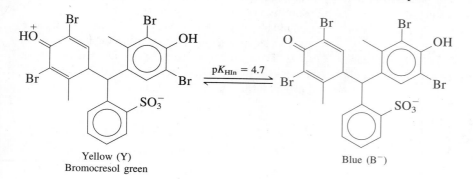

Yellow (Y)
Bromocresol green

Blue (B⁻)

pK_{HIn} for bromocresol green is 4.7. Below pH 4.7, the predominant species is yellow (Y); above pH 4.7, the predominant species is blue (B⁻).

The equilibrium between Y and B⁻ can be written

$$Y \rightleftharpoons B^- + H^+ \qquad K_{HIn} = \frac{[B^-][H^+]}{[Y]}$$

for which the Henderson-Hasselbalch equation is

$$pH = pK_{HIn} + \log\left(\frac{[B^-]}{[Y]}\right) \qquad (10\text{-}4)$$

TABLE 10-3 Common indicators

Indicator	Transition range (pH)	Acid color	Base color	Indicator	Transition range (pH)	Acid color	Base color
Methyl violet	0.0–1.6	Yellow	Violet	Litmus	5.0–8.0	Red	Blue
Cresol red	0.2–1.8	Red	Yellow	Bromothymol blue	6.0–7.6	Yellow	Blue
Thymol blue	1.2–2.8	Red	Yellow	Phenol red	6.4–8.0	Yellow	Red
Cresol purple	1.2–2.8	Red	Yellow	Neutral red	6.8–8.0	Red	Yellow
Erythrosine, disodium	2.2–3.6	Orange	Red	Cresol red	7.2–8.8	Yellow	Red
				α-Naphtholphthalein	7.3–8.7	Pink	Green
Methyl orange	3.1–4.4	Red	Yellow	Cresol purple	7.6–9.2	Yellow	Purple
Congo red	3.0–5.0	Violet	Red	Thymol blue	8.0–9.6	Yellow	Blue
Ethyl orange	3.4–4.8	Red	Yellow	Phenolphthalein	8.0–9.6	Colorless	Red
Bromocresol green	3.8–5.4	Yellow	Blue	Thymolphthalein	8.3–10.5	Colorless	Blue
Methyl red	4.8–6.0	Red	Yellow	Alizarin yellow	10.1–12.0	Yellow	Orange-red
Chlorophenol red	4.8–6.4	Yellow	Red	Nitramine	10.8–13.0	Colorless	Orange-brown
Bromocresol purple	5.2–6.8	Yellow	Purple	Tropaeolin O	11.1–12.7	Yellow	Orange
p-Nitrophenol	5.6–7.6	Colorless	Yellow				

Indicators and the Acidity of CO₂

This one is just plain fun. Fill two 1-L graduated cylinders with 900 mL of water and place a magnetic stirring bar in each. Add 10 mL of 1 M NH₃ to each. Then put 2 mL of phenolphthalein indicator solution in one and 2 mL of bromothymol blue indicator solution in the other. Both indicators will have the color of their basic species.

Drop a few chunks of Dry Ice (solid CO₂) into each cylinder. As the CO₂ bubbles through each cylinder, the solutions become more acidic.

First the pink phenolphthalein color disappears. After some time, the pH drops just low enough for bromothymol blue to change from blue to its pale green intermediate color. The pH does not go low enough to turn bromothymol blue to its yellow color.

Add about 20 mL of 6 M HCl to *the bottom* of each cylinder, using a length of Tygon tubing attached to a funnel. Then stir each solution for a few seconds on a magnetic stirrer. Explain what happens. The sequence of events is shown in Color Plate 8.

$$CO_2 + H_2O \rightleftharpoons H_2CO_3$$
Carbonic acid

pH	[B⁻]:[Y]	Color
3.7	1:10	Yellow
4.7	1:1	Green
5.7	10:1	**Blue**

At pH = pK_a = 4.7, there will be a 1:1 mixture of the yellow and blue species, which will appear green. As a crude rule of thumb, we can say that the solution will appear yellow when [Y]/[B⁻] ≳ 10/1 and blue when [B⁻]/[Y] ≳ 10/1. (The symbol ≳ means "is approximately greater than.") From Equation 10-4, we can see that the solution will be yellow when pH ≲ pK_{HIn} − 1 (= 3.7) and blue when pH ≳ pK_{HIn} + 1 (= 5.7). By comparison, Table 10-3 lists bromocresol green as yellow below pH 3.8 and blue above pH 5.4. Between pH 3.8 and 5.4, various shades of green are seen. Demonstration 10-2 illustrates indicator color changes.

Several indicators are listed twice in Table 10-3 with two different sets of colors. An example is thymol blue, which loses one proton with a pK of 1.7 and a second proton with a pK of 8.9:

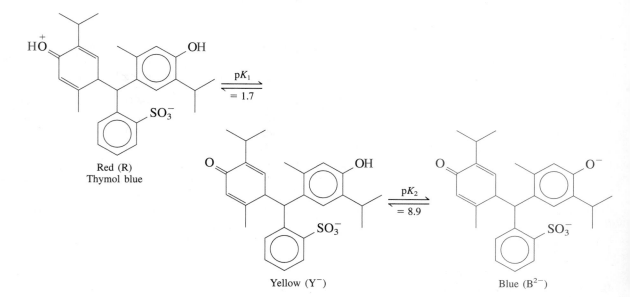

Red (R)
Thymol blue

Yellow (Y⁻)

Blue (B²⁻)

Below pH 1.7, the predominant species is red (R); between pH 1.7 and pH 8.9, the predominant species is yellow (Y^-); and above pH 8.9, the predominant species is blue (B^{2-}). The sequence of color changes for thymol blue is shown in Color Plate 9. The equilibrium between R and Y^- can be written

$$R \rightleftharpoons Y^- + H^+ \qquad K_1 = \frac{[Y^-][H^+]}{[R]}$$

$$pH = pK_1 + \log\left(\frac{[Y^-]}{[R]}\right) \qquad (10\text{-}5)$$

At pH = 1.7 ($= pK_1$), there will be a 1:1 mixture of the yellow and red species, which will appear orange. We crudely expect that the solution will appear red when $[R]/[Y^-] \gtrsim 10/1$ and yellow when $[Y^-]/[R] \gtrsim 10/1$. From Henderson-Hasselbalch equation 10-5, we can see that the solution will be red when pH $\leqslant pK_1 - 1$ ($= 0.7$) and yellow when pH $\geqslant pK_1 + 1$ ($= 2.7$). In Table 10-3, thymol blue is listed as red below pH 1.2 and yellow above pH 2.8. Thymol blue undergoes another transition, from yellow to blue, between pH 8.0 and pH 9.6. In this range, various shades of green are seen.

pH	$[Y^-]$:$[R]$	Color
0.7	1:10	Red
1.7	1:1	Orange
2.7	10:1	Yellow

Ask Yourself

10-E. (a) What is the origin of the rule of thumb that indicator color changes occur at $pK_{HIn} \pm 1$?

(b) What color do you expect to observe for cresol purple indicator (Table 10-3) at the following pH values: 1.0, 2.0, 3.0?

Key Equations

Henderson-Hasselbalch

$$pH = pK_a + \log\left(\frac{[A^-]}{[HA]}\right)$$

$$pH = pK_a + \log\left(\frac{[B]}{[BH^+]}\right) \quad \text{← } pK_a \text{ applies to } \textit{this} \text{ acid}$$

Important Terms

buffer   Henderson-Hasselbalch equation   indicator

Problems

10-1. A solution contains 63 different conjugate acid-base pairs. Among them is acrylic acid and acrylate ion, with the ratio [acrylate]/[acrylic acid] = 0.75. What is the pH of the solution?

$$H_2C=CHCO_2H \qquad pK_a = 4.25$$
Acrylic acid

10-2. Table 10-1 shows the relation of pH to the quotient $[A^-]/[HA]$.

(a) Use the Henderson-Hasselbalch equation to show that pH = pK_a + 2 when $[A^-]/[HA]$ = 100.

(b) Find the quotient $[A^-]/[HA]$ when pH = pK_a − 3.

(c) Find the pH when $[A^-]/[HA]$ = 10^{-4}.

10-3. Given that pK_b for iodate ion (IO_3^-) is 13.83, find the quotient $[HIO_3]/[IO_3^-]$ in a solution of sodium iodate at (a) pH 7.00; (b) pH 1.00.

10-4. Given that pK_b for nitrite ion (NO_2^-) is 10.85, find the quotient $[HNO_2]/[NO_2^-]$ in a solution of sodium nitrite at (a) pH 2.00; (b) pH 10.00.

10-5. Write the Henderson-Hasselbalch equation for a solution of methylamine. Calculate the quotient $[CH_3NH_2]/[CH_3NH_3^+]$ at (a) pH 4.00; (b) pH 10.64; (c) pH 12.00.

10-6. Why is the pH of a buffer nearly independent of concentration?

10-7. Find the pH of a solution prepared from 2.53 g of oxoacetic acid, 5.13 g of potassium oxoacetate, and 103 g of water. Use Appendix B to find pK_a for oxoacetic acid.

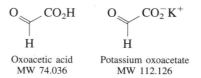

Oxoacetic acid   Potassium oxoacetate
MW 74.036   MW 112.126

10-8. (a) Find the pH of a solution prepared by dissolving 1.00 g of glycine amide hydrochloride plus 1.00 g of glycine amide in 0.100 L.

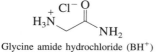

Glycine amide hydrochloride (BH^+)
MW 110.543, pK_a = 8.20

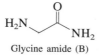

Glycine amide (B)
MW 74.083

(b) How many grams of glycine amide should be added to 1.00 g of glycine amide hydrochloride to give 100 mL of solution with pH 8.00?

(c) What would be the pH if the solution in (a) is mixed with 5.00 mL of 0.100 M HCl?

(d) What would be the pH if the solution in (c) is mixed with 10.00 mL of 0.100 M NaOH?

10-9. (a) Write the chemical reactions whose equilibrium constants are K_b and K_a for imidazole and imidazole hydrochloride, respectively.

(b) Calculate the pH of a 100-mL solution containing 1.00 g of imidazole (MW 68.078) and 1.00 g of imidazole hydrochloride (MW 104.539).

(c) Calculate the pH of the solution if 2.30 mL of 1.07 M $HClO_4$ is added to the solution.

(d) How many milliliters of 1.07 M $HClO_4$ should be added to 1.00 g of imidazole to give a pH of 6.993?

10-10. (a) Calculate the pH of a solution prepared by mixing 0.080 0 mol of chloroacetic acid plus 0.040 0 mol of sodium chloroacetate in 1.00 L of water.

(b) Using first your head, and then the Henderson-Hasselbalch equation, find the pH of a solution prepared by dissolving all of the following in a total volume of 1.00 L: 0.180 mol $ClCH_2CO_2H$, 0.020 mol $ClCH_2CO_2Na$, 0.080 mol HNO_3, and 0.080 mol $Ca(OH)_2$. Assume that $Ca(OH)_2$ dissociates completely.

10-11. How many milliliters of 0.246 M HNO_3 should be added to 213 mL of 0.006 66 M ethylamine to give a pH of 10.52?

10-12. Calculate how many milliliters of 0.626 M KOH should be added to 5.00 g of HEPES (Table 10-2) to give a pH of 7.40.

10-13. How many milliliters of 0.113 M HBr should be added to 52.2 mL of 0.013 4 M morpholine to give a pH of 8.00?

10-14. Which buffer system will have the greatest buffer capacity at pH 9.0?

(i) dimethylamine / dimethylammonium ion

(ii) ammonia / ammonium ion

(iii) hydroxylamine / hydroxylammonium ion

(iv) 4-nitrophenol / 4-nitrophenolate ion

10-15. (a) Describe the correct procedures for preparing 0.250 L of 0.050 0 M HEPES (Table 10-2), pH 7.45.

(b) Would you need NaOH or HCl to bring the pH to 7.45?

10-16. (a) Calculate how many milliliters of 0.100 M HCl should be added to how many grams of sodium acetate dihydrate (NaOAc · 2H$_2$O, FW 118.06) to prepare 250.0 mL of 0.100 M buffer, pH 5.00.

(b) If you mixed what you calculated, the pH would not be 5.00. Describe how you would actually prepare this buffer in the lab.

10-17. Consider the buffer in Figure 10-2 whose initial pH is 6.200 and whose pK_a is 5.000. It is made from p mol of HA plus q mol of A$^-$ such that $p + q = 1$ mol.

(a) Setting mol HA $= q$ and mol A$^- = 1 - q$, use the Henderson-Hasselbalch equation to find the values of p and q.

(b) Calculate the pH change when 0.010 0 mol of H$^+$ is added to the buffer. Did you get the same value shown in Figure 10-2? For the sake of this problem, use three decimal places for pH.

10-18. Cresol red has *two* transition ranges listed in Table 10-3. What color would you expect it to be at the following pH values?

    (a) 0   **(b)** 1   **(c)** 6   **(d)** 9

10-19. *Spectrophotometry with indicators.* Consider an indicator, HIn, which dissociates according to the equation

$$\text{HIn} \xrightleftharpoons{K_{\text{HIn}}} \text{H}^+ + \text{In}^-$$

Suppose that the molar absorptivity, ϵ, is 2 080 M^{-1} cm^{-1} for HIn and 14 200 M^{-1} cm^{-1} for In$^-$, at a wavelength of 440 nm.

(a) Use Beer's law to write an expression for the absorbance at 440 nm of a solution containing HIn at a concentration [HIn] and In$^-$ at a concentration [In$^-$]. Assume the cell pathlength is 1.00 cm. Note that absorbance is additive. The absorbance is the sum of absorbances of both components.

(b) A 1.84 $\times$ 10^{-4} M solution of HIn is adjusted to pH 6.23 to give a mixture of HIn and In$^-$ with a total concentration of 1.84 $\times$ 10^{-4} M. The measured absorbance of the solution at 440 nm is 0.868. Using your expression from **(a)** and the fact that [HIn] + [In$^-$] = 1.84 $\times$ 10^{-4} M, calculate pK_{HIn} for this indicator.

10-20. *Henderson-Hasselbalch spreadsheet.* Prepare a spreadsheet with the constant pK_a = 4 in column A. Type the heading [A$^-$]/[HA] for column B and enter values ranging from 0.001 to 1 000, with a selection of values in between. In column C, compute pH from the Henderson-Hasselbalch equation using pK_a from column A and [A$^-$]/[HA] from column B. In column D, compute log([A$^-$]/[HA]). Use the values from columns C and D to prepare a graph of pH versus log([A$^-$]/[HA]). Explain the shape of the resulting graph.

World Record Small Titration

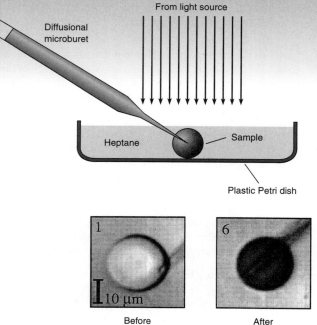

From light source

Diffusional microburet

Heptane

Sample

Plastic Petri dish

1

$\mathbf{I}$10 μm

Before end point

6

After end point

Titration of femtoliter samples. The frames at the bottom show an 8.7-pL droplet before and after the end point; the microburet is making contact at the right side.

The titration of 29 fmol (f = femto = 10^{-15}) of HNO_3 in a 1.9-pL (p = pico = 10^{-12}) water drop under a layer of hexane in a Petri dish was carried out with 2% precision by using KOH delivered by diffusion from the tip (1 μm in diameter) of a glass capillary tube. A mixture of the indicators bromothymol blue and bromocresol purple, with a distinct yellow-to-purple transition at pH 6.7, allowed the end point to be observed through a microscope by a video camera.

Acid-Base Titrations

$\mathbf{A}$s described in Chapter 6, a titration is a procedure in which we measure the quantity of a known reagent required to react with an unknown sample. From this quantity, we deduce the concentration of analyte in the unknown. Titrations involving the reaction between an acid and a base are among the most widespread in chemical analysis. For example, at the end of this chapter, we will see how acid-base titrations are used to measure the nitrogen content of foods.

11-1 Titration of Strong Acid with Strong Base

For each type of titration in this chapter, *our goal is to construct a graph showing how the pH changes as titrant is added.* If you can do this, then you understand what is happening during the titration, and you will be able to interpret an experimental titration curve.

The first step is to write the chemical reaction between titrant and analyte. Then use that reaction to calculate the composition and pH after each addition of titrant. Let's focus on the titration of 50.00 mL of 0.020 00 M KOH with 0.100 0 M HBr. The chemical reaction between titrant and analyte is merely

First write the reaction between *titrant* and *analyte.*

$$H^+ + OH^- \longrightarrow H_2O \qquad K = \frac{1}{K_w} = \frac{1}{10^{-14}} = 10^{14}$$

The titration reaction

Because the equilibrium constant for this reaction is 10^{14}, it is fair to say that it "goes to completion." Prior to the equivalence point, *any amount of H^+ added will consume a stoichiometric amount of OH^-.*

Equivalence point: when moles of added titrant are exactly sufficient for stoichiometric reaction with analyte.

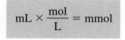

$$mL \times \frac{mol}{L} = mmol$$

A useful starting point is to calculate the volume of HBr (V_e) needed to reach the equivalence point:

$$\underbrace{(V_e(mL))\,(0.100\,0\,M)}_{\substack{\text{mmol of HBr} \\ \text{at equivalence point}}} = \underbrace{(50.00\,mL)\,(0.020\,00\,M)}_{\substack{\text{mmol of OH}^- \\ \text{being titrated}}} \Rightarrow V_e = 10.00\,mL$$

When 10.00 mL of HBr have been added, the titration is complete. Prior to V_e, excess, unreacted OH$^-$ is present. After V_e, there is excess H$^+$ in the solution.

In the titration of a strong base with a strong acid, three regions of the titration curve require different kinds of calculations:

1. Before the equivalence point, the pH is determined by excess OH$^-$ in the solution.

2. At the equivalence point, added H$^+$ is just sufficient to react with all of the OH$^-$ to make H$_2$O. The pH is determined by the dissociation of water.

3. After the equivalence point, pH is determined by excess H$^+$ in the solution.

We will do one sample calculation for each region. Complete results are shown in Table 11-1 and Figure 11-1.

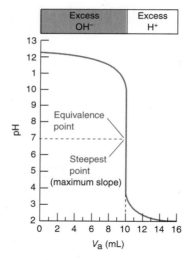

FIGURE 11-1 Calculated titration curve showing how pH changes as 0.100 0 M HBr is added to 50.00 mL of 0.020 00 M KOH. At the equivalence point, the curve is steepest; in other words, the first derivative reaches a maximum and the second derivative is 0. We will discuss derivatives of the titration curve as a method for finding the end point later in this chapter.

TABLE 11-1 Calculation of the titration curve for 50.00 mL of 0.020 00 M KOH treated with 0.100 0 M HBr

	mL HBr added (V_a)	Concentration of unreacted OH$^-$ (M)	Concentration of excess H$^+$ (M)	pH
	0.00	0.020 0		12.30
	1.00	0.017 6		12.24
	2.00	0.015 4		12.18
	3.00	0.013 2		12.12
	4.00	0.011 1		12.04
Region 1	5.00	0.009 09		11.95
(excess OH$^-$)	6.00	0.007 14		11.85
	7.00	0.005 26		11.72
	8.00	0.003 45		11.53
	9.00	0.001 69		11.22
	9.50	0.000 840		10.92
	9.90	0.000 167		10.22
	9.99	0.000 016 6		9.22
Region 2	10.00	—	—	7.00
	10.01		0.000 016 7	4.78
	10.10		0.000 166	3.78
	10.50		0.000 826	3.08
Region 3	11.00		0.001 64	2.79
(excess H$^+$)	12.00		0.003 23	2.49
	13.00		0.004 76	2.32
	14.00		0.006 25	2.20
	15.00		0.007 69	2.11
	16.00		0.009 09	2.04

184

Region 1: Before the Equivalence Point

Before adding any HBr titrant from the buret, the flask of analyte contains 50.00 mL of 0.020 00 M KOH, which amounts to $(50.00 \text{ mL})(0.020\ 00 \text{ M}) = 1.000$ mmol of OH^-. Remember that mL × (mol/L) = mmol.

If we add 3.00 mL of HBr from the buret, we are adding $(3.00 \text{ mL})(0.100\ 0 \text{ M}) = 0.300$ mmol of H^+, which consumes 0.300 mmol of OH^-.

$$OH^- \text{ remaining} = \underbrace{1.000 \text{ mmol}}_{\text{Initial } OH^-} - \underbrace{0.300 \text{ mmol}}_{\substack{OH^- \text{ consumed} \\ \text{by HBr}}} = 0.700 \text{ mmol}$$

The total volume in the flask is now 50.00 mL + 3.00 mL = 53.00 mL. Therefore the concentration of OH^- in the flask is

$$[OH^-] = \frac{0.700 \text{ mmol}}{53.00 \text{ mL}} = 0.013\ 2 \text{ M}$$

From the concentration of OH^-, it is easy to find the pH:

$$[H^+] = \frac{K_w}{[OH^-]} = \frac{1.0 \times 10^{-14}}{0.013\ 2} = 7.5_8 \times 10^{-13} \text{ M} \Rightarrow pH = 12.12$$

(If you were crazy) you could reproduce all of the calculations prior to the equivalence point in Table 11-1 in the same manner as the 3.00-mL point. The volume of acid added in Table 11-1 is designated V_a, and pH is expressed to the 0.01 decimal place, regardless of what is justified by significant figures. We do this for the sake of consistency and also because 0.01 is near the limit of accuracy in pH measurements.

Region 2: At the Equivalence Point

At the equivalence point, enough H^+ has been added to react with all the OH^-. We could prepare the same solution by dissolving KBr in water. The pH is determined by the dissociation of water:

$$\underset{x \quad\quad x}{H_2O \rightleftharpoons H^+ + OH^-}$$

$$K_w = x^2 = 1.0 \times 10^{-14} \Rightarrow x = 1.0 \times 10^{-7} \text{ M} \Rightarrow pH = 7.00$$

The pH at the equivalence point in the titration of any strong base (or acid) with strong acid (or base) will be 7.00 at 25°C.

As we will soon discover, *the pH is **not** 7.00 at the equivalence point in the titration of weak acids or bases*. The pH is 7.00 only if the titrant and analyte are both strong.

Region 3: After the Equivalence Point

Beyond the equivalence point, excess HBr is present. For example, at the point where 10.50 mL of HBr have been added, there is an excess of 10.50 − 10.00 = 0.50 mL.

11-1 Titration of Strong Acid with Strong Base

Before the equivalence point, there is excess OH^-.

$$\frac{\text{mmol}}{\text{mL}} = \frac{\text{mol}}{\text{L}} = \text{M}$$

Challenge Calculate $[OH^-]$ and pH when 6.00 mL of HBr have been added. Check your answers with Table 11-1.

At the equivalence point, pH = 7.00, but *only* in a strong acid–strong base reaction.

After the equivalence point, there is excess H^+.

The excess H^+ amounts to

$$\text{excess } H^+ = (0.50 \text{ mL})(0.100\ 0 \text{ M}) = 0.050 \text{ mmol}$$

Using the total volume of solution $(50.00 + 10.50 = 60.50 \text{ mL})$, we can find the pH:

$$[H^+] = \frac{0.050 \text{ mmol}}{60.50 \text{ mL}} = 8.2_6 \times 10^{-4} \text{ M} \Rightarrow \text{pH} = -\log(8.2_6 \times 10^{-4}) = 3.08$$

The Titration Curve

The titration curve in Figure 11-1 is a graph of pH versus V_a, the volume of acid added. The sudden change in pH near the equivalence point is characteristic of all analytically useful titrations. The curve is steepest at the equivalence point, which means that the slope is greatest. The pH at the equivalence point is 7.00 *only* in a strong acid-strong base titration. If one or both of the reactants is weak, the equivalence point pH is *not* 7.00.

EXAMPLE | **Titration of Strong Acid with Strong Base**

Find the pH when 12.74 mL of 0.087 42 M NaOH have been added to 25.00 mL of 0.066 66 M $HClO_4$.

SOLUTION The titration reaction is $H^+ + OH^- \rightarrow H_2O$. The equivalence point is

$$\underbrace{(V_e(\text{mL}))(0.087\ 42 \text{ M})}_{\substack{\text{mmol of NaOH} \\ \text{at equivalence point}}} = \underbrace{(25.00 \text{ mL})(0.066\ 66 \text{ M})}_{\substack{\text{mmol of } HClO_4 \\ \text{being titrated}}} \Rightarrow V_e = 19.06 \text{ mL}$$

At V_b (volume of base) $= 12.74$ mL, there is excess acid in the solution:

$$H^+ \text{ remaining} = \underbrace{(25.00 \text{ mL})(0.066\ 66 \text{ M})}_{\substack{\text{Initial mmol} \\ \text{of } HClO_4}} - \underbrace{(12.74 \text{ mL})(0.087\ 42 \text{ M})}_{\substack{\text{Added mmol} \\ \text{of NaOH}}} = 0.553 \text{ mmol}$$

$$[H^+] = \frac{0.553 \text{ mmol}}{(25.00 + 12.74) \text{ mL}} = 0.014\ 7 \text{ M}$$

$$\text{pH} = -\log(0.014\ 7) = 1.83$$

Ask Yourself

11-A. What is the equivalence volume in the titration of 50.00 mL of 0.010 0 M NaOH with 0.100 M HCl? Calculate the pH at the following points: $V_a =$ 0.00, 1.00, 2.00, 3.00, 4.00, 4.50, 4.90, 4.99, 5.00, 5.01, 5.10, 5.50, 6.00, 8.00, and 10.00 mL. Make a graph of pH versus V_a.

The titration of a weak acid with a strong base puts all of our knowledge of acid-base chemistry to work. The example we treat is the titration of 50.00 mL of 0.020 00 M MES with 0.100 0 M NaOH. MES is an abbreviation for 2-(*N*-morpholino)ethane-sulfonic acid, which is a weak acid with $pK_a = 6.15$. It is widely used in biochemistry as a buffer for the pH 6 region.

The *titration reaction* is

(11-1)

HA A^-
MES, $pK_a = 6.15$

Always start by writing the titration reaction.

Reaction 11-1 is the reverse of the K_b reaction for the base A^-. The equilibrium constant is $1/K_b = 1/(K_w/K_{HA}) = 7.1 \times 10^7$. The equilibrium constant is so large that we can say that the reaction goes "to completion" after each addition of OH^-. As we saw in Box 10-1, *strong + weak react completely.*

strong + weak → complete reaction

It is helpful first to calculate the volume of base needed to reach the equivalence point. Because one mole of OH^- reacts with one mole of MES, we can say

$$\underbrace{(V_e(mL))\ (0.100\ 0\ M)}_{mmol\ of\ base} = \underbrace{(50.00\ mL)\ (0.020\ 00\ M)}_{mmol\ of\ HA} \Rightarrow V_e = 10.00\ mL$$

The titration calculations for this problem are of four types:

1. Before any base is added, the solution contains just HA in water. This is a weak-acid problem in which the pH is determined by the equilibrium

$$HA \overset{K_a}{\rightleftharpoons} H^+ + A^-$$

2. From the first addition of NaOH until immediately before the equivalence point, there is a mixture of unreacted HA plus the A^- produced by Reaction 11-1. *Aha! A buffer!* We can use the Henderson-Hasselbalch equation to find the pH.

3. At the equivalence point, "all" of the HA has been converted to A^-. The problem is the same as if the solution had been made by merely dissolving A^- in water. We have a weak-base problem in which the pH is determined by the reaction

$$A^- + H_2O \overset{K_b}{\rightleftharpoons} HA + OH^-$$

4. Beyond the equivalence point, excess NaOH is being added to a solution of A^-. We calculate the pH as if we had simply added excess NaOH to water. We will neglect the very small effect from having A^- present as well.

Region 1: Before Base Is Added

Before adding any base, we have a solution of 0.020 00 M HA with $pK_a = 6.15$. This is simply a weak-acid problem.

The initial solution contains just the *weak acid* HA.

F is the formal concentration of HA, which is 0.020 00 M in this example.

$$HA \rightleftharpoons H^+ + A^- \qquad K_a = 10^{-6.15}$$
$$ F - x \qquad x \qquad x$$

$$\frac{x^2}{0.020\ 00 - x} = K_a \Rightarrow x = 1.19 \times 10^{-4} \Rightarrow pH = 3.93$$

Region 2: Before the Equivalence Point

Before the equivalence point, there is a mixture of HA and A^-, which is a *buffer*.

Once we begin to add OH^-, a mixture of HA and A^- is created by the titration reaction (11-1). This mixture is a buffer whose pH can be calculated with the Henderson-Hasselbalch equation (10-1) once we know the quotient $[A^-]/[HA]$.

Henderson-Hasselbalch equation:
$$pH = pK_a + \log\left(\frac{[A^-]}{[HA]}\right) \qquad (10\text{-}1)$$

Consider the point where 3.00 mL of OH^- have been added:

Titration reaction:	HA	+	OH^-	$\rightarrow$	A^-	+	H_2O
Initial mmol:	1.000		0.300		—		
Final mmol:	0.700		—		0.300		

Once we know the *quotient* $[A^-]/[HA]$ in any solution, we know its pH:

The Henderson-Hasselbalch equation needs only mmol because volumes cancel in the quotient $[A^-]/[HA]$.

$$pH = pK_a + \log\left(\frac{[A^-]}{[HA]}\right) = 6.15 + \log\left(\frac{0.300}{0.700}\right) = 5.78$$

The point at which the volume of titrant is $\frac{1}{2}V_e$ is a special one in any titration.

Titration reaction:	HA	+	OH^-	$\rightarrow$	A^-	+	H_2O
Initial mmol:	1.000		0.500		—		
Final mmol:	0.500		—		0.500		

Landmark point: When $V_b = \frac{1}{2}V_e$, pH = pK_a. (When activities are taken into account [see Box 5-2], this statement is not exactly true, but it is a good approximation.)

$$pH = pK_a + \log\left(\frac{0.500}{0.500}\right) = pK_a$$

When the volume of titrant is $\frac{1}{2}V_e$, pH = pK_a for the acid HA. From the experimental titration curve, you can find pK_a by reading the pH when $V_b = \frac{1}{2}V_e$, where V_b is the volume of added base.

Advice. As soon as you recognize a mixture of HA and A^- in any solution, *you have a buffer!* You can calculate the pH from the quotient $[A^-]/[HA]$.

Region 3: At the Equivalence Point

At the equivalence point, HA has been converted to A^-, a *weak base*.

At the equivalence point ($V_b = 10.00$ mL), the quantity of NaOH is exactly enough to consume the HA.

Titration reaction:	HA	+	OH^-	$\rightarrow$	A^-	+	H_2O
Initial mmol:	1.000		1.000		—		
Final mmol:	—		—		1.000		

The resulting solution contains "just" A^-. We could have prepared the same solution by dissolving the salt Na^+A^- in distilled water. *Na^+A^- is a weak base.*

To compute the pH of a weak base, we write the reaction of the weak base with water:

$$\underset{F'-x}{A^-} + H_2O \rightleftharpoons \underset{x}{HA} + \underset{x}{OH^-} \qquad K_b = \frac{K_w}{K_a} \qquad (11\text{-}2)$$

The only tricky point is that the formal concentration of A^- is no longer 0.020 00 M, which was the initial concentration of HA. The initial 1.000 mmol of HA in 50.00 mL has been diluted with 10.00 mL of titrant:

$$[A^-] = \frac{1.000 \text{ mmol}}{(50.00 + 10.00) \text{ mL}} = 0.016\ 67 \text{ M} \equiv F'$$

Designating the formal concentration of A^- as F', we can find the pH from Reaction (11-2):

$$\frac{x^2}{F'-x} = K_b = \frac{K_w}{K_a} = 1.43 \times 10^{-8} \Rightarrow x = 1.54 \times 10^{-5} \text{ M}$$

$$pH = -\log[H^+] = -\log\left(\frac{K_w}{x}\right) = 9.18$$

The pH at the equivalence point in this titration is 9.18. **It is not 7.00.** The equivalence point pH will *always* be above 7 for the titration of a weak acid with a strong base, because the acid is converted to its conjugate base at the equivalence point.

Region 4: After the Equivalence Point

Now we are adding NaOH to a solution of A^-. The base NaOH is so much stronger than the base A^- that it is a fair approximation to say that the pH is determined by the concentration of excess OH^- in the solution.

Let's calculate the pH when $V_b = 10.10$ mL. This is just 0.10 mL past V_e. The quantity of excess OH^- is (0.10 mL)(0.100 0 M) = 0.010 mmol, and the total volume of solution is 50.00 + 10.10 mL = 60.10 mL.

$$[OH^-] = \frac{0.010 \text{ mmol}}{50.00 + 10.10 \text{ mL}} = 1.66 \times 10^{-4} \text{ M}$$

$$pH = -\log\left(\frac{K_w}{[OH^-]}\right) = 10.22$$

The Titration Curve

A summary of the calculations for the titration of MES with NaOH is shown in Table 11-2. The titration curve in Figure 11-2 has two easily identified points. One is the equivalence point, which is the steepest part of the curve. The other landmark is the point where $V_b = \frac{1}{2}V_e$ and $pH = pK_a$. This latter point has the minimum slope, which means that the pH changes least for a given addition of NaOH. This is another way of saying that the maximum *buffer capacity* occurs when $pH = pK_a$ and $[HA] = [A^-]$.

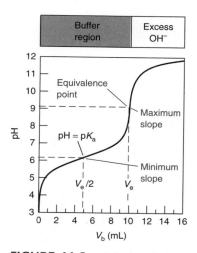

FIGURE 11-2 Calculated titration curve for the reaction of 50.00 mL of 0.020 00 M MES with 0.100 0 M NaOH. Landmarks occur at half of the equivalence volume ($pH = pK_a$) and at the equivalence point, which is the steepest part of the curve.

Challenge Compare the concentration of OH^- from excess titrant at $V_b = 10.10$ mL to the concentration of OH^- from hydrolysis of A^-. Satisfy yourself that it is fair to neglect the contribution of A^- to the pH after the equivalence point.

Landmarks in a titration:
At $V_b = V_e$, curve is steepest.
At $V_b = \frac{1}{2}V_e$, $pH = pK_a$ and the slope is minimal.

The *buffer capacity* measures the ability to resist changes in pH.

TABLE 11-2 **Calculation of the titration curve for 50.00 mL of 0.020 00 M MES treated with 0.100 0 M NaOH**

	mL base added (V_b)	pH
Region 1 (weak acid)	0.00	3.93
	0.50	4.87
	1.00	5.20
	2.00	5.55
	3.00	5.78
	4.00	5.97
	5.00	6.15
Region 2 (buffer)	6.00	6.33
	7.00	6.52
	8.00	6.75
	9.00	7.10
	9.50	7.43
	9.90	8.15
Region 3 (weak base)	10.00	9.18
	10.10	10.22
	10.50	10.91
	11.00	11.21
	12.00	11.50
Region 4 (excess OH⁻)	13.00	11.67
	14.00	11.79
	15.00	11.88
	16.00	11.95

Ask Yourself

11-B. Write the reaction between formic acid (Appendix B) and KOH. What is the equivalence volume (V_e) in the titration of 50.0 mL of 0.050 0 M formic acid with 0.050 0 M KOH? Calculate the pH at the points V_b = 0.0, 10.0, 20.0, 25.0, 30.0, 40.0, 45.0, 48.0, 49.0, 49.5, 50.0, 50.5, 51.0, 52.0, 55.0, and 60.0 mL. Draw a graph of pH versus V_b. Without doing any calculations, what should the pH be at $V_b = \frac{1}{2}V_e$? Does your calculated result agree with the prediction?

11-3 Titration of Weak Base with Strong Acid

The titration of a weak base with a strong acid is just the reverse of the titration of a weak acid with a strong base. The *titration reaction* is

$$B + H^+ \longrightarrow BH^+$$

Because the reactants are weak + strong, the reaction goes essentially to completion after each addition of acid. There are four distinct regions of the titration curve:

When $V_a = 0$, we have a *weak-base* problem.

1. Before acid is added, the solution contains just the weak base, B, in water. The pH is determined by the K_b reaction:

$$\underset{\substack{\text{B} \\ F - x}}{\text{B}} + H_2O \underset{}{\overset{K_b}{\rightleftharpoons}} \underset{x}{BH^+} + \underset{x}{OH^-}$$

2. Between the initial point and the equivalence point, there is a mixture of B and BH^+—*Aha! A buffer!* The pH is computed by using

$$pH = pK_a \text{ (for } BH^+) + \log\left(\frac{[B]}{[BH^+]}\right)$$

When $0 < V_a < V_e$, we have a buffer.

At the special point where $V_a = \frac{1}{2}V_e$, $pH = pK_a$ (for BH^+).

3. At the equivalence point, B has been converted into BH^+, a weak acid. The pH is calculated by considering the acid dissociation reaction of BH^+:

$$\underset{\substack{BH^+ \\ F' - x}}{BH^+} \rightleftharpoons \underset{x}{B} + \underset{x}{H^+} \qquad K_a = \frac{K_w}{K_b}$$

When $V_a = V_e$, the solution contains the weak acid BH^+.

The formal concentration of BH^+, F', is not the same as the original formal concentration of B, because some dilution has occurred. Because the solution contains BH^+ at the equivalence point, it is acidic. *The pH at the equivalence point must be below 7.*

4. After the equivalence point, there is excess H^+ in the solution. We treat this problem by considering only the concentration of excess H^+ and neglecting the contribution of weak acid, BH^+.

For $V_a > V_e$, there is excess strong acid.

EXAMPLE | **Titration of Pyridine with HCl**

Consider the titration of 25.00 mL of 0.083 64 M pyridine with 0.106 7 M HCl.

Titration reaction:

Pyridine (B)
$K_b = 1.69 \times 10^{-9}$

BH^+

$$\underset{\text{mmol of HCl}}{\underline{(V_e(\text{mL})) (0.106\ 7\ M)}} = \underset{\text{mmol of pyridine}}{\underline{(25.00\ \text{mL}) (0.083\ 64\ M)}} \Rightarrow V_e = 19.60\ \text{mL}$$

Find the pH when $V_a = 4.63$ mL, which is before the equivalence point.

SOLUTION Part of the pyridine has been neutralized, so there is a mixture of pyridine and pyridinium ion—*Aha! A buffer!* The initial millimoles of pyridine are $(25.00\ \text{mL})(0.083\ 64\ M) = 2.091$ mmol. The added H^+ is $(4.63\ \text{mL})(0.106\ 7\ M) = 0.494$ mmol. Therefore we can write

Titration reaction:	B	+	H^+	$\rightarrow$	BH^+
Initial mmol:	2.091		0.494		—
Final mmol:	1.597		—		0.494

$$pH = pK_{BH^+} + \log\left(\frac{[B]}{[BH^+]}\right) = 5.23 + \log\left(\frac{1.597}{0.494}\right) = 5.74$$
$$\underset{-\log(K_w/K_b)}{\uparrow}$$

11-4 *Finding the End Point*

The *equivalence point* in a titration is defined by the stoichiometry of the reaction. The *end point* is the abrupt change of a physical property (such as pH) that we measure to locate the equivalence point. Indicators and pH measurements are commonly used to find the end point in an acid-base titration.

Using Indicators to Find the End Point

Choose an indicator whose color change comes as close as possible to the theoretical pH of the equivalence point.

In Section 10-6 we learned that an indicator is an acid or base whose various protonated species have different colors. For the weak-acid indicator HIn, the solution takes on the color of HIn when $pH \lesssim pK_{HIn} - 1$ and has the color of In^- when $pH \gtrsim pK_{HIn} + 1$. In the interval $pK_{HIn} - 1 \lesssim pH \lesssim pK_{HIn} + 1$, a mixture of both colors is observed.

    A titration curve for which pH = 5.54 at the equivalence point is shown in Figure 11-3. The pH drops steeply (from 7 to 4) over a small volume interval. An indicator with a color change in this pH interval would provide a fair approximation to the equivalence point. The closer the point of color change is to pH 5.54, the more accurate will be the end point. The difference between the observed end point (color change) and the true equivalence point is called the **indicator error.**

One of the most common indicators is phenolphthalein, which changes from colorless in acid to pink in base:

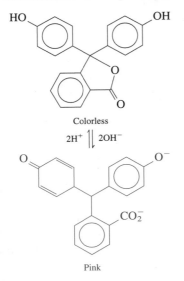

Colorless

$2H^+ \parallel 2OH^-$

Pink

    If you dump half a bottle of indicator into your reaction, you will introduce a different indicator error. Because indicators are acids or bases, they consume analyte or titrant. We use indicators under the assumption that the moles of indicator are negligible relative to the moles of analyte. Never use more than a few drops of dilute indicator solution.

    Many indicators in Table 10-3 would be useful for the titration in Figure 11-3. For example, if bromocresol purple were used, we would use the purple-to-yellow color change as the end point. The last trace of purple should disappear near pH 5.2, which is quite close to the true equivalence point in Figure 11-3. If bromocresol green were used as the indicator, a color change from blue to green (= yellow + blue) would mark the end point.

    In general, *we seek an indicator whose transition range overlaps the steepest part of the titration curve as closely as possible.* The steepness of the titration curve near the equivalence point in Figure 11-3 ensures that the indicator error caused by the noncoincidence of the color change and equivalence point will not be large. For example, if the indicator color change were at pH 6.4 (instead of 5.54), the error in V_e would be only 0.25% in this particular case.

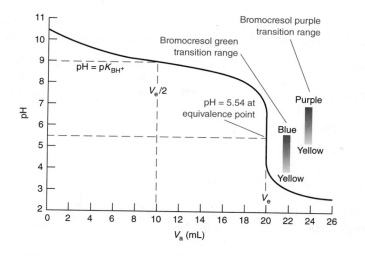

FIGURE 11-3 Calculated titration curve for the reaction of 100 mL of 0.010 0 M base ($pK_b = 5.00$) with 0.050 0 M HCl. As in the titration of HA with OH^-, $pH = pK_{BH^+}$ when $V_a = \frac{1}{2}V_e$.

Using a pH Electrode to Find the End Point

Figure 11-4 shows experimental results for the titration of the weak acid, H_6A, with NaOH. Because the compound is difficult to purify, only a tiny amount was available for titration. Just 1.430 mg was dissolved in 1.00 mL of water and titrated with microliter (μL) quantities of 0.065 92 M NaOH delivered with a Hamilton syringe.

When H_6A is titrated, we might expect to see an abrupt change in pH at all six equivalence points. The curve in Figure 11-4 shows two clear breaks, near 90 and 120 μL, which correspond to titration of the *third* and *fourth* protons of H_6A.

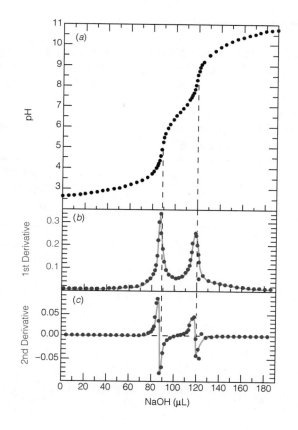

FIGURE 11-4 (*a*) Experimental points in the titration of 1.430 mg of xylenol orange, a hexaprotic acid, dissolved in 1.000 mL of 0.10 M $NaNO_3$. The titrant was 0.065 92 M NaOH. (*b*) The first derivative, $\Delta pH/\Delta V$, of the titration curve. (*c*) The second derivative, $\Delta(\Delta pH/\Delta V)/\Delta V$, which is the derivative of the middle curve. Derivatives for the first end point are calculated in Table 11-3. End points are taken as maxima in the derivative curve and zero crossings of the second derivative.

The end point has maximum slope.

$$H_4A^{2-} + OH^- \longrightarrow H_3A^{3-} + H_2O \quad (\sim 90 \ \mu L \text{ equivalence point})$$
$$H_3A^{3-} + OH^- \longrightarrow H_2A^{4-} + H_2O \quad (\sim 120 \ \mu L \text{ equivalence point})$$

The first two and last two equivalence points give unrecognizable end points, because they occur at pH values that are too low or too high.

The end point is where the slope of the titration curve is greatest. The slope is the change in pH (ΔpH) between two points divided by the change in volume (ΔV) between the points:

Slope of titration curve: $$\text{slope} = \frac{\Delta \text{pH}}{\Delta V} \quad (11\text{-}3)$$

The slope (which is also called the *first derivative*) displayed in the middle of Figure 11-4 is calculated in Table 11-3. The first two columns of this table give experimental volumes and pH measurements. (The pH meter was precise to three digits, even though accuracy ends in the second decimal place.) To compute the first derivative, each pair of volumes is averaged and the quantity ΔpH/ΔV is calculated.

The last two columns of Table 11-3 and Figure 11-4c give the slope of the slope (called the *second derivative*), computed as follows:

The slope of the slope (the second derivative) is 0 at the end point.

Second derivative: $$\frac{\Delta(\text{slope})}{\Delta V} = \frac{\Delta(\Delta \text{pH}/\Delta V)}{\Delta V} \quad (11\text{-}4)$$

The end point is the volume at which the second derivative is 0. A graph on the scale of Figure 11-5 allows us to make a good estimate of the end-point volume.

EXAMPLE Computing Derivatives of a Titration Curve

Let's see how the first and second derivatives in Table 11-3 are calculated.

SOLUTION The first number in the third column, 85.5, is the average of the first two volumes (85.0 and 86.0) in the first column. The slope (first derivative) ΔpH/ΔV is calculated from the first two pH values and the first two volumes:

TABLE 11-3 Computation of first and second derivatives for a titration curve

μL NaOH	pH	\multicolumn{2}{c}{First derivative}	\multicolumn{2}{c}{Second derivative}
		μL	$\frac{\Delta pH}{\Delta \mu L}$	μL	$\frac{\Delta(\Delta pH/\Delta \mu L)}{\Delta \mu L}$
85.0	4.245	85.5	0.155		
86.0	4.400			86.0	0.071 0
87.0	4.626	86.5	0.226	87.0	0.081 0
88.0	4.933	87.5	0.307	88.0	0.033 0
89.0	5.273	88.5	0.340	89.0	−0.083 0
90.0	5.530	89.5	0.257	90.0	−0.068 0
91.0	5.719	90.5	0.189	91.25	−0.039 0
93.0	5.980	92.0	0.130		

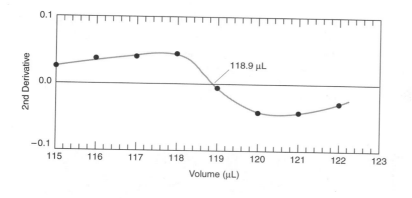

FIGURE 11-5 Enlargement of the second end point in the second derivative curve shown in Figure 11-4c.

$$\frac{\Delta pH}{\Delta V} = \frac{4.400 - 4.245}{86.0 - 85.0} = 0.155$$

The coordinates ($x = 85.5$, $y = 0.155$) are one point in the graph of the first derivative in Figure 11-4b.

The second derivative is computed from the first derivative. The first volume in the fifth column of Table 11-3 is 86, which is the average of 85.5 and 86.5. The second derivative is found as follows:

$$\frac{\Delta(\Delta pH/\Delta V)}{\Delta V} = \frac{0.226 - 0.155}{86.5 - 85.5} = 0.071$$

The coordinates ($x = 86$, $y = 0.071$) are plotted in the second derivative graph in Figure 11-4c. These calculations are tedious by hand, but they are not bad with a spreadsheet.

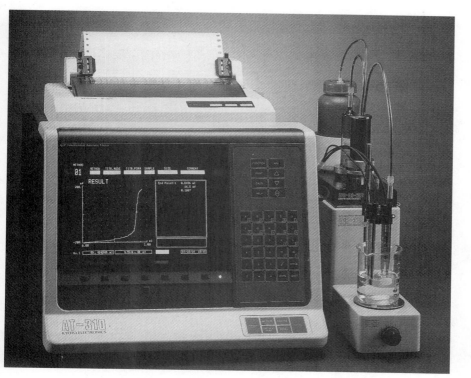

FIGURE 11-6 Autotitrator delivers titrant from the bottle at the back right to the beaker of analyte on the stirring motor at the front right. Electrodes immersed in the analyte solution monitor pH (or concentrations of specific ions) and display results on the screen during the titration.

Figure 11-6 shows an *autotitrator* in which the titration and calculations are performed automatically. Titrant from the plastic bottle at the rear is dispensed in small increments by a syringe pump, and the pH is measured by electrodes immersed in the beaker of analyte on the stirrer. The instrument waits for the pH to stabilize after each addition, before adding the next increment. pH is displayed on the screen, and the end point is computed automatically by finding the maximum slope in the titration curve.

Ask Yourself

11-D. (a) Select indicators from Table 10-3 that would be useful for the titrations in Figures 11-1 and 11-2, and the $pK_a = 8$ curve in Figure 11-10. Select a different indicator for each titration and state what color change you would use as the end point.

(b) Here are data points around the second apparent end point in Figure 11-4. Prepare a table analogous to Table 11-3 showing the first and second derivatives. Plot both derivatives versus V_b and locate the end points from each plot.

V_b (μL)	pH	V_b (μL)	pH
107	6.921	117	7.878
110	7.117	118	8.090
113	7.359	119	8.343
114	7.457	120	8.591
115	7.569	121	8.794
116	7.705	122	8.952

11-5 *Practical Notes*

NaOH and KOH must be standardized with primary standards.

Acids and bases listed in Table 11-4 can be purchased in forms pure enough to be used as *primary standards*. NaOH and KOH are not primary standards, because the reagent-grade materials contain carbonate (from reaction with atmospheric CO_2) and adsorbed water. Solutions of NaOH and KOH must be standardized against a primary standard. Potassium hydrogen phthalate is among the most convenient compounds for this purpose. Dilute solutions of NaOH for titrations are prepared by diluting a stock solution of 50 wt % aqueous NaOH. Sodium carbonate is relatively insoluble in this stock solution and settles to the bottom.

Alkaline solutions (e.g., 0.1 M NaOH) must be protected from the atmosphere because they absorb CO_2:

$$OH^- + CO_2 \longrightarrow HCO_3^-$$

CO_2 changes the concentration of base over a period of time and reduces the sharpness of the end point in the titration of weak acids. If base is kept in a tightly capped polyethylene bottle, it can be used for weeks with little change. Strong base attacks glass and should not be kept in a buret longer than necessary.

TABLE 11-4 Primary standards

Compound	Formula weight	Notes
ACIDS		
Potassium hydrogen phthalate (structure with CO_2H and CO_2K on benzene ring)	204.233	The pure solid is dried at 105°C and used to standardize base. A phenolphthalein end point is satisfactory.
$KH(IO_3)_2$ Potassium hydrogen iodate	389.912	This is a strong acid, so any indicator with an end point between ~5 and ~9 is adequate.
BASES		
$H_2NC(CH_2OH)_3$ Tris(hydroxymethyl)aminomethane (also called tris or tham)	121.136	The pure solid is dried at 100°–103°C and titrated with strong acid. The end point is in the range pH 4.5–5.
Na_2CO_3 Sodium carbonate	105.989	Primary standard grade Na_2CO_3 is titrated with acid to an end point of pH 4–5. Just before the end point, the solution is boiled to expel CO_2.
$Na_2B_4O_7 \cdot 10H_2O$ Borax	381.367	The recrystallized material is dried in a chamber containing an aqueous solution saturated with NaCl and sucrose. This procedure gives the decahydrate in pure form. The standard is titrated with acid to a methyl red end point.

(benzene ring with CO_2H and CO_2^-) $+ OH^- \longrightarrow$ (benzene ring with CO_2^- and CO_2^-) $+ H_2O$

$$H_2NC(CH_2OH)_3 + H^+ \longrightarrow H_3\overset{+}{N}C(CH_2OH)_3$$

$$\text{"}B_4O_7^{2-} \cdot 10H_2O\text{"} + 2H^+ \longrightarrow 4B(OH)_3 + 5H_2O$$

Ask Yourself

11-E. (a) Give the name and formula of a primary standard used to standardize **(i)** HCl and **(ii)** NaOH.

(b) In the reaction shown in Table 11-4, one mole of potassium hydrogen phthalate reacts with one mole of NaOH. How many grams of potassium hydrogen phthalate should be used to standardize ~0.05 M NaOH if you wish to use ~30 mL of base for the titration?

11-6 Kjeldahl Nitrogen Analysis

Developed in 1883, the **Kjeldahl nitrogen analysis** (pronounced KEL-dall) remains one of the most widely used methods for determining nitrogen in organic substances such as protein, cereal, and flour. The solid is first *digested* (decomposed and dis-

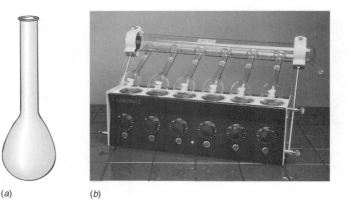

(a)   (b)

FIGURE 11-7 (a) The Kjeldahl digestion flask has a long neck to minimize loss by spattering. (b) Six-port manifold for multiple samples provides for exhaust of fumes.

solved) in boiling sulfuric acid, which converts nitrogen to ammonium ion, NH_4^+, and carbon to CO_2:

Each atom of nitrogen in the unknown is converted into one NH_4^+ ion.

Kjeldahl digestion: organic C, H, N $\xrightarrow[H_2SO_4]{\text{Boiling}}$ $NH_4^+ + CO_2 + H_2O$

Mercury, copper, and selenium compounds catalyze the digestion process. To speed the rate of reaction, the boiling point of concentrated (98 wt %) sulfuric acid (338°C) is raised by adding K_2SO_4. Digestion is carried out in a long-neck *Kjeldahl flask* (Figure 11-7) that prevents loss of sample by spattering. (A convenient alternative to the Kjeldahl flask is to use H_2SO_4 and H_2O_2 in a microwave bomb, as in Figure 2-19.)

After digestion is complete, the solution containing NH_4^+ is made basic, and the liberated NH_3 is distilled (with a large excess of steam) into a receiver containing a known amount of HCl (Figure 11-8). Excess, unreacted HCl is titrated with standard NaOH to determine how much HCl was consumed by NH_3.

Neutralization of NH_4^+: $NH_4^+ + OH^- \longrightarrow NH_3(g) + H_2O$ (11-5)

Distillation of NH_3 into standard HCl: $NH_3 + H^+ \longrightarrow NH_4^+$ (11-6)

Titration of unreacted HCl with NaOH: $H^+ + OH^- \longrightarrow H_2O$ (11-7)

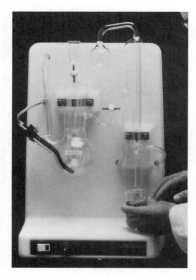

FIGURE 11-8 Kjeldahl distillation unit employs electric immersion heater in flask at left to carry out distillation in 5 min. Beaker at right collects liberated NH_3 in standard HCl.

EXAMPLE   Kjeldahl Analysis

A typical protein contains 16.2 wt % nitrogen. A 0.500-mL aliquot of protein solution was digested, and the liberated NH_3 was distilled into 10.00 mL of 0.021 40 M HCl. The unreacted HCl required 3.26 mL of 0.019 8 M NaOH for complete titration. Find the concentration of protein (mg protein/mL) in the original sample.

SOLUTION The original mmol of HCl in the receiver was (10.00 mL) (0.021 40 M) = 0.214 0 mmol. The NaOH required for titration of unreacted HCl in Reaction 11-7 was (3.26 mL)(0.019 8 M) = 0.064 5 mmol. The difference, 0.214 0 − 0.064 5 = 0.149 5 mmol, must equal the quantity of NH_3 produced in Reaction 11-5 and distilled into the HCl.

Because 1 mmol of nitrogen in the protein gives rise to 1 mmol of NH_3, there must have been 0.149 5 mmol of nitrogen in the protein, corresponding to

$$(0.149\ 5\ \text{mmol}) \left(14.006\ 74\ \frac{\text{mg N}}{\text{mmol}} \right) = 2.093\ \text{mg N}$$

If the protein contains 16.2 wt % N, there must be

$$\frac{2.093\ \text{mg N}}{0.162\ \text{mg N/mg protein}} = 12.9\ \text{mg protein} \Rightarrow \frac{12.9\ \text{mg protein}}{0.500\ \text{mL}} = 25.8\ \frac{\text{mg protein}}{\text{mL}}$$

Ask Yourself

11-F. The Kjeldahl procedure was used to analyze 256 μL of a solution containing 37.9 mg protein/mL. The liberated NH_3 was collected in 5.00 mL of 0.033 6 M HCl, and the remaining acid required 6.34 mL of 0.010 M NaOH for complete titration.
   (a) How many moles of NH_3 were liberated?
   (b) How many grams of nitrogen are contained in the NH_3 in (a)?
   (c) How many grams of protein were analyzed?
   (d) What is the weight percent of nitrogen in the protein?

11-7 *Putting Your Spreadsheet to Work*

In Sections 11-1 to 11-3, we calculated titration curves because they helped us understand the chemistry behind a titration curve. Now we will see how a spreadsheet and graphics program decrease the agony and mistakes of titration calculations. First we must derive equations relating pH to volume of titrant for use in the spreadsheet.

Charge Balance

The **charge balance** states that in any solution, the sum of positive charges must equal the sum of negative charges, because the solution must have zero net charge. For a solution of the weak acid HA plus NaOH, the charge balance is

Charge balance: $\qquad [H^+] + [Na^+] = [A^-] + [OH^-]$ $\qquad$ (11-8)

The sum of the positive charges of H^+ and Na^+ equals the sum of the negative charges of A^- and OH^-.

Titrating a Weak Acid with a Strong Base

Consider the titration of a volume V_a of acid HA (initial concentration C_a) with a volume V_b of NaOH of concentration C_b. The concentration of Na^+ is just the moles of NaOH ($C_b V_b$) divided by the total volume of solution ($V_a + V_b$):

$$[Na^+] = \frac{C_b V_b}{V_a + V_b}$$ $\qquad$ (11-9)

If the solution contained HA and $Ca(OH)_2$, the charge balance would be

$$[H^+] + 2[Ca^{2+}] = [A^-] + [OH^-]$$

because one mole of Ca^{2+} provides two moles of charge. If $[Ca^{2+}] = 0.1$ M, the positive charge it contributes is 0.2 M.

Similarly, the formal concentration of the weak acid is

$$F_{HA} = [HA] + [A^-] = \frac{C_a V_a}{V_a + V_b} \tag{11-10}$$

because we have diluted $C_a V_a$ moles of HA to a total volume of $V_a + V_b$.

Now we finally get to use the fractional composition equations from Chapter 9:

Fraction of weak acid in the form HA: $\quad \alpha_{HA} = \dfrac{[HA]}{F} = \dfrac{[H^+]}{[H^+] + K_a} \tag{11-11}$

Fraction of weak acid in the form A^-: $\quad \alpha_{A^-} = \dfrac{[A^-]}{F} = \dfrac{K_a}{[H^+] + K_a} \tag{11-12}$

Equations 11-11 and 11-12 say that if a weak acid has a formal concentration F, the concentration of HA is $\alpha_{HA} \cdot F$ and the concentration of A^- is $\alpha_{A^-} \cdot F$. The fractions must add up to 1.

$\alpha_{HA} + \alpha_{A^-} = 1$

Getting back to our titration, we can write an expression for the concentration of A^- by combining Equation 11-12 with Equation 11-10:

$$[A^-] = \alpha_{A^-} \cdot F_{HA} = \frac{\alpha_{A^-} \cdot C_a V_a}{V_a + V_b} \tag{11-13}$$

Substituting for $[Na^+]$ (Equation 11-9) and $[A^-]$ (Equation 11-13) in the charge balance (Equation 11-8) gives

$$[H^+] + \frac{C_b V_b}{V_a + V_b} = \frac{\alpha_{A^-} \cdot C_a V_a}{V_a + V_b} + [OH^-]$$

	A	B	C	D	E	F	G
1	Cb =	pH	[H+]	[OH–]	Alpha[A–]	Phi	Vb (mL)
2	0.1	3.00	1.00E-03	1.00E-11	7.075E-04	-4.880E-02	-0.488
3	Ca =	3.93	1.17E-04	8.51E-11	5.990E-03	1.153E-04	0.001
4	0.02	4.00	1.00E-04	1.00E-10	7.030E-03	2.028E-03	0.020
5	Va =	5.00	1.00E-05	1.00E-09	6.612E-02	6.561E-02	0.656
6	50	6.15	7.08E-07	1.41E-08	5.000E-01	5.000E-01	5.000
7	Ka =	7.00	1.00E-07	1.00E-07	8.762E-01	8.762E-01	8.762
8	7.08E-7	8.00	1.00E-08	1.00E-06	9.861E-01	9.861E-01	9.861
9	Kw =	9.18	6.61E-10	1.51E-05	9.991E-01	1.000E+00	10.000
10	1.E-14	10.00	1.00E-10	1.00E-04	9.999E-01	1.006E+00	10.059
11		11.00	1.00E-11	1.00E-03	1.000E+00	1.061E+00	10.606
12		12.00	1.00E-12	1.00E-02	1.000E+00	1.667E+00	16.667
13							
14	C2 = 10^-B2						
15	D2 = $A$10/C2						
16	E2 = $A$8/(C2+$A$8)						
17	F2 = (E2-(C2-D2)/$A$4)/(1+(C2-D2)/$A$2) [Equation 11-14]						
18	G2 = F2*$A$4*$A$6/$A$2						

FIGURE 11-9 Spreadsheet that uses Equation 11-14 to calculate the titration curve for 50 mL of the weak acid, 0.02 M MES (pK_a = 6.15), treated with 0.1 M NaOH. We provide pH as input in column B, and the spreadsheet tells us what volume of base in column G is required to generate that pH.

which can be rearranged to the form

Fraction of titration for weak acid by strong base:

$$\phi = \frac{C_b V_b}{C_a V_a} = \frac{\alpha_{A^-} - \frac{[H^+] - [OH^-]}{C_a}}{1 + \frac{[H^+] - [OH^-]}{C_b}} \qquad (11\text{-}14)$$

At last! Equation 11-14 is really useful. It relates the volume of titrant (V_b) to the pH and a bunch of constants. The quantity ϕ, which is the quotient $C_b V_b / C_a V_a$, gives the fraction of the way to the equivalence point, V_e. When $\phi = 1$, the volume of base added, V_b, is equal to V_e. Equation 11-14 works backward from the way you are accustomed to thinking, because you need to put in pH (on the right) to get out volume (on the left).

Let's set up a spreadsheet to use Equation 11-14 to calculate the titration curve for 50.00 mL of the weak acid 0.020 00 M MES with 0.100 0 M NaOH, which was shown in Figure 11-2 and Table 11-2. The equivalence volume is $V_e = 10.00$ mL. The quantities in Equation 11-14 are

$C_b = 0.1$

$C_a = 0.02$

$V_a = 50$

$K_a = 7.0_8 \times 10^{-7}$

$K_w = 10^{-14}$

$[H^+] = 10^{-pH}$

$[OH^-] = K_w / [H^+]$

$$\alpha_{A^-} = \frac{K_a}{[H^+] + K_a}$$

pH is the input

$V_b = \dfrac{\phi C_a V_a}{C_b}$ is the output

$\phi = C_b V_b / C_a V_a$ is the fraction of the way to the equivalence point:

ϕ	Volume of base
0.5	$V_b = \frac{1}{2} V_e$
1	$V_b = V_e$
2	$V_b = 2 V_e$

2-(*N*-Morpholino)ethanesulfonic acid
MES, $pK_a = 6.15$

The input to the spreadsheet in Figure 11-9 is pH in column B and the output is V_b in column G. From the pH, the values of $[H^+]$, $[OH^-]$, and α_{A^-} are computed in columns C, D, and E. Equation 11-14 is used in column F to find the fraction of titration, ϕ. From this value, we calculate the volume of titrant, V_b, in column G.

How do we know what pH values to put in? Trial-and-error allows us to find the starting pH, by putting in a pH and seeing if V_b is positive or negative. In a few tries, it is easy to home in on the pH at which $V_b = 0$. In Figure 11-9 we see that a pH of 3.00 is too low, because ϕ and V are both negative. Input values of pH are spaced as closely as you like, so that you can generate a smooth titration curve. To save space, we only show a few points in Figure 11-9, including the midpoint (pH $6.15 \Rightarrow V_b = 5.00$ mL) and the end point (pH $9.18 \Rightarrow V_b = 10.00$ mL). The spreadsheet reproduces Table 11-2 without dividing the titration into different regions that use different approximations.

The spreadsheet in Figure 11-9 can be used to find the pH of a weak acid. Just search for the pH at which $V_b = 0$.

To home in on an exact volume (such as V_e), set the spreadsheet to show extra digits in the cell of interest. Tables in this book have been formatted to reduce the number of digits.

The Power of a Spreadsheet

By changing K_a in cell A8 of Figure 11-9, we can calculate a whole family of curves for the titration of different acids. Figure 11-10 shows how the titration curve depends on the acid dissociation constant of HA. The strong-acid curve at the bottom of Figure 11-10 was computed with a large value of K_a ($K_a = 10^3$) in cell A8 of the spreadsheet in Figure 11-9. Figure 11-10 shows that as K_a decreases (pK_a increases),

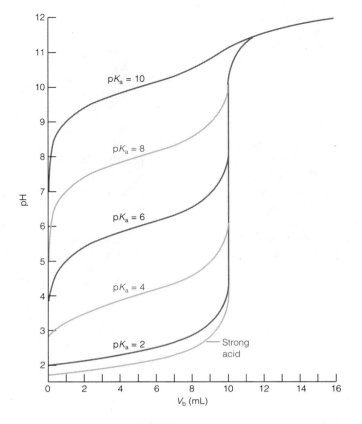

FIGURE 11-10 Calculated curves showing the titration of 50.0 mL of 0.020 0 M HA with 0.100 M NaOH. As the acid becomes weaker, the change in slope at the equivalence point becomes less abrupt.

the extent of the change near the equivalence point decreases, until the equivalence point becomes too shallow to detect. Similar behavior occurs as the concentrations of analyte and titrant decrease. *It is not practical to titrate an acid or a base when its strength is too weak or its concentration too dilute.*

Titrating a Weak Base with a Strong Acid

Following a similar derivation as that above, we can derive a spreadsheet equation to describe the titration of weak base B with strong acid:

Fraction of titration for weak base by strong acid:

$$\phi = \frac{C_a V_a}{C_b V_b} = \frac{\alpha_{BH^+} + \dfrac{[H^+] - [OH^-]}{C_b}}{1 - \dfrac{[H^+] - [OH^-]}{C_a}} \tag{11-15}$$

where C_a is the concentration of strong acid in the buret, V_a is the volume of acid added, C_b is the initial concentration of weak base being titrated, V_b is the initial volume of weak base being titrated, and α_{BH^+} is the fraction of base in the form BH^+:

Fraction of weak base in the form BH⁺:

$$\alpha_{BH^+} = \frac{[HA]}{F} = \frac{[H^+]}{[H^+] + K_{BH^+}} \tag{11-16}$$

where K_{BH^+} is the acid dissociation constant of BH^+.

Ask Yourself

11-G. (a) *Effect of pK$_a$ in the titration of weak acid with strong base.* Use the spreadsheet in Figure 11-9 to compute and plot the family of curves in Figure 11-10. For strong acid, use $K_a = 10^3$.

(b) *Effect of concentration in the titration of weak acid with strong base.* Use your spreadsheet to prepare a family of titration curves for pK$_a$ = 6, with the following combinations of concentrations: **(i)** C_a = 20 mM, C_b = 100 mM; **(ii)** C_a = 2 mM, C_b = 10 mM; and **(iii)** C_a = 0.2 mM, C_b = 1 mM.

Key Equations

Useful shortcut

$$mL \times \frac{mol}{L} = mmol$$

Equivalence volume (V_e)

$$\underbrace{C_aV_a = C_bV_e}_{\substack{\text{Titrating acid} \\ \text{with base}}} \qquad \text{or} \qquad \underbrace{C_aV_e = C_bV_b}_{\substack{\text{Titrating base} \\ \text{with acid}}}$$

C_a = acid concentration $\qquad C_b$ = base concentration

V_a = acid volume $\qquad\qquad V_b$ = base volume $\qquad V_e$ = equivalence volume

Titration of weak acid

1. Initial solution—weak acid

$$\underset{F-x}{HA} \overset{K_a}{\rightleftharpoons} \underset{x}{H^+} + \underset{x}{A^-} \qquad \frac{x^2}{F-x} = K_a$$

2. Before equivalence point—buffer

Titration reaction tells how much HA and A$^-$ are present

$$pH = pK_a + \log\left(\frac{[A^-]}{[HA]}\right)$$

3. Equivalence point—weak base—pH > 7

$$\underset{F'-x}{A^-} + H_2O \overset{K_b}{\rightleftharpoons} \underset{x}{HA} + \underset{x}{OH^-} \qquad K_b = \frac{K_w}{K_a}$$

F' is diluted concentration

4. After equivalence point—excess strong base

$$pH = -\log(K_w/[OH^-]_{excess})$$

Titration of weak base

1. Initial solution—weak base

$$\underset{F-x}{B} + H_2O \overset{K_b}{\rightleftharpoons} \underset{x}{BH^+} + \underset{x}{OH^-} \qquad \frac{x^2}{F-x} = K_b$$

2. Before equivalence point—buffer

Titration reaction tells us how much B and BH$^+$ are present

$$pH = pK_{BH^+} + \log\left(\frac{[B]}{[BH^+]}\right)$$

3. Equivalence point—weak acid—pH < 7

$$BH^+ \xrightleftharpoons{K_{BH^+}} B + H^+$$
$$\;\;F' - x \qquad\quad x \quad\; x$$

4. After equivalence point—excess strong acid

$$pH = -\log([H^+]_{excess})$$

Spreadsheet titration equations	Use Equations 11-14 and 11-15
	Input is pH and output is volume
Choosing indicator	Use indicator with color change close to theoretical pH at equivalence point of titration
Using electrodes for end point	End point has greatest slope: $\Delta pH/\Delta V$ is maximum
	End point has zero second derivative: $\dfrac{\Delta(\Delta pH/\Delta V)}{\Delta V} = 0$

Important Terms

charge balance	indicator error	Kjeldahl nitrogen analysis

Problems

11-1. Consider the titration of 100.0 mL of 0.100 M NaOH with 1.00 M HBr. What is the equivalence volume? Find the pH at the following volumes of HBr and make a graph of pH versus V_a: $V_a = 0$, 1.00, 5.00, 9.00, 9.90, 10.00, 10.10, and 12.00 mL.

11-2. Consider the titration of 25.0 mL of 0.050 0 M $HClO_4$ with 0.100 M KOH. Find the equivalence volume. Find the pH at the following volumes of KOH and plot pH versus V_b: $V_b = 0$, 1.00, 5.00, 10.00, 12.40, 12.50, 12.60, and 13.00 mL.

11-3. 50.0 mL of 0.050 0 M weak acid HA ($pK_a = 4.00$) was titrated with 0.500 M NaOH. Write the titration reaction and find V_e. Find the pH at $V_b = 0$, 1.00, 2.50, 4.00, 4.90, 5.00, 5.10, and 6.00 mL and plot pH versus V_b.

11-4. When methylammonium chloride is titrated with tetramethylammonium hydroxide, the titration reaction is

$$\underset{\substack{BH^+ \\ \text{Weak acid}}}{CH_3NH_3^+} + \underset{\substack{\text{From} \\ (CH_3)_4N^+OH^-}}{OH^-} \longrightarrow \underset{\substack{B \\ \text{Weak base}}}{CH_3NH_2} + H_2O$$

Find the equivalence volume in the titration of 25.0 mL of 0.010 0 M methylammonium chloride with 0.050 0 M tetramethylammonium hydroxide. Calculate the pH at $V_b = 0$, 2.50, 5.00, and 10.00 mL. Sketch the titration curve.

11-5. Write the reaction for the titration of 100 mL of 0.100 M anilinium bromide ("aminobenzene · HBr") with 0.100 M NaOH. Sketch the titration curve for the points $V_b = 0$, $0.100V_e$, $0.500V_e$, $0.900V_e$, V_e, and $1.200V_e$.

11-6. What is the pH at the equivalence point when 0.100 M hydroxyacetic acid is titrated with 0.050 0 M KOH?

11-7. When 16.24 mL of 0.064 3 M KOH was added to 25.00 mL of 0.093 8 M weak acid, HA, the observed pH was 3.62. Find pK_a for the acid.

11-8. When 22.63 mL of aqueous NaOH was added to 1.214 g of CHES (FW 207.29, structure in Table 10-2) dissolved in 41.37 mL of water, the pH was 9.24. Calculate the molarity of the NaOH.

11-9. (a) When 100.0 mL of weak acid HA was titrated with 0.093 81 M NaOH, 27.63 mL was required to reach the equivalence point. Find the molarity of HA.

(b) What is the formal concentration of A^- at the equivalence point?

(c) The pH at the equivalence point was 10.99. Find pK_a for HA.

(d) What was the pH when only 19.47 mL of NaOH had been added?

11-10. A 100.0-mL aliquot of 0.100 M weak base B ($pK_b = 5.00$) was titrated with 1.00 M $HClO_4$. Find V_e and calculate the pH at $V_a = 0$, 1.00, 5.00, 9.00, 9.90, 10.00, 10.10, and 12.00 mL and make a graph of pH versus V_a.

11-11. A solution of 100.0 mL of 0.040 0 M sodium propanoate (the sodium salt of propanoic acid) was titrated with 0.083 7 M HCl. Find V_e and calculate the pH at $V_a = 0$, $\frac{1}{4}V_e$, $\frac{1}{2}V_e$, $\frac{3}{4}V_e$, V_e, and 1.1 V_e.

11-12. A solution containing 50.0 mL of 0.031 9 M benzylamine was titrated with 0.050 0 M HCl.

(a) What is the equilibrium constant for the titration reaction?

(b) Find V_e and calculate the pH at $V_a = 0$, 12.0, $\frac{1}{2}V_e$, 30.0, V_e, and 35.0 mL.

11-13. Don't ever mix acid with cyanide (CN^-) because it liberates poisonous hydrogen cyanide (HCN). But, just for fun, calculate the pH of a solution made by mixing 50.00 mL of 0.100 M NaCN with

(a) 4.20 mL of 0.438 M $HClO_4$

(b) 11.82 mL of 0.438 M $HClO_4$

(c) What is the pH at the equivalence point with 0.438 M $HClO_4$?

11-14. Would the indicator bromocresol green, with a transition range of pH 3.8–5.4, ever be useful in the titration of a weak acid with a strong base? Why?

11-15. Consider the titration in Figure 11-2, for which the pH at the equivalence point is calculated to be 9.18. If thymol blue is used as an indicator, what color will be observed through most of the titration prior to the equivalence point? At the equivalence point? After the equivalence point?

11-16. Why would an indicator end point not be very useful in the titration curve for $pK_a = 10.00$ in Figure 11-10?

11-17. Phenolphthalein is used as an indicator for the titration of HCl with NaOH.

(a) What color change is observed at the end point?

(b) The basic solution just after the end point slowly absorbs CO_2 from the air and becomes more acidic by virtue of the reaction $CO_2 + OH^- \rightleftharpoons HCO_3^-$. This causes the color to fade from pink to colorless. If you carry out the titration too slowly, does this reaction lead to a systematic or a random error in finding the end point?

11-18. In the titration of 0.10 M pyridinium bromide (the salt of pyridine plus HBr) by 0.10 M NaOH, the pH at $0.99V_e$ is 7.22. At V_e, pH = 8.96, and at $1.01V_e$, pH = 10.70. Select an indicator from Table 10-3 that would be suitable for this titration and state what color change will be used.

11-19. Prepare a graph of the second derivative to find the end point from the titration data below:

mL NaOH	pH	mL NaOH	pH
10.679	7.643	10.729	5.402
10.696	7.447	10.733	4.993
10.713	7.091	10.738	4.761
10.721	6.700	10.750	4.444
10.725	6.222	10.765	4.227

11-20. Borax (Table 11-4) was used to standardize a solution of HNO_3. Titration of 0.261 9 g borax required 21.61 mL. What is the molarity of the HNO_3?

11-21. A 10.231-g sample of window cleaner containing ammonia was diluted with 39.466 g of water. Then 4.373 g of solution was titrated with 14.22 mL of 0.106 3 M HCl to reach a bromocresol green end point.

(a) What fraction of the 10.231-g sample of window cleaner is contained in the 4.373 g that was analyzed?

(b) How many grams of NH_3 (MW 17.031) were in the 4.373-g sample?

(c) Find the weight percent of NH_3 in the cleaner.

11-22. In the Kjeldahl nitrogen analysis, the final product is NH_4^+ in HCl solution. It is necessary to titrate the HCl without titrating the NH_4^+ ion.

(a) Calculate the pH of pure 0.010 M NH_4Cl.

(b) The steep part of the titration curve when HCl is titrated with NaOH runs from pH $\approx$ 4 up to pH $\approx$ 10. Select an indicator that would allow you to titrate HCl but not NH_4^+.

11-23. Prepare a spreadsheet like the one in Figure 11-9, but using Equation 11-15, to reproduce the titration curve in Figure 11-3.

11-24. *Effect of pK_b in the titration of weak base with strong acid.* Use the spreadsheet from the previous problem to compute and plot a family of curves analogous to Figure 11-10 for the titration of 50.0 mL of 0.020 0 M B ($pK_b = -2.00$, 2.00, 5.00, 8.00, and 10.00) with 0.100 M HCl. ($pK_b = -2.00$ corresponds to $K_b = 10^{+2.00}$, which represents a strong base.)

Disappearing Statues and Teeth

Erosion of carbonate stone. The columns at the left are on the inside of a structure. The columns at the right are on the outside and beginning to wear out from exposure to acid rain. Figure 9-1 shows rainfall pH in North America.

The main constituent of limestone and marble is calcite, a crystalline form of calcium carbonate. This mineral is insoluble in neutral or basic solution but dissolves in acid by virtue of two *coupled equilibria,* in which the product of one reaction is consumed in the next reaction:

$$CaCO_3(s) \rightleftharpoons Ca^{2+} + CO_3^{2-} \qquad CO_3^{2-} + H^+ \rightleftharpoons HCO_3^-$$

Calcite   Carbonate   Bicarbonate

Le Châtelier's principle tells us that if we remove a product of the first reaction, we will draw the reaction to the right, making calcite more soluble.

Tooth enamel contains the mineral hydroxyapatite, a calcium hydroxyphosphate. This mineral also dissolves in acid, because both PO_4^{3-} and OH^- react with H^+:

$$Ca_{10}(PO_4)_6(OH)_2 + 14H^+ \rightleftharpoons 10Ca^{2+} + 6H_2PO_4^- + 2H_2O$$

Hydroxyapatite

Bacteria residing on your teeth metabolize sugar into lactic acid, thereby creating a pH below 5 at the surface of a tooth. Acid dissolves the hydroxyapatite, creating tooth decay.

$$\overset{\displaystyle O}{\overset{\displaystyle \|}{CH_3CHCO_2H}} \quad \text{Lactic acid}$$

Chapter 12

Polyprotic Acids and Bases

Carbonic acid from marble, phosphoric acid from teeth, and amino acids from proteins are all **polyprotic acids**—those having more than one acidic proton. This chapter extends our discussions of acids, bases, and buffers to polyprotic systems encountered throughout nature.

12-1 Amino Acids Are Polyprotic

The **amino acids** from which proteins are built have an acidic carboxylic acid group, a basic amino group, and a variable substituent designated R in the structure below:

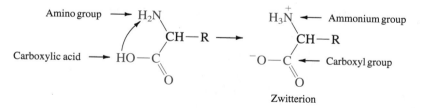

Zwitterion

A *zwitterion* is a molecule with positive and negative charges.

Because the amino group is more basic than the carboxyl group, the acidic proton resides on nitrogen of the amino group instead of oxygen from the carboxyl group. The resulting structure, with positive and negative sites, is called a **zwitterion.**

At low pH, both the ammonium group and the carboxyl group are protonated. At high pH, neither is protonated. The substituent may also have acidic or basic properties. Acid dissociation constants of the 20 common amino acids are given in

Table 12-1, in which each substituent (R) is shown in its fully protonated form. For example, Table 12-1 shows the amino acid cysteine, which has three acidic protons:

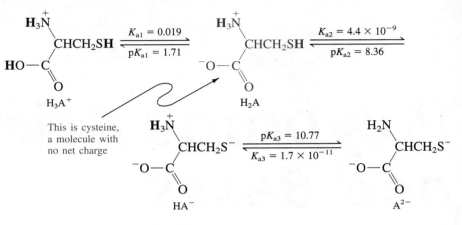

This is cysteine, a molecule with no net charge

K_{a1} refers to the *most acidic* proton. The subscript a in K_{a1} and K_{a2} is customarily omitted and we will write K_1 and K_2 in most of this chapter.

In general, a *diprotic* acid has two acid dissociation constants designated K_{a1} and K_{a2} (where $K_{a1} \geq K_{a2}$):

$$H_2A \xrightleftharpoons[]{K_{a1}} HA^- + H^+ \qquad HA^- \xrightleftharpoons[]{K_{a2}} A^{2-} + H^+$$

The two base association constants are designated K_{b1} and K_{b2} ($K_{b1} > K_{b2}$):

$$A^{2-} + H_2O \xrightleftharpoons[]{K_{b1}} HA^- + OH^- \qquad HA^- + H_2O \xrightleftharpoons[]{K_{b2}} H_2A + OH^-$$

Relation Between K_a and K_b

If you add the K_{a1} reaction to the K_{b2} reaction, the sum is $H_2O \rightleftharpoons H^+ + OH^-$—the K_w reaction. By this means, you can derive a most important set of relationships between the acid and base equilibrium constants:

Challenge Add the K_{a1} and K_{b2} reactions to prove that $K_{a1} \cdot K_{b2} = K_w$.

Relation between K_a and K_b for diprotic system:

$$K_{a1} \cdot K_{b2} = K_w$$
$$K_{a2} \cdot K_{b1} = K_w$$

(12-1)

For a *triprotic* system, with three acidic protons, the corresponding relationships are

Relation between K_a and K_b for triprotic system:

$$K_{a1} \cdot K_{b3} = K_w$$
$$K_{a2} \cdot K_{b2} = K_w$$
$$K_{a3} \cdot K_{b1} = K_w$$

(12-2)

The standard notation for successive acid dissociation constants of a polyprotic acid is K_1, K_2, K_3, and so on, with the subscript a usually omitted. We retain or omit the subscript a as dictated by clarity. For successive base hydrolysis constants, we retain the subscript b. K_{a1} *(or K_1) refers to the acidic species with the most protons, and K_{b1} refers to the basic species with no acidic protons.*

TABLE 12-1 Acid dissociation constants of amino acids

Amino acid[a]	Substituent	Carboxylic acid pK_a	Ammonium pK_a	Substituent pK_a	Molecular weight
Alanine (A)	$-CH_3$	2.348	9.867		89.09
Arginine (R)	$-CH_2CH_2CH_2NHC\overset{+NH_2}{\underset{NH_2}{}}$	1.823	8.991	12.48	174.20
Asparagine (N)	$-CH_2\overset{O}{\overset{\|}{C}}NH_2$	2.14	8.72		132.12
Aspartic acid (D)	$-CH_2CO_2H$	1.990	10.002	3.900	133.10
Cysteine (C)	$-CH_2SH$	1.71	10.77	8.36	121.16
Glutamic acid (E)	$-CH_2CH_2CO_2H$	2.23	9.95	4.42	147.13
Glutamine (Q)	$-CH_2CH_2\overset{O}{\overset{\|}{C}}NH_2$	2.17	9.01		146.15
Glycine (G)	$-H$	2.350	9.778		75.07
Histidine (H)	$-CH_2$ (imidazole)	1.7	9.08	6.02	155.16
Isoleucine (I)	$-CH(CH_3)(CH_2CH_3)$	2.319	9.754		131.17
Leucine (L)	$-CH_2CH(CH_3)_2$	2.329	9.747		131.17
Lysine (K)	$-CH_2CH_2CH_2CH_2NH_3^+$	2.04	9.08	10.69	146.19
Methionine (M)	$-CH_2CH_2SCH_3$	2.20	9.05		149.21
Phenylalanine (F)	$-CH_2$ (phenyl)	2.20	9.31		165.19
Proline (P)	HO_2C (pyrrolidine, H_2N^+) ← Structure of entire amino acid	1.952	10.640		115.13
Serine (S)	$-CH_2OH$	2.187	9.209		105.09
Threonine (T)	$-CH(CH_3)(OH)$	2.088	9.100		119.12
Tryptophan (W)	$-CH_2$ (indole)	2.35	9.33		204.23
Tyrosine (Y)	$-CH_2$—(phenyl)—OH	2.17	9.19	10.47	181.19
Valine (V)	$-CH(CH_3)_2$	2.286	9.718		117.15

a. Standard abbreviations are shown in parentheses. Acidic protons are **bold**. Each substituent is written in its fully protonated form.

Ask Yourself

12-A. (a) Each ion below has two consecutive acid-base reactions (called the stepwise acid-base reactions) when placed in water. Write the reactions and the correct symbol (e.g., K_2 or K_{b1}) for the equilibrium constant for each. Use Appendix B to find the numerical value of each acid or base equilibrium constant.

(i)   **(ii)**

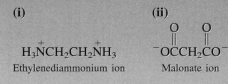

Ethylenediammonium ion   Malonate ion

   (b) Starting with the fully protonated species shown below, write the stepwise acid dissociation reactions of the amino acids aspartic acid and arginine. Be sure to remove the protons in the correct order based on the pK_a values in Table 12-1. Remember that the proton with the lowest pK_a (i.e., the greatest K_a) comes off first. Label the neutral molecules that we call aspartic acid and arginine.

(iii)   **(iv)**

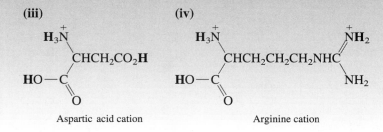

Aspartic acid cation   Arginine cation

12-2 Finding the pH in Diprotic Systems

Consider the amino acid leucine, designated HL:

The side chain in leucine is an isobutyl group: $R = -CH_2CH(CH_3)_2$

The equilibrium constants refer to the following reactions:

Diprotic acid: $H_2L^+ \rightleftharpoons HL + H^+$ $K_{a1} \equiv K_1$ (12-3)

                      $HL \rightleftharpoons L^- + H^+$ $K_{a2} \equiv K_2$ (12-4)

Diprotic base: $L^- + H_2O \rightleftharpoons HL + OH^-$ K_{b1} (12-5)

        $HL + H_2O \rightleftharpoons H_2L^+ + OH^-$ K_{b2} (12-6)

   We now set out to calculate the pH and composition of individual solutions of 0.050 0 M H_2L^+, 0.050 0 M HL, and 0.050 0 M L^-. Our methods do not depend on the charge of the acids and bases. We would use the same procedure to find the pH of the diprotic H_2A, where A is anything, or H_2L^+, where HL is leucine.

The Acidic Form, H_2L^+

Easy stuff.

A salt such as leucine hydrochloride contains the protonated species H_2L^+, which can dissociate twice, as indicated in Reactions 12-3 and 12-4. Because $K_1 = 4.69 \times 10^{-3}$, H_2L^+ is a weak acid. HL is an even weaker acid, because $K_2 = 1.79 \times 10^{-10}$. It appears that the H_2L^+ will dissociate only partly, and the resulting HL will hardly dissociate at all. For this reason, we make the (superb) approximation that a solution of H_2L^+ behaves as a monoprotic acid, with $K_a = K_1$.

With this approximation, the calculation of the pH of 0.050 0 M H_2L^+ is simple:

$$\underset{\substack{H_2L^+ \\ 0.050\,0 - x}}{\overset{\overset{\displaystyle R}{\overset{\displaystyle +\,|}{H_3NCHCO_2H}}}{}} \quad \overset{K_1 = 4.69 \times 10^{-3}}{\rightleftharpoons} \quad \underset{\substack{HL \\ x}}{\overset{\overset{\displaystyle R}{\overset{\displaystyle +\,|}{H_3NCHCO_2^-}}}{}} + \underset{x}{H^+}$$

H_2L^+ can be treated as monoprotic, with $K_a = K_1$.

$$\frac{x^2}{F - x} = K_1 \Rightarrow x = 1.31 \times 10^{-2}\ M$$

$$[HL] = x = 1.31 \times 10^{-2}\ M$$

$$[H^+] = x = 1.31 \times 10^{-2}\ M \Rightarrow pH = 1.88$$

$$[H_2L^+] = F - x = 3.69 \times 10^{-2}\ M$$

F is the formal concentration of H_2L^+ (= 0.050 0 M in this example).

What is the concentration of L^- in the solution? We have already assumed that it is very small, but it cannot be zero. We can now calculate $[L^-]$ from the K_2 equation (12-4):

$$HL \overset{K_2}{\rightleftharpoons} L^- + H^+ \qquad K_2 = \frac{[H^+][L^-]}{[HL]} \Rightarrow [L^-] = \frac{K_2[HL]}{[H^+]}$$

$$[L^-] = \frac{(1.79 \times 10^{-10})(1.31 \times 10^{-2})}{(1.31 \times 10^{-2})} = 1.79 \times 10^{-10}\ M\ (= K_2)$$

Our approximation that the second dissociation of a diprotic acid is much less than the first dissociation is confirmed by this last result. The concentration of L^- is about eight orders of magnitude smaller than that of HL. As a source of protons, the dissociation of HL is negligible relative to the dissociation of H_2L^+. For most diprotic acids, K_1 is sufficiently larger than K_2 for this approximation to be valid. Even if K_2 were just 10 times less than K_1, the value of $[H^+]$ calculated by ignoring the second ionization would be in error by only 4%. The error in pH would be only 0.01 pH unit. In summary, *a solution of a diprotic acid behaves like a solution of a monoprotic acid, with $K_a = K_1$.*

The Basic Form, L^-

More easy stuff.

The fully basic species, L^-, would be found in a salt such as sodium leucinate, which could be prepared by treating leucine with an equimolar quantity of NaOH. Dissolving sodium leucinate in water gives a solution of L^-, the fully basic species. The K_b values for this dibasic anion are

$$L^- + H_2O \rightleftharpoons HL + OH^- \qquad K_{b1} = K_w/K_{a2} = 5.59 \times 10^{-5}$$

$$HL + H_2O \rightleftharpoons H_2L^+ + OH^- \qquad K_{b2} = K_w/K_{a1} = 2.13 \times 10^{-12}$$

211

Hydrolysis is the reaction of anything with water. Specifically, the reaction $L^- + H_2O \rightleftharpoons HL + OH^-$ is called hydrolysis.

L^- can be treated as monobasic with $K_b = K_{b1}$.

K_{b1} tells us that L^- will not **hydrolyze** (react with water) very much to give HL. Furthermore, K_{b2} tells us that the resulting HL is such a weak base that hardly any further reaction to make H_2L^+ will occur.

We therefore treat L^- as a monobasic species, with $K_b = K_{b1}$. The results of this (fantastic) approximation can be outlined as follows:

$$\underset{\underset{0.050\,0 - x}{L^-}}{\underset{|}{\overset{\overset{R}{|}}{H_2NCHCO_2^-}}} + H_2O \underset{\xrightarrow{\quad K_{b1} = 5.59 \times 10^{-5} \quad}}{\rightleftharpoons} \underset{\underset{x}{HL}}{\overset{\overset{R}{\overset{+\,|}{}}}{H_3NCHCO_2^-}} + \underset{x}{OH^-}$$

$$\frac{x^2}{F - x} = 5.59 \times 10^{-5} \Rightarrow x = [OH^-] = 1.64 \times 10^{-3}\ M$$

$$[HL] = x = 1.64 \times 10^{-3}\ M$$

$$[H^+] = K_w/[OH^-] = K_w/x = 6.08 \times 10^{-12}\ M \Rightarrow pH = 11.22$$

$$[L^-] = F - x = 4.84 \times 10^{-2}\ M$$

The concentration of H_2L^+ can be found from the K_{b2} equilibrium (12-6).

$$HL + H_2O \xrightarrow{\quad K_{b2} \quad} H_2L^+ + OH^- \qquad K_{b2} = \frac{[H_2L^+][OH^-]}{[HL]} = \frac{[H_2L^+]x}{x} = [H_2L^+]$$

We find that $[H_2L^+] = K_{b2} = 2.13 \times 10^{-12}\ M$, and the approximation that $[H_2L^+]$ is insignificant relative to $[HL]$ is well justified. In summary, if there is any reasonable separation between K_1 and K_2 (and, therefore, between K_{b1} and K_{b2}), *the fully basic form of a diprotic acid can be treated as monobasic, with $K_b = K_{b1}$.*

A tougher problem.

The Intermediate Form, HL

A solution prepared from leucine, HL, is more complicated than one prepared from either H_2L^+ or L^-, because HL is both an acid and a base.

HL is both an acid and a base.

$$HL \rightleftharpoons H^+ + L^- \qquad K_a = K_2 = 1.79 \times 10^{-10} \qquad (12\text{-}7)$$
$$HL + H_2O \rightleftharpoons H_2L^+ + OH^- \qquad K_b = K_{b2} = 2.13 \times 10^{-12} \qquad (12\text{-}8)$$

A molecule that can both donate and accept a proton is said to be **amphiprotic.** The acid dissociation reaction (12-7) has a larger equilibrium constant than the base hydrolysis reaction (12-8), so we expect the solution of leucine to be acidic.

However, we cannot simply ignore Reaction 12-8, even if K_a and K_b differ by several orders of magnitude. Both reactions proceed to a nearly equal extent, because H^+ produced in Reaction 12-7 reacts with OH^- from Reaction 12-8, thereby driving Reaction 12-8 to the right.

To treat this case correctly, we resort to writing a *charge balance* (Section 11-7), which says that the sum of positive charges in solution must equal the sum of negative charges in solution. The procedure is applied to leucine, whose intermediate form (HL) has no net charge. However, the results apply to the intermediate form of *any* diprotic acid, regardless of its charge.

Our problem deals with 0.050 0 M leucine, in which both Reactions 12-7 and 12-8 can happen. The charge balance is

$$\underbrace{[H^+] + [H_2L^+]}_{\substack{\text{Sum of positive} \\ \text{charges}}} = \underbrace{[L^-] + [OH^-]}_{\substack{\text{Sum of negative} \\ \text{charges}}}$$

Charge balance for HL:

sum of positive charges = sum of negative charges

which can be rearranged to

$$[H_2L^+] - [L^-] + [H^+] - [OH^-] = 0 \tag{12-9}$$

Using acid dissociation equilibria (Equations 12-3 and 12-4), we replace $[H_2L^+]$ with $[HL][H^+]/K_1$, and $[L^-]$ with $[HL]K_2/[H^+]$. Also, we can always write $[OH^-] = K_w/[H^+]$. Putting these expressions into Equation 12-9 gives

$$\frac{[HL][H^+]}{K_1} - \frac{[HL]K_2}{[H^+]} + [H^+] - \frac{K_w}{[H^+]} = 0$$

which can be solved for $[H^+]$. First, multiply all terms by $[H^+]$:

$$\frac{[HL][H^+]^2}{K_1} - [HL]K_2 + [H^+]^2 - K_w = 0$$

Then factor out $[H^+]^2$ and rearrange:

$$[H^+]^2\left(\frac{[HL]}{K_1} + 1\right) = K_2[HL] + K_w$$

$$[H^+]^2 = \frac{K_2[HL] + K_w}{\dfrac{[HL]}{K_1} + 1}$$

Multiplying the numerator and denominator by K_1 and taking the square root of both sides give

$$[H^+] = \sqrt{\frac{K_1K_2[HL] + K_1K_w}{K_1 + [HL]}} \tag{12-10}$$

We solved for $[H^+]$ in terms of known constants plus the single unknown, $[HL]$. Where do we proceed from here?

Luckily, a chemist gallops down from the mountain mists on her snow-white unicorn to provide the missing insight: "The major species will be HL, because it is both a weak acid and a weak base. Neither Reaction 12-7 nor Reaction 12-8 goes very far. For the concentration of HL in Equation 12-10, you can simply substitute the value 0.050 0 M."

The missing insight!

Taking the chemist's advice, we rewrite Equation 12-10 as follows:

$$[H^+] \approx \sqrt{\frac{K_1K_2F + K_1K_w}{K_1 + F}} \tag{12-11}$$

where F is the formal concentration of HL (= 0.050 0 M). It turns out that Equation 12-11 can be further simplified under most conditions. The first term in the numerator is almost always much greater than the second term, so the second term can be dropped:

$$[H^+] \approx \sqrt{\frac{K_1 K_2 F + \cancel{K_1 K_w}}{K_1 + F}}$$

Then, if $K_1 \ll F$, the first term in the denominator can also be neglected.

$$[H^+] \approx \sqrt{\frac{K_1 K_2 F}{\cancel{K_1} + F}}$$

Canceling F in the numerator and denominator gives

$$[H^+] \approx \sqrt{K_1 K_2} = (K_1 K_2)^{1/2} \qquad (12\text{-}12)$$

Making use of the identity $\log(x^{1/2}) = \frac{1}{2}\log x$, we rewrite Equation 12-12 in the form

$$\log[H^+] \approx \tfrac{1}{2}\log(K_1 K_2)$$

Noting that $\log xy = \log x + \log y$, we can rewrite the equation once more:

$$\log[H^+] \approx \tfrac{1}{2}(\log K_1 + \log K_2)$$

To convert the left side of the preceding equation to pH (= $-\log$ [H$^+$]) and the right side to pK (= $-\log K$), multiply both sides by -1:

$$\underbrace{-\log[H^+]}_{\text{pH}} \approx \tfrac{1}{2}(\underbrace{-\log K_1}_{\text{p}K_1} \underbrace{-\log K_2}_{\text{p}K_2})$$

The pH of the intermediate form of a diprotic acid is close to midway between the two pK_a values and is almost independent of concentration.

Intermediate form of diprotic acid:

$$\boxed{\text{pH} \approx \tfrac{1}{2}(\text{p}K_1 + \text{p}K_2)} \qquad (12\text{-}13)$$

where K_1 and K_2 are the acid dissociation constants (= K_{a1} and K_{a2}) of the diprotic acid.

Equation 12-13 is a good one to keep in your head. It says that *the pH of a solution of the intermediate form of a diprotic acid is close to midway between pK$_1$ and pK$_2$, regardless of the formal concentration.*

For leucine, Equation 12-13 gives a pH of $\frac{1}{2}(2.329 + 9.747) = 6.038$, or $[H^+] = 10^{-\text{pH}} = 9.16 \times 10^{-7}$ M. The concentrations of H_2L^+ and L^- can be found from the K_1 and K_2 equilibria, using $[HL] = 0.050\ 0$ M.

$$[H_2L^+] = \frac{[H^+][HL]}{K_1} = \frac{(9.16 \times 10^{-7})(0.050\ 0)}{4.69 \times 10^{-3}} = 9.77 \times 10^{-6} \text{ M}$$

$$[L^-] = \frac{K_2[HL]}{[H^+]} = \frac{(1.79 \times 10^{-10})(0.050\ 0)}{9.16 \times 10^{-7}} = 9.77 \times 10^{-6} \text{ M}$$

Was the approximation $[HL] \approx 0.050\ 0$ M a good one? It certainly was, because $[H_2L^+]$ (= 9.77×10^{-6} M) and $[L^-]$ (= 9.77×10^{-6} M) are small in comparison with $[HL]$ ($\approx 0.050\ 0$ M). Nearly all the leucine remained in the form HL.

Potassium hydrogen phthalate, KHP, is a salt of the intermediate form of phathalic acid. Calculate the pH of 0.10 M KHP and 0.010 M KHP.

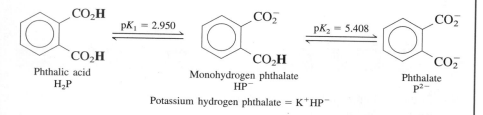

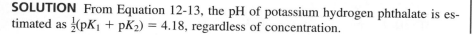

Phthalic acid
H_2P

Monohydrogen phthalate
HP^-

Phthalate
P^{2-}

Potassium hydrogen phthalate = K^+HP^-

SOLUTION From Equation 12-13, the pH of potassium hydrogen phthalate is estimated as $\frac{1}{2}(pK_1 + pK_2) = 4.18$, regardless of concentration.

Summary of Diprotic Acid Calculations

The fully protonated species H_2A is treated as a monoprotic acid with acid dissociation constant K_1. The fully basic species A^{2-} is treated as a monoprotic base, with base association constant $K_{b1} = K_w/K_{a2}$. For the intermediate form HA^-, use the equation pH $\approx \frac{1}{2}(pK_1 + pK_2)$, where K_1 and K_2 are the acid dissociation constants for H_2A. The same considerations apply to a diprotic base (B $\rightarrow$ BH^+ $\rightarrow$ BH_2^{2+}): B is treated as monobasic; BH_2^{2+} is treated as monoprotic; and BH^+ is treated as an intermediate with pH $\approx \frac{1}{2}(pK_1 + pK_2)$, where K_1 and K_2 are the acid dissociation constants of BH_2^{2+}.

Diprotic systems:

- Treat H_2A and BH_2^{2+} as monoprotic weak acids
- Treat A^{2-} and B as monoprotic weak bases
- Treat HA^- and BH^+ as intermediates: pH $\approx \frac{1}{2}(pK_1 + pK_2)$

Ask Yourself

12-B. Find the pH and the concentrations of H_2SO_3, HSO_3^-, and SO_3^{2-} in each of these solutions: **(a)** 0.050 M H_2SO_3; **(b)** 0.050 M $NaHSO_3$; **(c)** 0.050 M Na_2SO_3.

12-3 Which Is the Principal Species?

We are often faced with the problem of identifying which species of acid, base, or intermediate is predominant under given conditions. For example, what is the principal form of benzoic acid in an aqueous solution at pH 8?

Benzoic acid
$pK_a = 4.20$

The pK_a for benzoic acid is 4.20. This means that at pH 4.20 there is a 1:1 mixture of benzoic acid (HA) and benzoate ion (A^-). At pH = $pK_a + 1$ (= 5.20), the

At pH = pK_a, $[A^-] = [HA]$ because

$$pH = pK_a + \log\left(\frac{[A^-]}{[HA]}\right)$$
$$= pK_a + \log 1 = pK_a.$$

pH	Major species
$< pK_a$	HA
$> pK_a$	A^-

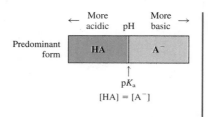

quotient $[A^-]/[HA]$ is 10:1. At $pH = pK_a + 2$ ($= 6.20$), the quotient $[A^-]/[HA]$ is 100:1. As the pH increases, the quotient $[A^-]/[HA]$ increases still further.

For a monoprotic system, the basic species, A^-, is the predominant form when $pH > pK_a$. The acidic species, HA, is the predominant form when $pH < pK_a$. The predominant form of benzoic acid at pH 8 is the benzoate anion, $C_6H_5CO_2^-$.

EXAMPLE **Principal Species—Which One and How Much?**

What is the predominant form of ammonia in a solution at pH 7.0? Approximately what fraction is in this form?

SOLUTION In Appendix B we find $pK_a = 9.24$ for the ammonium ion (NH_4^+, the conjugate acid of ammonia, NH_3). At $pH = 9.24$, $[NH_4^+] = [NH_3]$. Below pH 9.24, NH_4^+ will be the predominant form. Because $pH = 7.0$ is about 2 pH units below pK_a, the quotient $[NH_3]/[NH_4^+]$ will be around 1:100. Approximately 99% is in the form NH_4^+.

pH	Major species
$pH < pK_1$	H_2A
$pK_1 < pH < pK_2$	HA^-
$pH > pK_2$	A^{2-}

For diprotic systems, the reasoning is the same, but there are two pK_a's. Consider fumaric acid, H_2A, with $pK_1 = 3.053$ and $pK_2 = 4.494$. At $pH = pK_1$, $[H_2A] = [HA^-]$. At $pH = pK_2$, $[HA^-] = [A^{2-}]$. The chart in the margin shows the major species in each pH region. At pH values below pK_1, H_2A is dominant. At pH values above pK_2, A^{2-} is dominant. At pH values between pK_1 and pK_2, HA^- is dominant. Figure 12-1 shows the fraction of each species as a function of pH.

The diagram below shows the major species for a triprotic system and introduces an important extension of what we learned in the previous section.

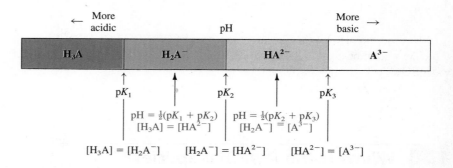

H_3A is the dominant species of a triprotic system in the most acidic solution at $pH < pK_1$. H_2A^- is dominant between pK_1 and pK_2. HA^{2-} is the major species between pK_2 and pK_3, and A^{3-} dominates in the most basic solution at $pH > pK_3$.

The diagram above shows that the pH of the first intermediate species, H_2A^-, is $\frac{1}{2}(pK_1 + pK_2)$. At this pH, the concentrations of H_3A and HA^{2-} are small and equal to each other. *The new feature of the diagram is that the pH of the second intermediate species, HA^{2-}, is $\frac{1}{2}(pK_2 + pK_3)$. At this pH, the concentrations of H_2A^- and A^{3-} are small and equal to each other.*

Triprotic systems:

- Treat H_3A as monoprotic weak acid
- Treat A^{3-} as monoprotic weak base
- Treat H_2A^- as intermediate: $pH \approx \frac{1}{2}(pK_1 + pK_2)$
- Treat HA^{2-} as intermediate: $pH \approx \frac{1}{2}(pK_2 + pK_3)$

The amino acid arginine has the following forms:

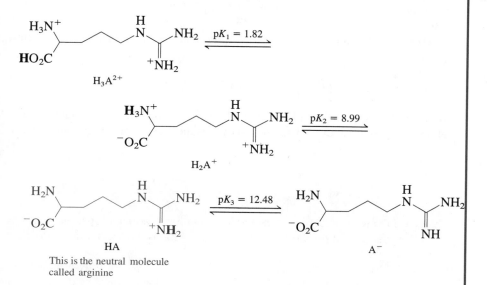

H₃A²⁺ $\xrightarrow{pK_1 = 1.82}$ H₂A⁺ $\xrightarrow{pK_2 = 8.99}$ HA $\xrightarrow{pK_3 = 12.48}$ A⁻

This is the neutral molecule called arginine

Note that the ammonium group next to the carboxyl group at the left is more acidic than the substituent ammonium group at the right. What is the principal form of arginine at pH 10.0? Approximately what fraction is in this form? What is the second most abundant form at this pH?

SOLUTION We know that at $pH = pK_2 = 8.99$, $[H_2A^+] = [HA]$. At $pH = pK_3 = 12.48$, $[HA] = [A^-]$. At $pH = 10.0$, the major species is HA. Because pH 10.0 is about one pH unit higher than pK_2, we can say that $[HA]/[H_2A^+] \approx 10:1$. About 90% of arginine is in the form HA. The second most important species is H_2A^+, which makes up about 10% of the arginine.

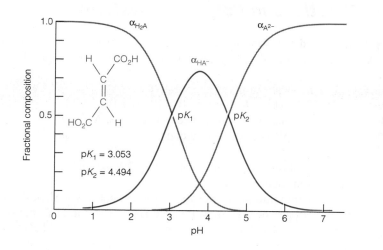

FIGURE 12-1 Fractional composition diagram for fumaric acid (*trans*-butenedioic acid). α_i is the fraction of species i at each pH. At low pH, H_2A is the dominant form. At intermediate pH, HA^- is dominant and at high pH, A^{2-} dominates. Because pK_1 and pK_2 are not very different, the fraction of HA^- never gets very close to unity. The procedure for computing these curve is analogous to that used for Figure 9-4 for a monoprotic system.

EXAMPLE **More on Polyprotic Systems**

In the pH range 1.82 to 8.99, H_2A^+ is the principal form of arginine. Which is the second most prominent species at pH 6.0? At pH 5.0?

SOLUTION We know that the pH of the pure intermediate (amphiprotic) species, H_2A^+, is

$$\text{pH of } H_2A^+ \approx \tfrac{1}{2}(pK_1 + pK_2) = 5.40$$

Above pH 5.40 (and below pH = pK_2), HA is the second most important species. Below pH 5.40 (and above pH = pK_1), H_3A^{2+} is the second most important species.

Ask Yourself

12-C. (a) Draw the structure of the predominant form (principal species) of 1,3-dihydroxybenzene at pH 9.00 and at pH 11.00. What is the second most prominent species at each pH?

(b) The amino acid cysteine is a triprotic system whose fully protonated form could be designated H_3C^+. Which form of cysteine is drawn below: H_3C^+, H_2C, HC^-, or C^{2-}? What would the pH of a 0.10 M solution of this form of cysteine be?

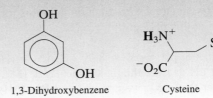

1,3-Dihydroxybenzene   Cysteine

12-4 *Titrations in Polyprotic Systems*

Figure 11-2 showed the titration curve for the monoprotic acid HA treated with OH^-. As a brief reminder, the pH at several critical points was computed as follows:

Initial solution:	Has the pH of the weak acid HA
$V_e/2$:	pH = pK_a because [HA] = [A^-]
V_e:	Has the pH of the conjugate base, A^-
Past V_e:	pH is governed by concentration of excess OH^-

The slope of the titration curve is minimum at $V_e/2$ when pH = pK_a. The slope is maximum at the equivalence point.

Before delving into titration curves for diprotic systems, you should realize that there are two buffer pairs derived from the acid H_2A. H_2A and HA^- constitute one buffer pair and HA^- and A^{2-} constitute a second pair. For the acid H_2A, there are *two* Henderson-Hasselbalch equations, both of which are *always* true. If you happen

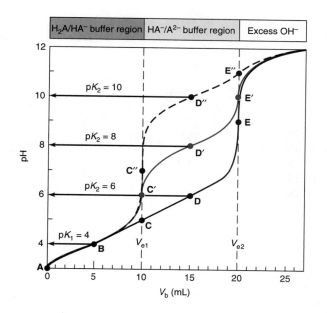

FIGURE 12-2 Calculated titration curves for 50.0 mL of 0.020 0 M H_2A titrated with 0.100 M NaOH. *Lowest curve:* $pK_1 = 4.00$ and $pK_2 = 6.00$. *Middle curve:* $pK_1 = 4.00$ and $pK_2 = 8.00$. *Upper curve:* $pK_1 = 4.00$ and $pK_2 = 10.00$.

to know the concentrations $[H_2A]$ and $[HA^-]$, then use the pK_1 equation. If you know $[HA^-]$ and $[A^{2-}]$, use the pK_2 equation.

$$pH = pK_1 + \log\left(\frac{[HA^-]}{[H_2A]}\right) \qquad pH = pK_2 + \log\left(\frac{[A^{2-}]}{[HA^-]}\right)$$

All Henderson-Hasselbalch equations are always true for a solution at equilibrium.

So now let us turn our attention to the titration of diprotic acids. Figure 12-2 shows calculated curves for 50.0 mL each of 0.020 0 M diprotic acids H_2A titrated with 0.100 M OH^-. For all three curves, H_2A has $pK_1 = 4.00$. In the lowest curve, pK_2 is 6.00. In the middle curve, pK_2 is 8.00, and in the upper curve pK_2 is 10.00. The first equivalence volume (V_{e1}) occurs when the moles of added OH^- equal the moles of H_2A. The second equivalence volume (V_{e2}) is always exactly twice as great as the first equivalence volume, because we must add the same amount of OH^- to convert HA^- to A^{2-}. Let's consider why pH varies as it does during these titrations.

$V_{e2} = 2V_{e1}$ (always!)

Point A has the same pH in all three cases. It is the pH of the acid H_2A, which is treated as a monoprotic acid with $pK_a = pK_1 = 4.00$ and formal concentration F.

$$HA \rightleftharpoons A^- + H^+ \qquad \frac{x^2}{F - x} = K_1 \qquad (12\text{-}14)$$
$$F-x \qquad x \qquad x$$

Point A: weak acid H_2A

$$\frac{x^2}{0.020\ 0 - x} = 10^{-4.00} \Rightarrow x = 1.37 \times 10^{-3}\ M \Rightarrow pH = -\log(x) = 2.86$$

Point B, which is halfway to the first equivalence point, has the same pH in all three cases. It is the pH of a 1:1 mixture of $H_2A:HA^-$, which is treated as a monoprotic acid with $pK_a = pK_1 = 4.00$ and formal concentration F:

$$pH = pK_1 + \log\left(\frac{[HA^-]}{[H_2A]}\right) = pK_1 + \log 1 = pK_1 = 4.00 \qquad (12\text{-}15)$$

Point B: buffer containing $H_2A + HA^-$

Because all three acids have the same pK_1, the pH at this point is the same in all three cases.

Point C (and C′ and C″) is the first equivalence point. H_2A has been converted to HA^-, the intermediate form of a diprotic acid. The pH is calculated from Equation 12-13:

Point C: intermediate form HA^-

$$pH \approx \tfrac{1}{2}(pK_1 + pK_2) = \begin{cases} 5.00 \text{ at C} \\ 6.00 \text{ at C′} \\ 7.00 \text{ at C″} \end{cases} \tag{12-13}$$

The three acids have the same pK_1, but different values of pK_2. Therefore the pH at the first equivalence point is different in all three cases.

Point D is halfway from the first equivalence point to the second equivalence point. Half of the HA^- has been converted to A^{2-}. The pH is

Point D: buffer containing $HA^- + A^{2-}$

$$pH = pK_2 + \log\left(\frac{[A^{2-}]}{[HA^-]}\right) = pK_2 + \log 1 = pK_2 = \begin{cases} 6.00 \text{ at D} \\ 8.00 \text{ at D′} \\ 10.00 \text{ at D″} \end{cases} \tag{12-16}$$

Look at Figure 12-2 and you will see that points D, D′, and D″ come at pK_2 for each of the acids.

Point E is the second equivalence point. All of the acid has been converted into A^{2-}, a weak base with concentration F'. We find the pH by treating A^{2-} as a monoprotic base with $K_{b1} = K_w/K_{a2}$:

Point E: weak base A^{2-}

$$A^{2-} + H_2O \rightleftharpoons HA^- + OH^- \qquad \frac{x^2}{F' - x} = K_{b1} \tag{12-17}$$
$$\phantom{A^{2-} + H_2O \rightleftharpoons} F' - x \qquad\quad x \qquad x$$

The pH is different for the three acids, because K_{b1} is different for all three. Here is how we find the pH at the second equivalence point:

$$F' = [A^{2-}] = \frac{\text{mmol } A^{2-}}{\text{total mL}} = \frac{(0.020\ 0 \text{ M})(50.0 \text{ mL})}{70.0 \text{ mL}} = 0.014\ 3 \text{ M}$$

$$\frac{x^2}{0.014\ 3 - x} = K_{b1} = \begin{cases} 10^{-8.00} \text{ at E} \\ 10^{-6.00} \text{ at E′} \\ 10^{-4.00} \text{ at E″} \end{cases} \Rightarrow x = \begin{cases} 1.20 \times 10^{-5} \text{ at E} \\ 1.19 \times 10^{-4} \text{ at E′} \\ 1.15 \times 10^{-3} \text{ at E″} \end{cases} = [OH^-]$$

$$\Rightarrow pH = -\log(K_w/x) = \begin{cases} 9.08 \text{ at E} \\ 10.08 \text{ at E′} \\ 11.06 \text{ at E″} \end{cases}$$

Beyond V_{e2}, the pH is governed by the concentration of excess OH^-. pH rapidly converges to the same value for all three titrations.

You can see in Figure 12-2 that in a favorable case (the middle curve), we observe two obvious, steep equivalence points in the titration curve. When the pK values are too close together, or when the pK values are too low or too high, there may not be a distinct break at each equivalence point.

EXAMPLE   **Titration of Sodium Carbonate**

Let's reverse the process of Figure 12-2 and calculate the pH at points A–E in the titration of a diprotic base. Figure 12-3 shows the calculated titration curve for 50.0

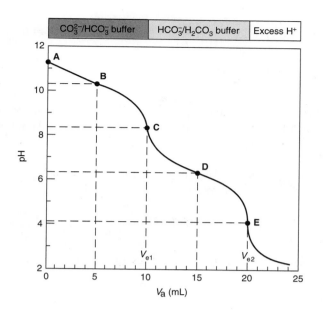

FIGURE 12-3 Calculated titration curve for 50.0 mL of 0.020 0 M Na_2CO_3 titrated with 0.100 M HCl.

mL of 0.020 0 M Na_2CO_3 titrated with 0.100 M HCl. The first equivalence point is at 10.0 mL and the second is at 20.0 mL. Find the pH at points A–E.

$$Na_2CO_3 \quad pK_1 = 6.352 \quad K_{a1} = 4.45 \times 10^{-7} \quad K_{b1} = K_w/K_{a2} = 2.13 \times 10^{-4}$$
$$pK_2 = 10.329 \quad K_{a2} = 4.69 \times 10^{-11} \quad K_{b2} = K_w/K_{a1} = 2.25 \times 10^{-8}$$

$$H_2CO_3 \xrightleftharpoons{pK_1} HCO_3^- \xrightleftharpoons{pK_2} CO_3^{2-}$$
Carbonic acid Carbonate

SOLUTION *Point A:* The initial point in the titration is just a solution of 0.020 0 M Na_2CO_3, which can be treated as a monoprotic base:

$$\underset{0.020\ 0 - x}{CO_3^{2-}} + H_2O \xrightleftharpoons{K_{b1}} \underset{x}{HCO_3^-} + \underset{x}{OH^-} \qquad \frac{x^2}{0.020\ 0 - x} = K_{b1}$$

$$\Rightarrow x = 1.96 \times 10^{-3} = [OH^-] \Rightarrow pH = -\log(K_w/x) = 11.29$$

Point B: Now we are halfway to the first equivalence point. Half of the carbonate has been converted to bicarbonate, so there is a 1:1 mixture of CO_3^{2-} and HCO_3^-—*Aha! A buffer!*

$$pH = pK_2 + \log\left(\frac{[CO_3^{2-}]}{[HCO_3^-]}\right) = pK_2 + \log 1 = pK_2 = 10.33$$

↑
Use pK_2 because the equilibrium
involves CO_3^{2-} and HCO_3^-

Point C: At the first equivalence point, we have a solution of HCO_3^-, the intermediate form of a diprotic acid. To a good approximation, the pH is independent of concentration and is given by

$$pH \approx \tfrac{1}{2}(pK_1 + pK_2) = \tfrac{1}{2}(6.352 + 10.329) = 8.34$$

Point D: We are halfway to the second equivalence point. Half of the bicarbonate has been converted to carbonic acid, so there is a 1:1 mixture of HCO_3^- and H_2CO_3—*Aha! Another buffer!*

$$pH = pK_1 + \log\left(\frac{[HCO_3^-]}{[H_2CO_3]}\right) = pK_1 + \log 1 = pK_1 = 6.35$$

Use pK_1 because the equilibrium
involves HCO_3^- and H_2CO_3

Point E: At the second equivalence point, all carbonate has been converted to carbonic acid, which has been diluted from its initial volume of 50.0 mL to a volume of 70.0 mL.

$$F' = [H_2CO_3] = \frac{\text{mmol } H_2CO_3}{\text{total mL}} = \frac{(0.020\ 0\ M)(50.0\ mL)}{70.0\ mL} = 0.014\ 3\ M$$

$$H_2CO_3 \overset{K_1}{\rightleftharpoons} HCO_3^- + H^+ \qquad \frac{x^2}{0.014\ 3 - x} = K_1$$
$$F' - x \qquad\quad x \qquad x$$

$$\Rightarrow x = 7.98 \times 10^{-5} = [H^+] \Rightarrow pH = -\log(x) = 4.10$$

Proteins Are Polyprotic Acids and Bases

Proteins are polymers made of amino acids:

$$\overset{+}{H_3N}-\underset{R_1}{\overset{H}{\underset{|}{\overset{|}{C}}}}-CO_2^- + \overset{+}{H_3N}-\underset{R_2}{\overset{H}{\underset{|}{\overset{|}{C}}}}-CO_2^- + \overset{+}{H_3N}-\underset{R_3}{\overset{H}{\underset{|}{\overset{|}{C}}}}-CO_2^- \qquad \text{Amino acids}$$

$-2\ H_2O$

$$\overset{+}{H_3N}-\underset{R_1}{\overset{H}{C}}-\overset{O}{\overset{\|}{C}}-\underset{H}{\overset{}{N}}-\underset{R_2}{\overset{H}{C}}-\overset{O}{\overset{\|}{C}}-\underset{H}{\overset{}{N}}-\underset{R_3}{\overset{H}{C}}-CO_2^-$$

A polypeptide
(A long polypeptide
is called a protein)

N-terminal residue · · · Peptide bond · · · C-terminal residue

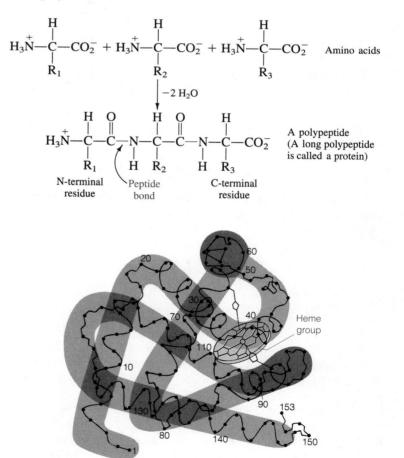

FIGURE 12-4 Structure showing folding of the amino acid backbone of myoglobin. Substituents (R groups from Table 12-1) are omitted for clarity. The flat *heme* group at the right side of the protein contains an iron atom that can bind O_2, CO, and other small molecules.

Myoglobin

Box 12-1 *Explanation*

What Is Isoelectric Focusing?

At its *isoelectric pH,* a protein has zero net charge and will therefore not migrate in an electric field. This effect is the basis of a very sensitive technique of protein separation called **isoelectric focusing.** A mixture of proteins is subjected to a strong electric field in a medium specifically designed to have a pH gradient. Positively charged molecules move toward the negative pole and negatively charged molecules move toward the positive pole. The proteins migrate in one direction or the other until the point where the pH is the same as their isoelectric pH. At this point, they have no net charge and no longer move. Thus, each protein in the mixture is focused in one small region at its isoelectric pH.

An example of isoelectric focusing is shown in the figure. A mixture of proteins was applied to a polyacrylamide gel containing a mixture of polyprotic compounds called *ampholytes.* Several hundred volts were applied across the length of the gel. The ampholytes migrated until they formed a stable pH gradient (ranging from about pH 3 at one end of the gel to pH 10 at the other). The proteins migrated until they reached regions with their isoelectric pH, at which point they had no net charge and ceased migrating. If a molecule diffuses out of its isoelectric region, it becomes charged and immediately migrates back to its isoelectric zone. When the proteins finished migrating, the electric field was removed and proteins were precipitated in place on the gel and stained with a dye to make them visible.

The stained gel is shown at the bottom of the figure. A spectrophotometer scan of the dye peaks is shown on the graph, and a profile of measured pH is also plotted. Each dark band of stained protein gives an absorbance peak.

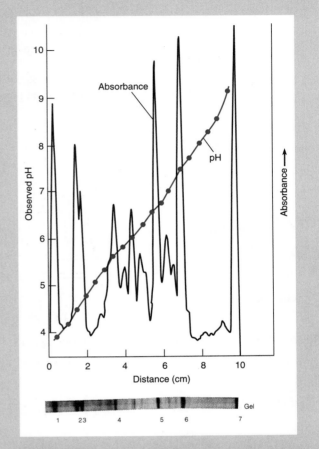

Isoelectric focusing of a mixture of proteins: (1) soybean trypsin inhibitor; (2) β-lactoglobulin A; (3) β-lactoglobulin B; (4) ovotransferrin; (5) horse myoglobin; (6) whale myoglobin; (7) cytochrome *c.*

Proteins have biological functions such as structural support, catalysis of chemical reactions, immune response to foreign substances, transport of molecules across membranes, and control of genetic expression. The three-dimensional structure and function of a protein are determined by the sequence of amino acids from which the protein is made. Figure 12-4 shows the general shape of the protein myoglobin, whose function is to store O_2 in muscle cells. Of the 153 amino acids in sperm-whale myoglobin, 35 have basic side groups and 23 are acidic.

At high pH, most proteins have lost so many protons that they have a negative charge. At low pH, most proteins have gained so many protons that they have a positive charge. At some intermediate pH, called the *isoelectric pH* (or isoelectric point), each protein has exactly zero net charge. Box 12-1 explains how proteins can be separated from each other on the basis of their different isoelectric points.

Ask Yourself

12-D. Consider the titration of 50.0 mL of 0.050 0 M malonic acid with 0.100 M NaOH.

    (a) How many milliliters of titrant are required to reach each equivalence point?

    (b) Calculate the pH at $V_b = 0.0, 12.5, 25.0, 37.5, 50.0,$ and 55.0 mL.

    (c) Put the points from **(b)** on a graph and sketch the titration curve.

Key Equations

Diprotic acid equilibria	$H_2A \rightleftharpoons HA^- + H^+ \qquad K_{a1} \equiv K_1$ $HA^- \rightleftharpoons A^{2-} + H^+ \qquad K_{a2} \equiv K_2$
Diprotic base equilibria	$A^{2-} + H_2O \rightleftharpoons HA^- + OH^- \qquad K_{b1}$ $HA^- + H_2O \rightleftharpoons H_2A + OH^- \qquad K_{b2}$

Relation between K_a and K_b

Monoprotic system	$K_a K_b = K_w$	Triprotic system	$K_{a1} K_{b3} = K_w$
Diprotic system	$K_{a1} K_{b2} = K_w$		$K_{a2} K_{b2} = K_w$
	$K_{a2} K_{b1} = K_w$		$K_{a3} K_{b1} = K_w$

pH of H_2A (or BH_2^{2+})

$$H_2A \overset{K_{a1}}{\rightleftharpoons} H^+ + HA^- \qquad \frac{x^2}{F - x} = K_{a1}$$
$$\;\; F-x \qquad\quad\; x \quad\; x$$

(This gives $[H^+]$, $[HA^-]$, and $[H_2A]$. You can solve for $[A^{2-}]$ from the K_{a2} equilibrium.)

pH of HA^- (or diprotic BH^+) $pH \approx \frac{1}{2}(pK_1 + pK_2)$

pH of A^{2-} (or diprotic B)

$$A^{2-} + H_2O \overset{K_{b1}}{\rightleftharpoons} HA^- + OH^- \qquad \frac{x^2}{F - x} = K_{b1} = \frac{K_w}{K_{a2}}$$
$$F-x \qquad\qquad\; x \quad\;\; x$$

(This gives $[OH^-]$, $[HA^-]$, and $[A^{2-}]$. You can find $[H^+]$ from the K_w equilibrium and $[H_2A]$ from the K_{a1} equilibrium.)

Diprotic buffer

$$pH = pK_1 + \log\left(\frac{[HA^-]}{[H_2A]}\right) \qquad\qquad pH = pK_2 + \log\left(\frac{[A^{2-}]}{[HA^-]}\right)$$

Both equations are always true and either can be used, depending on which set of concentrations you happen to know.

Titration of H_2A with OH^-

$V_b = 0$	Find pH of H_2A	$V_b = \frac{3}{2}V_{e1}$	$pH = pK_2$
$V_b = \frac{1}{2}V_{e1}$	$pH = pK_1$	$V_b = V_{e2}$	Find pH of A^{2-}
$V_b = V_{e1}$	$pH \approx \frac{1}{2}(pK_1 + pK_2)$	$V_b > V_{e2}$	Find concentration of excess OH^-

Titration of diprotic B with H^+ $V_a = 0$ Find pH of B $V_a = \frac{3}{2}V_{e1}$ pH $= pK_{a1}$ (for BH_2^{2+})

$V_a = \frac{1}{2}V_{e1}$ pH $= pK_{a2}$ (for BH_2^{2+}) $V_a = V_{e2}$ Find pH of BH_2^{2+}

$V_a = V_{e1}$ pH $\approx \frac{1}{2}(pK_{a1} + pK_{a2})$ $V_a > V_{e2}$ Find concentration of excess H^+

Principal species

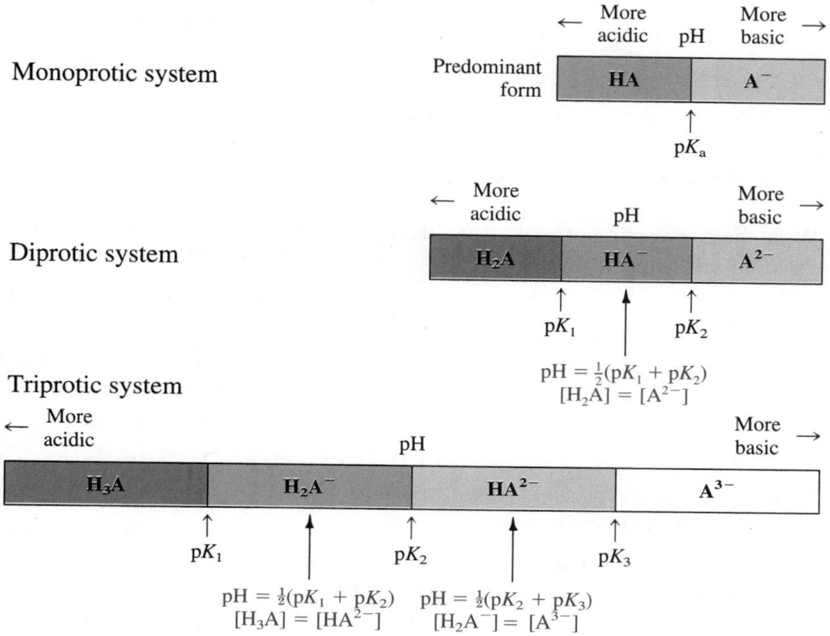

Monoprotic system

Diprotic system

Triprotic system

Important Terms

amino acid

amphiprotic

hydrolysis

isoelectric focusing

polyprotic acid

zwitterion

Problems

12-1. Write the K_{a2} reaction of sulfuric acid (H_2SO_4) and the K_{b2} reaction of disodium oxalate ($Na_2C_2O_4$) and find their numerical values.

12-2. The base association constants of phosphate are $K_{b1} = 0.014$, $K_{b2} = 1.58 \times 10^{-7}$, and $K_{b3} = 1.41 \times 10^{-12}$. From the K_b values, calculate K_{a1}, K_{a2}, and K_{a3} for H_3PO_4.

12-3. Write the general structure of an amino acid. Why do some amino acids in Table 12-1 have two pK values and others three?

12-4. Write the stepwise acid-base reactions for the following species in water. Write the correct symbol (e.g., K_{b1}) for the equilibrium constant for each reaction and find its numerical value.

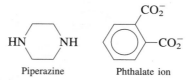

Piperazine   Phthalate ion

12-5. Write the K_{a2} reaction of proline and the K_{b2} reaction of the trisodium salt below.

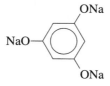

12-6. From the K_a values for citric acid in Appendix B, find K_{b1}, K_{b2}, and K_{b3} for trisodium citrate.

12-7. Write the chemical reactions whose equilibrium constants are K_{b1} and K_{b2} for the amino acid serine, and find their numerical values.

12-8. We will abbreviate malonic acid, $CH_2(CO_2H)_2$, as H_2M. Find the pH and concentrations of H_2M, HM^-, and M^{2-} in each of the following solutions: **(a)** 0.100 M H_2M; **(b)** 0.100 M NaHM; **(c)** 0.100 M Na_2M.

12-9. Consider the dibasic compound, B, that can form BH^+ and BH_2^{2+} with $K_{b1} = 1.00 \times 10^{-5}$ and $K_{b2} = 1.00 \times 10^{-9}$. Find the pH and concentrations of B, BH^+, and BH_2^{2+} in each of the following solutions: **(a)** 0.100 M B; **(b)** 0.100 M BH^+I^-; **(c)** 0.100 M $BH_2^{2+}(I^-)_2$.

12-10. Calculate the pH of a 0.300 M solution of the dibasic compound piperazine, which we will designate B. Calculate the concentration of each form of piperazine (B, BH^+, BH_2^{2+}) in this solution.

12-11. The salt piperazine monohydrochloride is formed when one mole of HCl is added to the dibasic compound piperazine:

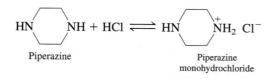

Piperazine   Piperazine
                              monohydrochloride

Find the pH of 0.150 M piperazine monohydrochloride and calculate the concentration of each form of piperazine in this solution.

12-12. Draw the structure of the amino acid glutamine and satisfy yourself that it is the intermediate form of a diprotic system. Find the pH of 0.050 M glutamine.

12-13. The acid HA has p$K_a = 7.00$.
   (a) Which is the principal species, HA or A^-, at pH 6.00?
   (b) Which is the principal species at pH 8.00?
   (c) What is the quotient $[A^-]/[HA]$ at pH 7.00? at pH 6.00?

12-14. The diprotic acid H_2A has p$K_1 = 4.00$ and p$K_2 = 8.00$.
   (a) At what pH is $[H_2A] = [HA^-]$?
   (b) At what pH is $[HA^-] = [A^{2-}]$?
   (c) Which is the principal species, H_2A, HA^-, or A^{2-}, at pH 2.00?
   (d) Which is the principal species at pH 6.00?
   (e) Which is the principal species at pH 10.00?

12-15. The base B has p$K_b = 5.00$.
   (a) What is the value of pK_a for the acid BH^+?
   (b) At what pH is $[BH^+] = [B]$?
   (c) Which is the principal species, B or BH^+, at pH 7.00?
   (d) What is the quotient $[B]/[BH^+]$ at pH 12.00?

12-16. Draw the structures of the predominant forms of glutamic acid and tyrosine at pH 9.0 and pH 10.0. What is the second most abundant species at each pH?

12-17. Calculate the pH of a 0.10 M solution of each amino acid in the form drawn below.

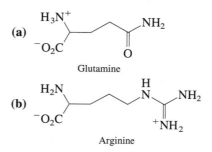

(a)
Glutamine

(b)
Arginine

12-18. Draw the structure of the predominant form of pyridoxal-5-phosphate at pH 7.00.

12-19. Find the pH and concentration of each species of arginine in 0.050 M arginine · HCl solution.

12-20. What is the charge of the predominant form of citric acid at pH 5.00?

12-21. A 100.0-mL aliquot of 0.100 M diprotic acid H_2A ($pK_1 = 4.00$, $pK_2 = 8.00$) was titrated with 1.00 M NaOH. At what volumes are the two equivalence points? Find the pH at the following volumes of base added and sketch a graph of pH versus V_b: $V_b = 0$, 5.0, 10.0, 15.0, 20.0, and 22.0 mL.

12-22. The dibasic compound B ($pK_{b1} = 4.00$, $pK_{b2} = 8.00$) was titrated with 1.00 M HCl. The initial solution of B was 0.100 M and had a volume of 100.0 mL. At what volumes are the two equivalence points? Find the pH at the following volumes of acid added and sketch a graph of pH versus V_a: $V_a = 0$, 5.0, 10.0, 15.0, 20.0, and 22.0 mL.

12-23. Select one indicator from Table 10-3 that would be useful for detecting each equivalence point shown in Figure 12-2 and listed below. State the color change you would observe in each case.

    (a) Second equivalence point of the lowest curve.

    (b) First equivalence point of the upper curve.

    (c) First equivalence point of the middle curve.

    (d) Second equivalence point of the middle curve.

12-24. Write the two consecutive reactions that occur when 40.0 mL of 0.100 M piperazine is titrated with 0.100 M HCl and indicate the equivalence volumes. Find the pH at $V_a = 0$, 20.0, 40.0, 60.0, 80.0, and 100.0 mL and sketch the titration curve.

12-25. Write the chemical reactions (including structures of reactants and products) that occur when the amino acid histidine is titrated with perchloric acid. (Histidine is a molecule with no net charge.) A solution containing 25.0 mL of 0.050 0 M histidine was titrated with 0.050 0 M $HClO_4$. List the equivalence volumes and calculate the pH at $V_a = 0$, 12.5, 25.0, and 50.0 mL.

12-26. An aqueous solution containing ~1 g of oxobutanedioic acid (MW 132.073) per 100 mL was titrated with 0.094 32 M NaOH to measure the acid molarity.

    (a) What will be the pH at each equivalence point?

    (b) Which equivalence point would be best to use in this titration?

    (c) You have the indicators erythrosine, ethyl orange, bromocresol green, bromothymol blue, thymolphthalein, and alizarin yellow. Which indicator will you use and what color change will you look for?

A Heparin Sensor Electrode

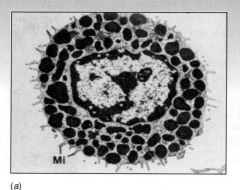

(a)

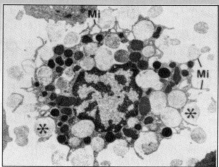

(b)

(*a*) Mast cell with numerous dark granules containing regulatory molecules such as heparin. (*b*) Granules expelled from the mast cell release heparin into a blood vessel. Two of the granules in the process of exiting the cell are denoted by asterisks. (Mi indicates details of a cell's surface called microvilli and microfolds.)

Blood clotting must be precisely regulated in the human body to minimize *hemorrhage* (bleeding) in an injury and *thrombosis* (uncontrolled clotting) in healthy tissue. The exquisitely complex clotting process is regulated by numerous substances, including *heparin,* a negatively charged molecule secreted by mast cells near the walls of blood vessels.

The concentration of heparin administered during surgery to prevent clotting must be monitored to prevent uncontrolled bleeding. Until recently, only clotting time could be measured; there was no direct, rapid determination of heparin at physiologic levels. Therefore an ion-selective electrode was designed to respond to heparin. In this chapter we will study basic principles of electrochemistry necessary to understand how electrodes work. In the next chapter we will see how electrodes are designed to respond selectively to particular analytes.

Chapter 13

Electrode Potentials

Measurements of the acidity of rainwater, the fuel-air mixture in an automobile engine, and the concentrations of gases and electrolytes in your bloodstream during surgery are all made with electrical sensors. This chapter provides the foundation needed to discuss some of the most common electrochemical sensors in the next chapter.

13-1 *Redox Chemistry and Electricity*

A **redox reaction** involves transfer of electrons from one species to another. A species is said to be **oxidized** when it *loses electrons*. It is **reduced** when it *gains electrons*. An **oxidizing agent,** also called an **oxidant,** takes electrons from another substance and becomes reduced. A **reducing agent,** also called a **reductant,** gives electrons to another substance and is oxidized in the process. In the reaction

$$\underset{\substack{\text{Oxidizing}\\\text{agent}}}{Fe^{3+}} + \underset{\substack{\text{Reducing}\\\text{agent}}}{V^{2+}} \longrightarrow Fe^{2+} + V^{3+}$$

Oxidation: loss of electrons
Reduction: gain of electrons

Oxidizing agent: takes electrons
Reducing agent: gives electrons

Fe^{3+} is the oxidizing agent because it takes an electron from V^{2+}. V^{2+} is the reducing agent because it gives an electron to Fe^{3+}. Fe^{3+} is reduced (its oxidation state goes down from $+3$ to $+2$), and V^{2+} is oxidized (its oxidation state goes up from $+2$ to $+3$) as the reaction proceeds from left to right. Appendix D discusses oxidation numbers and balancing redox equations. You should be able to write a balanced redox reaction as the sum of two *half-reactions,* one involving oxidation and the other involving reduction.

229

F is the *Faraday constant:*

$$F = 96\ 485.309\ \text{C/mol}$$

The *quantity* of electrons flowing from a reaction is proportional to the quantity of analyte that reacts.

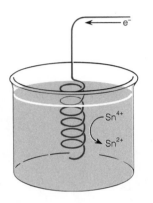

FIGURE 13-1 Electrons flowing into a coil of Pt wire at which Sn^{4+} ions in solution are reduced to Sn^{2+} ions. This process could not happen by itself because there is not a complete circuit. If Sn^{4+} is to be reduced at this Pt electrode, some other species must be oxidized at some other place.

$$1\ \text{A} = 1\ \text{C/s}$$

Chemistry and Electricity

Electric charge(q) is measured in **coulombs** (C). The magnitude of the charge of a single electron (or proton) is 1.602×10^{-19} C. One mole of electrons has a charge of 9.649×10^4 C, which is called the **Faraday constant** (*F*).

Relation between charge and moles:

$$\underset{\text{Coulombs}}{q} = \underset{\text{Moles}}{n} \cdot \underset{\frac{\text{Coulombs}}{\text{Mole}}}{F} \tag{13-1}$$

In the reaction $Fe^{3+} + V^{2+} \rightarrow Fe^{2+} + V^{3+}$, one electron is transferred to oxidize one atom of V^{2+} and to reduce one atom of Fe^{3+}. If we know how many moles of electrons are transferred from V^{2+} to Fe^{3+}, then we know how many moles of product have been formed.

EXAMPLE	**Relating Coulombs to Quantity of Reaction**

If 5.585 g of Fe^{3+} were reduced in the reaction $Fe^{3+} + V^{2+} \rightarrow Fe^{2+} + V^{3+}$, how many coulombs of charge must have been transferred from V^{2+} to Fe^{3+}?

SOLUTION First, we find that 5.585 g of Fe^{3+} equals 0.100 0 mol of Fe^{3+}. Because each Fe^{3+} ion requires one electron, 0.100 0 mol of electrons must have been transferred. Equation 13-1 relates coulombs of charge to moles of electrons:

$$q = nF = (0.100\ 0\ \text{mol e}^-)\left(9.649 \times 10^4\ \frac{\text{C}}{\text{mol e}^-}\right) = 9.649 \times 10^3\ \text{C}$$

Electric Current Is Proportional to the Rate of a Redox Reaction

Electric **current** (*I*) is the quantity of charge flowing each second past a point in an electric circuit. The unit of current is the **ampere** (A), which represents a flow of one coulomb per second.

Consider Figure 13-1 in which electrons flow into a Pt wire dipped into a solution in which the reduction of Sn^{4+} to Sn^{2+} occurs:

$$Sn^{4+} + 2e^- \longrightarrow Sn^{2+}$$

The Pt wire is an **electrode**—a device to conduct electrons into or out of the chemicals involved in the redox reaction. Platinum is an *inert* electrode; it does not participate in the redox reaction except as a conductor of electrons. We call a molecule that can donate or accept electrons at an electrode an **electroactive species.** The rate at which electrons flow into the electrode is a direct measure of the rate of reduction of Sn^{4+}.

| EXAMPLE | Relating Current to the Rate of Reaction |

Suppose that Sn^{4+} is reduced to Sn^{2+} at a constant rate of 4.24 mmol/h in Figure 13-1. How much current flows into the solution?

SOLUTION *Two* electrons are required to reduce *one* Sn^{4+} ion to Sn^{2+}. If Sn^{4+} is reacting at a rate of 4.24 mmol/h, electrons flow at a rate of $2(4.24) = 8.48$ mmol/h, which corresponds to

$$\frac{8.48 \text{ mmol/h}}{3\ 600 \text{ s/h}} = 2.356 \times 10^{-3} \text{ mmol/s} = 2.356 \times 10^{-6} \text{ mol/s}$$

To find the current, we convert moles of electrons per second to coulombs per second:

$$\text{current} = \frac{\text{coulombs}}{\text{second}} = \frac{\text{moles}}{\text{second}} \cdot \frac{\text{coulombs}}{\text{mole}}$$

$$= \left(2.356 \times 10^{-6} \frac{\text{mol}}{\text{s}}\right)\left(9.649 \times 10^{4} \frac{\text{C}}{\text{mol}}\right) = 0.227 \text{ C/s} = 0.227 \text{ A}$$

Voltage and Electrical Work

Because of their negative charge, electrons are attracted to positively charged regions and repelled from negatively charged regions. If electrons are attracted from one point to another, they can do useful work along the way. If we want to force electrons into a region from which they are repelled, we must do work on the electrons to push them along. Work has the dimensions of energy, whose units are *joules* (J).

The difference in **electric potential** between two points measures the work that is needed (or can be done) when electrons move from one point to another. Potential difference is measured in **volts** (V).

A good analogy for understanding current and potential is to think of water flowing through a garden hose. Electric current is the quantity of electric charge flowing per second through a wire. Electric current is analogous to the volume of water flowing per second through the hose. Electric potential is a measure of the force pushing on the electrons. The greater the force, the more current flows. Electric potential is analogous to the pressure on the water in the hose. The greater the pressure, the faster the water flows.

When a charge, q, moves through a potential difference, E, the work done is

*Relation between work
and voltage:*

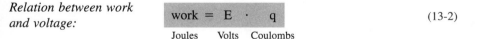

$$\underset{\text{Joules} \quad \text{Volts} \quad \text{Coulombs}}{\text{work} = E \cdot q} \tag{13-2}$$

One joule of energy is gained or lost when one coulomb of charge moves through a potential difference of one volt. Equation 13-2 tells us that the dimensions of volts are J/C.

The garden hose analogy for work is this: Suppose that one end of a hose is raised 1 m above the other end and a volume of 1 L of water flows through the hose. The water can be made to flow through a mechanical device to do a certain

It costs energy to move like charges toward each other. Energy is released when opposite charges move toward each other.

1 volt = 1 J/C

amount of work. If one end of the hose is raised 2 m above the other, the amount of work that can be done by the falling water is twice as great. The elevation difference between the ends of the hose is analogous to electric potential difference and the volume of water is analogous to electric charge. The greater the electric potential difference between two points in a circuit, the more work can be done by the charge flowing between those two points.

EXAMPLE Electrical Work

How much work can be done when 2.36 mmol of electrons moves "downhill" through a potential difference of 1.05 V? By "downhill," we mean that the flow of current is energetically favorable.

SOLUTION To use Equation 13-2, first convert moles of electrons to coulombs of charge:

$$q = nF = (2.36 \times 10^{-3} \text{ mol})(9.649 \times 10^4 \text{ C/mol}) = 2.277 \times 10^2 \text{ C}$$

The work that can be done by the moving electrons is

$$\text{work} = E \cdot q = (1.05 \text{ V})(2.277 \times 10^2 \text{ C}) = 239 \text{ J}$$

Electrolysis is a chemical reaction in which we apply a voltage to drive a redox reaction that would not otherwise occur. For example, electrolysis is used to make aluminum metal from Al^{3+} and to make chlorine gas (Cl_2) from Cl^- in seawater. Demonstration 13-1 is a great classroom illustration of electrolysis.

Electrolytic production of aluminum by the Hall-Heroult process consumes 4.5% of the electrical output of the United States! Al^{3+} in a molten solution of Al_2O_3 and cryolite (Na_3AlF_6) is reduced to aluminum metal at the cathode of a cell that typically draws 250 000 A. This process was invented by Charles Hall in 1886 when he was 22 years old, just after graduating from Oberlin College.

Charles Martin Hall. [Photo courtesy Alcoa]

Ask Yourself

13-A. Consider the redox reaction

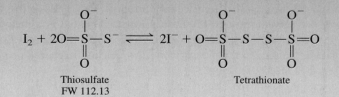

Thiosulfate
FW 112.13   Tetrathionate

   (a) Identify the oxidizing agent on the left side of the reaction and write a balanced oxidation half-reaction.
   (b) Identify the reducing agent on the left side of the reaction and write a balanced reduction half-reaction.
   (c) How many coulombs of charge are passed from reductant to oxidant when 1.00 g of thiosulfate reacts?
   (d) If the rate of reaction is 1.00 g of thiosulfate consumed per minute, what current (in amperes) flows from reductant to oxidant?
   (e) If the charge in **(c)** flows "downhill" through a potential difference of 0.200 V, how much work (in joules) can be done by the electric current?

Electrochemical Writing

Approximately 7% of electric power in the United States goes into electrolytic chemical production. The electrolysis apparatus below consists of a sheet of aluminum foil taped or cemented to a glass or wood surface. Any size will work, but an area about 15 cm on a side is convenient for a classroom demonstration. On the metal foil is taped (at one edge only) a sandwich consisting of filter paper, writing paper, and another sheet of filter paper. A stylus is prepared from a length of copper wire (18 gauge or thicker) looped at the end and passed through a length of glass tubing.

A fresh solution is prepared from 1.6 g of KI, 20 mL of water, 5 mL of 1 wt % starch solution, and 5 mL of phenolphthalein indicator solution. (If the solution darkens after standing for several days, it can be decolorized by adding a few drops of dilute $Na_2S_2O_3$.) The three layers of paper are soaked with the KI-starch-phenolphthalein solution. Connect the stylus and foil to a 12-V DC power source, and write on the paper with the stylus.

When the stylus is the cathode, pink color appears from the reaction of OH^- with phenolphthalein:

cathode: $H_2O + e^- \longrightarrow \frac{1}{2}H_2(g) + OH^-$

When the polarity is reversed and the stylus is the anode, a black (very dark blue) color appears from the reaction of I_2 with starch:

anode: $I^- \longrightarrow \frac{1}{2}I_2 + e^-$

Pick up the top sheet of filter paper and the typing paper, and you will discover that the writing appears in the opposite color on the bottom sheet of filter paper. This sequence is shown in Color Plate 10. Color Plate 11 shows the same oxidation reaction in a solution where you can clearly see the process in action.

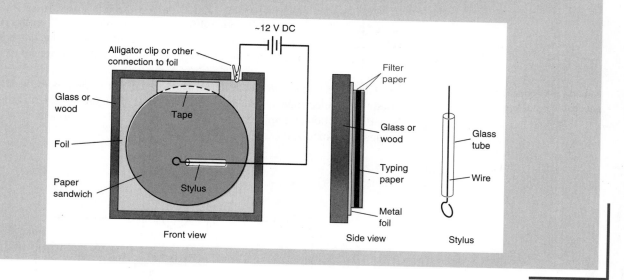

Front view — Alligator clip or other connection to foil; ~12 V DC; Glass or wood; Tape; Foil; Paper sandwich; Stylus

Side view — Filter paper; Glass or wood; Typing paper; Metal foil

Stylus — Glass tube; Wire

13-2 Galvanic Cells

A **galvanic cell** uses a *spontaneous* chemical reaction to generate electricity. To accomplish this, one reagent must be oxidized and another must be reduced. The two cannot be in contact, or electrons would flow directly from the reducing agent to the oxidizing agent without going through the external circuit. Therefore, the oxidizing and reducing agents are physically separated, and electrons are forced to flow through the circuit to go from one reactant to the other.

A spontaneous reaction is one in which it is energetically favorable for reactants to be converted to products.

233

A Cell in Action

For more on salt bridges, see Demonstration 13-2.

Figure 13-2 shows a galvanic cell consisting of two *half-cells* connected by a **salt bridge** through which ions can migrate to maintain electroneutrality in each vessel. The left half-cell has a metallic zinc electrode dipped into aqueous $ZnCl_2$. The right half-cell has a copper electrode immersed in aqueous $CuSO_4$. The salt bridge is filled with a gel containing saturated aqueous KCl. The electrodes are connected by a *potentiometer* (a voltmeter) to measure the voltage difference between the two half-cells.

The chemical reactions occurring in this cell are

reduction half-reaction: $\qquad Cu^{2+}(aq) + 2e^- \rightleftharpoons Cu(s)$

oxidation half-reaction: $\qquad\qquad\qquad Zn(s) \rightleftharpoons Zn^{2+}(aq) + 2e^-$

net reaction: $\qquad Cu^{2+}(aq) + Zn(s) \rightleftharpoons Cu(s) + Zn^{2+}(aq)$

The two half-reactions are always written with equal numbers of electrons so that their sum includes no free electrons.

Oxidation of $Zn(s)$ at the left side of Figure 13-2 produces $Zn^{2+}(aq)$. Electrons from the Zn metal flow through the potentiometer into the copper electrode, where $Cu^{2+}(aq)$ is reduced to $Cu(s)$. If there were no salt bridge, the left half-cell would soon build up positive charge (from excess Zn^{2+}) and the right half-cell would build up negative charge (by depletion of Cu^{2+}). In an instant, the charge buildup would oppose the driving force for the chemical reaction and the process would cease.

To maintain electroneutrality, Zn^{2+} ions on the left side diffuse into the salt bridge and Cl^- from the salt bridge diffuses into the left half-cell. In the right half-cell, SO_4^{2-} diffuses into the salt bridge and K^+ diffuses out of the salt bridge. The result is that the positive and negative charges in each half-cell remain exactly balanced.

cathode ↔ reduction
anode ↔ oxidation

Chemists define the electrode at which *reduction* occurs as the **cathode.** The **anode** is the electrode at which *oxidation* occurs. In Figure 13-2, Cu is the cathode because reduction takes place at its surface ($Cu^{2+} + 2e^- \rightarrow Cu$) and Zn is the anode because it is oxidized ($Zn \rightarrow Zn^{2+} + 2e^-$).

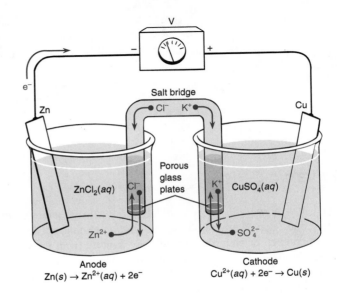

FIGURE 13-2 A galvanic cell consisting of two half-cells and a salt bridge through which ions can diffuse to maintain electroneutrality on each side.

The Human Salt Bridge

A salt bridge is an ionic medium through which ions diffuse to maintain electroneutrality in each compartment of an electrochemical cell. One way to prepare a salt bridge is to heat 3 g of agar (the stuff used to grow bacteria in a Petri dish) with 30 g of KCl in 100 mL of water until a clear solution is obtained. The solution is poured into a U-tube and allowed to gel. The bridge is stored in saturated aqueous KCl.

To conduct this demonstration, set up the galvanic cell in Figure 13-2 with 0.1 M $ZnCl_2$ on the left side and 0.1 M $CuSO_4$ on the right side. You can use a voltmeter or a pH meter to measure voltage. If you use a pH meter, the positive terminal is the socket to which the glass electrode goes and the negative terminal is the reference electrode socket.

You should be able to write the two half-reactions for this cell and use the Nernst equations (13-6 and 13-7) to calculate the theoretical voltage. Measure the voltage with a conventional salt bridge. Then replace the salt bridge with one made of filter paper freshly soaked in NaCl solution and measure the voltage again. Finally, replace the filter paper with two fingers of the same hand and measure the voltage again. Your body is really just a bag of salt housed in a *semipermeable membrane* (skin) through which ions can diffuse. Small differences in voltage observed when the salt bridge is replaced can be attributed to the junction potential discussed in Section 14-2.

Challenge One hundred and eighty students at Virginia Polytechnic Institute and State University made a salt bridge by holding hands. (Their electrical resistance was lowered by a factor of 100 by wetting everyone's hands.) Can your school beat this record?

Line Notation

A shorthand notation employing just two symbols is commonly used to describe electrochemical cells:

| phase boundary   ‖ salt bridge

The cell in Figure 13-2 is represented by the *line diagram*.

$$Zn(s) \,|\, ZnCl_2(aq) \,\|\, CuSO_4(aq) \,|\, Cu(s)$$

Each phase boundary is indicated by a vertical line. The electrodes are shown at the extreme left- and right-hand sides of the diagram. The contents of the salt bridge are not specified.

EXAMPLE Interpreting Line Diagrams of Cells

Write a line diagram for the cell in Figure 13-3. Write an anode reaction (an oxidation) for the left half-cell, a cathode reaction (a reduction) for the right half-cell, and the net cell reaction.

235

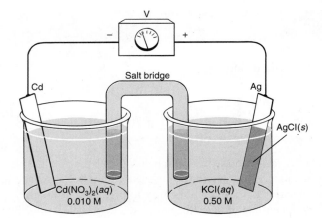

FIGURE 13-3 Another galvanic cell.

SOLUTION The right half-cell contains two solid phases (Ag and AgCl) and an aqueous phase. The left half-cell contains one solid phase and one aqueous phase. The line diagram is

$$Cd(s) \mid Cd(NO_3)_2(aq) \parallel KCl(aq) \mid AgCl(s) \mid Ag(s)$$

$Cd(NO_3)_2$ is a strong electrolyte, so it is completely dissociated to Cd^{2+} and NO_3^-; and KCl is dissociated into K^+ and Cl^-. In the left half-cell, we find cadmium in the oxidation states 0 and +2. In the right half-cell, we find silver in the 0 and +1 states. The Ag(I) is solid AgCl adhering to the Ag electrode. The electrode reactions are

anode (oxidation): $\qquad\qquad\qquad Cd(s) \rightleftharpoons Cd^{2+}(aq) + 2e^-$
cathode (reduction): $\qquad AgCl(s) + e^- \rightleftharpoons Ag(s) + Cl^-(aq)$

net reaction: $\qquad\qquad Cd(s) + AgCl(s) \rightleftharpoons Cd^{2+}(aq) + Ag(s) + Cl^-(aq) + e^-$

Oops! We didn't write both half-reactions with the same number of electrons, so the net reaction still has electrons in it. This is not allowed, so we will double the coefficients of the second half-reaction:

anode (oxidation): $\qquad\qquad\qquad Cd(s) \rightleftharpoons Cd^{2+}(aq) + 2e^-$
cathode (reduction): $\qquad 2AgCl(s) + 2e^- \rightleftharpoons 2Ag(s) + 2Cl^-(aq)$

net reaction: $\qquad\qquad Cd(s) + 2AgCl(s) \rightleftharpoons Cd^{2+}(aq) + 2Ag(s) + 2Cl^-(aq)$

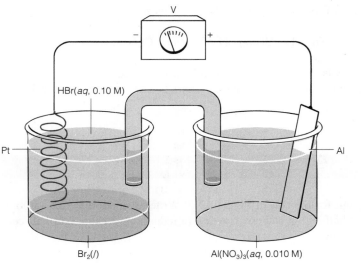

FIGURE 13-4 Cell for Ask Yourself 13-B.

Ask Yourself

13-B. (a) What is the line notation for the cell in Figure 13-4?
(b) Draw a picture of the following cell.

$$Au(s) \mid Fe(CN)_6^{4-}(aq), Fe(CN)_6^{3-}(aq) \parallel Ag(S_2O_3)_2^{3-}(aq), S_2O_3^{2-}(aq) \mid Ag(s)$$

Write an oxidation half-reaction for the left half-cell, a reduction half-reaction for the right half-cell, and a balanced net reaction.

13-3 *Standard Potentials*

The voltage measured in the experiment in Figure 13-2 is the difference in electric potential between the Cu electrode on the right and the Zn electrode on the left. The greater the voltage, the more favorable is the net cell reaction and the more work can be done by electrons flowing from one side to the other (Equation 13-2). The potentiometer terminals are labeled $+$ and $-$. The voltage is positive when electrons flow into the negative terminal, as in Figure 13-2. If electrons flow the other way, the voltage is negative. We will draw every cell with the negative terminal of the potentiometer at the left.

To predict the voltage when different half-cells are connected to each other, the **standard reduction potential** ($E°$) for each half-cell is measured by an experiment shown in idealized form in Figure 13-5. The half-reaction of interest in this diagram is

$$Ag^+ + e^- \rightleftharpoons Ag(s) \qquad (13\text{-}3)$$

occurring in the right-hand half-cell, which is connected to the *positive* terminal of the potentiometer. The term *standard* means that species are solids or liquids or their concentrations are 1 M or their pressures are 1 atm. We call these conditions the *standard states* of the reactants and products.

We oversimplify the standard half-cell in this text. The word *standard* really means that the *activities* of all species are 1. Activity (Box 5-2) is not the same as concentration, but we will consider *standard* to mean that concentrations are 1 M or 1 atm.

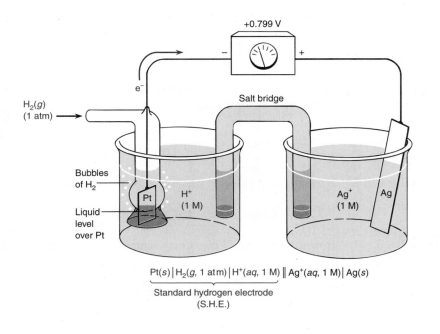

Pt(s) | H₂(g, 1 atm) | H⁺(aq, 1 M) ‖ Ag⁺(aq, 1 M) | Ag(s)

Standard hydrogen electrode
(S.H.E.)

FIGURE 13-5 Setup used to measure the standard reduction potential ($E°$) for the half-reaction $Ag^+ + e^- \rightleftharpoons Ag(s)$. The left half-cell is called the standard hydrogen electrode (S.H.E.).

Question What is the pH of the standard hydrogen electrode?

We will write all half-reactions as *reductions*. By convention, $E° = 0$ for S.H.E.

The left half-cell, connected to the *negative* terminal of the potentiometer, is called the **standard hydrogen electrode** (S.H.E.). It consists of a catalytic Pt surface in contact with an acidic solution in which $[H^+] = 1$ M. A stream of $H_2(g, 1$ atm$)$ is bubbled through the electrode. The reaction at the surface of the Pt electrode is

$$H^+(aq, 1 \text{ M}) + e^- \rightleftharpoons \tfrac{1}{2}H_2(g, 1 \text{ atm}) \qquad (13\text{-}4)$$

We *assign* a potential of 0 to the standard hydrogen electrode. The voltage measured in Figure 13-5 can therefore be *assigned* to Reaction 13-3, which occurs in the right half-cell. The measured value $E° = +0.799$ V is the standard reduction potential for Reaction 13-3. The positive sign tells us that electrons flow from left to right through the meter.

The line notation for the cell in Figure 13-5 is

$$Pt(s) \mid H_2(g, 1 \text{ atm}), H^+(aq, 1 \text{ M}) \parallel Ag^+(aq, 1 \text{ M}) \mid Ag(s)$$

which is abbreviated

$$\text{S.H.E.} \parallel Ag^+(aq, 1 \text{ M}) \mid Ag(s)$$

Measured voltage = right-hand electrode potential − left-hand electrode potential.

The standard reduction potential is really the *difference* between the standard potential of the reaction of interest and the potential of the S.H.E., which we have arbitrarily set to 0.

If we wanted to measure the standard potential of the half-reaction

$$Cd^{2+} + 2e^- \rightleftharpoons Cd(s) \qquad (13\text{-}5)$$

we would construct the cell

$$\text{S.H.E.} \parallel Cd^{2+}(aq, 1 \text{ M}) \mid Cd(s)$$

Challenge Draw a picture of the cell S.H.E. $\parallel Cd^{2+}(aq, 1$ M$) \mid Cd(s)$ and show the direction of electron flow.

with the cadmium half-cell *at the right*. In this case, we would observe a *negative* voltage of -0.402 V. The negative sign means that electrons flow from Cd to Pt, a direction opposite that of the cell in Figure 13-5.

Formal Potential

Appendix C lists many standard reduction potentials. Sometimes, multiple potentials are listed for one reaction, as for the $AgCl(s) \mid Ag(s)$ half-reaction:

Standard potential: 0.222 V
Formal potential for saturated KCl: 0.197 V

$$AgCl(s) + e^- \rightleftharpoons Ag(s) + Cl^- \quad \begin{cases} 0.222 \text{ V} \\ 0.197 \text{ V saturated KCl} \end{cases}$$

The value 0.222 V is the standard potential that would be measured in the cell

$$\text{S.H.E.} \parallel Cl^-(aq, 1 \text{ M}) \mid AgCl(s) \mid Ag(s)$$

The value 0.197 V is measured in a cell containing saturated KCl solution instead of 1 M Cl^-:

$$\text{S.H.E.} \parallel KCl (aq, \text{ saturated}) \mid AgCl(s) \mid Ag(s)$$

The potential for a cell containing a specified concentration of reagent other than 1 M is called the **formal potential.**

Ask Yourself

13-C. Draw a line diagram and a picture of the cell used to measure the standard potential of the reaction $Fe^{3+} + e^- \rightleftharpoons Fe^{2+}$, which comes to equilibrium at the surface of a Pt electrode. Use Appendix C to find the cell potential. Based on the cell potential, show the direction of electron flow in your picture.

13-4 *The Nernst Equation*

Le Châtelier's principle tells us that if we increase reactant concentrations, we drive a reaction to the right. Increasing the product concentrations drives a reaction to the left. The net driving force for a reaction is expressed by the **Nernst equation,** whose two terms include the driving force under standard conditions ($E°$, which applies when concentrations are 1 M or 1 atm) and a term that shows the dependence on concentrations.

Nernst Equation for a Half-Reaction

For the half-reaction

$$aA + ne^- \rightleftharpoons bB$$

the Nernst equation giving the half-cell potential, E, is

Nernst equation: $$E = E° - \frac{0.059\ 16}{n} \log\left(\frac{[B]^b}{[A]^a}\right) \qquad (13\text{-}6)$$

where $E°$ is the standard reduction potential that applies when $[A] = [B] = 1$ M.

The logarithmic term in the Nernst equation is the *reaction quotient, Q* ($= [B]^b/[A]^a$). Q has the same form as the equilibrium constant, but the concentrations need not be at their equilibrium values. Concentrations of solutes are expressed as moles per liter and concentrations of gases are expressed as pressures in atmospheres. Pure solids, pure liquids, and solvents are omitted from Q. When all concentrations are 1 M and all pressures are 1 atm, $Q = 1$ and $\log Q = 0$, thus giving $E = E°$.

In the Nernst equation:
- E and $E°$ are measured in volts
- solute concentration = mol/L
- gas concentration = atm
- solid, liquid, solvent omitted

EXAMPLE Writing the Nernst Equation for a Half-Reaction

Let's write the Nernst equation for the reduction of white phosphorus to phosphine gas:

$$\tfrac{1}{4}P_4(s,\ white) + 3H^+ + 3e^- \rightleftharpoons PH_3(g) \qquad E° = -0.046 \text{ V}$$
$$\text{Phosphine}$$

SOLUTION We omit solids from the reaction quotient, and the concentration of phosphine gas is expressed as its pressure in atm (P_{PH_3}):

$$E = -0.046 - \frac{0.059\ 16}{3}\ \log\left(\frac{P_{PH_3}}{[H^+]^3}\right)$$

EXAMPLE	Multiplication of a Half-Reaction

If you multiply a half-reaction by any factor, the value of $E°$ does not change. However, the factor n before the log term and the exponents in the reaction quotient do change. Write the Nernst equation for the reaction in the previous example, multiplied by 2:

$$\tfrac{1}{2}P_4(s, \text{white}) + 6H^+ + 6e^- \rightleftharpoons 2PH_3(g) \qquad E° = -0.046\ V$$

SOLUTION

$$E = -0.046 - \frac{0.059\ 16}{6}\ \log\left(\frac{P_{PH_3}^2}{[H^+]^6}\right)$$

$E°$ remains at -0.046 V, as in the previous example. However, the factor in front of the log term and the exponents in the log term have changed.

Nernst Equation for a Complete Reaction

Consider the cell in Figure 13-3 in which the negative terminal of the potentiometer is connected to the Cd electrode and the positive terminal is connected to the Ag electrode. The voltage, E, is the difference between the potentials of the two electrodes:

Nernst equation for a complete cell:

$$E = E_+ - E_-$$

(13-7)

where E_+ is the potential of the half-cell attached to the positive terminal of the potentiometer and E_- is the potential of the half-cell attached to the negative terminal. The potential of each half-reaction (*written as a reduction*) is governed by the Nernst equation 13-6.

Here is a procedure for writing a net cell reaction and finding its voltage:

Both half-reactions are written as *reductions* when we use Equation 13-7.

Step 1. Write *reduction* half-reactions for both half-cells and find $E°$ for each in Appendix C. Multiply the half-reactions as necessary so that they each contain the same number of electrons. When you multiply a reaction by any number, *do not* multiply $E°$.

Step 2. Write a Nernst equation for the half-reaction in the right half-cell, which is attached to the positive terminal of the potentiometer. This is E_+.

Step 3. Write a Nernst equation for the half-reaction in the left half-cell, which is attached to the negative terminal of the potentiometer. This is E_-.

Step 4. Find the net cell voltage by subtraction: $E = E_+ - E_-$.

Step 5. To write a balanced net cell reaction, subtract the left half-reaction from the right half-reaction. (*This is equivalent to reversing the left half-reaction and adding.*)

If the net cell voltage, $E (= E_+ - E_-)$, is positive, then the net cell reaction is spontaneous in the forward direction. If the net cell voltage is negative, then the reaction is spontaneous in the reverse direction.

EXAMPLE | Nernst Equation for a Complete Reaction

Find the voltage of the cell in Figure 13-3 if the right half-cell contains 0.50 M KCl(aq) and the left half-cell contains 0.010 M Cd(NO$_3$)$_2$(aq). Write the net cell reaction and state whether it is spontaneous in the forward or reverse direction.

SOLUTION

Step 1. right half-cell: $2AgCl(s) + 2e^- \rightleftharpoons 2Ag(s) + 2Cl^-$ $E_+^\circ = 0.222$ V

left half-cell: $Cd^{2+} + 2e^- \rightleftharpoons Cd(s)$ $E_-^\circ = -0.402$ V

Step 2. Nernst equation for right half-cell:

$$E_+ = E_+^\circ - \frac{0.059\ 16}{2} \log([Cl^-]^2) \qquad (13-8)$$

$$= 0.222 - \frac{0.059\ 16}{2} \log([0.50]^2) = 0.240\ \text{V}$$

Pure solids, pure liquids, and solvents are omitted from Q.

Step 3. Nernst equation for left half-cell:

$$E_- = E_-^\circ - \frac{0.059\ 16}{2}\log\left(\frac{1}{[Cd^{2+}]}\right) = -0.402 - \frac{0.059\ 16}{2}\log\left(\frac{1}{[0.010]}\right)$$

$$= -0.461\ \text{V}$$

Step 4. cell voltage: $E = E_+ - E_- = 0.240 - (-0.461) = 0.701$ V

Step 5. net cell reaction:

$$\begin{array}{rl}
 & 2AgCl(s) + 2e^- \rightleftharpoons 2Ag(s) + 2Cl^- \\
- & \underline{Cd^{2+} + 2e^- \rightleftharpoons Cd(s)} \\
 & Cd(s) + 2AgCl(s) \rightleftharpoons Cd^{2+} + 2Ag(s) + 2Cl^-
\end{array}$$

Subtracting a reaction is the same as *reversing the reaction and adding.*

Because the voltage is positive, the net reaction is spontaneous in the forward direction. Cd(s) is oxidized to Cd^{2+} and AgCl(s) is reduced to Ag(s). Electrons flow from the left electrode to the right electrode.

What if you had written the Nernst equation for the right half-cell with just one electron instead of two: $AgCl(s) + e^- \rightleftharpoons Ag(s) + Cl^-$? Try this and you will discover that the half-cell potential is unchanged. *Neither $E°$ nor E changes when you multiply a reaction.*

Different Descriptions of the Same Reaction

We know that the right half-cell in Figure 13-3 must contain some $Ag^+(aq)$ in equilibrium with $AgCl(s)$. Suppose that instead of writing the reaction $2AgCl(s) + 2e^- \rightleftharpoons 2Ag(s) + 2Cl^-$, a different, less handsome, author wrote the reaction

$$2Ag^+(aq) + 2e^- \rightleftharpoons 2Ag(s) \qquad E_+^\circ = 0.799 \text{ V}$$

$$E_+ = 0.799 - \frac{0.059\ 16}{2} \log\left(\frac{1}{[Ag^+]^2}\right) \qquad (13\text{-}9)$$

Both descriptions of the right half-cell are equally valid. In both cases, Ag(I) is reduced to Ag(0).

If the two descriptions are equal, then they should predict the same voltage. To use Equation 13-9, you must know the concentration of Ag^+ in the right half-cell, which is not obvious. But you are clever and cunning and realize that you can find $[Ag^+]$ from the solubility product for AgCl and the known concentration of Cl^- (which is 0.50 M).

$$K_{sp} = [Ag^+][Cl^-] \Rightarrow [Ag^+] = \frac{K_{sp}}{[Cl^-]} = \frac{1.8 \times 10^{-10}}{0.50} = 3.6 \times 10^{-10} \text{ M}$$

Putting this concentration into Equation 13-9 gives

$$E_+ = 0.799 - \frac{0.059\ 16}{2} \log\left(\frac{1}{(3.6 \times 10^{-10})^2}\right) = 0.240 \text{ V}$$

The cell voltage cannot depend on how we write the reaction!

which is exactly the same voltage computed in Equation 13-8! The two choices of half-reaction give the same voltage because they describe the same cell.

Advice for Finding Relevant Half-Reactions

When faced with a cell drawing or line diagram, the first step is to write reduction reactions for each half-cell. To do this, *look for elements in two oxidation states.* For the cell

$$Pb(s) \mid PbF_2(s) \mid F^-(aq) \parallel Cu^{2+}(aq) \mid Cu(s)$$

we see lead in the oxidation states 0 in Pb(s) and +2 in $PbF_2(s)$, and copper in the oxidation states 0 in Cu(s) and +2 in Cu^{2+}. Thus, the half-reactions are

Don't write a reaction such as $F_2(g) + 2e^- \rightleftharpoons 2F^-$, because $F_2(g)$ is not shown in the line diagram of the cell. $F_2(g)$ is neither a reactant nor a product.

right half-cell: $\qquad Cu^{2+} + 2e^- \rightleftharpoons Cu(s)$

left half-cell: $\qquad PbF_2(s) + 2e^- \rightleftharpoons Pb(s) + 2F^- \qquad (13\text{-}10)$

You might have chosen to write the lead half-reaction as

left half-cell: $\qquad Pb^{2+} + 2e^- \rightleftharpoons Pb(s) \qquad (13\text{-}11)$

because you know that if $PbF_2(s)$ is present, there must be some Pb^{2+} in the solution. Reactions 13-10 and 13-11 are equally valid descriptions of the cell, and each should predict the same cell voltage. Your choice of reaction depends on whether the F^- or Pb^{2+} concentration is more easily known to you.

Ask Yourself

13-D. (a) Arsine (AsH_3) is a poisonous gas used to make gallium arsenide for diode lasers. Find E for the half-reaction $As(s) + 3H^+ + 3e^- \rightleftharpoons AsH_3(g)$ if pH = 3.00 and P_{AsH_3} = 1.00 torr. (1 atm = 760 torr, so 1.00 torr = 1.00/760 atm.)

    (b) The cell in Demonstration 13-2 (and Figure 13-2) can be written

$$Zn(s) \mid Zn^{2+}(0.1 \text{ M}) \parallel Cu^{2+}(0.1 \text{ M}) \mid Cu(s)$$

Write a reduction half-reaction for each half-cell and use the Nernst equation to predict the cell voltage.

13-5 *E° and the Equilibrium Constant*

A *galvanic cell produces electricity because the cell reaction is not at equilibrium.* If the cell runs long enough, reactants are consumed until the reaction comes to equilibrium, and the cell voltage, E, reaches zero. This is what happens to a battery when it runs down.

At equilibrium, E (not $E°$) = 0.

    If E_+° is the standard reduction potential for the right half-cell and E_-° is the standard reduction potential for the left half-cell, $E°$ for the net cell reaction is

E° for net cell reaction: $$E° = E_+^\circ - E_-^\circ \qquad (13\text{-}12)$$

Now let's relate $E°$ to the equilibrium constant for the net cell reaction.

    As an example, consider the cell in Figure 13-6. The left half-cell contains a silver electrode coated with solid AgCl and dipped into saturated aqueous KCl. The

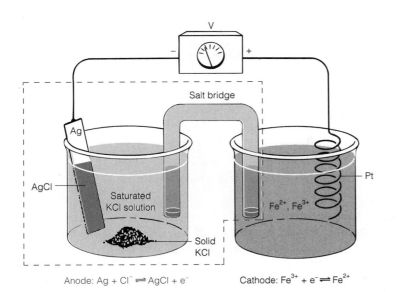

Anode: $Ag + Cl^- \rightleftharpoons AgCl + e^-$ Cathode: $Fe^{3+} + e^- \rightleftharpoons Fe^{2+}$

$Ag(s) \mid AgCl(s) \mid Cl^-(aq) \parallel Fe^{2+}(aq), Fe^{3+}(aq) \mid Pt(s)$

FIGURE 13-6 Cell used to illustrate the relation of $E°$ and the equilibrium constant. The dashed line encloses the part of the cell that we will call the *reference electrode* in Section 13-6.

E_+° and E_+ refer to the half-cell connected to the positive terminal of the potentiometer.

E_-° and E_- refer to the half-cell connected to the negative terminal.

right half-cell has a platinum wire dipped into an aqueous solution containing Fe^{2+} and Fe^{3+}.

The two half-reactions are

right: $\qquad Fe^{3+} + e^- \rightleftharpoons Fe^{2+} \qquad\qquad E_+^\circ = 0.771$ V $\quad$ (13-13)

left: $\qquad AgCl(s) + e^- \rightleftharpoons Ag(s) + Cl^- \qquad E_-^\circ = 0.222$ V $\quad$ (13-14)

The net cell reaction is found by subtracting the left half-reaction from the right half-reaction:

net reaction: $\qquad Fe^{3+} + Ag(s) + Cl^- \rightleftharpoons Fe^{2+} + AgCl(s)$

$$E^\circ = E_+^\circ - E_-^\circ = 0.771 - 0.222 = 0.549 \text{ V} \qquad (13\text{-}15)$$

The two electrode potentials are given by the Nernst equation 13-6:

$$E_+ = E_+^\circ - \frac{0.059\ 16}{n} \log\left(\frac{[Fe^{2+}]}{[Fe^{3+}]}\right) \qquad (13\text{-}16)$$

$$E_- = E_-^\circ - \frac{0.059\ 16}{n} \log([Cl^-]) \qquad (13\text{-}17)$$

where $n = 1$ for Reactions 13-13 and 13-14. The cell voltage is the difference $E_+ - E_-$:

$$E = E_+ - E_- = \left\{E_+^\circ - \frac{0.059\ 16}{n} \log\left(\frac{[Fe^{2+}]}{[Fe^{3+}]}\right)\right\} - \left\{E_-^\circ - \frac{0.059\ 16}{n} \log([Cl^-])\right\}$$

$$= (E_+^\circ - E_-^\circ) - \left\{\frac{0.059\ 16}{n} \log\left(\frac{[Fe^{2+}]}{[Fe^{3+}]}\right) - \frac{0.059\ 16}{n} \log([Cl^-])\right\} \quad (13\text{-}18)$$

Algebra of logarithms:

$\log x + \log y = \log(xy)$

$\log x - \log y = \log(x/y)$

To combine logarithms, we use the equality $\log x - \log y = \log(x/y)$:

$$E = \underbrace{(E_+^\circ - E_-^\circ)}_{E^\circ \text{ for net reaction}} - \frac{0.059\ 16}{n} \log\underbrace{\left(\frac{[Fe^{2+}]}{[Fe^{3+}][Cl^-]}\right)}_{\substack{\text{Reaction quotient } (Q) \\ \text{for net reaction}}} = E^\circ - \frac{0.059\ 16}{n} \log Q \quad (13\text{-}19)$$

Equation 13-19 is true at any time. *In the special case when the cell is at equilibrium, $E = 0$ and $Q = K$, the equilibrium constant.* Therefore,

$$0 = E^\circ - \frac{0.059\ 16}{n} \log K \Rightarrow E^\circ = \frac{0.059\ 16}{n} \log K \qquad (13\text{-}20)$$

To go from Equation 13-20 to 13-21:

$$\frac{0.059\ 16}{n} \log K = E^\circ$$

$$\log K = \frac{nE^\circ}{0.059\ 16}$$

$$10^{\log K} = 10^{nE^\circ/0.059\ 16}$$

$$K = 10^{nE^\circ/0.059\ 16}$$

Finding K from E°: $\qquad K = 10^{nE^\circ/0.059\ 16} \qquad$ (at 25°C) $\qquad$ (13-21)

Equation 13-21 gives the equilibrium constant for a net cell reaction from E°.

A positive value of E° means that $K > 1$ in Equation 13-21 and a negative value means that $K < 1$. We say that a reaction is spontaneous under standard conditions (i.e., when all concentrations of reactants and products are 1 M or 1 atm) if E° is positive. Biochemists use a different potential, called $E^{\circ\prime}$, which is described in Box 13-1.

Box 13-1 *Explanation*

Why Biochemists Use $E°'$

Redox reactions are essential for life. For example, the enzyme-catalyzed reduction of pyruvate to lactate is a step in the anaerobic fermentation of sugar by bacteria:

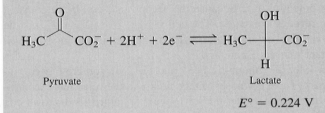

Pyruvate   Lactate

$$E° = 0.224 \text{ V}$$

The same reaction causes lactate to build up in your muscles and make you feel fatigued during intense exercise, when the flow of oxygen cannot keep pace with your requirement for energy.

$E°$ applies when the concentrations of reactants and products are 1 M. For a reaction involving H^+, this means that pH = 0 (because $[H^+] = 1$ M and log 1 = 0). Biochemists studying the energetics of fermentation or respiration are more interested in reduction potentials that apply near physiologic pH, not at pH 0. Therefore biochemists use a formal potential designated $E°'$, which applies at pH 7.

> $E°$ applies at pH 0
>
> $E°'$ applies at pH 7

For the conversion of pyruvate to lactate, the reactant and product are carboxylic acids at pH 0 and carboxylate anions at pH 7. The value of $E°'$ at pH 7 is −0.190 V, which is quite different from $E° = +0.224$ V at pH 0.

EXAMPLE   **Using $E°$ to Find the Equilibrium Constant**

Find the equilibrium constant for the reaction $Fe^{3+} + Ag(s) + Cl^- \rightleftharpoons Fe^{2+} + AgCl(s)$, which is the net cell reaction in Figure 13-6.

SOLUTION Equation 13-15 states that $E° = 0.549$ V. The equilibrium constant is computed with Equation 13-21, using $n = 1$, because one electron is transferred in each half-cell reaction:

$$K = 10^{nE°/0.059\ 16} = 10^{(1)(0.549)/(0.059\ 16)} = 1.9 \times 10^9$$

K has two significant figures, because $E°$ has three digits. One digit of $E°$ is used for the exponent (9), and the other two are left for the multiplier (1.9).

Ask Yourself

13-E. **(a)** Write the half-reactions for Figure 13-2. Calculate $E°$ and the equilibrium constant for the net cell reaction.

    (b) The solubility product reaction of AgBr is $AgBr(s) \rightleftharpoons Ag^+ + Br^-$. Use the reactions $Ag^+ + e^- \rightleftharpoons Ag(s)$ and $AgBr(s) + e^- \rightleftharpoons Ag(s) + Br^-$ to compute the solubility product of AgBr. Compare your answer to that in Appendix A.

13-6 *Reference Electrodes*

Imagine a solution containing an electroactive species whose concentration we wish to measure. We construct a half-cell by inserting an electrode (such as a Pt wire) into the solution to transfer electrons to or from the species of interest. Because this electrode responds directly to the analyte, it is called the **indicator electrode.** The potential of the indicator electrode is E_+. We then connect this half-cell to a second half-cell via a salt bridge. The second half-cell has a fixed composition that provides a known, constant potential, E_-. Because the second half-cell has a constant potential, it is called a **reference electrode.** The cell voltage ($E = E_+ - E_-$) is the difference between the variable potential that reflects changes in the analyte concentration and the constant reference potential.

Suppose you have a solution containing Fe^{2+} and Fe^{3+}. If you are clever, you can make this solution part of a cell whose voltage tells you the relative concentrations of these species—that is, the value of $[Fe^{2+}]/[Fe^{3+}]$. Figure 13-6 shows one way to do this. A Pt wire was inserted as an indicator electrode through which Fe^{3+} can receive electrons or Fe^{2+} can lose electrons. The left half-cell completes the galvanic cell and has a known, constant potential.

The two half-reactions were given in Reactions 13-13 and 13-14, and the two electrode potentials were given in Equations 13-16 and 13-17. The cell voltage is the difference $E_+ - E_-$:

Indicator electrode: responds to analyte concentration

Reference electrode: maintains a fixed (reference) potential

$$E = E_+ - E_-$$
$$= \underbrace{\left\{ E_+^\circ - 0.059\,16 \log\underbrace{\left(\frac{[Fe^{2+}]}{[Fe^{3+}]}\right)}_{\substack{\text{The variable} \\ \text{of interest}}} \right\}}_{\text{Constant}} - \underbrace{\{ E_-^\circ - 0.059\,16 \log([Cl^-]) \}}_{\substack{\text{Constant concentration} \\ \text{in left half-cell}}} \quad (13\text{-}22)$$

The cell voltage in Figure 13-6 responds only to changes in the quotient $[Fe^{2+}]/[Fe^{3+}]$. Everything else is constant.

Because $[Cl^-]$ in the left half-cell is constant (fixed by the solubility of KCl, with which the solution is saturated), the cell voltage changes only when the quotient $[Fe^{2+}]/[Fe^{3+}]$ changes.

The half-cell on the left in Figure 13-6 can be thought of as a *reference electrode.* We can picture the cell and salt bridge enclosed by the dashed line as a single unit dipped into the analyte solution, as shown in Figure 13-7. The Pt wire is

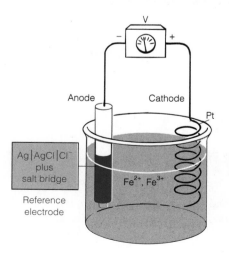

FIGURE 13-7 Another view of Figure 13-6. The contents of the dashed box in Figure 13-6 are now considered to be a reference electrode dipped into the analyte solution.

the indicator electrode, whose potential responds to changes in the quotient $[Fe^{2+}]/[Fe^{3+}]$. The reference electrode completes the redox reaction and provides a *constant potential* to the left side of the potentiometer. Changes in the cell voltage can be assigned to changes in the quotient $[Fe^{2+}]/[Fe^{3+}]$.

Silver-Silver Chloride Reference Electrode

The half-cell enclosed by the dashed line in Figure 13-6 is called a **silver-silver chloride electrode.** Figure 13-8 shows how the half-cell is reconstructed as a thin, glass-enclosed electrode that can be dipped into the analyte solution in Figure 13-7. The porous plug at the base of the electrode functions as a salt bridge. It allows electrical contact between solutions inside and outside the electrode with minimal physical mixing.

The standard reduction potential for AgCl | Ag is $+0.222$ V at 25°C. If the cell is saturated with KCl, the potential is $+0.197$ V. This is the value we will use for all problems involving the AgCl | Ag reference electrode. The advantage of a saturated KCl solution is that the concentration of chloride does not change if some of the liquid evaporates.

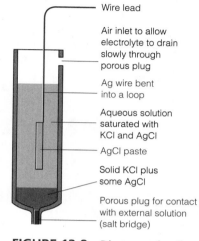

FIGURE 13-8   Diagram of a silver-silver chloride reference electrode.

Ag | AgCl electrode: $AgCl(s) + e^- \rightleftharpoons Ag(s) + Cl^-$ $E° = +0.222$ V
$E(\text{saturated KCl}) = +0.197$ V

EXAMPLE   Using a Reference Electrode

Calculate the cell voltage in Figure 13-7 if the reference electrode is a saturated silver-silver chloride electrode and $[Fe^{2+}]/[Fe^{3+}] = 10$.

SOLUTION We use Equation 13-22, noting that $E_- = 0.197$ V for a saturated silver-silver chloride electrode:

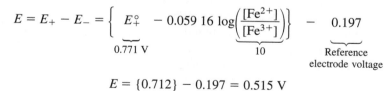

$$E = E_+ - E_- = \left\{ \underbrace{E°_+}_{0.771 \text{ V}} - 0.059\,16 \log \left(\underbrace{\frac{[Fe^{2+}]}{[Fe^{3+}]}}_{10} \right) \right\} - \underbrace{0.197}_{\substack{\text{Reference} \\ \text{electrode voltage}}}$$

$$E = \{0.712\} - 0.197 = 0.515 \text{ V}$$

Calomel Reference Electrode

The *calomel electrode* in Figure 13-9 is based on the reaction

Calomel electrode: $\frac{1}{2}Hg_2Cl_2(s) + e^- \rightleftharpoons Hg(l) + Cl^-$ $E° = +0.268$ V
Mercury(I) chloride $E(\text{saturated KCl}) = +0.241$ V
(calomel)

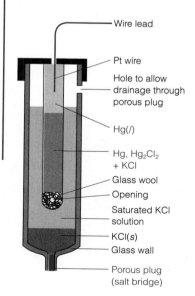

FIGURE   13-9   A   saturated calomel electrode (S.C.E.).

If the cell is saturated with KCl, it is called a **saturated calomel electrode** and the cell potential is +0.241 V at 25°C. This electrode is encountered so frequently that it is abbreviated **S.C.E.**

Voltage Conversions Between Different Reference Scales

It is sometimes necessary to convert potentials between different reference scales. If an electrode has a potential of −0.461 V with respect to a calomel electrode, what is the potential with respect to a silver-silver chloride electrode? What would be the potential with respect to the standard hydrogen electrode?

To answer these questions, Figure 13-10 shows the positions of the calomel and silver-silver chloride electrodes with respect to the standard hydrogen electrode.

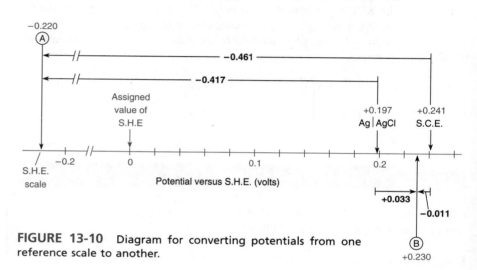

FIGURE 13-10 Diagram for converting potentials from one reference scale to another.

Point A, which is −0.461 V from S.C.E., is −0.417 V from the silver-silver chloride electrode and −0.220 V from S.H.E. What about point B, whose potential is +0.033 V with respect to silver-silver chloride? Its position is −0.011 V from S.C.E. and +0.230 V with respect to S.H.E. By keeping this diagram in mind, you can convert potentials from one scale to another.

Ask Yourself

13-F. (a) Find the concentration ratio $[Fe^{2+}]/[Fe^{3+}]$ for the cell in Figure 13-7 if the measured voltage is 0.703 V.

(b) Convert the potentials listed below. The Ag | AgCl and calomel reference electrodes are saturated with KCl.

       (i) 0.523 V versus S.H.E. = ? vs Ag | AgCl

       (ii) 0.222 V versus S.C.E. = ? vs S.H.E.

Key Equations

Definitions

Oxidizing agent—takes electrons

Reducing agent—gives electrons

Anode—where oxidation occurs

Cathode—where reduction occurs

Standard potential

Measured by the cell

S.H.E. ‖ half-reaction of interest

where S.H.E. is the standard hydrogen electrode and all reagents in the right half-cell are in their standard state (= 1 M, 1 atm, pure solid, or pure liquid)

Nernst equation

For the half-reaction $aA + ne^- \rightleftharpoons bB$

$$E = E^\circ - \frac{0.059\ 16}{n} \log\left(\frac{[B]^b}{[A]^a}\right) \quad \text{(at 25°C)}$$

Voltage of complete cell

$E = E_+ - E_-$

E_+ = voltage of electrode connected to + terminal of meter

E_- = voltage of electrode connected to − terminal of meter

E° for net cell reaction

$E^\circ = E^\circ_+ - E^\circ_-$

E°_+ = standard potential for half-reaction in cell connected to + terminal of meter

E°_- = standard potential for half-reaction in cell connected to − terminal of meter

Finding K from E°

$K = 10^{nE^\circ/0.059\ 16}$ (n = number of e^- in half-reaction)

Important Terms

ampere

anode

cathode

coulomb

current

electric potential

electroactive species

electrode

electrolysis

Faraday constant

formal potential

galvanic cell

indicator electrode

Nernst equation

oxidant

oxidation

oxidizing agent

redox reaction

reducing agent

reductant

reduction

reference electrode

salt bridge

saturated calomel electrode (S.C.E.)

silver-silver chloride electrode

standard hydrogen electrode (S.H.E.)

standard reduction potential

volt

Problems

13-1. (a) Explain the difference between electric charge (q, coulombs), electric current (I, amperes), and electric potential (E, volts).

(b) How many electrons are in one coulomb?

(c) How many coulombs are in one mole of charge?

13-2. (a) Identify the oxidizing and reducing agents among the reactants below and write a balanced half-reaction for each.

$$2S_2O_4^{2-} + TeO_3^{2-} + 2OH^- \rightleftharpoons 4SO_3^{2-} + Te(s) + H_2O$$

Dithionite   Tellurite   Sulfite

(b) How many coulombs of charge are passed from reductant to oxidant when 1.00 g of Te is deposited?

(c) If Te is created at a rate of 1.00 g/h, how much current is flowing?

13-3. Hydrogen ions can carry out useful work in a living cell (such as the synthesis of the molecule ATP that provides energy for chemical synthesis) when they pass from a region of high potential to a region of lower potential. How many joules of work can be done when 1.00 μmol of H^+ crosses a membrane and goes from a potential of +0.075 V to a potential of −0.090 V (i.e., through a potential difference of 0.165 V)?

13-4. The basal rate of consumption of O_2 by a 70-kg human is 16 mol O_2/day. This O_2 oxidizes food and is reduced to H_2O, thereby providing energy for the organism:

$$O_2 + 4H^+ + 4e^- \rightleftharpoons 2H_2O$$

(a) To what current (in amperes = C/s) does this respiration rate correspond? (Current is defined by the flow of electrons from food to O_2.)

(b) If the electrons flow from nicotinamide adenine dinucleotide (NADH) to O_2, they experience a potential drop of 1.1 V. How many joules of work can be done by 16 mol O_2?

13-5. Draw a picture of the following cell. Write an oxidation for the left half-cell and a reduction for the right half-cell.

$$Pt(s) \mid Hg(l) \mid Hg_2Cl_2(s) \mid KCl(aq) \parallel ZnCl_2(aq) \mid Zn(s)$$

13-6. Suppose that the concentrations of NaF and KCl were each 0.10 M in the cell

$$Pb(s) \mid PbF_2(s) \mid F^-(aq) \parallel Cl^-(aq) \mid AgCl(s) \mid Ag(s)$$

(a) Using the half-reactions $2AgCl(s) + 2e^- \rightleftharpoons 2Ag(s) + 2Cl^-$ and $PbF_2(s) + 2e^- \rightleftharpoons Pb(s) + 2F^-$, calculate the cell voltage.

(b) Now calculate the cell voltage by using the reactions $2Ag^+ + 2e^- \rightleftharpoons 2Ag(s)$ and $Pb^{2+} + 2e^- \rightleftharpoons Pb(s)$ and K_{sp} for AgCl and PbF$_2$ (see Appendix A).

13-7. Consider a circuit in which the left half-cell was prepared by dipping a Pt wire in a beaker containing an equimolar mixture of Cr^{2+} and Cr^{3+}. The right half-cell contained a Tl rod immersed in 1.00 M TlClO$_4$.

(a) Use line notation to describe this cell.

(b) Calculate the cell voltage.

(c) Write the spontaneous net cell reaction.

(d) When the two electrodes are connected by a salt bridge and a wire, which terminal (Pt or Tl) will be the anode?

13-8. Consider the cell

$$Pt(s) \mid H_2(g, 0.100 \text{ atm}) \mid H^+(aq, pH = 2.54)$$
$$\parallel Cl^-(aq, 0.200 \text{ M}) \mid Hg_2Cl_2(s) \mid Hg(l) \mid Pt(s)$$

(a) Write a reduction reaction and Nernst equation for each half-cell. For the Hg$_2$Cl$_2$ half-reaction, $E° = 0.268$ V.

(b) Find E for the net cell reaction and state whether reduction will occur at the left- or right-hand electrode.

13-9. (a) Calculate the cell voltage (E, not $E°$), and state the direction in which electrons will flow through the potentiometer in Figure 13-4. Write the spontaneous net cell reaction.

(b) The left half-cell was loaded with 14.3 mL of $Br_2(l)$ (density = 3.12 g/mL). The aluminum electrode contains 12.0 g of Al. Which element, Br_2 or Al, is the limiting reagent in this cell? (That is, which reagent will be used up first?)

(c) If the cell is somehow operated under conditions in which it produces a constant voltage of 1.50 V, how much electrical work will have been done when 0.231 mL of $Br_2(l)$ has been consumed?

(d) If the current is 2.89×10^{-4} A, at what rate (grams per second) is Al(s) dissolving?

13-10. Calculate $E°$ and K for each reaction.

(a) $Cu(s) + Cu^{2+} \rightleftharpoons 2Cu^+$

(b) $2F_2(g) + H_2O \rightleftharpoons F_2O(g) + 2H^+ + 2F^-$

13-11. From the half-reactions below, calculate the solubility product of Mg(OH)$_2$.

$$Mg^{2+} + 2e^- \rightleftharpoons Mg(s) \qquad E° = -2.360 \text{ V}$$
$$Mg(OH)_2(s) + 2e^- \rightleftharpoons Mg(s) + 2OH^- \quad E° = -2.690 \text{ V}$$

13-12. The cell in Ask Yourself 13-B(b) contains 1.3 mM Fe(CN)$_6^{4-}$, 4.9 mM Fe(CN)$_6^{3-}$, 1.8 mM Ag(S$_2$O$_3$)$_2^{3-}$, and 55 mM S$_2$O$_3^{2-}$.

(a) Find K for the net cell reaction.

(b) Find the cell voltage and state whether Ag(s) is oxidized or reduced by the spontaneous cell reaction.

13-13. From the standard potentials for reduction of $Br_2(aq)$ and $Br_2(l)$ in Appendix C, calculate the solubility of Br_2 in water at 25°C. Express your answer as g/L.

13-14. A solution contains 0.010 0 M IO_3^-, 0.010 0 M I^-, 1.00×10^{-4} M I_3^-, and pH 6.00 buffer. Consider the reactions

$$2IO_3^- + I^- + 12H^+ + 10e^- \rightleftharpoons I_3^- + 6H_2O$$
$$E° = 1.210 \text{ V}$$
$$I_3^- + 2e^- \rightleftharpoons 3I^- \qquad E° = 0.535 \text{ V}$$

(a) Write a balanced net reaction that can occur in this solution.

(b) Calculate $E°$ and K for the reaction.

(c) Calculate E for the conditions given above.

(d) At what pH would the concentrations of IO_3^-, I^-, and I_3^- listed above be in equilibrium?

13-15. (a) Write the half-reactions for the silver-silver chloride and calomel reference electrodes.

(b) Predict the voltage for the cell below:

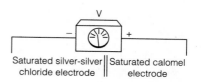

Saturated silver-silver ‖ Saturated calomel
chloride electrode ‖ electrode

13-16. Convert the potentials listed below. The Ag I AgCl and calomel reference electrodes are saturated with KCl.

(a) -0.111 V versus Ag I AgCl = ? vs S.H.E.

(b) 0.023 V versus Ag I AgCl = ? vs S.C.E.

(c) -0.023 V versus calomel = ? vs Ag I AgCl

13-17. Suppose that the silver-silver chloride electrode in Figure 13-7 is replaced by a saturated calomel electrode. Calculate the cell voltage if $[Fe^{2+}]/[Fe^{3+}] = 2.5 \times 10^{-3}$.

Measuring Ions in a Beating Heart

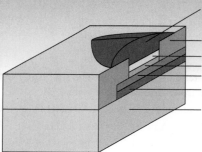

Liquid ion exchanger in polymer matrix—this layer is sensitive to K⁺ or H⁺
Polymer overlayer
AgCl ⎱ Reference electrode
Ag ⎰
Au (inert layer)
Cr (adhesion layer)
Kapton (flexible plastic substrate)

Structure of each ion-sensing dot in the array.

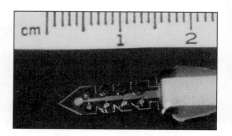

Flexible sensor array that can be inserted inside a beating heart. [R. P. Buck, University of North Carolina]

Cables to sensors   Heart

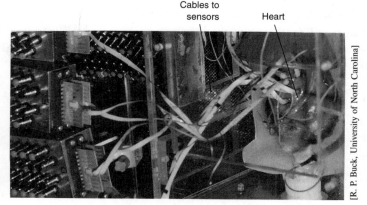

[R. P. Buck, University of North Carolina]

Multiple sensor arrays inserted into a beating pig heart monitor simultaneous changes in [K⁺] and [H⁺] when blood is denied to certain regions of the heart.

When blood flow to part of the heart is interrupted, potentially lethal changes in ionic concentrations occur. To study such an event, chemists, engineers, and doctors pooled their knowledge to create flexible, miniature ion-selective electrodes that could be inserted into a beating heart. In the device shown at the left above, four dots are sensors for K⁺ and four are sensors for H⁺. The dot at the front tip is a Ag | AgCl reference electrode for the eight sensors. The sensors provide a linear response to pH over the range 4–12 and a linear response to K⁺ over the concentration range 10^{-1} to 10^{-5} M. Fabrication of the device required photolithographic techniques from the microelectronics industry and polymer know-how to produce a mechanically flexible end product. Adhesion of the various layers to one another was a significant challenge in making a practical device.

Electrode Measurements

In the last chapter we learned that the voltage of an electrochemical cell is related to the concentrations of species in the cell. We saw that some cells could be divided into a *reference electrode* that provides a constant electric potential and an *indicator electrode* whose potential varies in response to analyte concentration. In this chapter we examine indicator electrodes that are widely used in chemical analysis. The use of voltage measurements to extract chemical information is called **potentiometry.**

Demonstration 14-1 uses a Pt indicator electrode to monitor a pretty amazing chemical reaction.

14-1 *The Silver Indicator Electrode*

Chemically inert platinum, gold, and carbon indicator electrodes are frequently used to conduct electrons to or from species in solution. In contrast to chemically inert elements, silver readily participates in the reaction $Ag^+ + e^- \rightleftharpoons Ag(s)$.

Figure 14-1 shows how a silver electrode can be used in conjunction with a calomel reference electrode to measure $[Ag^+]$ during the titration of halide ions by Ag^+ (as shown in Figures 6-3 and 6-4). The reaction at the silver indicator electrode is

$$Ag^+ + e^- \rightleftharpoons Ag(s) \qquad E_+^\circ = 0.799 \text{ V}$$

and the reference half-cell reaction is

$$Hg_2Cl_2(s) + 2e^- \rightleftharpoons 2Hg(l) + 2Cl^- \qquad E_- = 0.241 \text{ V}$$

The reference half-cell potential (E_-, not E_-°) is constant at 0.241 V because $[Cl^-]$ is fixed by the concentration of saturated KCl. The Nernst equation for the entire cell is therefore

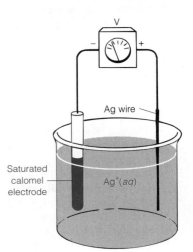

FIGURE 14-1 Use of silver and calomel electrodes to measure the concentration of Ag^+ in a solution.

253

Potentiometry with an Oscillating Reaction

Principles of potentiometry are illustrated in a fascinating manner by *oscillating reactions* in which chemical concentrations oscillate between high and low values. An example is the Belousov-Zhabotinskii reaction:

$$3CH_2(CO_2H)_2 + 2BrO_3^- + 2H^+ \xrightarrow{\text{Ce}^{3+/4+} \text{ catalyst}}$$

Malonic acid   Bromate

$$2BrCH(CO_2H)_2 + 3CO_2 + 4H_2O$$

Bromomalonic acid

During this reaction, the quotient $[Ce^{3+}]/[Ce^{4+}]$ oscillates by a factor of 10 to 100. When the Ce^{4+} concentration is high, the solution is yellow. When Ce^{3+} predominates, the solution is colorless.

To start the show, combine the following solutions in a 300-mL beaker:

160 mL of 1.5 M H_2SO_4

40 mL of 2 M malonic acid

30 mL of 0.5 M $NaBrO_3$ (or saturated $KBrO_3$)

4 mL of saturated ceric ammonium sulfate,
  $Ce(SO_4)_2 \cdot 2(NH_4)_2SO_4 \cdot 2H_2O$

After an induction period of 5 to 10 min with magnetic stirring, you can initiate oscillations by adding 1 mL of

ceric ammonium sulfate solution. The reaction is somewhat temperamental and may need more Ce^{4+} over a 5-min period to initiate oscillations.

A galvanic cell is built around the reaction as shown below. The $[Ce^{3+}]/[Ce^{4+}]$ ratio is monitored by Pt and calomel electrodes. You should be able to write the cell reactions and a Nernst equation for this experiment.

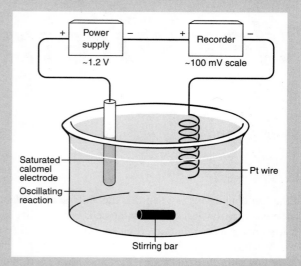

Apparatus used to monitor relative concentrations of Ce^{3+} and Ce^{4+} in an oscillating reaction.

E_- = reference electrode potential with actual concentrations in the reference cell

E_-° = standard potential of reference half-reaction when all species are in their standard states (pure solid, pure liquid, 1 M, or 1 atm)

$$E = E_+ - E_- = \underbrace{\left\{ 0.799 - 0.059\ 16 \log \left(\frac{1}{[Ag^+]} \right) \right\}}_{\substack{\text{Potential of } Ag \mid Ag^+ \\ \text{indicator electrode}}} - \underset{\substack{\uparrow \\ \text{Constant potential of} \\ \text{S.C.E. reference electrode}}}{\{0.241\}}$$

Noting that $\log (1/[Ag^+]) = -\log [Ag^+]$, we rewrite the expression above as

$$E = 0.558 + 0.059\ 16 \log [Ag^+] \tag{14-1}$$

The voltage changes by 0.059 16 V (at 25°C) for each factor-of-10 change in $[Ag^+]$.

The experiment in Figure 6-3 used a silver indicator electrode and a *glass* reference electrode. The glass electrode responds to the pH of the solution, which is held constant by a buffer. Therefore, the glass electrode remains at a constant potential.

In place of a potentiometer (a pH meter), we use a chart recorder to obtain a permanent record of the oscillations. Because the potential oscillates over a range of ~100 mV, but is centered near ~1.2 V, the cell voltage is offset by ~1.2 V with any available power supply.

Trace a (below) shows what is usually observed.

The potential changes rapidly during the abrupt colorless-to-yellow change and more gradually during the gentle yellow-to-colorless change. Trace b shows two different cycles superimposed in the same solution. This unusual event occurred in a reaction that had been oscillating normally for about 30 min.

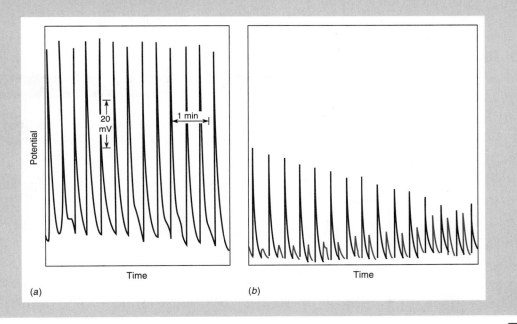

(a)   (b)

Titration of a Halide Ion with Ag$^+$

Let's consider how the concentration of Ag$^+$ varies during the titration of I$^-$ with Ag$^+$, as shown in Figure 6-4b. The titration reaction is

$$Ag^+ + I^- \longrightarrow AgI(s) \qquad K = \frac{1}{K_{sp}} = \frac{1}{8.3 \times 10^{-17}}$$

If you were monitoring the reaction with silver and calomel electrodes, you could use Equation 14-1 to compute the expected voltage at each point in the titration.

At any point prior to the equivalence point (V_e), there is a known excess of I$^-$ from which we can calculate [Ag$^+$]:

When adding Ag$^+$ to I$^-$:

- Before V_e, there is a known excess of I$^-$: [Ag$^+$] = K_{sp}/[I$^-$]
- At V_e, [Ag$^+$] = [I$^-$] = $\sqrt{K_{sp}}$
- After V_e, there is a known excess of Ag$^+$

Before V_e: $\qquad K_{sp} = [Ag^+][I^-] \Rightarrow [Ag^+] = K_{sp}/[I^-]$ $\qquad$ (14-2)

255

At the equivalence point, the quantity of Ag^+ added is exactly equal to the I^- that was originally present. We can imagine that $AgI(s)$ is made stoichiometrically and a little bit redissolves:

$$At\ V_e: \qquad K_{sp} = [\underset{x}{Ag^+}][\underset{x}{I^-}] \Rightarrow [Ag^+] = [I^-] = \sqrt{K_{sp}} \qquad (14\text{-}3)$$

Beyond the equivalence point, the quantity of excess Ag^+ added from the buret is known, and the concentration is just

$$After\ V_e: \qquad [Ag^+] = \frac{\text{moles of excess } Ag^+}{\text{total volume of solution}} \qquad (14\text{-}4)$$

EXAMPLE   **Potentiometric Precipitation Titration**

A 20.00-mL solution containing 0.100 4 M KI was titrated with 0.084 5 M $AgNO_3$, using the cell in Figure 14-1. Calculate the voltage at volumes $V_{Ag^+} = 15.00$, V_e, and 25.00 mL.

SOLUTION The titration reaction is $Ag^+ + I^- \rightarrow AgI(s)$, and the equivalence volume is

$$\underbrace{(V_e(\text{mL}))(0.084\ 5\ M)}_{\text{mmol } Ag^+} = \underbrace{(20.00\ \text{mL})(0.100\ 4\ M)}_{\text{mmol } I^-} \Rightarrow V_e = 23.76\ \text{mL}$$

15.00 mL: We began with $(20.00\ \text{mL})(0.100\ 4\ M) = 2.008$ mmol I^- and added $(15.00\ \text{mL})(0.084\ 5\ M) = 1.268$ mmol Ag^+. The concentration of unreacted I^- is

$$[I^-] = \frac{(2.008 - 1.268)\ \text{mmol}}{(20.00 + 15.00)\ \text{mL}} = 0.021\ 1\ M$$

The concentration of Ag^+ in equilibrium with the solid AgI is therefore

$$[Ag^+] = \frac{K_{sp}}{[I^-]} = \frac{8.3 \times 10^{-17}}{0.021\ 1} = 3.9 \times 10^{-15}\ M$$

The cell voltage is computed with Equation 14-1:

$$E = 0.558 + 0.059\ 16 \log (3.9 \times 10^{-15}) = -0.294\ V$$

At V_e: Equation 14-3 tells us that $[Ag^+] = \sqrt{K_{sp}} = 9.1 \times 10^{-9}\ M$, so

$$E = 0.558 + 0.059\ 16 \log (9.1 \times 10^{-9}) = 0.082\ V$$

25.00 mL: Now there is an excess of $25.00 - 23.76 = 1.24$ mL of 0.084 5 M $AgNO_3$ in a total volume of 45.00 mL.

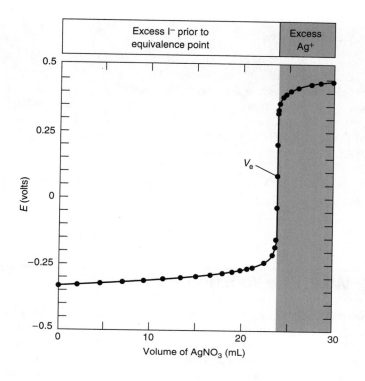

FIGURE 14-2 Calculated titration curve for the addition of 0.084 5 M Ag^+ to 20.00 mL of 0.100 4 M I^-, using the cell in Figure 14-1.

$$[Ag^+] = \frac{(1.24 \text{ mL})(0.084\ 5 \text{ M})}{45.00 \text{ mL}} = 2.33 \times 10^{-3} \text{ M}$$

and the cell voltage is

$$E = 0.558 + 0.059\ 16 \log (2.33 \times 10^{-3}) = 0.402 \text{ V}$$

The titration curve in Figure 14-2 barely changes prior to the equivalence point because the concentration of Ag^+ is very low and relatively constant until I^- is used up. When I^- has been consumed, $[Ag^+]$ suddenly increases and so does the voltage. Figure 14-2 is upside down compared to curve b in Figure 6-4. The reason is that in Figure 14-1 the indicator electrode is connected to the *positive* terminal of the potentiometer. In Figure 6-3 the indicator electrode is connected to the *negative* terminal because the glass pH electrode fits into the positive terminal of the meter. In addition to their opposite polarities, the voltages in Figures 14-2 and 6-4 are different because each experiment uses a different reference electrode.

Double-Junction Reference Electrode

If you tried to titrate I^- with Ag^+ by using the cell in Figure 14-1, KCl solution would slowly leak into the titration beaker from the porous plug at the base of the reference electrode (Figure 13-9). Cl^- introduces a titration error because it consumes Ag^+. The *double-junction reference electrode* in Figure 14-3 prevents the inner electrolyte solution from leaking directly into the titration vessel.

FIGURE 14-3 Double-junction reference electrode has an inner electrode identical to those in Figures 13-8 and 13-9. The outer compartment is filled with any desired electrolyte, such as KNO_3, that is compatible with the titration solution. KCl electrolyte from the inner electrode slowly leaks into the outer electrode, so the outer electrolyte should be changed periodically.

Ask Yourself

14-A. Consider the titration of 40.0 mL of 0.050 0 M NaCl with 0.200 M AgNO$_3$, using the cell in Figure 14-1. The equivalence volume is $V_e = 10.0$ mL.

(a) Prior to V_e, there is a known excess of Cl$^-$. Find [Cl$^-$] at the following volumes of added silver: $V_{Ag^+} = 0.10, 2.50, 5.00, 7.50,$ and 9.90 mL. From [Cl$^-$], use K_{sp} for AgCl to find [Ag$^+$] at each volume.

(b) Find [Cl$^-$] and [Ag$^+$] at $V_{Ag^+} = V_e = 10.00$ mL.

(c) After V_e, there is a known excess of Ag$^+$. Find [Ag$^+$] at $V_{Ag^+} = 10.10$ and 12.00 mL.

(d) Use Equation 14-1 to find the cell voltage at each volume in parts (a)–(c) and make a graph of the titration curve.

14-2 *What Is a Junction Potential?*

$E_{observed} = E_{cell} + E_{junction}$
Because the junction potential is usually unknown, E_{cell} is uncertain.

When two dissimilar electrolyte solutions are placed in contact, a voltage difference called the **junction potential** develops at the interface. This small, unknown voltage (usually a few millivolts) exists at each end of a salt bridge connecting two half-cells. *The junction potential puts a fundamental limitation on the accuracy of direct potentiometric measurements,* because we usually do not know the contribution of the junction to the measured voltage.

To see why a junction potential occurs, consider a solution containing NaCl in contact with distilled water (Figure 14-4). The Na$^+$ and Cl$^-$ ions diffuse from the NaCl solution into the water phase. However, Cl$^-$ ion has a greater *mobility* than Na$^+$. That is, Cl$^-$ diffuses faster than Na$^+$. As a result, a region rich in Cl$^-$, with excess negative charge, develops at the front. Behind it is a positively charged region depleted of Cl$^-$. The result is an electric potential difference at the junction of the NaCl and H$_2$O phases.

Mobilities of ions are shown in Table 14-1 and several junction potentials are listed in Table 14-2. Because K$^+$ and Cl$^-$ have similar mobilities, junction potentials of a KCl salt bridge are slight. This is why saturated KCl is used in salt bridges.

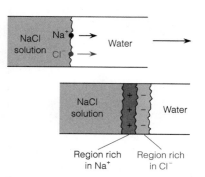

FIGURE 14-4 Development of the junction potential caused by unequal mobilities of Na$^+$ and Cl$^-$.

TABLE	14-1	Mobilities of ions in water at 25°C	
Ion	*Mobility [m^2/(s · V)][a]*	*Ion*	*Mobility [m^2/(s · V)][a]*
H$^+$	36.30×10^{-8}	OH$^-$	20.50×10^{-8}
K$^+$	7.62×10^{-8}	SO$_4^{2-}$	8.27×10^{-8}
NH$_4^+$	7.61×10^{-8}	Br$^-$	8.13×10^{-8}
La^{3+}	7.21×10^{-8}	I$^-$	7.96×10^{-8}
Ba^{2+}	6.59×10^{-8}	Cl$^-$	7.91×10^{-8}
Ag$^+$	6.42×10^{-8}	NO$_3^-$	7.40×10^{-8}
Ca^{2+}	6.12×10^{-8}	ClO$_4^-$	7.05×10^{-8}
Cu^{2+}	5.56×10^{-8}	F$^-$	5.70×10^{-8}
Na$^+$	5.19×10^{-8}	CH$_3$CO$_2^-$	4.24×10^{-8}
Li$^+$	4.01×10^{-8}		

a. The mobility of an ion is the velocity achieved in an electric field of 1 V/m. Mobility = velocity/field. The units of mobility are therefore (m/s)/(V/m) = m^2/(s · V).

Ask Yourself

14-B. A 0.1 M NaCl solution is placed in contact with a 0.1 M NaNO₃ solution. The concentration of Na$^+$ is the same on both sides of the junction, so there is no net diffusion of Na$^+$ from one side to the other. The mobility of Cl$^-$ is greater than that of NO$_3^-$, so Cl$^-$ diffuses away from the NaCl side faster than NO$_3^-$ diffuses away from NaNO₃. Which side of the junction will become positive and which will become negative? Explain your reasoning.

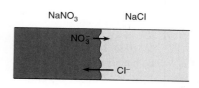

14-3 *How Ion-Selective Electrodes Work*

An **ion-selective electrode** responds preferentially to one species in a solution. Differences in concentration of the selected ion across the membrane produce an electric potential difference across the membrane.

To understand how ion-selective electrodes work, imagine a membrane separating two solutions of CaCl₂ (Figure 14-5). The membrane contains a molecule that binds and transports Ca^{2+}, but *not* Cl$^-$. Ca^{2+} can enter and cross the membrane, but Cl$^-$ cannot. Initially the potential difference across the membrane is zero, because both solutions are neutral (Figure 14-5a). After some time, there is a net diffusion of Ca^{2+} from the right (high concentration) to the left (low concentration).

TABLE 14-2

Liquid junction potentials at 25°C

Junction	Potential (mV)
0.1 M NaCl | 0.1 M KCl	−6.4
0.1 M NaCl | 3.5 M KCl	−0.2
1 M NaCl | 3.5 M KCl	−1.9
0.1 M HCl | 0.1 M KCl	+27
0.1 M HCl | 3.5 M KCl	+3.1

Note: A positive sign means that the right side of the junction becomes positive with respect to the left side.

Mechanism of ion-selective electrode:

1. A specific ion crosses the membrane, creating charge imbalance
2. Charge buildup opposes further movement of ion
3. Result: potential difference across membrane is related to difference in concentrations of specific ion on either side

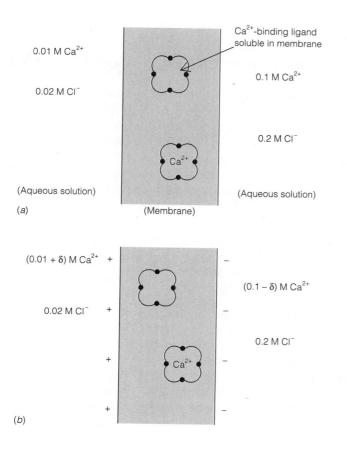

FIGURE 14-5 Mechanism of ion-selective electrode. (*a*) Initial conditions prior to Ca^{2+} migration across membrane. (*b*) After δ moles of Ca^{2+} per liter have crossed the membrane, giving the left side a charge of +2δ mol/L and the right side a charge of −2δ mol/L.

The factor in Equation 14-5 is 0.054 20 V at 0°C and 0.061 54 V at 37°C. Equation 14-5 should really be written in terms of activities (Box 5-2) instead of concentrations.

As Ca^{2+} migrates from right to left, positive charge builds up on the left (Figure 14-5b). Eventually the excess positive charge, which repels Ca^{2+}, prevents further migration of Ca^{2+} to the positively charged side. At equilibrium, the potential difference across the membrane is

Electric potential difference due to concentration difference:

$$E = \left(\frac{0.059\ 16}{n} \right) \log \left(\frac{[Ca^{2+}]_{\text{right}}}{[Ca^{2+}]_{\text{left}}} \right) \quad \text{(volts at 25°C)} \quad (14\text{-}5)$$

where n is the charge on the species of interest. Because the charge of Ca^{2+} is $n = 2$, a potential difference of $0.059\ 16/2 = 0.029\ 58$ V is expected for every factor-of-10 difference in concentration of Ca^{2+} across the membrane. No charge buildup results from the Cl^- because Cl^- has no way to cross the membrane.

The key to designing an ion-selective electrode is to fabricate a membrane across which only the desired species can migrate. The approach is to select soluble molecules or solid crystals that selectively interact with the desired analyte. No membrane is perfectly selective, so there is always some interference from unintended species.

Ask Yourself

14-C. Why doesn't Ca^{2+} continue to diffuse across the membrane in Figure 14-5 until the concentration is the same on both sides?

14-4 *pH Measurement with a Glass Electrode*

The most widely employed ion-selective electrode is the **glass electrode** for measuring pH. A pH electrode responds selectively to H^+, building up a potential difference of 0.059 16 V for every factor-of-10 difference in $[H^+]$ across the electrode membrane. A factor-of-10 difference in $[H^+]$ is one pH unit, so a change of, say, 4.00 pH units leads to a change in electrode potential of $4.00 \times 0.059\ 16 = 0.237$ V.

A typical **combination electrode,** incorporating both glass and reference electrodes in one body, is shown in Figure 14-6. The line diagram of this cell is

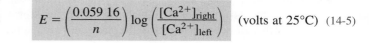

Glass membrane

$Ag(s) \mid AgCl(s) \mid Cl^-(aq) \parallel H^+(aq, \text{outside}) \vdots H^+(aq, \text{inside}), Cl^-(aq) \mid AgCl(s) \mid Ag(s)$

| Outer reference electrode | H^+ outside glass electrode (analyte solution) | H^+ inside glass electrode | Inner reference electrode |

The pH-sensitive part of the electrode is the thin glass membrane in the shape of a bulb at the bottom of the electrodes in Figures 14-6 and 14-7.

The glass membrane at the bottom of the pH electrode consists of an irregular network of SiO_4 tetrahedra through which Na^+ ions move sluggishly. Studies with tritium (the radioactive isotope ^3H) show that H^+ does *not* diffuse through the membrane. The glass surface contains exposed $-O^-$ groups that can bind H^+ from the solutions on either side of the membrane (Figure 14-8). H^+ equilibrates with the glass surface, thereby giving the side of the membrane exposed to the higher con-

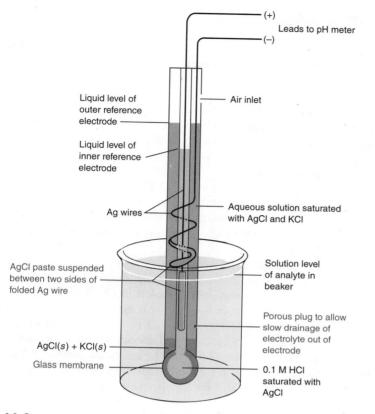

FIGURE 14-6 Glass combination electrode with a silver-silver chloride reference electrode. The glass electrode is immersed in a solution of unknown pH so that the porous plug on the lower right is below the surface of the liquid. The two Ag electrodes measure the voltage across the glass membrane.

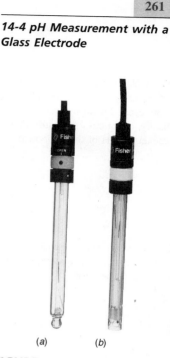

(a) (b)

FIGURE 14-7 Left: Glass-body combination electrode with pH-sensitive glass bulb at the bottom. The circle at the side of the electrode near the bottom is the porous junction salt bridge to the reference electrode compartment. Two silver wires coated with AgCl are visible inside the electrode. Right: Polymer body surrounds glass electrode to protect the delicate bulb.

centration of H^+ the more positive charge. Na^+ ions that are already in the glass migrate across the membrane from the positive side to the negative side, so the potential changes by 0.059 16 V for a unit change in pH. Because the electrical resistance of the glass membrane is high, little current actually flows across it.

The potential difference between the inner and outer silver-silver chloride electrodes in Figure 14-6 depends on $[Cl^-]$ in each electrode compartment and on the potential difference across the glass membrane. Because $[Cl^-]$ is fixed, and because $[H^+]$ is constant inside the glass electrode, the only variable is the pH of the analyte solution outside the glass membrane.

Real glass electrodes are described by the equation

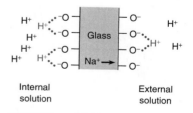

FIGURE 14-8 Ion-exchange equilibria on the inner and outer surfaces of the glass membrane. The pH of the internal solution is fixed. As the pH of the external solution (the sample) changes, the electric potential difference across the glass membrane changes.

Response of glass electrode:

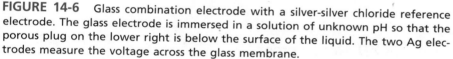

$$E = \text{constant} + \beta(0.059\ 16)\Delta pH \qquad \text{(at 25°C)} \qquad (14\text{-}6)$$

where ΔpH is the difference in pH between the analyte solution and the solution inside the glass bulb. The factor β, which is ideally 1, is typically 0.98–1.00. The constant term, called the *asymmetry potential*, arises because no two sides of a real object are identical, so a small voltage exists even if the pH is the same on both sides of the membrane. We correct for asymmetry by calibrating the electrode in solutions of known pH.

A pH electrode *must* be calibrated before use. It should be calibrated about every 2 h during sustained use. Ideally, calibration standards should bracket the pH of the unknown.

Don't leave a glass electrode out of water (or in a nonaqueous solvent) longer than necessary.

Calibrating a Glass Electrode

A pH electrode should be calibrated with two (or more) standard buffers selected so that the pH of the unknown lies within the range of the standards. Before calibrating the electrode, wash it with distilled water and gently *blot* it dry with a tissue. Don't *wipe* it, because this action might produce a static charge on the glass. Dip the electrode in a standard buffer whose pH is near 7 and allow the electrode to equilibrate for at least a minute. For best results, solutions should be stirred during testing. Adjust the meter reading (usually with a knob labeled "Calibrate") to indicate the pH of the standard buffer. Wash the electrode, blot it dry, and immerse it in the second standard buffer. If the electrode response were perfect, the voltage would change by 0.059 16 V per pH unit at 25°C. The actual change may be slightly less, so these two measurements establish the value of β in Equation 14-6. The pH of the second buffer is set on the meter with a knob that may be labeled "Slope" or "Temperature." Finally, dip the electrode in the unknown and read the pH on the meter.

Store the glass electrode in aqueous solution to prevent dehydration of the glass. If the electrode has dried out, it should be reconditioned in water for several hours. If the electrode is to be used above pH 9, soak it in a high-pH buffer.

If electrode response becomes sluggish, or if the electrode cannot be calibrated properly, try soaking it in 6 M HCl, followed by water. As a last resort, dip the electrode in 20 wt % aqueous ammonium bifluoride, NH_4HF_2, (in a plastic beaker) for 1 min. This reagent dissolves a little of the glass on the outside of the glass bulb and exposes a fresh surface. Wash the electrode with water and try calibrating it again. *Ammonium bifluoride must not contact your skin, because it produces HF burns.*

Errors in pH Measurement

To make intelligent use of a glass electrode, it is important to understand its limitations:

1. A pH measurement cannot be more accurate than our standards, which are typically ±0.01–0.02 pH units.

2. A *junction potential* exists at the porous plug near the bottom of the electrode in Figure 14-6. If the ionic composition of the analyte solution is different from that of the standard buffer, the junction potential will change *even if the pH of the two solutions is the same.* This factor gives an uncertainty of at least ~0.01 pH unit. Box 14-1 describes how junction potentials affect the measurement of the pH of rainwater.

3. When $[H^+]$ is very low and $[Na^+]$ is high, the electrode responds to Na^+ as if Na^+ were H^+, and the apparent pH is lower than the true pH. This is called the *alkaline error,* or *sodium error* (Figure 14-9).

4. In strong acid, the measured pH is higher than the actual pH, for reasons that are not well understood.

5. The electrode needs time to equilibrate with each solution. In a well-buffered solution, equilibration takes just seconds with adequate stirring. In a poorly buffered solution (such as near the equivalence point of a titration), it could take minutes.

FIGURE 14-9 Acid and alkaline errors of some glass electrodes. A: Corning 015, H_2SO_4. B: Corning 015, HCl. C: Corning 015, 1 M Na^+. D: Beckman-GP, 1 M Na^+. E: L & N Black Dot, 1 M Na^+. F: Beckman Type E, 1 M Na^+.

Systematic Error in Rainwater pH Measurement: The Effect of Junction Potential

Figure 9-1 shows the pH of rainfall over North America. Acidity in rainfall is partly a result of human activities and is slowly changing the nature of many ecosystems as we know them. Monitoring the pH of rainwater is a critical component of programs to reduce the production of acid rain.

To identify and correct systematic errors in the measurement of pH of rainwater, eight samples were provided to each of 17 laboratories, along with explicit instructions for how to conduct the measurements. Each lab used two standard buffers to calibrate its pH meter.

The figure below shows typical results for the pH of rainwater. The average of the 17 measurements is given by the horizontal line at pH 4.14 and the letters s, t, u, v, w, x, y, and z identify the types of pH electrodes used for the measurements. Laboratories using electrode types s and w had relatively large systematic errors. The type s electrode was a combination electrode (Figure 14-6) containing a reference electrode liquid junction with an exceptionally large area. Electrode type w had a reference electrode filled with a gel.

It was hypothesized that variability in the liquid junction potential (Section 14-2) led to the variability of the pH measurements. Standard buffers used for pH meter calibration typically have ion concentrations of ~0.05 M, whereas rainwater samples have ion concentrations two or more orders of magnitude lower. To test the hypothesis that junction potential caused systematic errors, a pure HCl solution with a concentration near 2×10^{-4} M was used as a pH calibration standard in place of high ionic strength buffers. The following data were obtained, with good results from all but the first lab. The standard deviation of all 17 measurements was reduced from 0.077 pH unit with the standard buffer to 0.029 pH unit with the HCl standard. It was concluded that junction potential is the cause of most of the variability between labs and that a low ionic strength standard is appropriate for rainwater pH measurements.

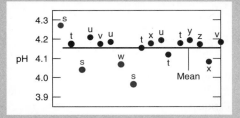

pH of rainwater from identical samples measured at 17 different labs, which all used the same standard calibration buffers. Letters designate different types of pH electrodes.

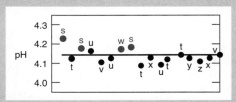

Rainwater pH measured after using low ionic strength HCl for calibration.

6. A dry electrode requires several hours of soaking before it responds to H^+ correctly.

7. A pH meter should be calibrated at the same temperature at which the measurement will be made. You should not calibrate your equipment at one temperature and then expect to make accurate measurements at a second temperature.

Errors 1 and 2 limit the accuracy of pH measurements with the glass electrode to ±0.02 pH unit, at best. Measurement of pH differences *between* solutions can be accurate to about ±0.002 pH unit, but knowledge of the true pH will still be at least an order of magnitude more uncertain. An uncertainty of ±0.02 pH unit corresponds to an uncertainty of ±5% in $[H^+]$.

Challenge Show that the potential of the glass electrode changes by 1.3 mV when the analyte H^+ concentration changes by 5.0%. Because 59 mV ≈ 1 pH unit, 1.3 mV = 0.02 pH unit.

Moral: A small uncertainty in voltage (1.3 mV) or pH (0.02 unit) corresponds to a large uncertainty (5%) in H^+ concentration. Similar uncertainties arise in other potentiometric measurements.

(a)

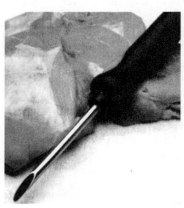

(b)

FIGURE 14-10 (a) pH-sensitive field effect transistor. (b) Rugged field effect transistor mounted in a steel shaft can be inserted into meat, poultry, or other damp solids to measure pH.

Electrode response really depends on $\log\left(\dfrac{[F^-]_{outside}}{[F^-]_{inside}}\right)$. The constant value of $[F^-]_{inside}$ is incorporated into the constant term in Equation 14-7.

Solid-State pH Sensors

There are pH sensors that do not depend on a fragile glass membrane. The *field effect transistor* in Figure 14-10 is a tiny semiconductor device whose surface binds H^+ from the medium in which the transistor is immersed. The higher the concentration of H^+ in the external medium, the more positively charged is the transistor's surface. The surface charge regulates the flow of current through the transistor, which therefore behaves as a pH sensor.

Ask Yourself

14-D. (a) List the sources of error associated with pH measurement made with the glass electrode.

(b) When the difference in pH across the membrane of a glass electrode at 25°C is 4.63 pH units, how much voltage is generated by the pH gradient? Assume that the constant β in Equation 14-6 is 1.00.

(c) Why do glass pH electrodes tend to indicate a pH lower than the actual pH in strongly basic solution?

14-5 Ion-Selective Electrodes

The glass pH electrode is an example of a *solid-state ion-selective electrode* whose operation depends on (1) an ion-exchange reaction of H^+ between the glass surface and the analyte solution and (2) transport of an ion (Na^+) across the glass membrane. A calcium ion-selective electrode whose membrane contains a solution of Ca^{2+}-binding compound (Figure 14-5) is an example of a *liquid-based ion-selective electrode*. We now examine several very useful ion-selective electrodes.

Solid-State Electrodes

The ion-sensitive component of a fluoride **solid-state ion-selective electrode** is a crystal of LaF_3 doped with EuF_2 (Figure 14-11a). *Doping* is the intentional addition of a small amount of an impurity (EuF_2 in this case) into the solid crystal (LaF_3). The inner surface of the crystal is exposed to filler solution containing a constant concentration of F^-. The outer surface is exposed to a variable concentration of F^- in the unknown. F^- from solution is selectively adsorbed on each surface of the crystal. Doping LaF_3 with EuF_2 creates anion vacancies that allow F^- to jump from one site to the next. Unlike H^+ in a pH electrode, F^- itself migrates across the LaF_3 crystal, as shown in Figure 14-11b.

The response of the electrode to F^- is given by

Response of F⁻ electrode: $$E = \text{constant} - \beta(0.059\,16) \log [F^-]_{outside} \qquad (14\text{-}7)$$

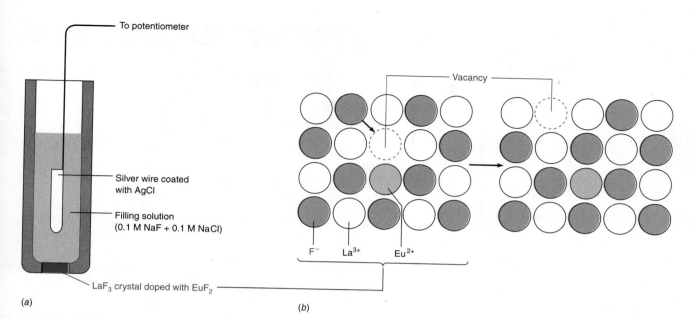

To potentiometer

Silver wire coated with AgCl

Filling solution (0.1 M NaF + 0.1 M NaCl)

LaF_3 crystal doped with EuF_2

Vacancy

F^- La^{3+} Eu^{2+}

(a)

(b)

FIGURE 14-11 (a) Fluoride ion-selective electrode employing a crystal of LaF_3 doped with EuF_2 as the ion-selective membrane. (b) Migration of F^- through the doped crystal: For charge conservation, every Eu^{2+} is accompanied by an anion vacancy in the crystal. When a neighboring F^- jumps into the vacancy, another site becomes vacant. Repetition of this process moves F^- through the lattice.

where $[F^-]_{outside}$ is the concentration of F^- in the analyte solution and β is close to 1.00. The electrode response is close to 59 mV per decade over a F^- concentration range from about 10^{-6} M to 1 M. The electrode is more responsive to F^- than to most other ions by a factor greater than 1 000. (Response to OH^- is one-tenth as great as the response to F^-.) At low pH, F^- is converted to HF ($pK_a = 3.17$), to which the electrode is insensitive. The F^- electrode is used to monitor and control the fluoridation of municipal water supplies. Fluoride is added to drinking water to help prevent tooth decay. Several other solid-state ion-selective electrodes are listed in Table 14-3.

TABLE 14-3 Solid-state ion-selective electrodes

Ion	Concentration range (M)	Membrane crystal[a]	pH range	Interfering species
F^-	10^{-6}–1	LaF_3	5–8	OH^-
Cl^-	10^{-4}–1	AgCl	2–11	CN^-, S^{2-}, I^-, $S_2O_3^{2-}$, Br^-
Br^-	10^{-5}–1	AgBr	2–12	CN^-, S^{2-}, I^-
I^-	10^{-6}–1	AgI	3–12	S^{2-}
CN^-	10^{-6}–10^{-2}	AgI	11–13	S^{2-}, I^-
S^{2-}	10^{-5}–1	Ag_2S	13–14	

a. Electrodes containing silver-based crystals such as Ag_2S should be stored in the dark and protected from light during use to prevent light-induced chemical degradation.

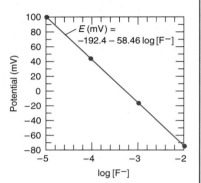

FIGURE 14-12 Calibration curve for fluoride ion-selective electrode.

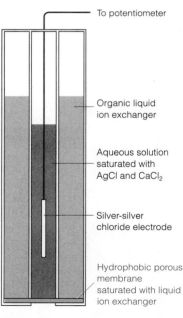

To potentiometer

Organic liquid ion exchanger

Aqueous solution saturated with AgCl and $CaCl_2$

Silver-silver chloride electrode

Hydrophobic porous membrane saturated with liquid ion exchanger

FIGURE 14-13 Schematic diagram of a Ca^{2+} ion-selective electrode based on a liquid ion exchanger.

EXAMPLE Calibration Curve for an Ion-Selective Electrode

A fluoride electrode immersed in standard solutions gave the following potentials:

$[F^-]$ (M)	$\log [F^-]$	E (mV vs S.C.E.)
1.00×10^{-5}	5.00	100.0
1.00×10^{-4}	4.00	41.4
1.00×10^{-3}	3.00	-17.0
1.00×10^{-2}	2.00	-75.4

(a) What potential is expected if $[F^-] = 5.00 \times 10^{-5}$ M? (b) What concentration of F^- will give a potential of 0.0 V?

SOLUTION (a) Our strategy is to fit the calibration data with Equation 14-7 and then to substitute the concentration of F^- into the equation to find the potential:

$$E = \underbrace{\text{constant}}_{\substack{y \\ \text{Intercept}}} - \underbrace{\text{m}}_{\text{Slope}} \cdot \underbrace{\log [F^-]}_{x}$$

Using the method of least squares in Section 4-6, we plot E versus $\log [F^-]$ to find a straight line with a slope of -58.46 mV and an intercept of -192.4 mV (Figure 14-12). Setting $[F^-] = 5.00 \times 10^{-5}$ M gives

$$E = -192.4 - 58.46 \log [5.00 \times 10^{-5}] = 59.0 \text{ mV}$$

(b) If $E = 0.0$ mV, we can solve for the concentration of $[F^-]$:

$$0.0 = -192.4 - 58.46 \log [F^-] \Rightarrow [F^-] = 5.1 \times 10^{-4} \text{ M}$$

Liquid-Based Ion-Selective Electrodes

The principle of a **liquid-based ion-selective electrode** was described in Figure 14-5. Figure 14-13 shows a Ca^{2+} ion-selective electrode, at the base of which is a membrane saturated with a liquid ion exchanger (a Ca^{2+}-binding compound dissolved in organic solvent). Ca^{2+} is selectively transported across the membrane to establish a voltage related to the difference in concentrations of Ca^{2+} between the sample and internal solution:

Response of Ca^{2+} *electrode:* $\quad E = \text{constant} + \beta\left(\dfrac{0.059\ 16}{2}\right) \log [Ca^{2+}]_{\text{outside}}$ (14-8)

where β is close to 1.00. Equations 14-8 and 14-7 have different signs before the log term because one involves an anion and the other a cation. The charge of the calcium ion requires a factor of 2 in the denominator before the logarithm. Figure 14-14 shows the variability of results when identical blood serum samples were analyzed with 14 different Ca^{2+} ion-selective electrodes.

The most serious interference with the Ca^{2+} electrode comes from Zn^{2+}, Fe^{2+}, Pb^{2+}, and Cu^{2+}, but high concentrations of Sr^{2+}, Mg^{2+}, Ba^{2+}, and Na^+ also interfere. Interference from H^+ is substantial below pH 4. pH and the ionic strength of standards and unknowns must generally be held constant when using ion-selective electrodes.

The heparin-sensitive electrode at the beginning of Chapter 13 is similar to the Ca^{2+} electrode. Figure 14-15 shows exchange of chloride ions in the membrane with the negatively charged chain of the heparin molecule.

Selectivity Coefficient

No electrode responds exclusively to one kind of ion, but the glass electrode is among the most selective. A high-pH glass electrode responds to Na^+ only when $[H^+] \leq 10^{-12}$ M and $[Na^+] \geq 10^{-2}$ M (Figure 14-9).

If an electrode that measures ion X also responds to ion Y, the **selectivity coefficient** is defined as

Selectivity coefficient:
$$k_{X,Y} = \frac{\text{response to Y}}{\text{response to X}} \qquad (14-9)$$

Ideally, the selectivity coefficient should be very small ($k \ll 1$).

The behavior of most ion-selective electrodes is described by the equation

Response of ion-selective electrode:
$$E = \text{constant} \pm \beta\left(\frac{0.059\ 16}{n_X}\right) \log\left\{[X] + \sum_Y (k_{X,Y}[Y]^{n_X/n_Y})\right\} \qquad (14-10)$$

where [X] is the concentration of the ion intended to be measured and [Y] is the concentration of an interfering species. The magnitude of the charge of each species is n_X or n_Y, and $k_{X,Y}$ is the selectivity coefficient. If the ion-selective electrode is connected to the positive terminal of the potentiometer, the sign before the log term is positive if X is a cation and negative if X is an anion. The value of β is near 1.00.

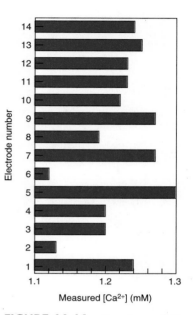

FIGURE 14-14 Response of 14 different Ca^{2+} ion-selective electrodes to identical human blood serum samples. The mean value is 1.22 ±0.05 mM.

Equation 14-10 describes the response of an electrode to its primary ion, X, and to all interfering ions, Y.

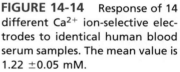

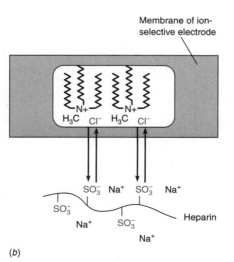

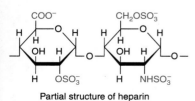

Partial structure of heparin

(a)

(b)

FIGURE 14-15 (a) Partial structure of heparin molecule, showing negatively charged substituents. Heparin belongs to a class of molecules called *proteoglycans* and contain about 95% polysaccharide (sugar) and 5% protein. (b) Ion exchange between the negatively charged heparin and Cl^- ions associated with tetraalkylammonium ions dissolved in the membrane of the ion-selective electrode.

EXAMPLE **Using the Selectivity Coefficient**

A Tl^{3+} ion-selective electrode has a selectivity coefficient $k_{Tl^{3+},K^+} = 3.4 \times 10^{-4}$. What is the change in electrode potential when 10^{-1} M K^+ is added to 10^{-5} M Tl^{3+}?

SOLUTION Using Equation 14-10 with $\beta = 1.00$, $n_X = +3$, and $n_Y = +1$, the potential without K^+ is

$$E = \text{constant} + \left(\frac{0.059\,16}{3}\right)\log[10^{-5}] = \text{constant} - 98.60 \text{ mV}$$

Addition of 10^{-1} M K^+ gives an electrode potential of

$$E = \text{constant} + \left(\frac{0.059\,16}{3}\right)\log[10^{-5} + (3.4 \times 10^{-4})(10^{-1})^{3/1}]$$

$$= \text{constant} - 98.31 \text{ mV}$$

The change is $(-98.31) - (-98.60) = 0.29$ mV, barely enough to measure.

A particular Ca^{2+} ion-selective electrode has selectivity coefficients $k_{Ca^{2+},Fe^{2+}} = 0.80$ and $k_{Ca^{2+},Mg^{2+}} = 0.01$. These coefficients mean that the response to Fe^{2+} is 80% as great as the response to Ca^{2+}, while the response to Mg^{2+} is only 1% as great as the response to Ca^{2+}.

Compound Electrodes

A **compound electrode** is a conventional electrode surrounded by a membrane that isolates (or generates) the analyte to which the electrode responds. The CO_2 gas-sensing electrode in Figure 14-16 is an ordinary glass pH electrode surrounded by

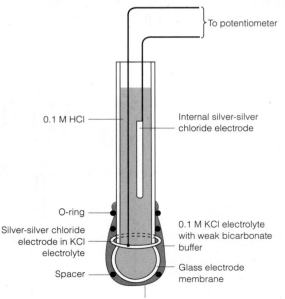

To potentiometer

0.1 M HCl

Internal silver-silver chloride electrode

O-ring

Silver-silver chloride electrode in KCl electrolyte

Spacer

0.1 M KCl electrolyte with weak bicarbonate buffer

Glass electrode membrane

CO_2 permeable membrane

FIGURE 14-16 Schematic diagram of a CO_2 gas-sensing electrode.

electrolyte solution enclosed in a semipermeable membrane made of rubber, Teflon, or polyethylene. A silver-silver chloride reference electrode is immersed in the electrolyte solution. When CO_2 diffuses through the semipermeable membrane, it lowers the pH in the electrolyte compartment. The response of the glass electrode to the change in pH is measured.

Other acidic or basic gases, including NH_3, SO_2, H_2S, NO_x (nitrogen oxides), and HN_3 (hydrazoic acid) can be detected in the same manner. These electrodes can be used to measure gases *in the gas phase* or dissolved in solution. Some ingenious compound electrodes contain a conventional electrode coated with an enzyme that catalyzes a reaction of the analyte. The product of the reaction is detected by the electrode. Compound electrodes based on enzymes are among the most selective because enzymes tend to be extremely specific in their reactivity with just the species of interest.

Ask Yourself

14-E. In Chapters 6 and 8 we discussed the measurement of nitrite in a saltwater aquarium at Georgia Tech. Now we consider the measurement of ammonia in the fishtank, using an ammonia-selective compound electrode. The procedure is to mix 100.0 mL of unknown or standard with 1.0 mL of 10 M NaOH and then to measure NH_3 with an electrode. The purpose of the NaOH is to raise the pH above 11, so that ammonia is almost all in the form NH_3, not NH_4^+.

(a) A series of standards gave the readings below. Prepare a calibration curve of potential (mV) versus log (ppm) and determine the equation of the straight line by the method of least squares. (Calibration is done in terms of nitrogen in the original standards. There is no dilution factor to consider from the NaOH.)

NH_3 nitrogen concentration (ppm)	log (ppm)	Electrode potential (mV vs S.C.E.)
0.100	−1.000	72
0.500	−0.301	42
1.000	0.000	25

Linda A. Hughes

(b) Two students measured NH_3 in the aquarium and observed values of 106 and 115 mV. What NH_3 nitrogen concentration (in ppm) should be reported by each student?

(c) Artificial seawater for the aquarium is prepared by adding a commercial seawater salt mix to the correct volume of distilled water. There is an unhealthy level of NH_4Cl impurity in the salt mix, so the instructions call for several hours of aerating freshly prepared seawater to remove $NH_3(g)$ before adding the water to a tank containing live fish. A student measured the concentration of NH_3 in freshly prepared seawater prior to aeration and observed a potential of 56 mV. What is the concentration of NH_3 in the freshly prepared seawater?

Key Equations

Voltage of complete cell

$$E = E_+ - E_- \quad \text{(repeated from Chapter 13)}$$
$E_+ =$ voltage of electrode connected to + terminal of meter
$E_- =$ voltage of electrode connected to − terminal of meter

Titration of X^- with M^+

Before V_e: $[M^+] = K_{sp}/[X^-]$

At V_e: $[M^+] = [X^-] = \sqrt{K_{sp}}$

After V_e: $[M^+] = \dfrac{\text{mol excess } M^+}{\text{total volume}}$

Response of glass pH electrodes

$E = \text{constant} + \beta(0.059\ 16)\Delta\text{pH}$

$\Delta\text{pH} = (\text{analyte pH}) - (\text{pH of internal solution})$

$\beta\ (\approx 1.00)$ is measured with standard buffers

constant = asymmetry potential (measured by calibration)

Ion-selective electrode response

$E = \text{constant} \pm \beta\left(\dfrac{0.059\ 16}{n_X}\right) \log\left\{[X] + \sum_Y (k_{X,Y}[Y]^{n_X/n_Y})\right\}$

X = analyte ion with charge n_X

Use + sign if X is cation and − sign if X is anion

Y = interfering ion with charge n_Y

$k_{X,Y}$ = selectivity coefficient

Important Terms

combination electrode

compound electrode

glass electrode

ion-selective electrode

junction potential

liquid-based ion-selective electrode

potentiometry

selectivity coefficient

solid-state ion-selective electrode

Problems

14-1. A cell was prepared by dipping a Cu wire and a saturated Ag | AgCl electrode into 0.10 M $CuSO_4$ solution. The Cu wire was attached to the positive terminal of a potentiometer and the reference electrode was attached to the negative terminal.

(a) Write a half-reaction for the Cu electrode.

(b) Write the Nernst equation for the Cu electrode.

(c) Calculate the cell voltage.

14-2. Pt and saturated calomel electrodes are dipped into a solution containing 0.002 17 M $Br_2(aq)$ and 0.234 M Br^-.

(a) Write the reaction that occurs at Pt and find the half-cell potential E_+.

(b) Find the net cell voltage, E.

14-3. A 50.0-mL solution of 0.100 M NaSCN was titrated with 0.200 M $AgNO_3$ in the cell in Figure 14-1. Find $[Ag^+]$ and E at $V_{Ag^+} = 0.1$, 10.0, 25.0, and 30.0 mL and sketch the titration curve.

14-4. A 10.0-mL solution of 0.050 0 M $AgNO_3$ was titrated with 0.025 0 M NaBr in the cell in Figure 14-1. Find the cell voltage at $V_{Br^-} = 0.1$, 10.0, 20.0, and 30.0 mL and sketch the titration curve.

14-5. Which side of the liquid junction 0.1 M KNO_3 | 0.1 M NaCl will be negative? Explain your answer.

14-6. If electrode C in Figure 14-9 is placed in a solution of pH 11.0, what will the pH reading be?

14-7. Suppose that the Ag | AgCl outer electrode in Figure 14-6 is filled with 0.1 M NaCl instead of saturated KCl. Suppose that the electrode is calibrated in a dilute buffer containing 0.1 M KCl at pH 6.54 at 25°C. The electrode is then dipped in a second buffer *at the same pH* and same temperature, but containing 3.5 M KCl.

(a) Use Table 14-2 to estimate the change in junction potential and how much the indicated pH will change.

(b) Suppose that a change in junction potential causes the apparent pH to change from 6.54 to 6.60. By what percentage does $[H^+]$ appear to change?

14-8. Explain the principle of operation of ion-selective electrodes. How does a compound electrode differ from a simple ion-selective electrode?

14-9. What does the selectivity coefficient tell us? Is it better to have a large or a small selectivity coefficient?

14-10. By how many volts will the potential of a Mg^{2+} ion-selective electrode change if the electrode is transferred from 1.00×10^{-4} M $MgCl_2$ to 1.00×10^{-3} M $MgCl_2$?

14-11. A cyanide ion-selective electrode obeys the equation

$$E = \text{constant} - (0.059\ 16) \log [CN^-]$$

The electrode potential was -0.230 V when the electrode was immersed in 1.00×10^{-3} M NaCN.

(a) Evaluate the constant in the equation above.

(b) Using the result from part **(a)**, find the concentration of CN^- if $E = -0.300$ V.

14-12. The selectivity coefficient, $k_{Li^+,Ca^{2+}}$, for a lithium electrode is 5×10^{-5}. When this electrode is placed in 3.44×10^{-4} M Li^+ solution, the potential is -0.333 V versus S.C.E. What would the potential be if Ca^{2+} were added to give 0.100 M Ca^{2+}?

14-13. An ammonia gas-sensing electrode gave the following calibration points when all solutions contained 1 M NaOH:

NH_3 (M)	E (mV)	NH_3 (M)	E (mV)
1.00×10^{-5}	268.0	5.00×10^{-4}	368.0
5.00×10^{-5}	310.0	1.00×10^{-3}	386.4
1.00×10^{-4}	326.8	5.00×10^{-3}	427.6

A dry food sample weighing 312.4 mg was digested by the Kjeldahl procedure (Section 11-6) to convert all the nitrogen to NH_4^+. The digestion solution was diluted to 1.00 L, and 20.0 mL was transferred to a 100-mL volumetric flask. The 20.0-mL aliquot was treated with 10.0 mL of 10.0 M NaOH plus enough NaI to complex the Hg catalyst from the digestion, and diluted to 100.0 mL. When measured with the ammonia electrode, this solution gave a reading of 339.3 mV.

(a) From the calibration data, find $[NH_3]$ in the 100-mL solution.

(b) Calculate the wt % of nitrogen in the food sample.

14-14. The selectivities of a lithium ion-selective electrode are indicated in the diagram below. Which alkali metal (Group 1) ion causes the most interference?

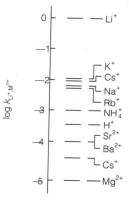

14-15. **(a)** Write an expression analogous to Equation 14-8 to describe the response of a La^{3+}-selective electrode to La^{3+} ion.

(b) If $\beta \approx 1.00$, by how many mV will the potential change when the electrode is removed from 1.00×10^{-4} M $LaClO_4$ and placed in 1.00×10^{-3} M $LaClO_4$?

(c) By how many mV will the potential of the electrode change if it is removed from 2.36×10^{-4} M $LaClO_4$ and placed in 4.44×10^{-3} M $LaClO_4$?

(d) The electrode potential is $+100$ mV in 1.00×10^{-4} M $LaClO_4$ and the selectivity coefficient $k_{La^{3+},Fe^{3+}}$ is $\frac{1}{1\ 200}$. What will the potential be when 0.010 M Fe^{3+} is added?

14-16. The following data were obtained when a Ca^{2+} ion-selective electrode was immersed in a series of standard solutions.

Ca^{2+} (M)	E (mV)
3.38×10^{-5}	-74.8
3.38×10^{-4}	-46.4
3.38×10^{-3}	-18.7
3.38×10^{-2}	$+10.0$
3.38×10^{-1}	$+37.7$

(a) Calculate the slope and the y-intercept (and their standard deviations) of the best straight line through the points using your least-squares spreadsheet from Chapter 4.

(b) Calculate the concentration of a sample that gave a reading of -22.5 mV.

(c) Your spreadsheet gives the uncertainty in $\log [Ca^{2+}]$. Using the upper and lower limits for $\log [Ca^{2+}]$, express the Ca^{2+} concentration as $[Ca^{2+}] = x \pm y$.

An Antibiotic Chelate Captures Its Prey

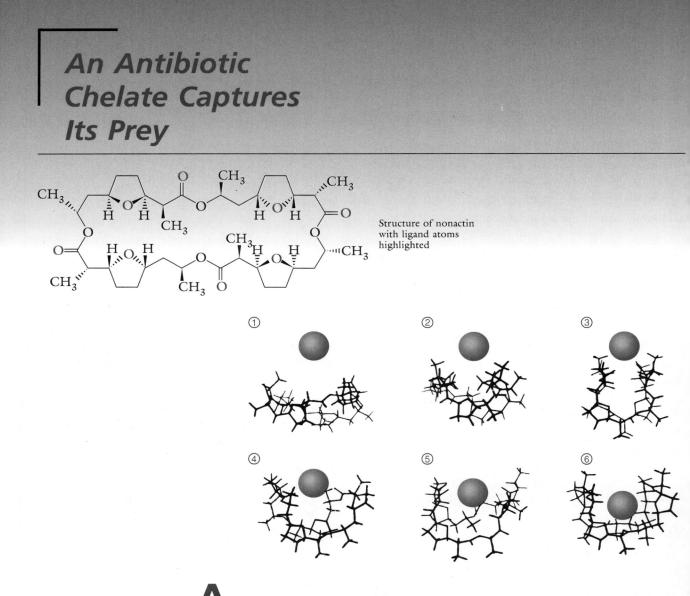

Structure of nonactin with ligand atoms highlighted

A s a nerve impulse passes along an axon, the *hydrophilic* ("water-loving") ions, K$^+$ and Na$^+$, must cross the *hydrophobic* ("water-hating") cell membrane. Carrier molecules with *polar* interiors and *nonpolar* exteriors perform this function. (Polar compounds have positive and negative regions that attract neighboring molecules by electrostatic forces. Nonpolar compounds do not have charged regions and are soluble inside the nonpolar cell membrane.)

A class of antibiotics called *ionophores,* including nonactin, gramicidin, and nigericin, alter the permeability of bacterial cells to metal ions and thus disrupt their metabolism. Ionophores are said to be *chelates* (pronounced KEE-lates), which means that they bind metal ions through more than one atom of the ionophore. One molecule of nonactin locks onto a K$^+$ ion through eight oxygen atoms. The sequence above shows how the conformation of nonactin changes as it wraps itself around a K$^+$ ion. The outside of the K$^+$-nonactin complex is mainly hydrocarbon, so the complex easily passes through the nonpolar cell membrane.

15

EDTA and Iodine Titrations

This chapter describes two reagents that are widely used for titrations in analytical chemistry. EDTA is a merciful abbreviation for *ethylenediaminetetraacetic acid,* a compound that can be used to titrate most metal ions by forming strong 1:1 complexes (Figure 15-1). Convenient redox titrations are available for numerous analytes with iodine (I_2, a mild oxidizing agent) or iodide (I^-, a mild reducing agent). For example, vitamin C in foods is measured by its reaction with iodine.

15-1 Metal-Chelate Complexes

An atom or group of atoms bound to whatever atom you are interested in is called a **ligand.** A ligand with a pair of electrons to share can bind to a metal ion that can accept a pair of electrons. Electron pair acceptors are called **Lewis acids** and electron

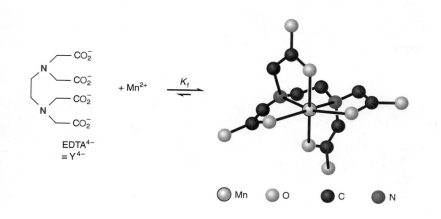

FIGURE 15-1 EDTA forms strong 1:1 complexes with most metal ions, binding through four oxygen and two nitrogen atoms. The six-coordinate structure of Mn^{2+}-EDTA is found in the compound $KMnEDTA \cdot 2H_2O$.

273

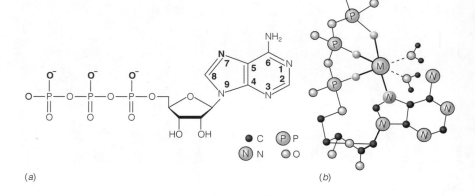

FIGURE 15-2 (*a*) Structure of adenosine triphosphate (ATP), with ligand atoms shown in **color**. (*b*) Possible structure of a metal-ATP complex, with four bonds to ATP and two bonds to H₂O ligands.

(*a*)   (*b*)

Lewis acid: electron pair acceptor
Lewis base: electron pair donor

pair donors are called **Lewis bases.** Cyanide is said to be a **monodentate** ("one-toothed") **ligand** because it binds to a metal ion through only one atom (the carbon atom):

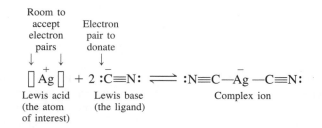

A **multidentate ligand** (also called a *polydentate ligand,* a **chelating ligand,** or just a *chelate*), binds to a metal ion through more than one ligand atom. EDTA in Figure 15-1 is a chelating ligand.

Most transition metal ions bind six ligand atoms. An important *tetradentate* ligand is adenosine triphosphate (ATP), which binds to divalent metal ions (such as Mg^{2+}, Mn^{2+}, Co^{2+}, and Ni^{2+}) through four of their six coordination positions (Figure 15-2). The fifth and sixth positions are occupied by water molecules. The biologically active form of ATP is generally the Mg^{2+} complex.

The aminocarboxylic acids in Figure 15-3 are synthetic chelating agents whose nitrogen and carboxylate oxygen atoms can lose protons and bind to metal ions. The chelating agents in Figure 15-3 form strong 1:1 complexes with all metal ions, except univalent ions such as Li^+, Na^+, and K^+. *The stoichiometry is 1:1 regardless of the charge on the ion.* A titration based on complex formation is called a **complexometric titration.** Box 15-1 describes an important use of chelating agents in medicine.

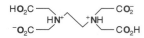

EDTA
Ethylenediaminetetraacetic acid
(also called ethylenedinitrilotetraacetic acid)

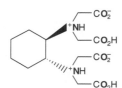

DCTA
trans-1,2-Diaminocyclohexanetetraacetic acid

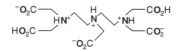

DTPA
Diethylenetriaminepentaacetic acid

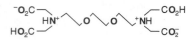

EGTA
bis-(Aminoethyl)glycolether-*N,N,N',N'*-tetraacetic acid

FIGURE 15-3 Analytically useful synthetic chelating agents that form strong 1:1 complexes with most metal ions.

Ask Yourself

15-A. What is the difference between a monodentate and a multidentate ligand? Is a chelating ligand monodentate or multidentate?

Box 15-1 *Informed Citizen*

Chelation Therapy and Thalassemia

Oxygen is carried in the human circulatory system by the iron-containing protein hemoglobin, which consists of two pairs of subunits, designated α and β. β-Thalassemia major is a genetic disease in which the β subunits of hemoglobin are not synthesized in adequate quantities. Children afflicted with this disease survive only with frequent transfusions of normal red blood cells. However, the patient accumulates 4–8 g of iron per year from hemoglobin in the transfused cells. The body has no mechanism for excreting such large quantities of iron, and most patients die by age 20 from the toxic effects of iron overload.

To enhance iron excretion, intensive chelation therapy is used. The most successful drug is *desferrioxamine B*, whose iron complex (ferrioxamine B) is shown here along with a graph illustrating its life-prolonging effects. The formation constant for ferrioxamine B is $10^{30.6}$.

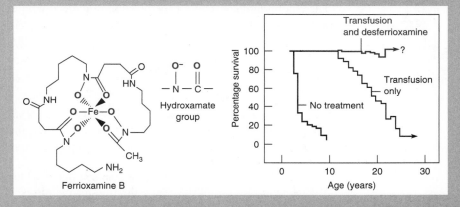

Ferrioxamine B

Used in conjunction with ascorbic acid—vitamin C, a reducing agent that reduces Fe^{3+} to the more soluble Fe^{2+}—desferrioxamine clears several grams of iron per year from an overloaded patient. The ferrioxamine complex is excreted in the urine.

Clinical trials demonstrate that desferrioxamine reduces the incidence of heart and liver disease in thalassemia patients and maintains approximately correct iron balance. However, desferrioxamine is expensive and must be taken by continuous injection. It is not absorbed through the intestine. Many potent iron chelators have been tested to find an effective one that can be taken orally. In the long term, bone marrow transplants or gene therapy might be effective cures for the disease.

15-2 *EDTA*

EDTA is, by far, the most widely used chelator in analytical chemistry. By direct titration or through an indirect sequence of reactions, virtually every element of the periodic table can be analyzed with EDTA.

EDTA is a hexaprotic system, designated H_6Y^{2+}. The highlighted acidic hydrogen atoms are the ones that are lost upon metal-complex formation:

One mole of EDTA reacts with *one* mole of metal ion.

$$HO_2CCH_2 \diagdown \diagup CH_2CO_2H$$
$$\overset{+}{H}NCH_2CH_2\overset{+}{N}H$$
$$HO_2CCH_2 \diagup \diagdown CH_2CO_2H$$
$$H_6Y^{2+}$$

$pK_1 = 0.0\ (CO_2H)$
$pK_2 = 1.5\ (CO_2H)$
$pK_3 = 2.0\ (CO_2H)$
$pK_4 = 2.66\ (CO_2H)$
$pK_5 = 6.16\ (NH^+)$
$pK_6 = 10.24\ (NH^+)$

Only some of the EDTA is in the form Y^{4-}.

The first four pK values apply to carboxyl protons, and the last two are for the ammonium protons. Below a pH of 10.24, most EDTA is protonated and is not in the form Y^{4-} that binds to metal ions (Figure 15-1).

Neutral EDTA is tetraprotic, with the formula H_4Y. A common reagent is the disodium salt, $Na_2H_2Y \cdot 2H_2O$, which attains the dihydrate composition upon heating at 80°C.

The equilibrium constant for the reaction of a metal with a ligand is called the **formation constant, K_f,** or the *stability constant:*

Formation constant:
$$M^{n+} + Y^{4-} \rightleftharpoons MY^{n-4} \qquad K_f = \frac{[MY^{n-4}]}{[M^{n+}][Y^{4-}]} \qquad (15\text{-}1)$$

Table 15-1 shows that formation constants for EDTA complexes are large and tend to be larger for more positively charged metal ions. Note that K_f is defined for reaction of the species Y^{4-} with the metal ion. At low pH, most EDTA is in one of its protonated forms, not Y^{4-}.

A metal-EDTA complex becomes less stable at lower pH because H^+ competes with the metal ion for the EDTA. For a titration reaction to be effective, it must go

TABLE 15-1 Formation constants for metal-EDTA complexes

Ion	log K_f	Ion	log K_f	Ion	log K_f
Li^+	2.79	Fe^{2+}	14.32	Pd^{2+}	18.5[a]
Na^+	1.66	Co^{2+}	16.31	Zn^{2+}	16.50
K^+	0.8	Ni^{2+}	18.62	Cd^{2+}	16.46
Be^{2+}	9.2	Cu^{2+}	18.80	Hg^{2+}	21.7
Mg^{2+}	8.79	Ti^{3+}	21.3[a]	Sn^{2+}	18.3
Ca^{2+}	10.69	V^{3+}	26.0	Pb^{2+}	18.04
Sr^{2+}	8.73	Cr^{3+}	23.4	Al^{3+}	16.3
Ba^{2+}	7.86	Mn^{3+}	25.3[a]	Ga^{3+}	20.3
Ra^{2+}	7.1	Fe^{3+}	25.1	In^{3+}	25.0
Sc^{3+}	23.1	Co^{3+}	41.4[a]	Tl^{3+}	37.8
Y^{3+}	18.09	Zr^{4+}	29.5	Bi^{3+}	27.8
La^{3+}	15.50	VO^{2+}	18.8	Ce^{3+}	15.98
V^{2+}	12.7	VO_2^+	15.55	Gd^{3+}	17.37
Cr^{2+}	13.6	Ag^+	7.32	Th^{4+}	23.2
Mn^{2+}	13.87	Tl^+	6.54	U^{4+}	25.8

Note: The stability constant is the equilibrium constant for the reaction $M^{n+} + Y^{4-} \rightleftharpoons MY^{n-4}$. Values in table generally apply at 20°C and ionic strength 0.1 M.

a. 25°C.

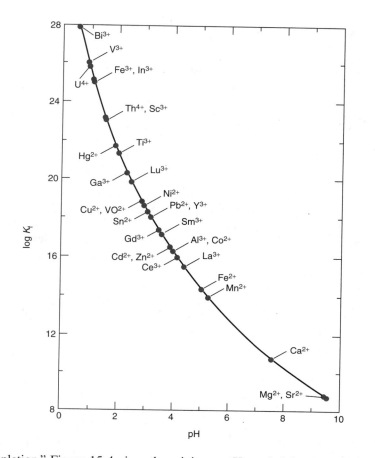

FIGURE 15-4 Minimum pH for effective titration of various metal ions by EDTA. pH was chosen such that complex formation is 99.9% complete at the equivalence point for the titration of 0.01 M metal ion.

"to completion." Figure 15-4 gives the minimum pH needed for titrations of metal ions to be 99.9% complete at a formal concentration of 0.01 M. The figure provides a strategy for the selective titration of one ion in the presence of another. For example, a solution containing both Fe^{3+} and Ca^{2+} could be titrated with EDTA at pH 4. At this pH, Fe^{3+} is titrated without interference from Ca^{2+}.

Ask Yourself

15-B. (a) Write the reaction whose equilibrium constant is the formation constant for EDTA complex formation and write the algebraic form of K_f.

    (b) Why is EDTA complex formation more complete at high pH than at low pH?

    (c) The following diagram is analogous to those in Section 12-3 for polyprotic acids. It shows the pH at which each species of EDTA is predominant. Fill in the pH at each arrow.

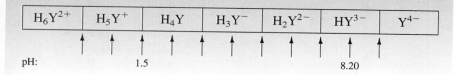

| H_6Y^{2+} | H_5Y^+ | H_4Y | H_3Y^- | H_2Y^{2-} | HY^{3-} | Y^{4-} |

pH: 1.5   8.20

15-3 Metal Ion Indicators

A **metal ion indicator** is a compound whose color changes when it binds to a metal ion. Two common indicators are shown in Table 15-2. *For an indicator to be useful, it must bind metal less strongly than EDTA does.*

A typical analysis is illustrated by the titration of Mg^{2+} with EDTA, using Eriochrome black T as the indicator:

The indicator must release its metal to EDTA.

$$MgIn + EDTA \longrightarrow MgEDTA + In \qquad (15\text{-}2)$$
$$\text{(red)} \quad \text{(colorless)} \qquad \text{(colorless)} \quad \text{(blue)}$$

At the start of the experiment, a small amount of indicator (In) is added to the colorless solution of Mg^{2+} to form a red complex. As EDTA is added, it reacts first with free, colorless Mg^{2+}. When free Mg^{2+} is used up, the last EDTA added before the equivalence point displaces indicator from the red MgIn complex. The change from the red of MgIn to the blue of unbound In signals the end point of the titration (Demonstration 15-1).

Most metal ion indicators are also acid-base indicators. Because the color of free indicator is pH dependent, most indicators can be used only in certain pH ranges. For example, xylenol orange (pronounced ᴢʏ-leen-ol) in Table 15-2 changes from yellow to red when it binds to a metal ion at pH 5.5. This is an easy color change to observe. At pH 7.5, the change is from violet to red and is rather difficult to see.

For an indicator to be useful in an EDTA titration, the indicator must give up its metal ion to EDTA. If metal does not freely dissociate from the indicator, the metal is said to **block** the indicator. Eriochrome black T is blocked by Cu^{2+}, Ni^{2+}, Co^{2+}, Cr^{3+}, Fe^{3+}, and Al^{3+}. It cannot be used for the direct titration of any of these metals. It can be used for a back titration, however. For example, excess standard EDTA can be added to Cu^{2+}. Then indicator is added and excess EDTA is back-titrated with Mg^{2+}.

Question What will the color change be when the back titration is performed?

TABLE 15-2   Some metal ion indicators

Name	Structure	pK_a		Color of free indicator	Color of metal ion complex
Eriochrome black T	(H₂In⁻)	$pK_2 = 6.3$ $pK_3 = 11.6$	H_2In^- HIn^{2-} In^{3-}	red blue orange	Wine red
Xylenol orange	(H₃In³⁻)	$pK_2 = 2.32$ $pK_3 = 2.85$ $pK_4 = 6.70$ $pK_5 = 10.47$ $pK_6 = 12.23$	H_5In^- H_4In^{2-} H_3In^{3-} H_2In^{4-} HIn^{5-} In^{6-}	yellow yellow yellow violet violet violet	Red

Demonstration 15-1

Metal Ion Indicator Color Changes

This demonstration illustrates the color change associated with Reaction 15-2 and shows how a second dye can be added to a solution to produce a more easily detected color change.

Stock solutions:

Eriochrome black T : dissolve 0.1 g of the solid in 7.5 mL of triethanolamine plus 2.5 mL of absolute ethanol

Methyl red: dissolve 0.02 g in 60 mL of ethanol; then add 40 mL of water

Buffer (pH 10): add 142 mL of concentrated (14.5 M) aqueous ammonia to 17.5 g of ammonium chloride and dilute to 250 mL with water

$MgCl_2$: 0.05 M

EDTA: 0.05 M $Na_2H_2EDTA \cdot 2H_2O$

Prepare a solution containing 25 mL of $MgCl_2$, 5 mL of buffer, and 300 mL of water. Add 6 drops of Eriochrome black T indicator and titrate with EDTA. Note the color change from wine red to pale blue at the end point (Color Plate 12a).

For some people, the change of indicator color is not so easy to see. Adding 3 mL of methyl red (or other yellow dyes) produces an orange color prior to the end point and a green color after it. This sequence of colors is shown in Color Plate 12b.

Ask Yourself

15-C. (a) Explain why the change from red to blue in Reaction 15-2 occurs suddenly at the equivalence point instead of gradually throughout the entire titration.

(b) EDTA buffered to pH 5 was titrated with standard Pb^{2+} using xylenol orange as indicator (Table 15-2).

   (i) Which is the principal species of the indicator at pH 5?

   (ii) What color was observed before the equivalence point?

   (iii) What color was observed after the equivalence point?

   (iv) What would the color change be if the titration were conducted at pH 8 instead of pH 5?

15-4 EDTA Titration Techniques

Chemists have developed procedures in which EDTA can be used directly or indirectly to analyze most elements of the periodic table. In this section we discuss several important techniques.

Direct Titration

In a **direct titration,** analyte is titrated with standard EDTA. The analyte is buffered to an appropriate pH, at which the reaction with EDTA is essentially complete and the free indicator has a color distinctly different from that of the metal-indicator complex.

An **auxiliary complexing agent,** such as ammonia (Color Plate 13), tartrate, citrate, or triethanolamine, prevents the metal ion from precipitating in the absence

The larger the effective formation constant, the more abrupt is the change in metal ion concentration at the end point.

279

of EDTA. For example, the direct titration of Pb^{2+} is carried out at pH 10 in the presence of tartrate, which complexes the metal ion and does not allow $Pb(OH)_2$ to precipitate. The lead-tartrate complex must be less stable than the lead-EDTA complex, or the titration would not be feasible.

Back Titration

In a **back titration,** a known excess of EDTA is added to the analyte. Excess EDTA is then titrated with a standard solution of metal ion. A back titration is necessary if the analyte precipitates in the absence of EDTA, if analyte reacts too slowly with EDTA, or if analyte blocks the indicator. The metal used in the back titration must not displace analyte from EDTA.

EXAMPLE A Back Titration

Ni^{2+} can be analyzed by a back titration with standard Zn^{2+} at pH 5.5 and xylenol orange indicator. A solution containing 25.00 mL of Ni^{2+} in dilute HCl was treated with 25.00 mL of 0.052 83 M Na_2EDTA. The solution was neutralized with NaOH, and the pH was adjusted to 5.5 with acetate buffer. The solution turned yellow when a few drops of indicator were added. Titration with 0.022 99 M Zn^{2+} required 17.61 mL to reach the red end point. What was the molarity of Ni^{2+} in the unknown?

SOLUTION The unknown was treated with 25.00 mL of 0.052 83 M EDTA, which contains (25.00 mL)(0.052 83 M) = 1.320 8 mmol of EDTA. Back titration required (17.61 mL)(0.022 99 M) = 0.404 9 mmol of Zn^{2+}. Because 1 mol of EDTA reacts with 1 mol of any metal ion, there must have been

$$1.320\ 8\ \text{mmol EDTA} - 0.404\ 9\ \text{mmol Zn}^{2+} = 0.915\ 9\ \text{mmol Ni}^{2+}$$

The concentration of Ni^{2+} is 0.915 9 mmol/25.00 mL = 0.036 64 M.

An EDTA back titration can prevent precipitation of analyte. For example, $Al(OH)_3$ precipitates at pH 7 in the absence of EDTA. An acidic solution of Al^{3+} can be treated with excess EDTA, adjusted to pH 7 with sodium acetate, and boiled to ensure complete complexation. The Al^{3+}-EDTA complex is stable at pH 7. The solution is then cooled; Eriochrome black T indicator is added; and back titration with standard Zn^{2+} is performed.

Displacement Titration

For metal ions that do not have a satisfactory indicator, a **displacement titration** may be feasible. In this procedure the analyte usually is treated with excess $Mg(EDTA)^{2-}$ to displace Mg^{2+}, which is later titrated with standard EDTA.

Challenge Calculate the equilibrium constant for Reaction 15-3 if $M^{n+} = Hg^{2+}$. Why is $Mg(EDTA)^{2-}$ used for a displacement titration?

$$M^{n+} + MgY^{2-} \longrightarrow MY^{n-4} + Mg^{2+} \tag{15-3}$$

Hg^{2+} is determined in this manner. The formation constant of $Hg(EDTA)^{2-}$ must be greater than the formation constant of $Mg(EDTA)^{2-}$, or else the displacement of Mg^{2+} from $Mg(EDTA)^{2-}$ does not occur.

There is no suitable indicator for Ag^+. However, Ag^+ will displace Ni^{2+} from the tetracyanonickelate(II) ion:

Box 15-2 *Explanation*

What Is Hard Water?

Hardness refers to the total concentration of alkaline earth ions in water. Because the concentrations of Ca^{2+} and Mg^{2+} are usually much greater than the concentrations of other Group 2 ions, hardness can be equated to $[Ca^{2+}] + [Mg^{2+}]$. Hardness is commonly expressed as the equivalent number of milligrams of $CaCO_3$ per liter. Thus, if $[Ca^{2+}] + [Mg^{2+}] = 1$ mM, we would say that the hardness is 100 mg $CaCO_3$ per liter because 100 mg $CaCO_3 = 1$ mmol $CaCO_3$. Water whose hardness is less than 60 mg $CaCO_3$ per liter is considered to be "soft."

Hard water reacts with soap to form insoluble curds:

$$Ca^{2+} + 2RSO_3^- \longrightarrow Ca(RSO_3)_2(s) \qquad (A)$$

Soap   Precipitate

Enough soap to consume the Ca^{2+} and Mg^{2+} must be used before the soap is useful for cleaning. Hard water is not thought to be unhealthy. Hardness is beneficial in irrigation water because the alkaline earth ions tend to *flocculate* (cause to aggregate) *colloidal* particles in soil and thereby increase the permeability of the soil to water.

Colloids are soluble particles that are 1–100 nm in diameter (see Demonstration 19-1). Such small particles tend to plug the paths by which water can drain through soil.

To measure hardness, water is treated with ascorbic acid to reduce Fe^{3+} to Fe^{2+} and with cyanide to mask Fe^{2+}, Cu^+, and other minor metal ions. Titration with EDTA at pH 10 in ammonia buffer gives $[Ca^{2+}] + [Mg^{2+}]$. $[Ca^{2+}]$ can be determined separately if the titration is carried out at pH 13 without ammonia. At this pH, $Mg(OH)_2$ precipitates and is inaccessible to the EDTA.

Insoluble carbonates are converted to soluble bicarbonates by excess carbon dioxide:

$$CaCO_3(s) + CO_2 + H_2O \longrightarrow Ca(HCO_3)_2(aq)$$

Calcium carbonate   Calcium bicarbonate

(B)

Heating reverses Reaction B to form a solid scale of $CaCO_3$ that clogs boiler pipes. The fraction of hardness due to $Ca(HCO_3)_2(aq)$ is called *temporary hardness* because this calcium is lost (by precipitation of $CaCO_3$) upon heating. Hardness arising from other salts (mainly dissolved $CaSO_4$) is called *permanent hardness*, because it is not removed by heating.

$$2Ag^+ + Ni(CN)_4^{2-} \longrightarrow 2Ag(CN)_2^- + Ni^{2+}$$

The liberated Ni^{2+} can then be titrated with EDTA to find out how much Ag^+ was added.

Indirect Titration

Anions that precipitate metal ions can be analyzed with EDTA by **indirect titration.** For example, sulfate can be analyzed by precipitation with excess Ba^{2+} at pH 1. The $BaSO_4(s)$ is filtered, washed, and boiled with excess EDTA at pH 10 to bring Ba^{2+} back into solution as $Ba(EDTA)^{2-}$. The excess EDTA is back-titrated with Mg^{2+}.

Alternatively, an anion can be precipitated with excess metal ion. The precipitate is filtered and washed, and the excess metal ion in the filtrate is titrated with EDTA. CO_3^{2-}, CrO_4^{2-}, S^{2-}, and SO_4^{2-} can be determined in this manner.

Masking

A **masking agent** is a reagent that protects some component of the analyte from reaction with EDTA. For example, Mg^{2+} in a mixture of Mg^{2+} and Al^{3+} can be

Masking prevents one element from interfering in the analysis of another element. Box 15-2 describes an application of masking.

N(CH$_2$CH$_2$OH)$_3$

Triethanolamine

SH
|
HOCH$_2$CHCH$_2$SH

2,3-Dimercaptopropanol

titrated by first masking the Al^{3+} with F$^-$, thereby leaving only the Mg^{2+} to react with EDTA.

Cyanide is a masking agent that forms complexes with Cd^{2+}, Zn^{2+}, Hg^{2+}, Co^{2+}, Cu$^+$, Ag$^+$, Ni^{2+}, Pd^{2+}, Pt^{2+}, Fe^{2+}, and Fe^{3+}, but not with Mg^{2+}, Ca^{2+}, Mn^{2+}, or Pb^{2+}. When CN$^-$ is added to a solution containing Cd^{2+} and Pb^{2+}, only the Pb^{2+} can react with EDTA. (***Caution:*** Cyanide forms toxic gaseous HCN below pH 11. Cyanide solutions should be strongly basic and handled in a hood.) Fluoride masks Al^{3+}, Fe^{3+}, Ti^{4+}, and Be^{2+}. (***Caution:*** HF formed by F$^-$ in acidic solution is extremely hazardous and should not contact skin or eyes. It may not be immediately painful, but the affected area should be flooded with water and then treated with calcium gluconate gel that you have on hand *before* the accident. First aiders must wear rubber gloves to protect themselves.) Triethanolamine masks Al^{3+}, Fe^{3+}, and Mn^{2+}; and 2,3-dimercaptopropanol masks Bi^{3+}, Cd^{2+}, Cu^{2+}, Hg^{2+}, and Pb^{2+}. Selectivity afforded by masking and pH control allows individual components of complex mixtures to be analyzed by EDTA titration.

Ask Yourself

15-D. (a) A 50.0-mL sample containing Ni^{2+} was treated with 25.0 mL of 0.050 0 M EDTA to complex all the Ni^{2+} and leave excess EDTA in solution. How many mmol of EDTA are contained in 25.0 mL of 0.050 0 M EDTA?

(b) The excess EDTA in part **(a)** was then back-titrated, requiring 5.00 mL of 0.050 0 M Zn^{2+}. How many mmol of Zn^{2+} are in 5.00 mL of 0.050 0 M Zn^{2+}?

(c) The mmol of Ni^{2+} in the unknown is the difference between the EDTA added in step **(a)** and the Zn^{2+} required in step **(b)**. Find the mmol of Ni^{2+} and the concentration of Ni^{2+} in the unknown.

15-5 *Titrations Involving Iodine*

Titrations that create or consume I$_2$ are widely used in quantitative analysis. For example, iodine titrations provide the best method to measure the oxidation states of metal ions in high-temperature superconductors (Figure 15-5 and Box 15-3). When a reducing analyte is titrated with iodine (to produce I$^-$), the method is called *iodimetry. Iodometry* is the titration of iodine produced when an oxidizing analyte is added to excess I$^-$. The iodine is usually titrated with standard thiosulfate solution.

I$_2$ is only slightly soluble in water (1.3×10^{-3} M at 20°C), but its solubility is enhanced by complexation with iodide:

$$I_2(aq) + I^- \rightleftharpoons I_3^- \qquad K = 7 \times 10^2$$

Iodine   Iodide   Triiodide

A typical 0.05 M solution of I$_3^-$ for titrations is prepared by dissolving 0.12 mol of KI plus 0.05 mol of I$_2$ in 1 L of water. When we speak of using iodine as a titrant, we almost always mean that we are using a solution of I$_2$ plus excess I$^-$.

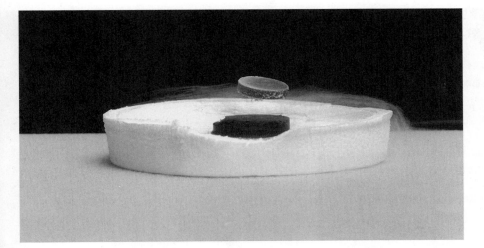

FIGURE 15-5 Permanent magnet levitates above superconducting disk cooled in a pool of liquid nitrogen. Redox titrations are crucial in measuring the chemical composition of a superconductor.

Starch Indicator

Starch is the indicator of choice for iodine because it forms an intense blue complex with iodine. The active fraction of starch is amylose, a polymer of the sugar α-D-glucose (Figure 15-6). The polymer coils into a helix, inside of which chains of I_6 (made from $3I_2$) fit and form an intense blue color.

$$\cdots [\text{I-I-I-I-I-I}] \cdots [\text{I-I-I-I-I-I}] \cdots$$

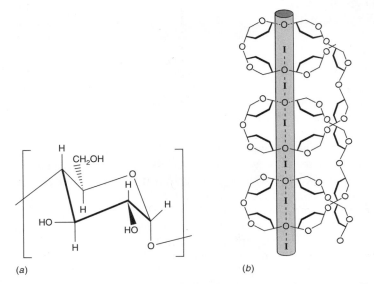

(a)   (b)

FIGURE 15-6 (a) Structure of the repeating unit of the sugar amylose found in starch. (b) Schematic structure of the starch-iodine complex. The sugar chain forms a helix around nearly linear I_6 chains.

Box 15-3 *Explanation*

How Iodine Is Used to Analyze High-Temperature Superconductors

Superconductors lose all electric resistance when cooled below a critical temperature. Prior to 1987, all known superconductors required cooling to temperatures near that of liquid helium (4 K), which is costly and impractical for most applications. In 1987 a giant step was taken when "high-temperature" superconductors that retain superconductivity above the boiling point of liquid nitrogen (77 K) were discovered.

The most startling characteristic of a superconductor is magnetic levitation, shown in Figure 15-5. When a magnetic field is applied to a superconductor, current flows in the outer skin of the material such that the applied magnetic field is exactly canceled by the induced magnetic field, and the net field inside the specimen is zero. Expulsion of a magnetic field from a superconductor is called the *Meissner effect.*

An important application of superconductors is in electromagnets for medical magnetic resonance imaging. Ordinary electromagnets require a huge amount of electrical power, but a superconducting electromagnet oper-

ates on *zero* electrical power because electricity moves through a superconductor with no resistance. Magnetic resonance imaging today requires liquid helium cooling, but advances in high-temperature superconducting wire may allow units to operate with liquid nitrogen in the future.

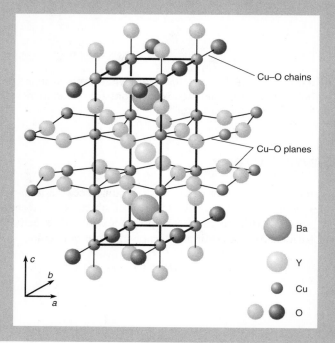

Structure of $YBa_2Cu_3O_7$, showing one-dimensional Cu–O chains (with darker oxygen atoms) and two-dimensional Cu–O planes. Loss of oxygen from the chains at elevated temperature results in $YBa_2Cu_3O_6$.

An alternative to the use of starch is the addition of a few milliliters of *p*-xylene to the vigorously stirred titration vessel. After each addition of reagent near the end point, stop stirring long enough to examine the color of the organic phase. I_2 is 400 times more soluble in *p*-xylene than in water, and its color is readily detected in the organic phase.

In a solution with no other colored species, it is possible to see the color of $\sim 5 \times 10^{-6}$ M I_3^-. With a starch indicator, the limit of detection is extended by about a factor of 10.

Starch is readily biodegraded, so either it should be freshly dissolved or the solution should contain a preservative, such as HgI_2 or thymol. A hydrolysis product of starch is glucose, which is a reducing agent. A partially hydrolyzed solution of starch could thus be a source of error in a redox titration.

In iodimetry (titration *with* I_3^-), starch can be added at the beginning of the titration. The first drop of excess I_3^- after the equivalence point causes the solution to turn dark blue. In iodometry (titration *of* I_3^-), I_3^- is present throughout the reaction up to the equivalence point. *Starch should not be added to such a reaction until immediately before the equivalence point* (as detected visually, by fading of the I_3^-; see Color Plate 14). Otherwise some iodine tends to remain bound to starch particles after the equivalence point is reached.

A prototypical high-temperature superconductor is yttrium barium copper oxide, $YBa_2Cu_3O_7$. When heated, the material readily loses oxygen atoms from the Cu–O chains, and compositions between $YBa_2Cu_3O_7$ and $YBa_2Cu_3O_6$ are observed.

When high-temperature superconductors were discovered, the oxygen content in the formula $YBa_2Cu_3O_x$ was unknown. Common oxidation states of yttrium and barium are Y^{3+} and Ba^{2+}, and common states of copper are Cu^{2+} and Cu^+. If copper were Cu^{2+}, the formula of the superconductor would be $(Y^{3+})(Ba^{2+})_2(Cu^{2+})_3(O^{2-})_{6.5}$, with a cation charge of $+13$ and an anion charge of -13. The composition $YBa_2Cu_3O_7$ requires Cu^{3+}, which is rather rare. Formally, $YBa_2Cu_3O_7$ can be thought of as $(Y^{3+})(Ba^{2+})_2(Cu^{2+})_2(Cu^{3+})(O^{2-})_7$, with a cation charge of $+14$ and an anion charge of -14.

Redox titrations provide the most reliable way to measure the oxidation state of copper and thereby deduce the oxygen content of $YBa_2Cu_3O_x$. An iodometric method involves two experiments. In *Experiment 1*, $YBa_2Cu_3O_x$ is dissolved in dilute acid, in which Cu^{3+} is converted to Cu^{2+}. For simplicity, we write the equations for the formula $YBa_2Cu_3O_7$, but you could balance these equations for $x \neq 7$:

$$YBa_2Cu_3O_7 + 13H^+ \longrightarrow$$
$$Y^{3+} + 2Ba^{2+} + 3Cu^{2+} + \tfrac{13}{2}H_2O + \tfrac{1}{4}O_2 \quad (A)$$

The total copper content is measured by treatment with iodide:

$$3Cu^{2+} + \tfrac{15}{2}I^- \longrightarrow 3CuI(s) + \tfrac{3}{2}I_3^- \quad (B)$$

and titration of the liberated triiodide with standard thiosulfate by Reaction 15-5. Each mole of Cu in $YBa_2Cu_3O_7$ is equivalent to 1 mol of $S_2O_3^{2-}$ in Experiment 1.

In *Experiment 2*, $YBa_2Cu_3O_x$ is dissolved in dilute acid containing I^-. Each mole of Cu^{3+} produces 1 mol of I_3^-, and each mole of Cu^{2+} produces 0.5 mol of I_3^-:

$$Cu^{3+} + 4I^- \longrightarrow CuI(s) + I_3^- \quad (C)$$
$$Cu^{2+} + \tfrac{5}{2}I^- \longrightarrow CuI(s) + \tfrac{1}{2}I_3^- \quad (D)$$

The moles of thiosulfate required in Experiment 1 equal the total moles of Cu in the superconductor. The difference in thiosulfate required between Experiments 2 and 1 gives the Cu^{3+} content. From this difference, it is possible to calculate the value of x in the formula $YBa_2Cu_3O_x$.

Although we can balance cation and anion charges in the formula $YBa_2Cu_3O_7$ by including Cu^{3+} in the formula, there is no evidence for discrete Cu^{3+} ions in the crystal. The best description of the valence state in the solid crystal involves delocalized charge in the Cu–O planes and chains. Nonetheless, the formal designation of Cu^{3+} and the chemistry in Equations A through D accurately describe the redox chemistry of $YBa_2Cu_3O_7$.

Preparation and Standardization of I_3^- Solutions

Triiodide (I_3^-) is prepared by dissolving solid I_2 in excess KI. I_2 is seldom used as a primary standard because some *sublimes* (evaporates) during weighing. Instead, an approximate amount is rapidly weighed, and the solution of I_3^- is standardized with a pure sample of the intended analyte or with As_4O_6 or $Na_2S_2O_3$.

Acidic solutions of I_3^- are unstable because the excess I^- is slowly oxidized by air:

$$6I^- + O_2 + 4H^+ \longrightarrow 2I_3^- + 2H_2O$$

In neutral solutions, oxidation is insignificant in the absence of heat, light, and metal ions. Above pH 11, triiodide disproportionates to hypoiodous acid (HOI), iodate (IO_3^-), and iodide.

There is a significant vapor pressure of toxic I_2 above solid I_2 and aqueous I_3^-. Vessels containing I_2 or I_3^- should be sealed and kept in a fume hood. Waste solutions of I_3^- should not be dumped into a sink in the open lab.

Disproportionation means that an element in one oxidation state reacts to give the same element in both higher and lower oxidation states.

Triiodide can be standardized by reaction with primary standard grade arsenic(III) oxide, As_4O_6 (Figure 15-7), at pH 7–8 in bicarbonate buffer:

$$As_4O_6(s) \ + \ 6H_2O \rightleftharpoons 4H_3AsO_3$$

Arsenic(III) oxide   Arsenious acid

$$H_3AsO_3 + I_3^- + H_2O \rightleftharpoons H_3AsO_4 + 3I^- + 2H^+$$

Standard I_3^- can be made by adding a weighed quantity of pure potassium iodate to a small excess of KI. Addition of excess strong acid (to give pH $\approx$ 1) produces I_3^-:

KIO$_3$ is a primary standard for the generation of I_3^-.

Preparing
standard I_3^-:
$$IO_3^- + 8I^- + 6H^+ \rightleftharpoons 3I_3^- + 3H_2O \qquad (15\text{-}4)$$
    Iodate

Freshly acidified iodate plus iodide can be used to standardize thiosulfate. The I_3^- reagent must be used immediately, for it is soon oxidized by air. The only disadvantage of KIO$_3$ is its low molecular weight relative to the number of electrons it accepts. This property leads to a larger than desirable relative weighing error in preparing solutions.

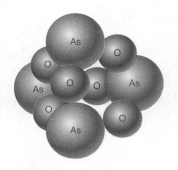

FIGURE 15-7 The As$_4$O$_6$ molecule consists of an As$_4$ tetrahedron with a bridging oxygen atom on each edge.

Use of Sodium Thiosulfate

Sodium thiosulfate is the almost universal titrant for triiodide. In neutral or acidic solution (pH < 9), triiodide oxidizes thiosulfate cleanly to tetrathionate:

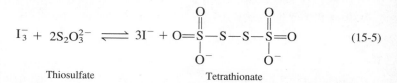

$$I_3^- + 2S_2O_3^{2-} \rightleftharpoons 3I^- + O{=}S{-}S{-}S{-}S{=}O \qquad (15\text{-}5)$$

      Thiosulfate   Tetrathionate

The common form of thiosulfate, $Na_2S_2O_3 \cdot 5H_2O$, is not pure enough to be a primary standard. Instead, thiosulfate is standardized by reaction with a fresh solution of I_3^- prepared from KIO$_3$ plus KI or a solution of I_3^- standardized with As$_4$O$_6$.

A stable solution of $Na_2S_2O_3$ is prepared by dissolving the reagent in high-quality, freshly boiled distilled water. Dissolved CO_2 promotes disproportionation of $S_2O_3^{2-}$:

$$S_2O_3^{2-} + H^+ \rightleftharpoons HSO_3^- + S(s) \qquad (15\text{-}6)$$

               Bisulfate   Sulfur

and metal ions catalyze atmospheric oxidation of thiosulfate. Thiosulfate solutions are stored in the dark with 0.1 g of Na_2CO_3 per liter to maintain optimum pH. Three drops of chloroform should be added to a thiosulfate solution to prevent bacterial growth. Although an acidic solution of thiosulfate is unstable, the reagent can be used to titrate I_3^- in acid because Reaction 15-5 is faster than Reaction 15-6.

Analytical Applications of Iodine

reducing agent + $I_3^- \rightarrow 3I^-$

Examples are given in Tables 15-3 and 15-4. Reducing agents such as vitamin C are titrated directly with standard I_3^- in the presence of starch, until reaching the intense blue end point:

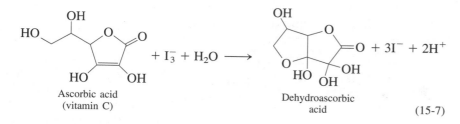

Ascorbic acid
(vitamin C)

Dehydroascorbic
acid

(15-7)

Oxidizing agents are treated with excess I^- to produce I_3^-. The iodometric analysis is completed by titrating the liberated I_3^- with standard thiosulfate. Starch is not added until just before the end point.

Reaction 15-7 has an important implication for backpackers who carry saturated aqueous I_2 to sterilize water from streams. I keep a few large crystals of I_2 in a 60-mL glass bottle with a Teflon-lined cap. Water from this bottle, which is always saturated with I_2, is measured out in the bottle cap and added to a 1-L bottle of water from a stream or lake. The required volume of saturated aqueous I_2 is shown in the margin. For example, I use 4 caps of iodine solution when the air temperature is near 20°C to deliver approximately 13 mL of disinfectant to my 1-L water bottle. It is important to use just the supernatant liquid, not the crystals of solid iodine, because too much iodine is harmful to humans. After allowing 30 min for the iodine to kill any critters, the water is safe to drink. Every time iodine solution is used, refill the small bottle with water so that aqueous I_2 is always available at the next water stop. *Don't put Tang or other beverages in the water until the 30-min period is over.* Tang, especially, is loaded with vitamin C, which destroys I_2 by Reaction 15-7 before it can disinfect the drinking water.

oxidizing agent $+ 3I^- \rightarrow I_3^-$.

Ask Yourself

15-E. (a) Potassium iodate solution was prepared by dissolving 1.022 g of KIO_3 (FW 214.00) in a 500-mL volumetric flask. Then 50.00 mL of the solution was pipetted into a flask and treated with excess KI (2 g) and acid (10 mL of 0.5 M H_2SO_4) to drive Reaction 15-4 to completion. How many moles of I_3^- are created by the reaction?

  (b) The triiodide from part **(a)** required 37.66 mL of sodium thiosulfate solution for Reaction 15-5. What is the concentration of the sodium thiosulfate solution?

  (c) A 1.223-g sample of solid containing ascorbic acid and inert ingredients was dissolved in dilute H_2SO_4 and treated with 2 g of KI and 50.00 mL of potassium iodate solution from part **(a)**. After Reactions 15-4 and 15-7 went to completion, the excess, unreacted triiodide required 14.22 mL of sodium thiosulfate solution from part **(b)** for complete titration. Find the moles of ascorbic acid and weight percent of ascorbic acid (MW 176.13) in the unknown.

Here's a recipe for disinfecting drinking water:

Temperature of saturated $I_2(aq)$	Volume to add to 1 L
3°C (37°F)	20 mL
20°C (68°F)	13 mL
25°C (77°F)	12.5 mL
40°C (104°F)	10 mL

TABLE 15-3   Titrations with standard triiodide (iodimetric titrations)

Species analyzed	Oxidation reaction	Notes
SO_2	$SO_2 + H_2O \rightleftharpoons H_2SO_3$ $H_2SO_3 + H_2O \rightleftharpoons SO_4^{2-} + 4H^+ + 2e^-$	Add SO_2 (or H_2SO_3 or HSO_3^- or SO_3^{2-}) to excess standard I^- in dilute acid and back-titrate unreacted I_3^- with standard thiosulfate.
H_2S	$H_2S \rightleftharpoons S(s) + 2H^+ + 2e^-$	Add H_2S to excess I_3^- in 1 M HCl and back-titrate with thiosulfate.
$Zn^{2+}, Cd^{2+}, Hg^{2+}, Pb^{2+}$	$M^{2+} + H_2S \longrightarrow MS(s) + 2H^+$ $MS(s) \rightleftharpoons M^{2+} + S + 2e^-$	Precipitate and wash metal sulfide. Dissolve in 3 M HCl with excess standard I_3^- and back-titrate with thiosulfate.
Cysteine, glutathione, mercaptoethanol	$2RSH \rightleftharpoons RSSR + 2H^+ + 2e^-$	Titrate the sulfhydryl compound at pH 4–5 with I_3^-.
$H_2C{=}O$	$H_2CO + 3OH^- \rightleftharpoons HCO_2^- + 2H_2O + 2e^-$	Add excess I_3^- plus NaOH to the unknown. After 5 min, add HCl and back-titrate with thiosulfate.
Glucose (and other reducing sugars)	$\underset{\displaystyle RCH}{\overset{\displaystyle O \atop \displaystyle \parallel}{}} + 3OH^- \rightleftharpoons RCO_2^- + 2H_2O + 2e^-$	Add excess I_3^- plus NaOH to the sample. After 5 min, add HCl and back-titrate with thiosulfate.

TABLE 15-4   Titration of I_3^- produced by analyte (iodometric titrations)

Species analyzed	Reaction	Notes
HOCl	$HOCl + H^+ + 3I^- \rightleftharpoons Cl^- + I_3^- + H_2O$	Reaction in 0.5 M H_2SO_4.
Br_2	$Br_2 + 3I^- \rightleftharpoons 2Br^- + I_3^-$	Reaction in dilute acid.
IO_3^-	$2IO_3^- + 16I^- + 12H^+ \rightleftharpoons 6I_3^- + 6H_2O$	Reaction in 0.5 M HCl.
IO_4^-	$2IO_4^- + 22I^- + 16H^+ \rightleftharpoons 8I_3^- + 8H_2O$	Reaction in 0.5 M HCl.
O_2	$O_2 + 4Mn(OH)_2 + 2H_2O \rightleftharpoons 4Mn(OH)_3$ $2Mn(OH)_3 + 6H^+ + 6I^- \rightleftharpoons$ $\qquad\qquad\qquad 2Mn^{2+} + 2I_3^- + 6H_2O$	The sample is treated with Mn^{2+}, NaOH, and KI. After 1 min, it is acidified with H_2SO_4, and the I_3^- is titrated.
H_2O_2	$H_2O_2 + 3I^- + 2H^+ \rightleftharpoons I_3^- + 2H_2O$	Reaction in 1 M H_2SO_4 with NH_4MoO_3 catalyst.
O_3	$O_3 + 3I^- + 2H^+ \rightleftharpoons O_2 + I_3^- + H_2O$	O_3 is passed through neutral 2 wt % KI solution. Add H_2SO_4 and titrate.
NO_2^-	$2HNO_2 + 2H^+ + 3I^- \rightleftharpoons 2NO + I_3^- + 2H_2O$	The nitric oxide is removed (by bubbling CO_2 generated in situ) prior to titration of I_3^-.
$S_2O_8^{2-}$	$S_2O_8^{2-} + 3I^- \rightleftharpoons 2SO_4^{2-} + I_3^-$	Reaction in neutral solution. Then acidify and titrate.
Cu^{2+}	$2Cu^{2+} + 5I^- \rightleftharpoons 2CuI(s) + I_3^-$	NH_4HF_2 is used as a buffer.
MnO_4^-	$2MnO_4^- + 16H^+ + 15I^- \rightleftharpoons 2Mn^{2+} + 5I_3^- + 8H_2O$	Reaction in 0.1 M HCl.
MnO_2	$MnO_2(s) + 4H^+ + 3I^- \rightleftharpoons Mn^{2+} + I_3^- + 2H_2O$	Reaction in 0.5 M H_3PO_4 or HCl.

Key Equation

Formation constant $\qquad M^{n+} + Y^{4-} \rightleftharpoons MY^{n-4} \qquad K_f = \dfrac{[MY^{n-4}]}{[M^{n+}][Y^{4-}]}$

Important Terms

auxiliary complexing agent

back titration

blocking

chelating ligand

complexometric titration

direct titration

displacement titration

formation constant

indirect titration

Lewis acid

Lewis base

ligand

masking agent

metal ion indicator

monodentate ligand

multidentate ligand

Problems

15-1. How many milliliters of 0.050 0 M EDTA are required to react with 50.0 mL of 0.010 0 M Ca^{2+}? with 50.0 mL of 0.010 0 M Al^{3+}?

15-2. Give three circumstances in which an EDTA back titration might be necessary.

15-3. Describe what is done in a displacement titration and give an example.

15-4. Give an example of the use of a masking agent.

15-5. What is meant by water hardness? Explain the difference between temporary and permanent hardness.

15-6. A 25.00-mL sample containing Fe^{3+} was treated with 10.00 mL of 0.036 7 M EDTA to complex all the Fe^{3+} and leave excess EDTA in solution. The excess EDTA was then back-titrated, requiring 2.37 mL of 0.046 1 M Mg^{2+}. What was the concentration of Fe^{3+} in the original solution?

15-7. A 50.0-mL solution containing Ni^{2+} and Zn^{2+} was treated with 25.0 mL of 0.045 2 M EDTA to bind all the metal. The excess unreacted EDTA required 12.4 mL of 0.012 3 M Mg^{2+} for complete reaction. An excess of the reagent 2,3-dimercapto-1-propanol was then added to displace the EDTA from zinc. Another 29.2 mL of Mg^{2+} was required for reaction with the liberated EDTA. Calculate the molarities of Ni^{2+} and Zn^{2+} in the original solution.

15-8. Sulfide ion was determined by indirect titration with EDTA. To a solution containing 25.00 mL of 0.043 32 M $Cu(ClO_4)_2$ plus 15 mL of 1 M acetate buffer (pH 4.5) was added 25.00 mL of unknown sulfide solution with vigorous stirring. The CuS precipitate was filtered and washed with hot water. Ammonia was added to the filtrate (which contains excess Cu^{2+}) until the blue color of $Cu(NH_3)_4^{2+}$ was observed. Titration with 0.039 27 M EDTA required 12.11 mL to reach the end point with the indicator murexide. Find the molarity of sulfide in the unknown.

15-9. The potassium ion in a 250.0 ($\pm$0.1)-mL water sample was precipitated with sodium tetraphenylborate:

$$K^+ + (C_6H_5)_4B^- \longrightarrow KB(C_6H_5)_4(s)$$

The precipitate was filtered, washed, and dissolved in an organic solvent. Treatment of the organic solution with an excess of Hg^{2+}-EDTA then gave the following reaction:

$$4HgY^{2-} + (C_6H_5)_4B^- + 4H_2O \longrightarrow$$
$$H_3BO_3 + 4C_6H_5Hg^+ + 4HY^{3-} + OH^-$$

The liberated EDTA was titrated with 28.73 ($\pm$0.03) mL of 0.043 7 ($\pm$0.000 1) M Zn^{2+}. Find the concentration (and absolute uncertainty) of K^+ in the original sample.

15-10. A 25.00-mL sample of unknown containing Fe^{3+} and Cu^{2+} required 16.06 mL of 0.050 83 M EDTA for complete titration. A 50.00-mL sample of the unknown was treated with NH_4F to protect the Fe^{3+}. Then the Cu^{2+} was reduced and masked by addition of thiourea. Upon addition of 25.00 mL of 0.050 83 M EDTA, the Fe^{3+} was liberated from its fluoride complex and formed an EDTA complex. The excess EDTA required 19.77 mL of 0.018 83 M Pb^{2+} to reach an end point, using xylenol orange. Find the concentration of Cu^{2+} in the unknown.

15-11. Cyanide recovered from refining of gold ore can be determined indirectly by EDTA titration. A known excess of Ni^{2+} is added to the cyanide to form tetracyanonickelate(II):

$$4CN^- + Ni^{2+} \longrightarrow Ni(CN)_4^{2-}$$

When the excess Ni^{2+} is titrated with standard EDTA, $Ni(CN)_4^{2-}$ does not react. In a cyanide analysis, 12.7 mL of cyanide solution was treated with 25.0 mL of standard solution containing excess Ni^{2+} to form tetracyanonickelate. The excess Ni^{2+} required 10.1 mL of 0.013 0 M EDTA for complete reaction. In a separate experiment, 39.3 mL of 0.013 0 M EDTA was required to react with 30.0 mL of the standard Ni^{2+} solution. Calculate the molarity of CN^- in the 12.7-mL sample of unknown.

15-12. A mixture of Mn^{2+}, Mg^{2+}, and Zn^{2+} was analyzed as follows: The 25.00-mL sample was treated with 0.25 g of $NH_3OH^+Cl^-$ (hydroxylammonium chloride, a reducing agent that maintains manganese in the +2 state), 10 mL of ammonia buffer (pH 10), and a few drops of Eriochrome black T indicator and then diluted to 100 mL. It was warmed to 40°C and titrated with 39.98 mL of 0.045 00 M EDTA to the blue end point. Then 2.5 g of NaF was added to displace Mg^{2+} from its EDTA complex. The liberated EDTA required 10.26 mL of standard 0.020 65 M Mn^{2+} for complete titration. After this second end point was reached, 5 mL of 15 wt % aqueous KCN was added to displace Zn^{2+} from its EDTA complex. This time the liberated EDTA required 15.47 mL of standard 0.020 65 M Mn^{2+}. Calculate the number of milligrams of each metal (Mn^{2+}, Zn^{2+}, and Mg^{2+}) in the 25.00-mL sample of unknown.

15-13. Why is iodine almost always used in a solution containing excess I^-?

15-14. A solution of I_3^- was standardized by titrating freshly dissolved arsenic(III) oxide (As_4O_6, FW 395.68). The titration of 25.00 mL of a solution prepared by dissolving 0.366 3 g of As_4O_6 in a volume of 100.0 mL required 31.77 mL of I_3^-.

(a) Calculate the molarity of the I_3^- solution.

(b) Does it matter whether starch indicator is added at the beginning or near the end point in this titration? Why?

15-15. A 128.6-mg sample of a protein (MW 58 600) was treated with 2.000 mL of 0.048 7 M $NaIO_4$ to react with all the serine and threonine residues.

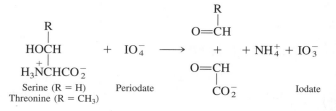

The solution was then treated with excess iodide to convert the unreacted periodate to iodine:

$$IO_4^- + 3I^- + H_2O \rightleftharpoons IO_3^- + I_3^- + 2OH^-$$

Titration of the iodine required 823 μL of 0.098 8 M thiosulfate.

(a) Calculate the number of serine plus threonine residues per molecule of protein. Answer to the nearest integer.

(b) How many milligrams of As_4O_6 (FW 395.68) would be required to react with the I_3^- liberated in this experiment?

15-16. The Kjeldahl analysis in Section 11-6 is used to measure the nitrogen content of organic compounds, which are digested in boiling sulfuric acid to decompose to ammonia, which, in turn, is distilled into standard acid. The remaining acid is then back-titrated with base. Kjeldahl himself had difficulty discerning by lamplight in 1880 the methyl red indicator end point in the back titration. He could have refrained from working at night, but instead he chose to complete the analysis differently. After distilling the ammonia into standard sulfuric acid, he added a mixture of KIO_3 and KI to the acid. The liberated iodine was then titrated with thiosulfate, using starch for easy end-point detection—even by lamplight. Explain how the thiosulfate titration is related to the nitrogen content of the unknown. Derive a relationship between moles of NH_3 liberated in the digestion and moles of thiosulfate required for titration of iodine.

15-17. From the reduction potentials below

$$I_2(s) + 2e^- \rightleftharpoons 2I^- \qquad E° = 0.535 \text{ V}$$
$$I_2(aq) + 2e^- \rightleftharpoons 2I^- \qquad E° = 0.620 \text{ V}$$
$$I_3^- + 2e^- \rightleftharpoons 3I^- \qquad E° = 0.535 \text{ V}$$

(a) Calculate the equilibrium constant for the reaction $I_2(aq) + I^- \rightleftharpoons I_3^-$.

(b) Calculate the equilibrium constant for the reaction $I_2(s) + I^- \rightleftharpoons I_3^-$.

(c) Calculate the solubility (g/L) of $I_2(s)$ in water.

15-18. *Iodometric analysis of high-temperature superconductor.* The procedure in Box 15-3 was carried out to find the effective copper oxidation state, and therefore the number of oxygen atoms, in the formula $YBa_2Cu_3O_{7-z}$, where z ranges from 0 to 0.5.

(a) In Experiment 1 of Box 15-3, 1.00 g of superconductor required 4.55 mmol of $S_2O_3^{2-}$. In Experiment 2, 1.00 g of superconductor required 5.68 mmol of $S_2O_3^{2-}$. Calculate the value of z in the formula $YBa_2Cu_3O_{7-z}$ (FW $666.246 - 15.9994\,z$).

(b) *Propagation of uncertainty.* In several replications of Experiment 1, the thiosulfate required was 4.55 (±0.10) mmol of $S_2O_3^{2-}$ per gram of $YBa_2Cu_3O_{7-z}$. In Experiment 2, the thiosulfate required was 5.68 (±0.05) mmol of $S_2O_3^{2-}$ per gram. Calculate the uncertainty of x in the formula $YBa_2Cu_3O_x$.

Katia and Dante

Courageous Katia Krafft samples vapors from a volcano.
[From Frank Press and Raymond Siever, *Understanding Earth* (New York: W. H. Freeman, 1994), p. 107.]

Robot Dante was equipped with a gas chromatograph to sample emissions from volcanic Mt. Erebus in Antarctica.

Over billions of years, volcanoes released gases that created today's atmosphere and condensed into the earth's oceans. Generally, 70–95% of volcanic gas is H_2O, followed by smaller volumes of CO_2 and SO_2, and traces of N_2, H_2, CO, sulfur, Cl_2, HF, HCl, and H_2SO_4. To reduce wear and tear on brave scientists who have ventured near a volcano to collect gases, the eight-legged robot Dante was designed to rappel 250 m into the sheer crater of Mt. Erebus in Antarctica in 1992 to sample volcanic emissions with a gas chromatograph and return information via a fiber-optic cable. Dante also had a quartz crystal microbalance (page 18) to measure aerosols, an infrared thermometer, and a γ-ray spectrometer to measure K, Th, and U in the crater's soil. Alas, Dante was not as reliable as Katia. The optical fiber broke after just 6 m of descent into the crater. In 1994 an improved Dante II successfully probed a volcano, Mt. Spurr, in Alaska.

Chapter 16

Principles of Chromatography

Chromatography is the most powerful tool in an analytical chemist's arsenal for separating and measuring the components of a complex mixture. In conjunction with mass spectrometry and infrared spectroscopy, chromatography can identify the components as well. This chapter discusses principles of chromatography, and the next two chapters take up specific methods of separation.

Chromatography is widely used for

quantitative analysis: How much of a component is present?

qualitative analysis: What is the identity of the component?

16-1 What Is Chromatography?

Chromatography is a process in which we separate compounds from one another by passing them all through a column that retains some compounds longer than others. In Figure 16-1, a solution containing compounds A and B is placed on top of a column previously packed with solid particles and filled with solvent. When the outlet is opened, A and B flow down into the column and are washed through with fresh solvent applied to the top of the column. If solute A is more strongly *adsorbed* than solute B on the solid particles, then A spends a smaller fraction of the time free in solution. Solute A moves down the column more slowly than B and emerges at the bottom after B.

Adsorption means sticking to the surface of the solid particles. Color Plate 15 illustrates adsorption chromatography.

The **mobile phase** (solvent moving through the column) in chromatography is either a liquid or a gas. The **stationary phase** (the substance that stays fixed inside the column) is either a solid or a viscous (syrupy) liquid coated onto solid particles or onto the inside wall of a hollow capillary column. Partitioning of solutes between the mobile and stationary phases gives rise to separation. In **gas chromatography** the mobile phase is a gas, and in **liquid chromatography** the mobile phase is a liquid.

293

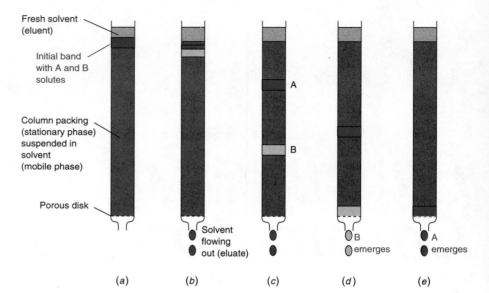

FIGURE 16-1 Separation by chromatography. Solute A has a greater affinity than solute B for the stationary phase, so A remains on the column longer than B. The term *chromatography* is derived from experiments in 1903 by M. Tswett, who separated plant pigments with a column containing solid $CaCO_3$ particles (the stationary phase) washed by hydrocarbon solvent (the mobile phase). The separation of colored bands led to the name *chromatography,* from the Greek word *chromatos,* meaning "color."

Fluid entering the column is called **eluent.** Fluid exiting the column is called **eluate.** The process of passing liquid or gas through a chromatography column is called **elution.**

Chromatography can be classified by the type of interaction of the solute with the stationary phase, as shown in Figure 16-2.

Adsorption chromatography uses a solid stationary phase and a liquid or gaseous mobile phase. Solute is adsorbed on the surface of the solid particles.

Partition chromatography involves a thin liquid stationary phase coated on the surface of a solid support. Solute equilibrates between the stationary liquid and the mobile phase.

Ion-exchange chromatography features ionic groups such as $-SO_3^-$ or $-N(CH_3)_3^+$ covalently attached to the stationary solid phase, which is usually a *resin.* Solute ions are attracted to the stationary phase by electrostatic forces. The mobile phase is a liquid.

Molecular exclusion chromatography (also called *gel filtration, gel permeation,* or *molecular sieve* chromatography) separates molecules by size, with larger molecules passing through most quickly. Unlike other forms of chromatography, there is no attractive interaction between the "stationary phase" and the solute. The stationary phase has pores small enough to exclude large molecules, but not small ones. Large molecules stream past without entering the pores. Small molecules take longer to pass through the column because they enter the pores and therefore must flow through a larger volume before leaving the column.

Affinity chromatography, the most selective kind of chromatography, employs specific interactions between one kind of solute molecule and a second molecule

Eluent
in

C
o
l
u
m
n

Eluate
out

For their pioneering work on liquid-liquid partition chromatography in 1941, A. J. P. Martin and R. L. M. Synge received the Nobel Prize in 1952.

B. A. Adams and E. L. Holmes developed the first synthetic ion-exchange resins in 1935. *Resins* are relatively hard, amorphous (noncrystalline) organic solids. *Gels* are relatively soft.

Large molecules pass through the column *faster* than small molecules.

that is covalently attached (immobilized) to the stationary phase. For example, the immobilized molecule might be an antibody to a particular protein. When a mixture containing a thousand different proteins is passed through the column, only that one protein binds to the antibody on the column. After washing all other proteins from the column, the desired protein is dislodged by changing the pH or ionic strength.

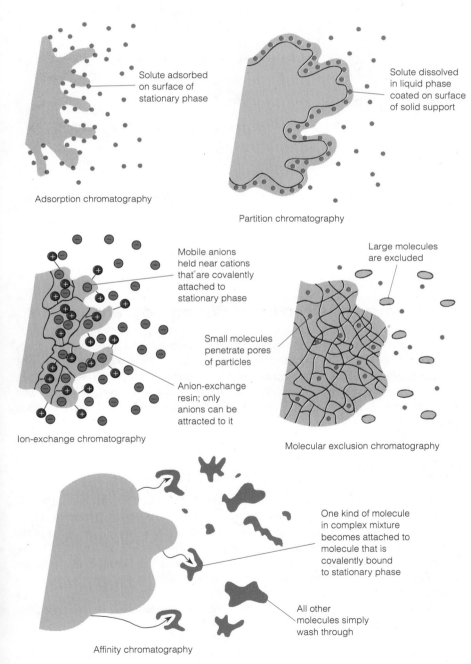

Solute adsorbed
on surface of
stationary phase

Adsorption chromatography

Solute dissolved
in liquid phase
coated on surface
of solid support

Partition chromatography

Mobile anions
held near cations
that are covalently
attached to
stationary phase

Anion-exchange
resin; only
anions can be
attracted to it

Ion-exchange chromatography

Large molecules
are excluded

Small molecules
penetrate pores
of particles

Molecular exclusion chromatography

One kind of molecule
in complex mixture
becomes attached to
molecule that is
covalently bound
to stationary phase

All other
molecules simply
wash through

Affinity chromatography

FIGURE 16-2 Types of chromatography.

16-A. Match the terms with their definitions:

1. Adsorption chromatography
2. Partition chromatography
3. Ion-exchange chromatography
4. Molecular exclusion chromatography
5. Affinity chromatography

A. Mobile-phase ions are attracted to ions covalently attached to stationary phase.

B. Solute is attracted to specific groups covalently attached to stationary phase.

C. Solute equilibrates between mobile phase and surface of stationary phase.

D. Solute equilibrates between mobile phase and stationary liquid film.

E. Solute penetrate voids in stationary phase. Largest solutes are eluted first.

16-2 *How We Describe a Chromatogram*

Detectors discussed in the next chapter respond to solutes as they exit the chromatography column. A **chromatogram** shows detector response as a function of time (or elution volume) in a chromatography experiment (Figure 16-3). Each peak in Figure 16-3 corresponds to a different substance eluted from the column. The **retention time,** t_r, is the time needed after injection for an individual solute to reach the detector.

Theoretical Plates

An ideal chromatographic peak has a Gaussian shape, like that in Figure 16-3. If the height of the peak is *h,* the *width at half-height,* $w_{1/2}$, is measured at $\frac{1}{2}h$. For a Gaussian peak, $w_{1/2}$ is equal to 2.35σ, where σ is the standard deviation of the peak. The width of the peak at the baseline, as shown in Figure 16-3, is 4σ.

In days of old, distillation was the most powerful means for separating volatile compounds from one another. A distillation column was divided into sections (*plates*) in which liquid and vapor equilibrated with each other. The more plates on a column, the more equilibration steps, and the better the separation between compounds with different boiling points.

Nomenclature from distillation has carried over into chromatography. We speak of a chromatography column as if it were divided into discrete sections (called **theoretical plates**) in which a solute molecule equilibrates between the mobile and sta-

Even though we speak of discrete theoretical plates, chromatography is really a continuous process. Theoretical plates are a purely imaginary way to describe the process.

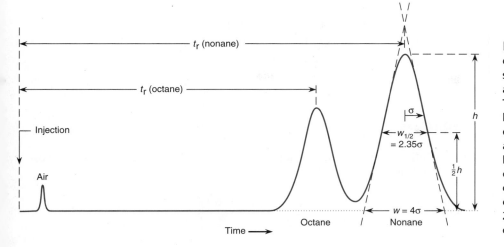

FIGURE 16-3 Schematic gas chromatogram, showing measurement of retention time (t_r) and width at half-height ($w_{1/2}$). The width at the base (w) is found by drawing tangents to the steepest parts of the Gaussian curve and extrapolating down to the baseline. The standard deviation of the Gaussian curve is σ. In gas chromatography, a small volume of air injected with the 0.1- to 2-μL sample is usually the first component to be eluted.

tionary phases. The retention of a compound on a column can be described by the number of theoretical equilibration steps that occur between injection and elution. The more equilibration steps (the more theoretical plates), the narrower the bandwidth when the compound emerges.

For any peak in Figure 16-3, the number of theoretical plates is computed by measuring the retention time and the width at half-height:

Number of plates on column: $$N = \frac{5.55\, t_r^2}{w_{1/2}^2} \qquad (16\text{-}1)$$

Retention and width can be measured in units of time or volume (such as mL of eluate). Both t_r and $w_{1/2}$ must be expressed in the same units in Equation 16-1.

If a column is divided into N theoretical plates (only in our minds), then the **plate height,** H, is the length of one plate. H is calculated by dividing the length of the column (L) by the number of theoretical plates on the column:

Plate height: $$H = L/N \qquad (16\text{-}2)$$

The smaller the plate height, the narrower is each eluted peak. The ability of a column to separate components of a mixture is improved by decreasing plate height. An efficient column has more theoretical plates than an inefficient column. Different solutes behave as if the column has somewhat different plate heights, because different compounds equilibrate between the mobile and stationary phases at different rates. Plate heights are in the neighborhood of ~0.1 to 1 mm in gas chromatography, ~10 μm in high-performance liquid chromatography, and <1 μm in capillary electrophoresis.

You can test for degradation of a column by injecting a test compound periodically and looking for peak asymmetry and changes in the number of plates.

small plate height
⇒ narrow peaks
⇒ better separations

EXAMPLE **Measuring Plates**

A solute with a retention time of 407 s has a width at half-height of 8.1 s on a column 12.2 m long. Find the number of plates and the plate height.

297

SOLUTION

$$N = \frac{5.55\, t_r^2}{w_{1/2}^2} = \frac{5.55 \cdot 407^2}{8.1^2} = 1.40 \times 10^4$$

$$H = \frac{L}{N} = \frac{12.2\ \text{m}}{1.40 \times 10^4} = 0.87\ \text{mm}$$

Resolution

The **resolution** between neighboring peaks is equal to the peak separation (Δt_r) divided by the average peak width (w_{av}, measured at the base, as in Figure 16-3):

Resolution:
$$\text{resolution} = \frac{\Delta t_r}{w_{av}} \tag{16-3}$$

The better the resolution, the more complete is the separation between neighboring peaks. Figure 16-4 shows that when two peaks have a resolution of 0.50, the overlap of area is 16%. When the resolution is 1.00, the overlap is 2.3%. If you double the length of an ideal chromatography column, you will improve resolution by $\sqrt{2}$.

Scaling Up a Separation

Analytical chromatography: small-scale analysis
Preparative chromatography: large-scale separation

Analytical chromatography is conducted on a small scale to separate, identify, and measure components of a mixture. *Preparative* chromatography is carried out on a large scale to isolate a significant quantity of one or more components of a mixture. Analytical chromatography typically uses long, thin columns to obtain good resolution. Preparative chromatography usually employs short, fat columns that can handle larger quantities of material but give inferior resolution. (Long, fat columns may be prohibitively expensive to buy or operate.)

If you have developed a chromatographic procedure to separate 2 mg of a mixture on a column with a diameter of 1.0 cm, what size column should you use to separate 20 mg of the mixture? The most straightforward way to scale up is to maintain the same column length and increase the cross-sectional area to maintain a constant ratio of unknown to column volume. Because area is πr^2, where r is the column radius, the desired radius is given by

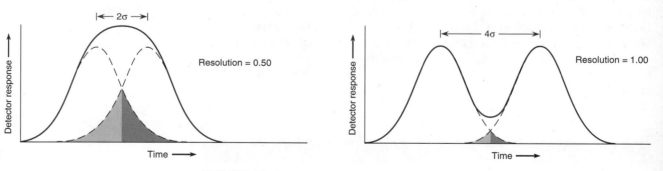

FIGURE 16-4 Resolution of Gaussian peaks of equal area and amplitude. Dashed lines show individual peaks; solid lines are the sum of two peaks. Overlapping area is shaded.

Scaling equation: $\dfrac{\text{large load (g)}}{\text{small load (g)}} = \left(\dfrac{\text{large column radius}}{\text{small column radius}}\right)^2$ (16-4)

$$\frac{20 \text{ mg}}{2 \text{ mg}} = \left(\frac{\text{large column radius}}{0.50 \text{ cm}}\right)^2$$

$$\text{large column radius} = 1.58 \text{ cm}$$

A column with a diameter near 3 cm would be appropriate.

To reproduce the conditions of the smaller column in the larger column, the *volume flow rate* (mL/min) should be increased in proportion to the cross-sectional area of the column. If the area of the large column is 10 times greater than that of the small column, the volume flow rate should be 10 times greater. If the small sample is dissolved in a volume V before it is applied to the small column, the large sample should be dissolved in a volume $10V$.

Rules for scaling up without reducing separation:

- cross-sectional area of column $\propto$ mass of analyte
- keep column length constant
- volume flow rate $\propto$ cross-sectional area of column
- sample volume $\propto$ mass of analyte

(The symbol $\propto$ means "is proportional to.")

Ask Yourself

16-B. (a) Use a ruler to measure retention times and widths at the base (w) of octane and nonane in Figure 16-3 to the nearest 0.1 mm.

   (b) Calculate the number of theoretical plates for octane and nonane.

   (c) If the column is 1.00 m long, find the plate height for octane and nonane.

   (d) Use your measurements from part **(a)** to compute the resolution between octane and nonane.

   (e) Suppose that the sample size for Figure 16-3 was 3.0 mg and the column dimensions were 4.0 mm diameter $\times$ 1.00 m length and the flow rate was 7.0 mL/min. What size column and what flow rate should be used to obtain the same quality of separation of 27.0 mg of sample?

16-3 *Why Do Bands Spread?*

Each band of solute becomes broader as it moves through a chromatography column (Figure 16-5). Broadening occurs because of diffusion, slow equilibration of solute between the mobile and stationary phases, and irregular flow paths on the column.

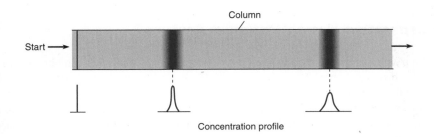

FIGURE 16-5 Broadening of an initially sharp band of solute as it moves through a chromatography column.

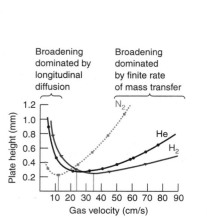

Zone of solute after short
time on column

Longitudinal
diffusion (B/u_x)

Zone of solute after longer
time on column

Direction of travel

FIGURE 16-6 In longitudinal diffusion, solute continuously diffuses away from the concentrated zone at the center of its band. The greater the flow rate, the less time is spent on the column and the less longitudinal diffusion occurs.

FIGURE 16-8 Optimum resolution (i.e., minimum plate height) occurs at an intermediate flow rate. Curves show measured plate height in gas chromatography of n-$C_{17}H_{36}$ at 175°C using N_2, He, or H_2 mobile phase.

Bands Diffuse

An infinitely narrow band of solute that is stationary inside the column slowly broadens because solute molecules diffuse away from the center of the band in both directions. This inescapable process, called *longitudinal diffusion,* begins at the moment solute is injected into the column. In chromatography, the further a band has traveled, the more time it has had to diffuse and the broader it becomes (Figure 16-6).

The faster the flow rate, the less time a band spends in the column, and the less time there is for diffusion to occur. The faster the flow, the sharper the peaks. Broadening by longitudinal diffusion is inversely proportional to flow rate:

*Broadening by
longitudinal diffusion:* $$\text{broadening} \propto \frac{1}{u} \qquad (16\text{-}5)$$

where u is the flow rate (mL/min).

Solute Requires Time to Equilibrate Between Phases

Imagine a solute that is distributed between the mobile and stationary phases at some moment in time at some position in a column with zero flow rate. Now set the flow into motion. If the solute cannot equilibrate rapidly enough between the phases, then solute in the stationary phase tends to lag behind solute in the mobile phase (Figure 16-7). This broadening due to the *finite rate of mass transfer* between phases becomes worse as the flow rate increases:

*Broadening by finite
rate of mass transfer:* $$\text{broadening} \propto u \qquad (16\text{-}6)$$

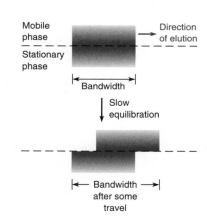

FIGURE 16-7 Solute requires a finite time to equilibrate between the mobile and stationary phases. If equilibration is slow, solute in the stationary phase tends to lag behind that in the mobile phase, causing the band to spread. The slower the flow rate, the less the zone broadening by this mechanism.

A Separation Has an Optimum Rate

When trying to separate closely spaced bands, we want to minimize band broadening during chromatography. If the bands broaden too much, we would not be able to resolve them from each other. Because broadening by longitudinal diffusion decreases with increasing flow rate (Equation 16-5), but broadening by the finite rate of mass transfer increases with increasing flow rate (Equation 16-6), there is some intermediate flow rate that gives minimum broadening and optimum resolution (Figure 16-8).

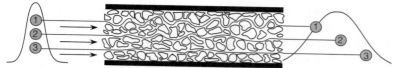

FIGURE 16-9 Band spreading from multiple flow paths. The smaller the stationary phase particles, the less serious is this problem. This process is absent in an open tubular column.

Part of the science and art of chromatography is to experiment with conditions such as flow rate and solvent composition to obtain the best separation (or adequate separation) between components of a mixture.

The rate of mass transfer between phases increases with temperature. Raising the column temperature might improve the resolution or allow faster separations without reducing resolution.

Some Band Broadening Is Independent of Flow Rate

There are some mechanisms of band broadening that are independent of flow rate. Figure 16-9 shows a mechanism that is called *multiple paths* and occurs in any column packed with solid particles. Because some of the random flow paths are longer than others, solute molecules entering the column at the same time on the left are eluted at different times on the right.

Open Tubular Columns

In contrast to a **packed column** that is filled with solid particles coated with stationary phase, an **open tubular column** is a hollow capillary whose inner wall is coated with a thin layer of stationary phase (Figure 16-10). In an open tubular column, there is no broadening from multiple paths, because there are no particles of stationary phase in the flow path. Therefore a given length of open tubular column generally gives better resolution than the same length of packed column. We say that the open tubular column has more theoretical plates (or a smaller plate height) than the packed column.

A packed gas chromatography column has greater resistance to gas flow than does an open tubular column. Therefore an open tubular column can be made much longer than a packed column with the same operating pressure. Because of its resis-

Open tubular columns give better separations than packed columns because

- there is no multiple path band broadening
- the open tubular column can be much longer

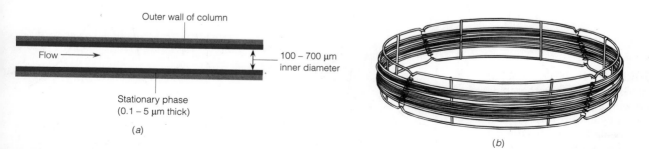

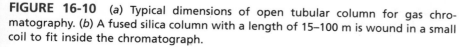

FIGURE 16-10 (a) Typical dimensions of open tubular column for gas chromatography. (b) A fused silica column with a length of 15–100 m is wound in a small coil to fit inside the chromatograph.

tance to gas flow, a packed gas chromatography column is usually just 2–3 m in length, whereas an open tubular column can be 100 m in length. The greater length and the smaller plate height of the open tubular column provide much better resolution than a packed column does.

An open tubular column cannot handle as much solute as a packed column, because there is less stationary phase in the open tubular column. Therefore open tubular columns are useful for analytical separations but not for preparative separations.

Sometimes Bands Have Funny Shapes

When a column is overloaded by too much solute in one band, the band emerges from the column with a gradually rising front and an abruptly cut off back (Figure 16-11a). The reason for this behavior is that a compound is most soluble in itself. As the concentration rises from zero at the front of the band to some high value inside the band, *overloading* can occur. When overloaded, the solute is so soluble in the concentrated part of the band that little trails behind the concentrated region.

Tailing is an asymmetric peak shape in which the trailing part of the chromatographic band is highly elongated (Figure 16-11b). Such a peak occurs when there are strongly polar, highly adsorptive sites (such as exposed –OH groups) in the stationary phase that retain solute more strongly than other sites. In a chemical treatment called *silanization,* polar –OH groups are converted to nonpolar –OSi(CH$_3$)$_3$ groups to eliminate tailing.

Box 16-1 discusses *polarity.*

Ask Yourself

16-C. (a) Why is longitudinal diffusion a more serious problem in gas chromatography than in liquid chromatography? (Think about this one; the answer is not in the text.)

(b) (i) In Figure 16-8, what is the optimal flow rate for best separation of solutes with He mobile phase?

(ii) Why does plate height increase at high flow rate?

(iii) Why does plate height increase at low flow rate?

(c) Why does an open tubular gas chromatography column give better resolution than the same length of packed column?

(d) Why is it desirable to use a very long open tubular column? What difference between packed and open tubular columns allows much longer open tubular columns to be used?

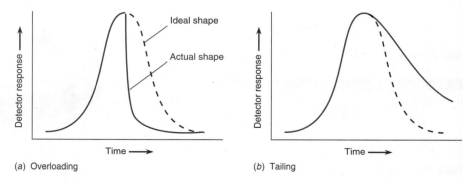

FIGURE 16-11 (*a*) Overloading gives rise to a chromatographic band with an ordinary front and an abruptly cut off rear. (*b*) Tailing is a peak shape with a normal front and a grossly elongated rear.

(*a*) Overloading

(*b*) Tailing

Box 16-1 *Explanation*

Polarity

Polar compounds have positive and negative regions that attract neighboring molecules by electrostatic forces.

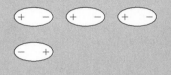

Polar molecules attract each other by electrostatic forces. Positive attracts negative.

Polarity arises because different atoms have different electronegativity–different abilities to attract electrons from the chemical bonds. For example, oxygen is more electronegative than carbon. In acetone, the oxygen attracts electrons from the C=O double bond and takes on a partial negative charge, designated $\delta-$. This shift leaves a partial positive charge ($\delta+$) on the neighboring carbon.

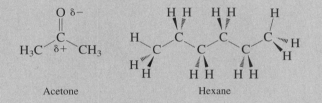

Acetone   Hexane

In contrast to acetone, hexane is considered to be a **nonpolar compound**, because there is little charge separation within a hexane molecule.

Water is a very polar molecule that forms hydrogen bonds between the electronegative oxygen atom in one molecule and the electropositive hydrogen atom of another molecule:

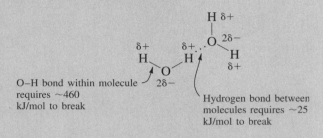

O–H bond within molecule requires ~460 kJ/mol to break

Hydrogen bond between molecules requires ~25 kJ/mol to break

Ionic compounds usually dissolve in water. In general, polar organic compounds tend to be most soluble in polar solvents and least soluble in nonpolar solvents. Nonpolar compounds tend to be most soluble in nonpolar solvents. "Like dissolves like."

Typical nonpolar and weakly polar compounds		Typical polar compounds	
	octane (C_8H_{18})	CH_3OH	methanol
	benzene	CH_3CH_2OH	ethanol
	toluene	$CHCl_3$	chloroform
	carbon tetrachloride	$CH_3-C(=O)OH$	acetic acid
	diethyl ether ($C_4H_{10}O$)	$CH_3C\equiv N$	acetonitrile

Box 16-2 *Explanation*

What Is Mass Spectrometry?

A **mass spectrometer** is a device that ionizes molecules in the gas phase, accelerates the ions, and separates them according to their mass. The output of the instrument, called a *mass spectrum,* is a graph showing the relative abundance of each charged species versus mass number (see Figures 16-12b and 16-12c). (The horizontal axis of a mass spectrum is really m/z, where m is the mass of the ion and z is the number of charges it carries. Most species in a mass spectrum carry one charge.)

The diagram below shows a quadrupole mass spectrometer connected to the exit of a capillary gas chromatograph to record a spectrum of each peak as it is eluted. The chromatography column passes through a heated, gas-tight connector into the ionization chamber of the mass spectrometer, which is pumped rapidly to maintain a vacuum. Electrons from a hot filament are accelerated through 70 V and collide with gaseous eluate molecules. Some molecules are ionized by collisions with the energetic electrons to make cations that have enough energy to decompose into smaller molecular fragments.

An ionized molecule (designated M in Figures 16-12b and 16-12c) and its ionized fragments are accelerated through 15 V and enter the quadrupole mass separator. The separator consists of four parallel metal cylinders to which are applied both a constant voltage

and a radio frequency alternating voltage. The electric field causes gaseous cations to travel in complex trajectories as they migrate from the ionization chamber toward the ion detector. The electric field allows cations with only one particular mass (resonant ions) to reach the detector. Cations with all other masses (nonresonant ions) collide with the cylinders and are lost before they reach the detector. By rapidly varying the applied voltages, ions of different mass are selected to generate the mass spectra in Figure 16-12. Quadrupole mass separators are scanned fast enough to record two to eight spectra per second, covering 800 mass units.

We can understand the pattern of fragments of 1-bromobutane in Figure 16-12c without too much difficulty. The peak at 136 mass units is the unfragmented ion called the *parent ion* (M). Its formula is $C_4H_9{}^{79}Br^+$. The peak at 138 mass units with almost equal intensity has the formula $C_4H_9{}^{81}Br^+$. Natural bromine consists of 50.5% ^{79}Br and 49.9% ^{81}Br, which explains why the peak at 136 mass units is slightly taller than the peak at 138 mass units. Small peaks at 107 and 109 mass units have lost C_2H_5 fragments. They are labeled M − C_2H_5 and have the formula $C_2H_4Br^+$. The peak at 57 mass units is $C_4H_9^+$, which has no bromine and therefore has no isotopic partner at 59 mass units. The peak at 41 mass units is $C_3H_5^+$.

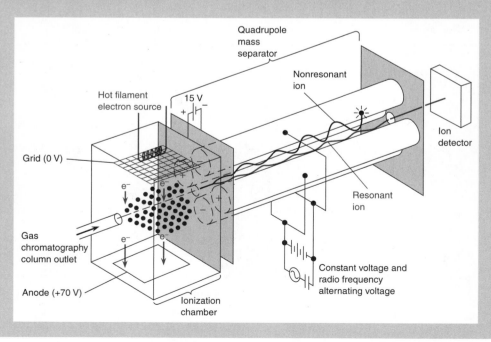

Chromatography is powerful for both qualitative and quantitative analysis. The simplest way to identify a chromatographic peak is to compare its retention time with that of an authentic sample of the suspected compound. The most reliable way to do this is by **co-chromatography,** in which a known analyte is added to the unknown sample. If the added compound is identical to one component of the unknown, then the relative size of that one peak will increase. Identification is only tentative when carried out with one column, but it is nearly conclusive when carried out with several different kinds of columns.

For qualitative analysis, each chromatographic peak can be directed into a mass spectrometer (Box 16-2) or infrared spectrophotometer to record a spectrum as the substance is eluted from the column. The spectrum can be identified by comparison with a library of spectra stored in a computer. Figure 16-12a shows a chromatogram from a student experiment in which 1-bromobutane was synthesized from 1-butanol:

Two compounds might have the same retention time on a particular column. It is unlikely that they will have the same retention time on several columns with different stationary phases.

$$CH_3CH_2CH_2CH_2OH + Br^- \longrightarrow CH_3CH_2CH_2CH_2Br + OH^- \quad (16\text{-}7)$$
$$\text{1-Butanol} \qquad\qquad\qquad \text{1-Bromobutane}$$

Peaks in the chromatogram are identified by recording the mass spectrum of each as it is eluted. Figure 16-12b shows the spectrum of Peak 1 and identifies it as starting material by comparison with a spectral library of known compounds. Figure 16-12c shows that Peak 2 is product. The progress of the reaction can be followed by chromatography, and additional products (if any) could be identified.

For quantitative analysis, we normally choose conditions under which *the area of a chromatographic peak is proportional to the quantity of that component.* Peak areas are measured automatically if the chromatograph is computer controlled. If you must make a measurement by hand, a triangle can be drawn with two sides tangent to the steepest points on each side of the peak (as in Figure 16-3). When the third side is drawn along the baseline, the area of the triangle ($= \frac{1}{2}$(base × height)) is 96% of the area of a Gaussian peak. Box 16-3 describes an application of quantitative analysis by chromatography to the measurement of prehistoric ocean temperatures.

FIGURE 16-12 (*a*) Gas chromatogram of a student sample from Reaction 16-7. The detector is the mass spectrometer set to record the total current from all ions of all masses. Dichloromethane solvent eluted prior to 2 min is not shown. (*b*) Mass spectrum of peak 1, which is identical to the spectrum of 1-butanol. (*c*) Mass spectrum of peak 2, which is identical to the spectrum of 1-bromobutane. In (*b*) and (*c*), M stands for the parent ion.

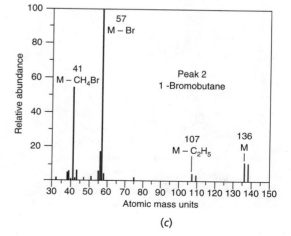
(a)   (b)   (c)

Ask Yourself

16-D. The mass spectrum of 1-bromobutane in Figure 16-12c has a pair of nearly equal-intensity peaks at 136 and 138 mass units and another pair at 107 and 109. The peaks at 57 and 41 have no similar equal-intensity partners. After reading Box 16-2, explain why this is so.

16-5 *Internal Standards*

An **internal standard** is a known amount of a compound, different from analyte, that is added to the unknown for the purpose of measuring the unknown. The signal due to analyte (X) is compared with that of the internal standard (S). A known mixture of standard and analyte is prepared beforehand to measure the relative response of the analytical method to the two species. The key relationship is

Box 16-3 *Informed Citizen*

How a Chemist Finds a Date: Molecular Stratigraphy in Geology

Phytoplankton living in the top layer of the ocean maintain the fluidity of their cell membranes by altering their lipid (fat) composition when the temperature changes. As ocean temperature decreases, these organisms synthesize membrane lipids that are more fluid at the lower temperature. Three membrane lipids specific to just a few species of phytoplankton are shown below. The compound designated 37:2Me increases relative to the compound 37:4Me in phytoplankton membranes as the ocean temperature increases.

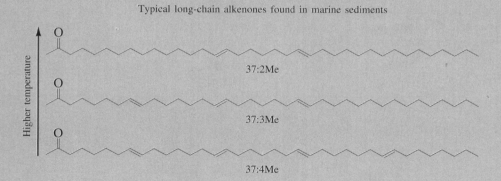

Typical long-chain alkenones found in marine sediments

Dead marine organisms eventually end up buried in the sediment at the bottom of the ocean. The deeper you sample a sediment, the further back into time you delve. By measuring the relative content of the three alkenones above at different depths in the sediment, you can infer the temperature of the ocean long ago.

The relevant index is called U_{37}^K, which stands for unsaturated ketones with 37 carbon atoms and is defined as

$$U_{37}^K = \frac{[37\!:\!2\text{Me}] - [37\!:\!4\text{Me}]}{[37\!:\!2\text{Me}] + [37\!:\!3\text{Me}] + [37\!:\!4\text{Me}]}$$

Internal standard equation:

$$\frac{\text{concentration ratio (X/S) in unknown}}{\text{concentration ratio in standard mixture}} = \frac{\text{signal ratio (X/S) in unknown}}{\text{signal ratio in standard mixture}} \quad (16\text{-}8)$$

Equation 16-8 assumes that the analytical method responds linearly to both the analyte and the standard. To use Equation 16-8, we normally measure the denominators on both sides in a preliminary experiment. Then we measure both numerators with the unknown.

Internal standards are desirable if loss of sample is likely to occur during handling. If a known quantity of standard is added to the unknown prior to any loss, the ratio of standard to analyte remains constant because the same fraction of each is lost in any operation.

As an example, consider the chromatogram in Figure 16-13. The area under each peak is proportional to the quantity of compound passing through the detector. By knowing the relative responses to the two compounds and how much standard S was added to the unknown X, we can deduce the concentration of X.

Linear response means that peak area is proportional to analyte concentration. For narrow peaks, height is often substituted for area.

U_{37}^{K} is the fraction of alkenones that are 37:2Me minus the fraction in the form 37:4Me. The higher the ocean temperature, the greater the value of U_{37}^{K}. Alkenones are extracted from ~1 g of sediment with organic solvents, and their concentrations are measured by gas chromatography with mass spectrometric detection.

The graph shows ocean temperature as a function of sediment age (depth) at a point in the Atlantic Ocean off the northwest coast of Africa where the sedimentation rate is 15–20 cm per 1 000 years. The major valleys in the graph are ice ages and the major peaks are interglacial periods.

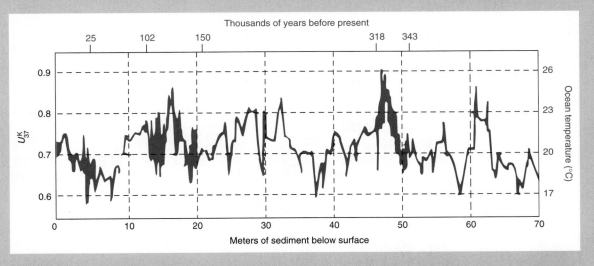

Ocean temperature inferred by measuring long-chain alkenones in sediment, using gas chromatography. Measuring concentrations of molecules in different layers of sediment is called *molecular stratigraphy*.

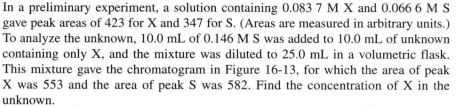

Detector signal →

Time (min)

FIGURE 16-13 Chromatogram illustrating the use of an internal standard. A known amount of standard S is added to unknown X. By measuring the area of both peaks in this chromatogram and the relative response of the detector to a known amount of each compound in another chromatogram (not shown), we can tell how much X is in the unknown.

In a preliminary experiment, a solution containing 0.083 7 M X and 0.066 6 M S gave peak areas of 423 for X and 347 for S. (Areas are measured in arbitrary units.) To analyze the unknown, 10.0 mL of 0.146 M S was added to 10.0 mL of unknown containing only X, and the mixture was diluted to 25.0 mL in a volumetric flask. This mixture gave the chromatogram in Figure 16-13, for which the area of peak X was 553 and the area of peak S was 582. Find the concentration of X in the unknown.

SOLUTION For the standard mixture, we can write

Standard mixture:
$$\frac{\text{area of X}}{\text{area of S}} = \frac{423}{347} \quad \text{when} \quad \frac{[X]}{[S]} = \frac{0.083\ 7\ \text{M}}{0.066\ 6\ \text{M}}$$

Information from the standard mixture allows us to write the denominators of Equation 16-8:

$$\frac{\text{concentration ratio (X/S) in unknown}}{0.083\ 7\ \text{M} / 0.066\ 6\ \text{M}} = \frac{\text{signal ratio (X/S) in unknown}}{423 / 347}$$

In the mixture of unknown plus standard, the concentration of S is

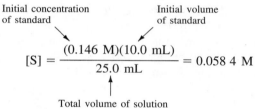

Initial concentration of standard

Initial volume of standard

$$[S] = \frac{(0.146\ \text{M})(10.0\ \text{mL})}{25.0\ \text{mL}} = 0.058\ 4\ \text{M}$$

Total volume of solution

This mixture gave the chromatogram in Figure 16-13, for which

Unknown mixture:
$$\frac{\text{area of X}}{\text{area of S}} = \frac{553}{582} \quad \text{when} \quad \frac{[X]}{[S]} = \frac{\text{unknown}}{0.058\ 4\ \text{M}}$$

We can now write the numerators of Equation 16-8 and solve for the unknown concentration [X]:

$$\frac{\text{concentration ratio (X/S) in unknown}}{\text{concentration ratio in standard mixture}} = \frac{\text{signal ratio (X/S) in unknown}}{\text{signal ratio in standard mixture}}$$

$$\frac{[X] / 0.058\ 4\ \text{M}}{0.083\ 7\ \text{M} / 0.066\ 6\ \text{M}} = \frac{553 / 582}{423 / 347}$$

$$\Rightarrow [X] = 0.057\ 2\ \text{M in mixture with S}$$

But X was diluted from 10.0 to 25.0 mL when the mixture with S was prepared, so the original concentration of X in the unknown is computed with the dilution equation 1-5:

$$M_{\text{conc}} \cdot V_{\text{conc}} = M_{\text{dil}} \cdot V_{\text{dil}}$$

$$[X] \cdot (10.0\ \text{mL}) = (0.057\ 2\ \text{M}) \cdot (25.0\ \text{mL}) \Rightarrow [X] = 0.143\ \text{M}$$

Ask Yourself

16-E. A mixture of 52.4 nM iodoacetone (X) and 38.9 nM *p*-dichlorobenzene (S) gave the relative response (peak area of X)/(peak area of S) = 0.644. A second solution containing an unknown quantity of X plus 742 nM S gave a response (peak area of X)/(peak area of S) = 1.093. Find [X] in the second solution.

Key Equations

Theoretical plates
$$N = \frac{5.55\, t_r^2}{w_{1/2}^2}$$

t_r = retention time of analyte
$w_{1/2}$ = peak width at half-height $\Big\}$ Both must be measured in the same units.

Plate height
$$H = L/N \qquad (L = \text{length of column})$$

Resolution
$$\text{resolution} = \frac{\Delta t_r}{w_{av}}$$

Δt_r = difference in retention time between two peaks $\Big\}$ Both must be measured
w_{av} = average width of two peaks at baseline $\quad$ in the same units.

Scaling equation
$$\frac{\text{large load}}{\text{small load}} = \left(\frac{\text{large column radius}}{\text{small column radius}} \right)^2$$

Internal standard equation
$$\frac{\text{concentration ratio (X/S) in unknown}}{\text{concentration ratio in standard mixture}} = \frac{\text{signal ratio (X/S) in unknown}}{\text{signal ratio in standard mixture}}$$

Important Terms

adsorption chromatography

affinity chromatography

chromatogram

chromatography

co-chromatography

eluate

eluent

elution

gas chromatography

internal standard

ion-exchange chromatography

liquid chromatography

mass spectrometer

mobile phase

molecular exclusion chromatography

nonpolar compound

open tubular column

packed column

partition chromatography

plate height

polar compound

resolution

retention time

stationary phase

theoretical plate

Problems

16-1. What is the difference between eluent and eluate?

16-2. Which column is more efficient: plate height = 0.1 mm or 1 mm?

16-3. Explain why a chromatographic separation normally has an optimum flow rate that gives the best separation.

16-4. Why does silanization reduce tailing of chromatographic peaks?

16-5. What kind of information does a mass spectrometer detector give in gas chromatography that is useful for qualitative analysis? For quantitative analysis?

16-6. Explain how co-chromatography is used in qualitative analysis. Why are several different types of columns necessary to make a convincing case for the identity of a compound?

16-7. (a) How many theoretical plates produce a chromatography peak eluting at 12.83 min with a width at half-height of 8.7 s?

(b) The length of the column is 15.8 cm. Find the plate height.

16-8. A gas chromatogram of a mixture of toluene and ethyl acetate is shown here.

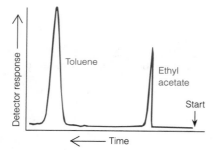

(a) Measure $w_{1/2}$ for each peak to the nearest 0.1 mm. When the thickness of the pen trace is significant relative to the length being measured, it is important to take the pen width into account. It is best to measure from the edge of one trace to the corresponding edge of the other trace, as shown here.

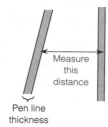

(b) Find the number of theoretical plates and the plate height for each peak.

16-9. Two components of a 12-mg sample are adequately separated by chromatography through a column 1.5 cm in diameter and 25 cm long at a flow rate of 0.8 mL/min. What size column and what flow rate should be used to obtain similar separation with a 250-mg sample?

16-10. The chromatogram here has a peak for isooctane at 13.81 min. The column is 30.0 m long.

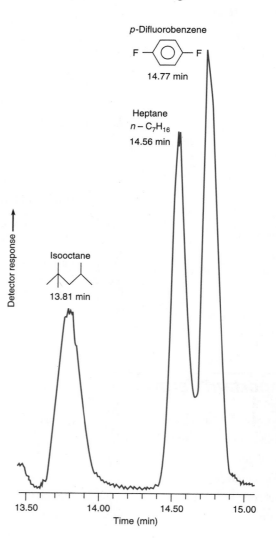

(a) Measure $w_{1/2}$ and find the number of theoretical plates for this peak.

(b) Find the plate height.

(c) Measure the width at the base, w, as defined in Figure 16-3. Figure 16-3 tells us that the expected ratio $w/w_{1/2}$ for a Gaussian peak is $4 \sigma/2.35 \ \sigma = 1.70$. Compare the measured ratio $w/w_{1/2}$ to the theoretical ratio.

16-11. Consider the peaks for heptane (14.56 min) and *p*-difluorobenzene (14.77 min) in the chromatogram for the previous problem. The column is 30.0 m long.

(a) Find the number of plates and the plate height for the two compounds.

(b) Measure w for each peak and find the resolution between the two peaks.

16-12. A column 3.00 cm in diameter and 32.6 cm long gives adequate resolution of a 72.4-mg mixture of unknowns, initially dissolved in 0.500 mL.

(a) If you wish to scale down to 10.0 mg of the same mixture with minimum use of chromatographic stationary phase and solvent, what length and diameter column would you use?

(b) In what volume would you dissolve the sample?

(c) If the flow rate in the large column is 1.85 mL/min, what should be the flow rate in the small column?

16-13. In Figure 16-8, flow rate is expressed as gas velocity in cm/s. The gas is flowing through an open tubular column with an inner diameter of 0.25 mm. What volume flow rate (mL/min) corresponds to a gas velocity of 50 cm/s? (Note that the volume of a cylinder is $\pi r^2 \times$ length, where r is the radius.)

16-14. As described in the section on internal standards, a solution containing 3.47 mM X and 1.72 mM S gave peak areas of 3 473 and 10 222, respectively, in a chromatographic analysis. Then 1.00 mL of 8.47 mM S was added to 5.00 mL of unknown X, and the mixed solution was diluted to 10.0 mL. This solution gave peak areas of 5 428 and 4 431 for X and S, respectively.

(a) Find the concentration of S (mM) in the 10.0 mL of mixed solution.

(b) Find the concentration of X (mM) in the 10.0 mL of mixed solution.

(c) Find the concentration of X in the original unknown.

16-15. Compounds C and D gave the following chromatography results:

Compound	Concentration (μg/mL in) mixture	Peak area (cm²)
C	236	4.42
D	337	5.52

A solution was prepared by mixing 1.23 mg of D in 5.00 mL with 10.00 mL of unknown containing just C and diluting to 25.00 mL. Peak areas of 3.33 and 2.22 cm² were observed for C and D, respectively. Find the concentration of C (μg/mL) in the unknown.

16-16. When 1.06 mmol of 1-pentanol and 1.53 mmol of 1-hexanol were separated by gas chromatography, they gave relative peak areas of 922 and 1 570 units, respectively. When 0.57 mmol of pentanol was added to an unknown containing hexanol, the relative chromatographic peak areas were 843:816 (pentanol:hexanol). How much hexanol did the unknown contain?

16-17. Suggest a molecular formula for the major peaks at 31, 41, 43, and 56 mass units in the mass spectrum of 1-butanol (C_4H_9OH) in Figure 16-12b.

In Vivo Microdialysis for Measuring Drug Metabolism

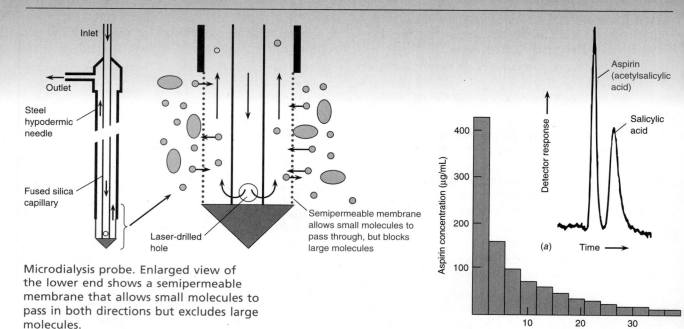

Microdialysis probe. Enlarged view of the lower end shows a semipermeable membrane that allows small molecules to pass in both directions but excludes large molecules.

Inlet
Outlet
Steel hypodermic needle
Fused silica capillary
Laser-drilled hole
Semipermeable membrane allows small molecules to pass through, but blocks large molecules

Aspirin (acetylsalicylic acid)
Salicylic acid
Detector response
(a) Time

Aspirin concentration (μg/mL)
400
300
200
100
10 20 30
(b) Time (min)

(a) Chromatogram of blood microdialysis sample drawn 5 min after intravenous injec of aspirin. (b) Aspirin concentration in bloo versus time after injection of aspirin.

Dialysis is the process in which small molecules diffuse across a *semipermeable membrane* that has pore sizes large enough to pass small but not large molecules. A *microdialysis probe* has a semipermeable membrane attached to the shaft of a hypodermic needle, which can be inserted into an animal. Fluid is pumped through the probe from the inlet to the outlet. Small molecules from the animal diffuse into the probe and are rapidly transported to the outlet. Fluid exiting the probe (*dialysate*) can be analyzed by liquid chromatography.

An in vivo (inside the living organism) study of aspirin metabolism in a rat is shown above. Aspirin is converted to salicylic acid by enzymes in the bloodstream. To measure the conversion rate, aspirin was injected into a rat and dialysate from a microdialysis probe in a vein of the rat was monitored by chromatography. If you simply withdrew blood for analysis, aspirin would continue to be metabolized by enzymes in the blood. Microdialysis separates the small aspirin molecule from large enzyme molecules.

Chapter 17

Gas and Liquid Chromatography

High-resolution capillary gas chromatography and high-performance liquid chromatography are workhorses of the modern analytical chemistry laboratory. This chapter describes the equipment and basic techniques of gas and liquid chromatography.

17-1 Gas Chromatography

In **gas chromatography,** a gaseous mobile phase transports a gaseous solute (or vapor from a volatile liquid) through a long, thin column containing stationary phase. Figure 17-1 shows the main components of a gas chromatograph. We begin the experiment by injecting a volatile liquid through a rubber *septum* (a thin disk) into

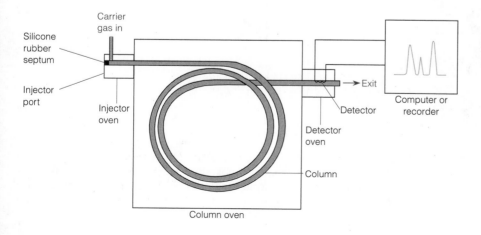

FIGURE 17-1 Schematic diagram of a gas chromatograph.

313

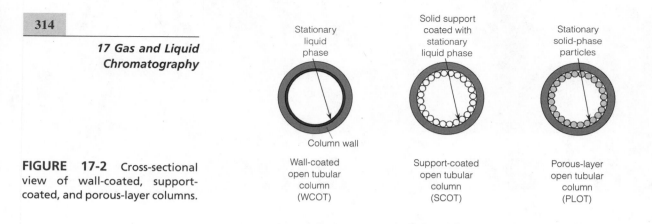

FIGURE 17-2 Cross-sectional view of wall-coated, support-coated, and porous-layer columns.

Wall-coated
open tubular
column
(WCOT)

Support-coated
open tubular
column
(SCOT)

Porous-layer
open tubular
column
(PLOT)

a heated port, which vaporizes the sample. The sample is swept through the column by He, N_2, or H_2 *carrier gas,* and the separated solutes flow through a detector, whose response is displayed on a recorder or computer. The column must be hot enough to produce sufficient vapor pressure for each solute to be eluted in a reasonable time. The detector is maintained at a higher temperature than the column, so that all solutes are gaseous. The injected sample size for a liquid is typically 0.1–2 μL for analytical chromatography, whereas preparative columns can handle 20–1 000 μL. Gases can be introduced in volumes of 0.5–10 mL by a gas-tight syringe or a gas-sampling valve.

Columns

Open tubular columns, such as the one shown in Figure 16-10, have a liquid or solid stationary phase coated on the inside wall (Figure 17-2). For example, a solid, porous carbon stationary phase, shown in Figure 17-3a, can be used to separate gases found in beer, as shown in Figure 17-3b.

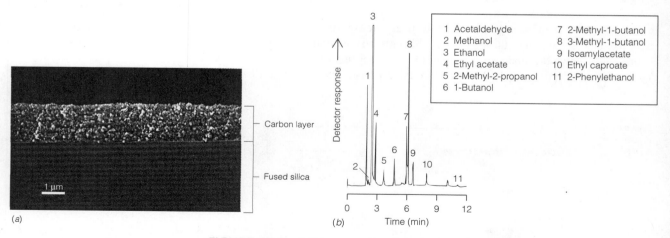

1 Acetaldehyde	7 2-Methyl-1-butanol
2 Methanol	8 3-Methyl-1-butanol
3 Ethanol	9 Isoamylacetate
4 Ethyl acetate	10 Ethyl caproate
5 2-Methyl-2-propanol	11 2-Phenylethanol
6 1-Butanol	

FIGURE 17-3 (a) Porous carbon stationary phase (2 μm thick) on the inside wall of a fused silica open tubular capillary gas chromatography column. (b) Chromatogram of compounds in the headspace (gas phase) of a beer can, obtained with a porous carbon column (0.25-mm diameter × 30 m long) operated at 30°C for 2 min and then ramped up to 160°C at 20°/min. Drink up!

TABLE 17-1 Common stationary phases in capillary gas chromatography

Structure	Polarity		Temperature range

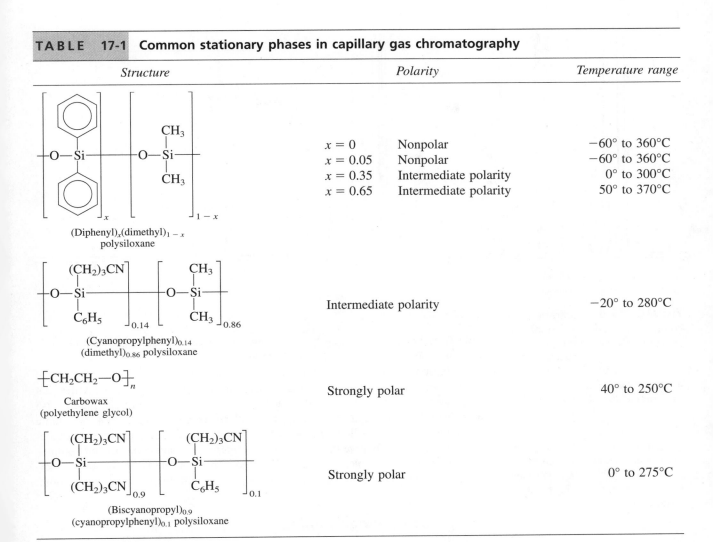

(Diphenyl)$_x$(dimethyl)$_{1-x}$ polysiloxane	$x = 0$ Nonpolar $x = 0.05$ Nonpolar $x = 0.35$ Intermediate polarity $x = 0.65$ Intermediate polarity		$-60°$ to $360°$C $-60°$ to $360°$C $0°$ to $300°$C $50°$ to $370°$C
(Cyanopropylphenyl)$_{0.14}$ (dimethyl)$_{0.86}$ polysiloxane	Intermediate polarity		$-20°$ to $280°$C
Carbowax (polyethylene glycol)	Strongly polar		$40°$ to $250°$C
(Biscyanopropyl)$_{0.9}$ (cyanopropylphenyl)$_{0.1}$ polysiloxane	Strongly polar		$0°$ to $275°$C

Open tubular columns are usually made of fused silica (SiO_2). As the column ages, stationary phase bakes off and exposes silanol groups (Si–O–H) on the silica surface. The silanol groups strongly retain some polar compounds by hydrogen bonding, causing *tailing* (Figure 16-11b) of chromatographic peaks. To reduce the tendency of a stationary phase to bleed from the column at high temperature, the stationary phase can be *bonded* (chemically attached) to the silica surface or *cross-linked* to itself with covalent chemical bridges.

Liquid stationary phases in Table 17-1 have a range of polarities. The choice of liquid phase for a given problem is based on the rule "like dissolves like." Nonpolar columns are usually best for nonpolar solutes, and polar solutes are usually best for polar solutes.

You can see the effects of column polarity on a separation in Figure 17-4. In Figure 17-4a, 10 compounds are eluted nearly in order of increasing boiling point from a nonpolar stationary phase: The higher the vapor pressure, the faster the compound is eluted. In Figure 17-4b, the strongly polar stationary phase strongly retains the polar solutes. The three alcohols (with –OH groups) are the last to be eluted, following the three ketones (with C=O groups), which follow four alkanes (having

Polarity was discussed in Box 16-1.

315

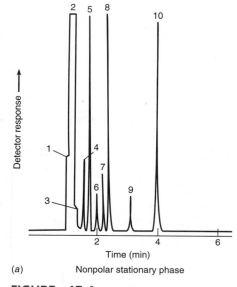

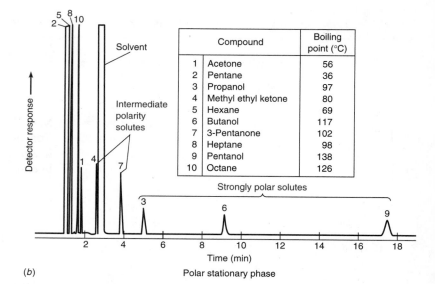

	Compound	Boiling point (°C)
1	Acetone	56
2	Pentane	36
3	Propanol	97
4	Methyl ethyl ketone	80
5	Hexane	69
6	Butanol	117
7	3-Pentanone	102
8	Heptane	98
9	Pentanol	138
10	Octane	126

FIGURE 17-4 Separation of compounds on (a) nonpolar poly(dimethylsiloxane) and (b) strongly polar polyethylene glycol stationary phases (1 μm thick) in open tubular columns (0.32-mm diameter × 30 m long) at 70°C.

Molecular sieves are also widely used to dry gases because sieves strongly retain water. Sieves are regenerated (freed of water) by heating to 300°C in vacuum.

Raising column temperature

- decreases retention time
- sharpens peaks

We refer to constant-temperature conditions as *isothermal* conditions.

only C–H bonds). Hydrogen bonding between solute and the stationary phase is probably the strongest force causing retention.

Common solid stationary phases include porous carbon (Figure 17-3a) and *molecular sieves,* which are inorganic materials with nanometer-size cavities that retain and separate small molecules such as H_2, O_2, N_2, CO_2, and CH_4. Figure 17-5 compares the separation of gases by molecular sieves in an open tubular column and a *packed column* filled with particles of the solid stationary phase. Open tubular columns typically give better separations (narrower peaks), but packed columns can handle larger samples. In Figure 17-5 the sample injected into the packed column was 250 times larger than the sample injected into the open tubular column.

Temperature Programming

If you increase the column oven temperature in Figure 17-1, solute vapor pressure increases and retention times decrease. To separate compounds with a wide range of boiling points or polarities, we usually raise the column temperature *during* the separation, a technique called **temperature programming.** Figure 17-6 shows the effect of temperature programming on the separation of nonpolar compounds with a range of boiling points from 69° for C_6H_{14} to 356°C for $C_{21}H_{44}$. At a constant column temperature of 150°C, the lower boiling compounds emerge close together, and the higher boiling compounds may not be eluted from the column. If the temperature is programmed to increase from 50° to 250°C, all compounds are eluted and the separation of peaks is fairly uniform.

Carrier Gas

Figure 16-8 showed that H_2 and He give better resolution (smaller plate height) than N_2 at high flow rate. This difference exists because solutes diffuse more rapidly through H_2 and He and therefore can equilibrate between the mobile and stationary phases more rapidly than they can in N_2. H_2 is explosive when mixed with air, so He is commonly used.

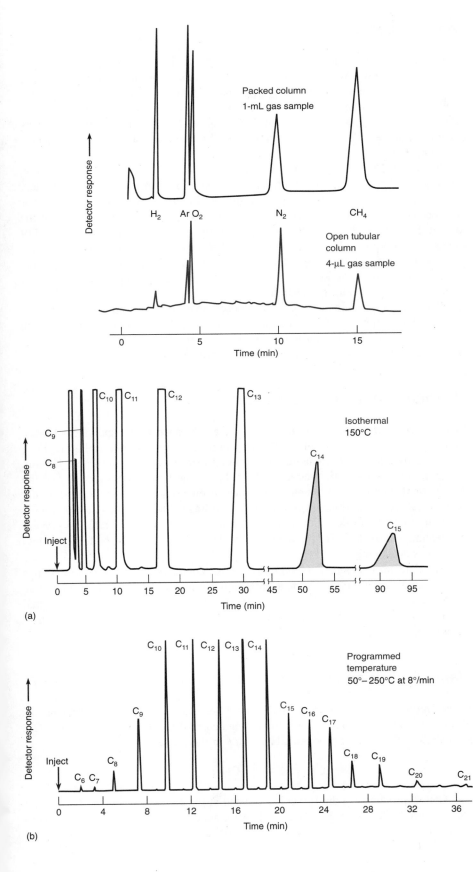

FIGURE 17-5 Gas chromatography with 5A molecular sieves. Upper chromatogram was obtained with a packed column (3.2-mm diameter × 4.6 m long) at 40°C, using 1 mL of sample containing 2 ppm (by volume) of each analyte in helium. Lower chromatogram was obtained with an open tubular column (0.32-mm diameter × 30 m long) at 30°C using 4 µL of the same sample.

FIGURE 17-6 Comparison of (*a*) isothermal and (*b*) programmed temperature chromatography of linear alkanes through a packed column with a nonpolar stationary phase. Detector sensitivity is 16 times greater in (*a*) than in (*b*).

Sample Injection

In Figure 17-1, liquid samples are injected through a rubber septum into a heated glass port. Carrier gas sweeps the vaporized sample from the port into the chromatography column. A complete injection usually contains too much material for a capillary column. In *split injection,* only 0.1–10% of the injected sample reaches the column. The remainder is blown out to waste. If the entire sample is not vaporized during injection, however, the higher boiling components will not be completely injected and there will be errors in quantitative analysis.

For quantitative analysis and for analysis of trace components of a mixture, *splitless injection* is appropriate. (*Trace components* are those present at extremely low concentrations.) For this purpose, a dilute sample in a low-boiling solvent is injected while the column temperature is 25°C lower than the boiling point of the solvent. Solvent condenses at the beginning of the column and traps a thin band of solute. After purging additional vapors from the injection port, the column temperature is raised and chromatography is begun. In splitless injection, ~80% of the sample is applied to the column, and little fractionation (selective evaporation of components) occurs during injection.

For sensitive compounds that decompose above their boiling temperature, we use *on-column injection* of solution directly into the column, without going through a hot port. The initial column temperature is low enough to condense solutes in a narrow zone. Warming the column initiates chromatography.

Injection into open tubular columns:

split: routine method
splitless: best for quantitative analysis
on-column: best for thermally unstable solutes

Thermal Conductivity Detector

Thermal conductivity measures the ability of a substance to transport heat from a hot region to a cold region. In the **thermal conductivity detector** in Figure 17-7, gas emerging from the chromatography column flows over a hot tungsten-rhenium filament. When solute emerges from the column, the thermal conductivity of the gas stream decreases, the filament gets hotter, its electrical resistance increases, and the voltage across the filament increases. The voltage change is the detector signal. Thermal conductivity detection is more sensitive at a lower flow rate. To prevent overheating and oxidation of the filament, the detector should never be left on unless carrier gas is flowing.

The detector responds to *changes* in thermal conductivity, so the conductivities of solute and carrier gas should be as different as possible. Because H_2 and He have the highest thermal conductivity, these two are the carriers of choice for thermal conductivity detection. A thermal conductivity detector responds to every substance except the carrier gas.

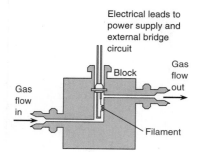

FIGURE 17-7 Thermal conductivity detector.

Electrical leads to power supply and external bridge circuit

Block

Gas flow out

Gas flow in

Filament

Flame Ionization Detector

In the **flame ionization detector** in Figure 17-8, eluate is burned in a mixture of H_2 and air. Carbon atoms (except carbonyl and carboxyl carbon atoms) produce CH radicals, which go on to produce CHO^+ ions in the flame:

$$CH + O \longrightarrow CHO^+ + e^-$$

The CHO^+ produced in the flame is collected at the cathode above the flame. The electric current that flows between the detector anode and cathode is proportional to the number of CHO^+ ions. Only about 1 in 10^5 carbon atoms produces an ion, but

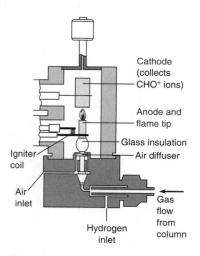

FIGURE 17-8 Flame ionization detector.

Cathode (collects CHO^+ ions)

Anode and flame tip

Glass insulation
Air diffuser

Igniter coil

Air inlet

Gas flow from column

Hydrogen inlet

ion production is strictly proportional to the number of susceptible carbon atoms entering the flame. The flame ionization detector is insensitive to O_2, CO_2, H_2O, and NH_3.

Response to organic compounds is directly proportional to solute mass over seven orders of magnitude. The detection limit is 100 times smaller than that of the thermal conductivity detector and is best with N_2 carrier gas. For open tubular columns, chromatography is conducted with H_2 or He at a low flow rate and N_2 *makeup gas* is added to the stream before it enters the detector. The makeup gas provides the higher flow rate needed by the detector and improves sensitivity.

Other Detectors

The *electron capture detector* is particularly sensitive to halogen-containing molecules, such as pesticides, but relatively insensitive to hydrocarbons, alcohols, and ketones. Gas entering the detector is ionized by high-energy electrons ("β-rays") emitted from a foil containing radioactive ^{63}Ni. Electrons liberated from the gas are attracted to an anode, producing a small, steady current. When analyte molecules with a high electron affinity enter the detector, they capture some of the electrons and reduce the current. The decrease in electric current is the analytical signal. Electron capture detectors are sensitive to as little as 5 fg (femtograms, 10^{-15} g) of analyte eluted per second. Pesticide residues in soil or produce can be detected at extremely low levels.

A *flame photometric detector* measures optical emission from phosphorus and sulfur compounds. When eluate passes through a H_2-air flame, excited sulfur- and phosphorus-containing species emit characteristic radiation, which is detected with a photomultiplier tube. The radiant emission is proportional to analyte concentration.

The *alkali flame detector* is a modified flame ionization detector that is selectively sensitive to phosphorus and nitrogen. It is especially important for drug analysis. When ions produced by these elements contact a Rb_2SO_4-containing glass bead at the burner tip, they create the electric current that is measured. N_2 carrier gas cannot be used for nitrogen-containing samples.

A *sulfur chemiluminescence detector* mixes the exhaust from a flame ionization detector with O_3 to form an excited state of SO_2 that emits light, which is detected. The *mass spectrometer* (Box 16-2) is a sensitive detector that provides qualitative information about the structure of the analyte, as well as a quantitative measure.

Gas chromatography detectors:

thermal conductivity: responds to everything
flame ionization: responds to compounds with C–H
electron capture: halogens, conjugated C=O, –C≡N, –NO_2
flame photometer: P and S
alkali flame: P and N
sulfur chemiluminescence: S
mass spectrometer: responds to everything

Ask Yourself

17-A. (a) What is the advantage of temperature programming in gas chromatography?

(b) What is the advantage of an open tubular column over a packed column? What is the advantage of a packed column over an open tubular column?

(c) Why do H_2 and He allow more rapid linear flow rates in gas chromatography than does N_2, without loss of column efficiency (Figure 16-8)?

(d) When would you use split, splitless, or on-column injection in gas chromatography?

(e) To which kinds of analytes do the following gas chromatography detectors respond: **(i)** thermal conductivity, **(ii)** flame ionization, **(iii)** electron capture, **(iv)** flame photometer, **(v)** alkali flame, **(vi)** sulfur chemiluminescence, and **(viii)** mass spectrometer?

17-2 *Classical Liquid Chromatography*

Modern chromatography evolved from the experiment in Figure 16-1 in which sample is applied to the top of an open, gravity-feed column containing stationary phase. The next section describes high-performance liquid chromatography, which uses closed columns under high pressure and is probably the most common form of chromatography practiced today. However, open columns are used for preparative separations in biochemistry and chemical synthesis.

There is an art to pouring uniform columns, applying samples evenly, and obtaining symmetric elution bands. Stationary solid phase is normally poured into a column by first making a *slurry* (a mixture of solid and liquid) and pouring the slurry gently down the wall of the column. Try to avoid creating distinct layers, which form when some of the slurry is allowed to settle before more is poured in. Don't drain solvent below the top of the stationary phase, because this creates air spaces and irregular flow patterns. Solvent should be directed gently down the wall of the column. In *no* case should the solvent be allowed to dig a channel into the stationary phase. Maximum resolution demands a slow flow rate.

The Stationary Phase

For adsorption chromatography, *silica gel* ($SiO_2 \cdot xH_2O$, also called silicic acid) is a common stationary phase whose active adsorption sites are Si–O–H (silanol) groups, which are slowly deactivated by adsorption of water from the air. Silica is activated by heating to 200°C to drive off the water. *Alumina* ($Al_2O_3 \cdot xH_2O$) is the other most common adsorbent. Preparative chromatography in the biochemistry lab is most often based on molecular exclusion and ion exchange, which are described in the next chapter.

TABLE 17-2 Eluotropic series

Solvent	Eluent strength	Solvent	Elluent strength
Fluoroalkanes	−0.25	Tetrahydrofuran	0.45
n-Pentane	0.00	2-Butanone	0.51
i-Octane	0.01	Acetone	0.56
n-Decane	0.04	Dioxane	0.56
Cyclohexane	0.04	Ethyl acetate	0.58
Carbon disulfide	0.15	1-Pentanol	0.61
Carbon tetrachloride	0.18	Dimethyl sulfoxide	0.62
1-Chloropentane	0.26	Nitromethane	0.64
i-Propyl ether	0.28	Acetonitrile	0.65
Toluene	0.29	Pyridine	0.71
Chlorobenzene	0.30	2-Propanol	0.82
Benzene	0.32	Ethanol	0.88
Diethyl ether	0.38	Methanol	0.95
Chloroform	0.40	1,2-Ethanediol	1.11
Dichloromethane	0.42	Acetic acid	Large

Solvents

In adsorption chromatography, solvent competes with the solute for adsorption sites on the stationary phase. *The relative abilities of different solvents to elute a given solute from the column are nearly independent of the nature of the solute.* Elution can be described as a displacement of solute from the adsorbent by solvent.

An *eluotropic series* ranks solvents by their relative abilities to displace solutes from a given adsorbent. **Eluent strength** is a measure of the solvent adsorption energy, with the value for pentane defined as 0. The more polar the solvent, the greater is its eluent strength. Table 17-2 gives eluent strengths for solvents on alumina, but a similar ranking applies to silica gel. The greater the eluent strength, the more rapidly will solutes be eluted from the column.

A *gradient* (steady change) of eluent strength is used for many separations. First, weakly retained solutes are eluted with a solvent of low eluent strength. Then a second solvent is mixed with the first, either in discrete steps or continuously, to increase eluent strength and elute more strongly adsorbed solutes. A small amount of polar solvent markedly increases the eluent strength of a nonpolar solvent.

Gradient elution in liquid chromatography is analogous to temperature programming in gas chromatography. Increased eluent strength is required to elute more strongly retained solutes.

Ask Yourself

17-B. Why are the relative eluent strengths of solvents in adsorption chromatography fairly independent of solute?

17-3 High-Performance Liquid Chromatography

High-performance liquid chromatography (HPLC) uses high pressure to force eluent through a closed column packed with micron-size particles that provide exquisite separations. The analytical HPLC equipment in Figure 17-9 uses columns with

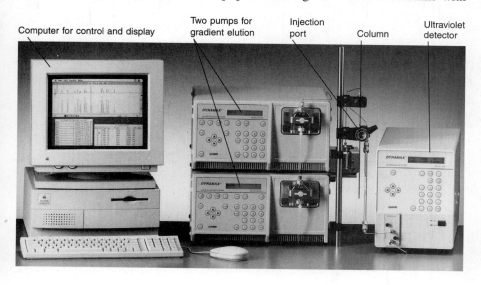

Computer for control and display | Two pumps for gradient elution | Injection port | Column | Ultraviolet detector

FIGURE 17-9 Typical laboratory equipment for high-performance liquid chromatography (HPLC).

Question According to the scaling rules in Section 16-2, if the column diameter is increased from 4 mm to 100 mm and all else remains the same, how much larger can the sample be to achieve the same resolution?

diameters of 1–5 mm and lengths of 5–30 cm, yielding 50 000 to 100 000 plates per meter. Essential components include a solvent delivery system, a sample injection valve, a detector, and a computer to display results. The industrial preparative column in Figure 17-10, which is 100 mm in diameter and 50 cm long, can handle up to 1 kg of sample.

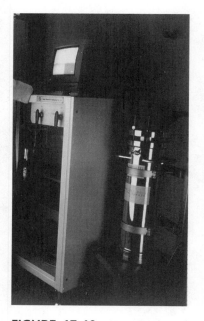

FIGURE 17-10 Up to 1 kg of sample can be applied to an industrial preparative HPLC column that is 100 mm in diameter.

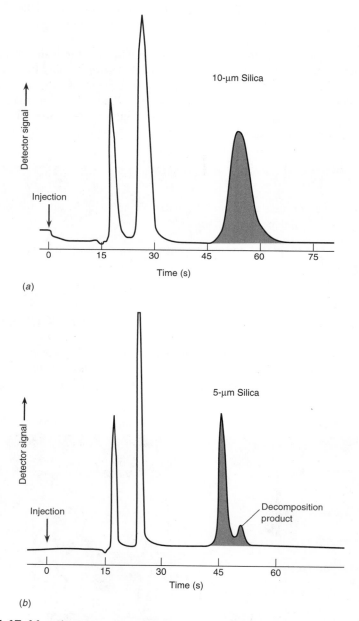

FIGURE 17-11 Chromatograms of the same sample run on columns packed with (a) 10-μm and (b) 5-μm particle diameter silica.

Figures 17-11 and 17-12 show that resolution increases when the stationary phase particle size decreases. Notice how much sharper the peaks become in Figure 17-11 and how a decomposition product is resolved from the slow-moving component. The penalty for using fine particles is resistance to solvent flow. Pressures of ~7–40 MPa (70–400 atm) are required for flow rates of ~0.5–5 mL/min. Smaller particles of stationary phase improve resolution by allowing solute to diffuse faster between the stationary and mobile phases. Also, smaller particles reduce the magnitude of irregular flow path (Figure 16-9).

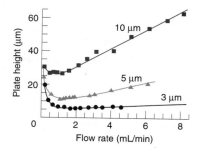

FIGURE 17-12 Plate height as a function of flow rate for stationary phase particle sizes of 10, 5, and 3 μm. The smaller the plate height, the sharper are the chromatographic peaks.

Stationary Phase

Normal-phase chromatography uses a polar stationary phase and a less polar solvent. *Eluent strength is increased by adding a more polar solvent.* **Reverse-phase chromatography** is the more common scheme in which the stationary phase is nonpolar or weakly polar and the solvent is more polar. *Eluent strength is increased by adding a less polar solvent.* Reverse-phase chromatography eliminates tailing arising from adsorption of polar compounds by polar packings (Figure 16-11b). Reverse-phase chromatography is also relatively insensitive to polar impurities (such as water) in the eluent.

Microporous particles of silica with diameters of 3–10 μm are the most common solid stationary phase support. These particles are permeable to solvent and have a surface area up to 500 m^2 per gram of silica. Adsorption chromatography occurs directly on the silica surface.

More commonly, liquid-liquid partition chromatography is conducted with a **bonded stationary phase** covalently attached to silanol groups on the silica surface:

Normal-phase chromatography:
  polar stationary phase and less polar solvent
Reverse-phase chromatography:
  low-polarity stationary phase and polar solvent

Bonded polar phases		Bonded nonpolar phases	
R = (CH$_2$)$_3$NH$_2$	Amino	R = (CH$_2$)$_{17}$CH$_3$	Octadecyl
R = (CH$_2$)$_3$C≡N	Cyano	R = (CH$_2$)$_7$CH$_3$	Octyl
R = (CH$_2$)$_2$OCH$_2$CH(OH)CH$_2$OH	Diol	R = (CH$_2$)$_3$C$_6$H$_5$	Phenyl

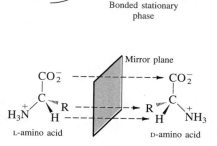

Bonded stationary phase

The octadecyl (C$_{18}$) stationary phase is by far the most common in HPLC. The Si–O–Si bond that attaches the stationary phase to the silica is stable only over the pH range 2–8. Strongly acidic or basic eluents generally cannot be used with silica.

Optical isomers are mirror image compounds such as D- and L-amino acids. Most compounds with four different groups attached to one tetrahedral carbon atom exist in two mirror image (optical) isomers. Solute optical isomers can be separated from each other by chromatography on a stationary phase containing just one optical isomer of the bonded phase. A major push for separating optical isomers has been made by the drug industry, which often needs to separate one isomer that is biologically active from the other, which might be inactive or toxic. Figure 17-13 shows excellent separation of the two optical isomers of the anti-inflammatory drug Naproxen.

Optical isomers

The Column

High-performance liquid chromatography columns are expensive and easily degraded by irreversible adsorption of impurities from samples and solvents. Therefore, the

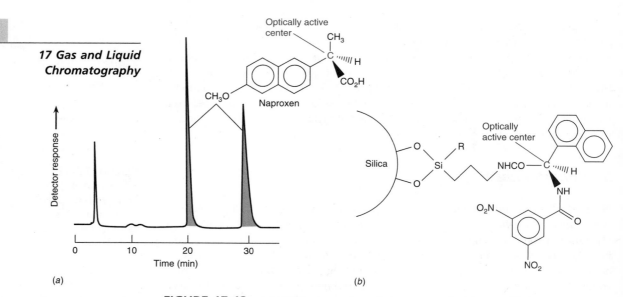

(a)   (b)

FIGURE 17-13 (a) HPLC separation of the two optical isomers (mirror image isomers) of the drug Naproxen eluted with 0.05 M ammonium acetate in methanol. Naproxen is the active ingredient of the anti-inflammatory drug Aleve. (b) Structure of the bonded stationary phase.

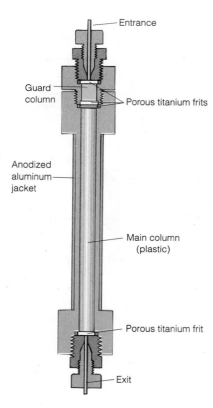

entrance to the main column is protected by a short, disposable **guard column** that contains the same stationary phase as the main column (Figure 17-14). The guard column collects irreversibly adsorbed solutes and is periodically replaced.

Because the column is under high pressure, a special technique is required to inject sample. The *injection valve* in Figure 17-15 has interchangeable steel sample loops that hold fixed volumes from 2 to 1 000 μL. In the load position, a syringe is used to wash and load the loop with fresh sample at atmospheric pressure. When the valve is rotated 60° counterclockwise, the content of the sample loop is injected into the column at high pressure. Samples should be passed through a 0.5- to 2-μm filter prior to injection, to avoid contaminating the column with particles, plugging the tubing, and damaging the pump.

Solvents

Elution with a single solvent or a constant solvent mixture is called **isocratic elution.** If one solvent does not discriminate adequately between the components of a mixture or if the solvent does not provide sufficiently rapid elution of all components, then **gradient elution** can be used. Figure 17-16 shows how increasing the fraction of acentonitrile in a water-acetonitrile mixture elutes solutes that were initially strongly retained by a C_{18}-silica column.

FIGURE 17-14 HPLC column with replaceable guard column to collect irreversibly adsorbed impurities. Titanium frits distribute the liquid evenly over the diameter of the column.

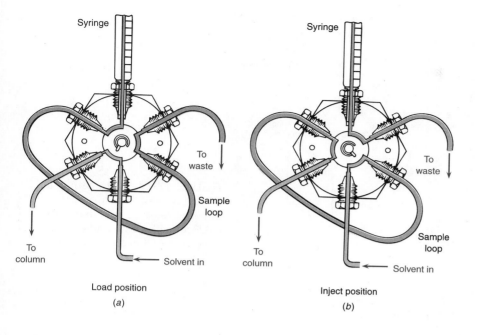

Syringe

To
waste

Sample
loop

To
column

Solvent in

Load position
(a)

Syringe

To
waste

Sample
loop

To
column

Solvent in

Inject position
(b)

FIGURE 17-15 Injection valve for HPLC. Replaceable sample loop comes in various fixed-volume sizes.

Pure HPLC solvents are expensive, and most organic solvents are very expensive to dispose of in an environmentally sound manner. Therefore solvent recycling is practiced. Commercially available equipment discards solvent from a column when it is contaminated with solutes but recycles pure solvent emerging between or after the chromatographic peaks.

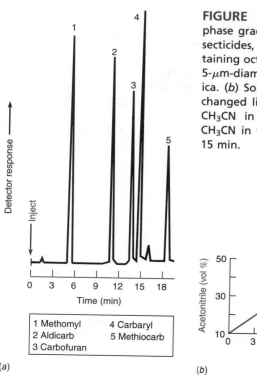

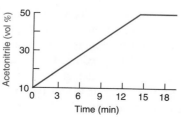

Detector response

Inject

Time (min)

1 Methomyl   4 Carbaryl
2 Aldicarb   5 Methiocarb
3 Carbofuran

(a)

FIGURE 17-16 (a) Reverse-phase gradient separation of insecticides, using a column containing octadecyl (C_{18}) groups on 5-μm-diameter microporous silica. (b) Solvent composition was changed linearly from 10 vol % CH_3CN in water to 50 vol % CH_3CN in water during the first 15 min.

Acetonitrile (vol %)

Time (min)

(b)

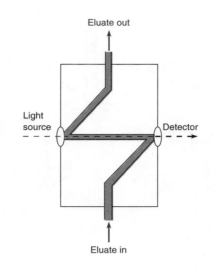

Eluate out

Light source

Detector

Eluate in

FIGURE 17-17 Light path in a micro flow cell of a spectrophotometric detector. Cells that have a 0.5-cm pathlength and contain only 10 μL of liquid are available.

Detectors

An **ultraviolet detector** is most common, with a flow cell such as that in Figure 17-17. Simple systems employ the intense 254-nm emission of a mercury lamp. More versatile instruments use a deuterium or xenon lamp and monochromator, so you can choose the optimum wavelength for your analytes. In some systems, the ultraviolet-visible absorption spectrum of each solute can be recorded as it is eluted.

A **refractive index detector** responds to almost every solute, but its detection limit is about 1 000 times poorer than that of the ultraviolet detector, and it is not useful for gradient elution because the baseline changes as the solvent changes. In general, a solute has a different refractive index (ability to deflect a ray of light) from that of the solvent. The detector measures the deflection of a light ray by the eluate.

An *electrochemical detector* responds to analytes that can be oxidized or reduced (losing or gaining electrons) at an electrode over which the eluate passes. Electric current through the electrode increases in proportion to the concentration of solute in the eluate.

Fluorescence detectors are especially sensitive but respond only to the few analytes that fluoresce. As in Figure 8-5, the fluorescence detector works by irradiating the eluate at one wavelength and monitoring emission at a different (longer) wavelength. The emission intensity is proportional to solute concentration.

Derivatization

Sometimes the analyte we wish to measure has no properties that make it easy to detect as it emerges from a chromatography column. To make analytes more sensitive to detection, we sometimes attach ultraviolet-absorbing, fluorescent, or electroactive groups (which can be oxidized or reduced at an electrode) to the analyte. The process of attaching an easily detected group to the analyte is called **derivatization,** and it can be performed prior to separation or later, by addition of reagents to the eluate between the column and the detector.

Figure 17-18 shows an example in which chromatography was used to monitor foods for highly toxic, carcinogenic (cancer-causing) aflatoxins produced by fungi that can grow in the food. Four major aflatoxins are designated B_1, B_2, G_1, and G_2. Of these, B_2 and G_2 are highly fluorescent and can be observed at parts-per-billion levels with a fluorescence detector (Figure 17-18b). To convert the weakly fluorescent aflatoxins B_1 and G_1 into highly fluorescent products that are readily observed, eluate is exposed to an ultraviolet lamp prior to reaching the detector. Absorption of ultraviolet radiation converts B_1 and G_1 into highly fluorescent products that are easily observed in Figure 17-18c.

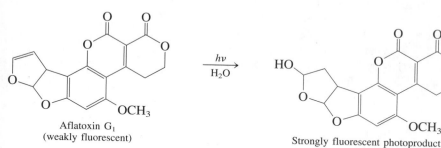

Aflatoxin G_1
(weakly fluorescent)

$\xrightarrow[H_2O]{h\nu}$

Strongly fluorescent photoproduct

Solid-Phase Extraction

Solid-phase extraction uses a chromatographic solid phase in a short, open column to separate one or more analytes from a mixture. For example, C_{18}-silica is used to isolate nonpolar organic substances from aqueous solution.

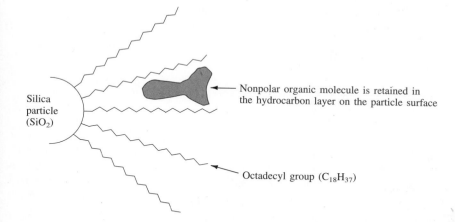

Silica particle (SiO_2)

Nonpolar organic molecule is retained in the hydrocarbon layer on the particle surface

Octadecyl group ($C_{18}H_{37}$)

When urine containing traces of steroid drugs is passed through a short, open column of C_{18}-silica, the steroids are retained in the C_{18} coating. Upon washing the column with a small volume of nonpolar or weakly polar solvent such as hexane or dichloromethane, the steroid is released from the column into the solvent, where it can be analyzed by chromatography.

Sample cleanup refers to the removal of undesirable components of the unknown that might interfere with measurement of the analyte. In the steroid analysis, the polar molecules in urine that are not retained by the C_{18}-silica wash right through the column and are separated from the analyte.

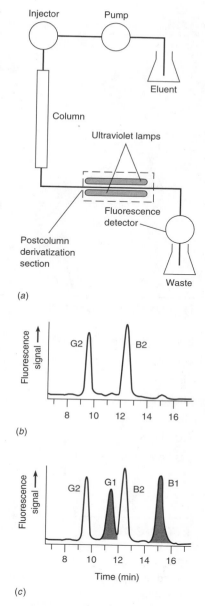

(a)

(b)

Time (min)

(c)

FIGURE 17-18 (a) Arrangement of photochemical reactor for postcolumn photochemical derivatization of aflatoxins. (b) Chromatogram of aflatoxins with ultraviolet lamps turned off. (c) Chromatogram with lamps turned on to convert B_1 and G_1 into fluorescent products.

Ask Yourself

17-C. (a) What is the difference between normal-phase and reverse-phase chromatography?

(b) What is the difference between isocratic and gradient elution?

(c) Why does eluent strength increase in normal-phase chromatography when a more polar solvent is added?

(d) Why does eluent strength increase in reverse-phase chromatography when a less polar solvent is added?

(e) What is the purpose of a guard column?

(f) What is the purpose of solid-phase extraction?

Important Terms

bonded stationary phase

derivatization

dialysis

eluent strength

flame ionization detector

gas chromatography

gradient elution

guard column

high-performance liquid chromatography

isocractic elution

normal-phase chromatography

refractive index detector

reverse-phase chromatography

sample cleanup

solid-phase extraction

temperature programming

thermal conductivity detector

ultraviolet detector

Problems

17-1. (a) Explain the difference between wall-coated, support-coated, and porous-layer open tubular columns for gas chromatography.

(b) What is the advantage of a bonded stationary phase in gas chromatography?

(c) Why do we use a makeup gas for some gas chromatography detectors?

17-2. Explain why plate height increases at **(a)** very low and **(b)** very high flow rates in Figure 17-12. (*Hint:* Refer to Section 16-3.)

17-3. Nonpolar aromatic compounds were separated by HPLC on a bonded phase containing octadecyl groups $[-(CH_2)_{17}CH_3]$ covalently attached to silica particles. The eluent was 65 vol % methanol in water. How would the retention times be affected if 90% methanol were used instead?

17-4. Polar solutes were separated by HPLC, using a bonded phase containing polar diol substituents $[-CH(OH)CH_2OH]$. How would the retention times be affected if the eluent were changed from 40 vol % to 60 vol % acetonitrile in water? Acetonitrile ($CH_3C\equiv N$) is less polar than water.

17-5. Draw the chemical structures of two nonpolar bonded phases and two polar bonded phases in HPLC. Begin with a silicon atom at the surface of a silica particle.

17-6. (a) Why is high pressure needed in HPLC?

(b) Why does the efficiency (decreased plate height) of liquid chromatography increase as the stationary phase particle size is reduced?

17-7. Octanoic acid and 1-aminooctane were separated by HPLC on a bonded phase containing octadecyl groups $[-(CH_2)_{17}CH_3]$. The eluent was 20 vol % methanol in water adjusted to pH 3.0 with HCl.

$$CH_3(CH_2)_6CO_2H \qquad CH_3(CH_2)_7NH_2$$

Octanoic acid   1-Aminooctane

(a) Draw the predominant form (neutral or ionic) of a carboxylic acid and an amine at pH 3.0.

(b) State which compound is expected to be eluted first and why.

17-8. A known mixture of compounds C and D gave the following HPLC results:

Compound	Concentration (mg/mL) in mixture	Peak area (cm^2)
C	1.03	10.86
D	1.16	4.37

A solution was prepared by mixing 12.49 mg of D plus 10.00 mL of unknown containing just C, and diluting to 25.00 mL. Peak areas of 5.97 and 6.38 cm^2 were observed for C and D, respectively. Find the concentration of C (mg/mL) in the unknown. (*Hint:* Review Section 16-5 on internal standards.)

17-9. Spherical, microporous silica particles used in chromatography have a density of 2.20 g/mL, a diameter of 10.0 μm, and a measured surface area of 300 m^2/g.

(a) The volume of a spherical particle is $\frac{4}{3}\pi r^3$, where r is the radius. The mass of the sphere is volume $\times$ density (= mL $\times$ g/mL). How many particles are in 1.00 g of silica?

(b) The surface area of a sphere is $4\pi r^2$. Calculate the surface area of 1.00 g of solid, spherical silica particles.

(c) Comparing the calculated and measured surface area, what can you say about the porosity of the particles?

Measuring the Contents of Single Cells

(a)

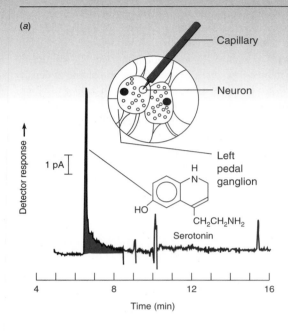

Capillary

Neuron

Left pedal ganglion

Serotonin

(b)

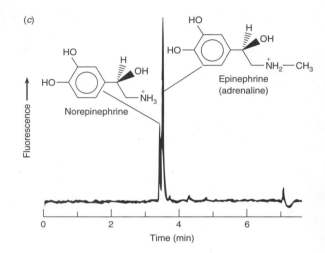

Red blood cell

Capillary

Cell sucked into capillary bursts open and contents begin to migrate in electric field

(c)

Norepinephrine

Epinephrine (adrenaline)

Capillary electrophoresis separates and measures ions on the basis of their different migration rates in an electric field. The tiny amount of sample required for an analysis is in the picoliter to nanoliter (10^{-12} to 10^{-9} L) range. Figure (a) shows a silica capillary with an inner diameter of 5 μm drawn to a fine point and inserted into a giant nerve cell of a pond snail. About 1% of the cell volume is then drawn into the capillary and the contents are separated with a 25-kV electric field and detected with an electrochemical detector. The first peak to be eluted is the neurotransmitter serotonin, whose concentration is 3 μM inside the cell.

Smaller cells, in Figure (b), are studied by sucking an entire cell into the end of the capillary, lysing the cell in place (bursting it open by osmotic pressure), and separating the components with an electric field. Figure (c) shows the measurement of hormones from a single cell of the adrenal gland of a cow. The separated hormones are detected by their fluorescence when stimulated with an ultraviolet laser.

330

Chromatography and Capillary Electrophoresis

Liquid chromatography, discussed in the last chapter, separates solutes by *adsorption* or *partition* mechanisms. Now we describe separations by *ion-exchange, molecular exclusion,* and *affinity* chromatography, which were all illustrated in Figure 16-2. We also consider *capillary electrophoresis,* which separates species on the basis of their different rates of migration in an electric field. Capillary electrophoresis is especially suited for extremely small volume samples. In studies of the enzyme lactate dehydrogenase in single red blood cells, quantities of enzyme as small as 780 molecules can be detected.

Question How many zeptomoles is 780 molecules?

18-1 Ion-Exchange Chromatography

Ion-exchange chromatography is based on the attraction between solute ions and charged sites in the stationary phase (Figure 16-2 and Color Plate 16). **Anion exchangers** have positively charged groups on the stationary phase that attract solute anions. **Cation exchangers** contain negatively charged groups that attract solute cations.

The stationary phase for ion-exchange chromatography is usually a *resin,* such as polystyrene, which consists of amorphous (noncrystalline) particles. Polystyrene is made into a cation exchanger when negative sulfonate ($-SO_3^-$) or carboxylate ($-CO_2^-$) groups are attached to the benzene rings (Figure 18-1). Polystyrene is an

Anion exchangers contain bound *positive* groups.
Cation exchangers contain bound *negative* groups.

331

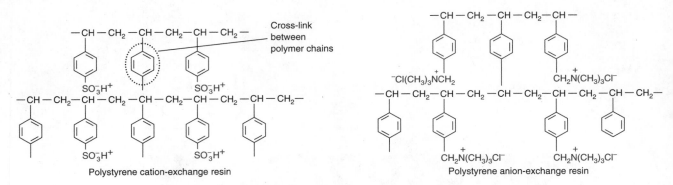

-Cl(CH$_3$)$_3$NCH$_2$ CH$_2$N(CH$_3$)$_3$Cl$^-$

CH$_2$N(CH$_3$)$_3$Cl$^-$ CH$_2$N(CH$_3$)$_3$Cl$^-$

Polystyrene cation-exchange resin   Polystyrene anion-exchange resin

FIGURE 18-1   Structures of polystyrene ion-exchange resins. *Cross-links* are covalent bridges between polymer chains.

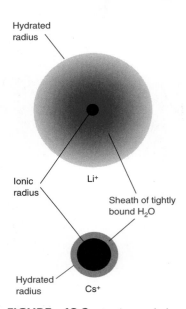

Hydrated radius

Ionic radius   Li$^+$

Sheath of tightly bound H$_2$O

Hydrated radius   Cs$^+$

FIGURE 18-2 Ionic and hydrated radii of Li$^+$ and Cs$^+$. The ionic radius is the size of the bare cation. The hydrated radius includes the effective size of the ion plus its sheath of water molecules, which are tightly bound by electrostatic attraction. Smaller bare ions have larger hydrated radii because they attract water molecules more strongly.

Electrostatic attraction

A large excess of one ion will displace another ion from the resin.

anion exchanger if ammonium groups (–NR$_3^+$) are attached. Cross-links (covalent bonds) between polystyrene chains in Figure 18-1 control the pore sizes into which solutes can diffuse.

Gel particles are softer than resin particles. Cellulose and dextran ion-exchange gels, which are polymers of the sugar glucose, possess larger pore sizes and lower charge densities. Gels are better suited than resins to ion exchange of macromolecules, such as proteins.

Ion-Exchange Selectivity

Consider the competition of Cs$^+$ and Li$^+$ for sites on the cation-exchange resin, R$^-$:

$$R^-Cs^+ + Li^+ \rightleftharpoons R^-Li^+ + Cs^+ \qquad K = \frac{[R^-Li^+][Cs^+]}{[R^-Cs^+][Li^+]} \qquad (18\text{-}1)$$

The equilibrium constant is called the *selectivity coefficient*, because it describes the relative affinities of the resin for Li$^+$ and Cs$^+$. Discrimination between different ions tends to increase with the extent of cross-linking, because the resin pore size shrinks as cross-linking increases.

The **hydrated radius** of an ion is the effective size of the ion plus its tightly bound sheath of water molecules, which are attracted by the positive or negative charge of the ion (Figure 18-2). The large species Li(H$_2$O)$_x^+$ does not have as much access to the resin as the smaller species Cs(H$_2$O)$_y^+$.

More highly charged ions bind more tightly to ion-exchange resins. For ions of the same charge, the larger the hydrated radius, the less tightly the ion is bound. An approximate order of selectivity for some cations is

$$Pu^{4+} \gg La^{3+} > Y^{3+} > Sc^{3+} > Al^{3+} \gg Ba^{2+} > Pb^{2+} > Sr^{2+} >$$
$$Ca^{2+} > Ni^{2+} > Cd^{2+} > Cu^{2+} > Co^{2+} > Zn^{2+} > Mg^{2+} \gg$$
$$Tl^+ > Ag^+ > Cs^+ > Rb^+ > K^+ > NH_4^+ > Na^+ > H^+ > Li^+$$

Reaction 18-1 can be driven in either direction. Washing a column containing Cs$^+$ with a substantial excess of Li$^+$ will replace Cs$^+$ with Li$^+$. Washing a column in the Li$^+$ form with excess Cs$^+$ will convert it to the Cs$^+$ form.

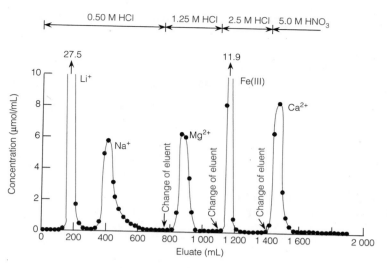

FIGURE 18-3 Separation of Na^+, Mg^{2+}, Ca^{2+}, Fe^{3+}, and Li^+ by using a cation-exchange resin with a gradient of increasing H^+ (shown at top). Low concentrations of H^+ elute Li^+ and Na^+, but high concentrations are required to elute Ca^{2+}. At sufficiently high Cl^- concentration, Fe^{3+} is converted to $FeCl_4^-$, which is not retained by a cation exchanger.

An ion exchanger loaded with one ion will bind a small amount of a different ion nearly quantitatively. A resin loaded with Cs^+ binds small amounts of Li^+ quantitatively, even though the selectivity is greater for Cs^+. The same resin binds large quantities of Ni^{2+} because the selectivity for Ni^{2+} is greater than that for Cs^+. Even though Fe^{3+} is bound more tightly than H^+, Fe^{3+} can be quantitatively removed from the resin by washing with excess acid.

To help separate one ion from another by ion-exchange chromatography, *gradient elution* with increasing ionic strength (ionic concentration) in the eluent is extremely valuable. Consider a column that binds Mg^{2+} more tightly than it binds Na^+. We might separate Mg^{2+} from Na^+ by elution with H^+, which is less tightly bound than either Mg^{2+} or Na^+. As the concentration of H^+ in the eluent is increased, Na^+ is displaced and moves down the column. At a still higher concentration of H^+, Mg^{2+} is also eluted (Figure 18-3).

"Quantitative" is chemists' jargon for "complete."

An ionic strength gradient in ion-exchange chromatography is analogous to a solvent gradient in HPLC or a temperature gradient in gas chromatography.

What Is Deionized Water?

Deionized water is prepared by passing water through an anion-exchange resin loaded with OH^- and a cation-exchange resin loaded with H^+. Suppose, for example, that $Cu(NO_3)_2$ is present in the water. The cation-exchange resin binds Cu^{2+} and replaces it with $2H^+$. The anion-exchange resin binds NO_3^- and replaces it with OH^-. The H^+ and OH^- combine, so the eluate is pure water:

$$\left. \begin{array}{l} Cu^{2+} \xrightarrow{\ H^+ \text{ ion exchange}\ } 2H^+ \\ 2NO_3^- \xrightarrow{\ OH^- \text{ ion exchange}\ } 2OH^- \end{array} \right\} \longrightarrow \text{pure } H_2O$$

Home water softeners use ion exchange to remove Ca^{2+} and Mg^{2+} from "hard" water (Box 15-2).

Charge is conserved during ion exchange. One Cu^{2+} displaces $2H^+$ from a cation-exchange column. It takes $3H^+$ to displace one Fe^{3+}. One SO_4^{2-} displaces $2OH^-$ from an anion-exchange column.

Preconcentration

Measuring extremely low levels of analyte is called **trace analysis.** Trace analysis is especially important for environmental problems in which low concentrations of substances, such as mercury in fish, can become concentrated over many years in people who eat much fish. For trace analysis, analyte concentration may be so low that it cannot be measured without **preconcentration,** in which analyte is brought to a higher concentration prior to analysis.

Metals in natural waters can be preconcentrated with a cation-exchange column:

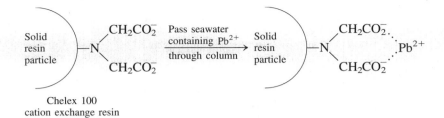

Chelex 100
cation exchange resin

When a large volume of water is passed through a small volume of resin, the cations are concentrated into the small column. The cations can then be displaced into a small volume of solution by eluting the column with concentrated acid:

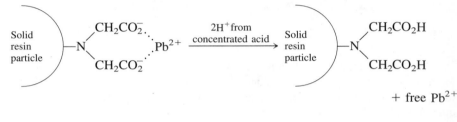

$$+ \text{ free } Pb^{2+}$$

When lead in seawater was preconcentrated by a factor of 50, the detection limit was reduced from 15 ng of lead per liter of seawater down to 0.3 ng/L. How many grams is 1 ng?

Ask Yourself

18-A. (a) What is deionized water? What kind of impurities are not removed by deionization?

(b) Why is gradient elution used in Figure 18-3?

(c) Based on Figure 18-3, which do you think has a smaller hydrated radius: Ca^{2+} or Mg^{2+}? Explain your reasoning.

18-2 | Ion Chromatography

Ion chromatography is a high-performance version of ion-exchange chromatography, with a key modification that removes eluent ions before detecting analyte ions. Ion chromatography has become the method of choice for anion analysis. It is used in the semiconductor industry to monitor anions and cations at 0.1 ppb levels in deionized water. Figure 18-4 shows an example of anion chromatography in environmental analysis.

In ion chromatography, anions are separated by ion exchange and detected by their electrical conductivity. The conductivity of the electrolyte in the eluent is ordinarily high enough to make it difficult or impossible to detect the conductivity change when analyte ions are eluted. Therefore, the key feature of *suppressed-ion* chromatography is removal of unwanted electrolyte prior to conductivity measurement.

The separator column separates the analytes, and the suppressor replaces the ionic eluent with a nonionic species.

Consider the sample containing KNO_3 and $CaSO_4$ shown in Figure 18-5, which is injected into the *separator column* (an anion-exchange column in the carbonate

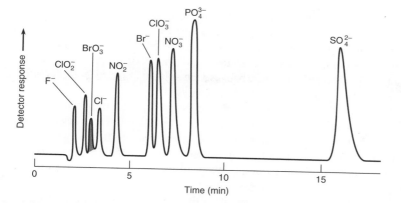

FIGURE 18-4 Converting one human hazard into another. Chlorination of drinking water converts some organic compounds into potential carcinogens, such as CHCl$_3$. To reduce this risk, ozone (O$_3$) has replaced Cl$_2$ in some municipal purification systems. Unfortunately, O$_3$ converts bromide (Br$^-$) into bromate (BrO$_3^-$), another carcinogen that must be monitored. The figure shows an anion chromatographic separation of ions found in drinking water. With preconcentration of the water, the detection limit for bromate is ~2 ppb.

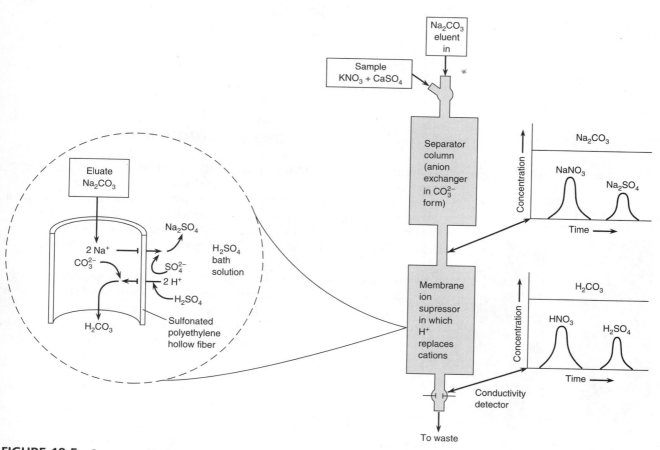

FIGURE 18-5 Suppressed-ion anion chromatography.

form) and eluted with Na_2CO_3. NO_3^- and SO_4^{2-} equilibrate with the resin and are slowly displaced by the CO_3^{2-} eluent. K^+ and Ca^{2+} cations are not retained and simply wash through. After a period of time, $NaNO_3$ and Na_2SO_4 are eluted from the separator column, as shown in the upper graph of Figure 18-5. These species are not easily detected, however, because the solvent contains Na_2CO_3, whose electrical conductivity obscures that of the analyte.

To remedy this problem, the solution is next passed through a *membrane ion suppressor* that replaces cations with H^+. The suppressor is a narrow channel enclosed by a thin, semipermeable cation-exchange membrane. Sulfonate groups ($-SO_3^-$) in the membrane repel anions but allow cations to pass freely. The outside of the membrane is bathed in H_2SO_4 solution. When $NaNO_3$ and Na_2CO_3 from the separator column pass through the suppressor, Na^+ is replaced by H^+, making a solution of HNO_3 and H_2CO_3. Na_2SO_4 formed in the outside bath is washed away. In the absence of analyte, only H_2CO_3, which has very low conductivity, emerges from the suppressor. When analyte is present, HNO_3 and H_2SO_4 with high conductivity are produced and detected.

Ask Yourself

18-B. What are the purposes of the separator column and the suppressor in suppressed-ion chromatography?

18-3 *Molecular Exclusion Chromatography*

In **molecular exclusion chromatography** (also called *gel filtration* or *gel permeation chromatography*), molecules are separated according to their size. Small molecules enter the small pores in the stationary phase, but large molecules do not (Figure 16-2). Because small molecules must pass through an effectively larger volume in the column, large molecules are eluted first (Figure 18-6). This technique is widely used in biochemistry and polymer chemistry to purify macromolecules and to measure molecular weights (Figure 18-7).

In molecular exclusion chromatography, the volume of mobile phase (the solvent) in the column outside the stationary phase is called the *void volume, V_0*. Large molecules that are excluded from the stationary phase are eluted in the void volume. Void volume is measured by passing through the column a large molecule that we know is too large to enter the pores. Blue Dextran, a dye of molecular weight 2×10^6, is commonly used.

Large molecules pass through the column faster than small molecules.

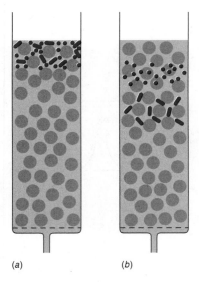

(a)   *(b)*

FIGURE 18-6 (a) A mixture of large and small molecules is applied to the top of a molecular exclusion chromatography column. (b) Large molecules cannot penetrate the pores of the stationary phase, but small molecules can. Therefore less of the volume is available to large molecules and they move down the column faster.

Molecular Weight Determination

Retention volume is the volume of mobile phase required to elute a particular solute from the column. Each stationary phase has a range over which there is a logarithmic relation between molecular weight and retention volume. We can estimate the molecular weight of an unknown by comparing its retention volume with those of standards. For proteins, it is important to use eluent with a high enough ionic strength (such as 0.05 M NaCl) to eliminate electrostatic adsorption of solute by occasional charged sites on the gel.

18-3 Molecular Exclusion Chromatography

1 Glutamate dehydrogenase (MW 290 000)
2 Lactate dehydrogenase (MW 140 000)
3 Enolase kinase (MW 67 000)
4 Adenylate kinase (MW 32 000)
5 Cytochrome *c* (MW 12 400)

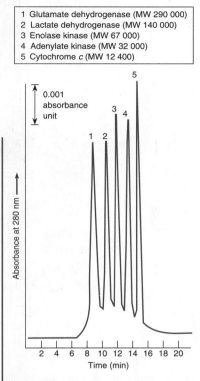

FIGURE 18-7 Separation of proteins by molecular exclusion chromatography using a TSK 3000SW HPLC column. The highest molecular weight molecules are eluted first.

EXAMPLE	Molecular Weight Determination by Gel Filtration

The proteins below were chromatographed on a gel-filtration column and retention volumes (V_r) were measured. Estimate the molecular weight of the unknown.

Compound	V_r (mL)	Molecular weight	Log (molecular weight)
Blue Dextran 2000	17.7	2×10^6	6.301
Aldolase	35.6	158 000	5.199
Catalase	32.3	210 000	5.322
Ferritin	28.6	440 000	5.643
Thyroglobulin	25.1	669 000	5.825
Unknown	30.3	?	

SOLUTION Figure 18-8 plots V_r versus log (MW). The least-squares fit to the five calibration standards is shown. Putting the retention volume of unknown protein into the least-squares equation allows us to solve for the molecular weight of the unknown:

$$V_r \text{ (mL)} = -15.75 \,(\log (\text{MW})) + 117.0$$
$$30.3 = -15.75 \,(\log (\text{MW})) + 117.0$$
$$\Rightarrow \log (\text{MW}) = 5.505 \Rightarrow \text{MW} = 10^{5.505} = 320\,000$$

Ask Yourself

18-C. A gel-filtration column has a radius (r) of 0.80 cm and a length (l) of 20.0 cm.

(a) Calculate the total volume of the column, which is equal to $\pi r^2 l$.

(b) Blue Dextran was eluted in a volume of 18.2 mL. What volume is occupied by the stationary phase plus the solvent inside the pores of the stationary phase?

(c) Suppose that the pores occupy 60.0% of the stationary-phase volume. Over what volume range (*x* mL for the largest molecules to *y* mL for the smallest molecules) are all solutes expected to be eluted?

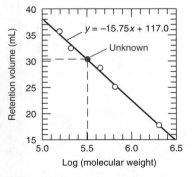

FIGURE 18-8 Calibration curve used to estimate the molecular weight of an unknown protein by molecular exclusion chromatography.

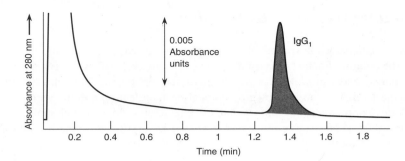

FIGURE 18-9 Purification of monoclonal antibody IgG by affinity chromatography on a column (4.6 mm in diameter × 5 cm long) containing protein A covalently attached to a polymer support. Other proteins in the sample are eluted from 0 to 0.3 min at pH 7.6. When the eluent pH is lowered to 2.6, IgG is freed from protein A and emerges from the column.

18-4 *Affinity Chromatography*

Affinity chromatography is used to isolate a single compound from a complex mixture. The technique is based on specific binding of that one compound to the stationary phase (Figure 16-2). When sample is passed through the column, only one solute is bound. After everything else has washed through, the one adhering solute is eluted by changing conditions such as pH or ionic strength to weaken its binding. Affinity chromatography is especially applicable in biochemistry and is based on specific interactions between enzymes and substrates, antibodies and antigens, or receptors and hormones.

Figure 18-9 shows the isolation of the protein immunoglobulin G (IgG) by affinity chromatography on a column containing covalently bound *protein A*. Protein A binds to one specific region of IgG at pH >7.2. When a crude mixture containing IgG and other proteins was passed through the column at pH 7.6, everything except IgG was eluted within 0.3 min. At 1 min, the eluent pH was lowered to 2.6 and IgG was cleanly eluted at 1.3 min.

18-5 *What Is Capillary Electrophoresis?*

Cations are attracted to the negative terminal (the cathode).
Anions are attracted to the positive terminal (the anode).

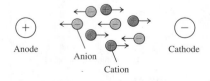

Background electrolyte, also called *run buffer,* is the solution in the electrode reservoirs. It controls pH and ionic composition in the capillary.

Electrophoresis is the migration of ions in an electric field. Anions are attracted to the anode and cations are attracted to the cathode. Different ions migrate at different speeds. **Capillary electrophoresis** is an extremely high resolution separation technique conducted with solutions of ions in a narrow capillary tube. As we shall see shortly, a clever modification of the technique allows us to separate neutral analytes as well as ions. Capillary electrophoresis applies with equal ease to the separation of macromolecules, such as proteins and DNA (deoxyribonucleic acid), and to the separation of small species such as Na^+ and benzene. The opening of this chapter showed how capillary electrophoresis could analyze the contents of a single cell.

The typical experiment in Figure 18-10 features a fused silica (SiO_2) capillary that is 50 cm long and has an inner diameter of 25–75 μm. The capillary tube is immersed in *background electrolyte* solution at each end. At the start of the exper-

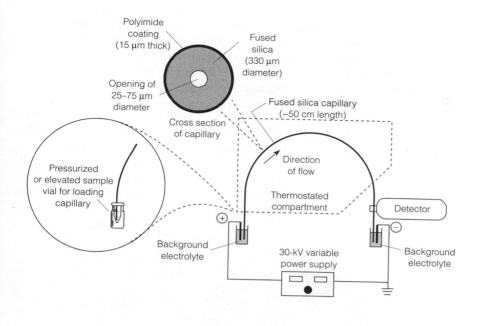

FIGURE 18-10 Capillary electrophoresis. Sample is injected by placing the capillary in the sample vial and elevating the vial or applying pressure to the vial or applying suction to the outlet of the capillary.

iment, one end of the capillary is dipped into a sample vial and ~10 nL (nanoliters, 10^{-9} L) of sample is introduced by raising the vial to siphon liquid into the capillary. After the capillary is placed back into the electrolyte solution, 20–30 kV is applied to the electrodes to cause ions in the capillary to migrate. Different ions migrate at different speeds, so they separate from one another as they travel through the capillary. Solutes are detected inside the capillary near the far end with an ultraviolet absorbance monitor (or some other detector).

Figure 18-11 shows the analysis of nanoliter quantities of surface fluid from the airway (trachea) of a rat. A layer of surface fluid propelled by tiny cilia (hairs) lining the trachea ejects mucus and particles from the lungs, thereby helping to prevent infection. To analyze ions in the fluid, a plastic capillary with an internal diameter of 280 μm was inserted into the trachea of a sedated rat and 50–200 nL of fluid was collected by *capillary action* in 2 min. A fraction of the collected fluid was then transferred to an even smaller silica capillary for electrophoresis. Comparison of the peak areas in Figure 18-11 with peaks from standard solutions gave the cation concentrations.

Capillary electrophoresis is not as sensitive as ion chromatography, but electrophoresis is more sensitive than ion-selective electrodes. For example, the working range for Cl^- analysis is 5 μg/L to 50 mg/L for ion chromatography, 50 μg/L to 50 mg/L for electrophoresis, and 1 000 μg/L to 6 000 mg/L for ion-selective electrodes. In each case, the lower number is the detection limit, defined as the concentration of analyte at which the signal is three times greater than the background noise. The upper limit is the highest concentration at which the detector response

The greater the charge on the ion, the faster it migrates in the electric field. The greater the size of the molecule, the slower it migrates.

Capillary action is based on the attraction between a liquid and the inside wall of a capillary tube. If you dip a glass capillary into an aqueous solution (such as a drop of blood), liquid is drawn spontaneously into the capillary.

FIGURE 18-11 Cation analysis of airway surface fluid from a rat by capillary electrophoresis at 20 kV. Multiple analyses were obtained from a single 100-nL fluid sample. The *indirect spectrophotometric detection* is described later in this chapter. A graph of detector response versus time in capillary electrophoresis is called an *electropherogram*. (In chromatography, we call the same graph a *chromatogram*.)

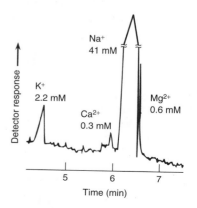

is still proportional to analyte concentration. The concentration range for capillary electrophoresis is intermediate, but the required volume of sample is smallest.

Capillary electrophoresis can provide extremely narrow peaks. Section 16-3 explained why bands are broadened during chromatography. The three mechanisms are longitudinal diffusion, the finite rate of mass transfer between the stationary and mobile phases, and multiple flow paths around solid particles. An open tubular column in chromatography or electrophoresis reduces peak broadening (relative to that of a packed column) by eliminating multiple flow paths. Capillary electrophoresis further reduces broadening by eliminating the mass transfer problem because *there is no stationary phase.* The only source of broadening under ideal conditions is longitudinal diffusion of solute as it migrates through the capillary. Capillary electrophoresis routinely achieves 50 000 to 500 000 theoretical plates, which is an order of magnitude higher than chromatography.

Ask Yourself

18-D. Capillary electrophoresis is noteworthy for analyzing very small volumes of sample and for producing very high resolution separations.

   (a) A typical injected sample occupies a 5-mm length of the capillary. What volume of sample is this if the inside diameter of the capillary is 25 μm? 50 μm? (The volume of a cylinder of radius r is $\pi r^2 \times$ length.)

   (b) Which mechanisms of peak broadening that operate in chromatography are absent in capillary electrophoresis?

18-6 *How Capillary Electrophoresis Works*

Two processes operate in capillary electrophoresis:

- *electrophoresis:* migration of cations to the cathode and anions to the anode
- *electroosmosis:* migration of the entire fluid toward the cathode

Capillary electrophoresis involves two simultaneous processes called *electrophoresis* and *electroosmosis.* Electrophoresis is the migration of ions in an electric field. Electroosmosis pumps the entire solution through the capillary from the anode toward the cathode. Superimposed on this one-way flow, cations are attracted to the cathode and anions are attracted to the anode. In Figure 18-10, cations swim from the injection end at the left toward the detection end at the right. Anions swim toward the left in Figure 18-10. Both cations and anions are swept from left to right by electroosmosis. Cations arrive at the detector before anions. Neutral molecules swept along by electroosmosis arrive at the detector after the cations and before the anions.

Electroosmosis

Electroosmosis is a pumping action caused by the applied electric field that propels the fluid inside a fused silica capillary from the anode toward the cathode. To understand why electroosmosis occurs, we need to consider what happens at the inside wall of the capillary. The wall is covered with silanol (Si–OH) groups that are negatively charged (Si–O⁻) above pH 2. Figure 18-12a shows that the capillary wall and the solution immediately adjacent to the wall form an *electric double layer.* The

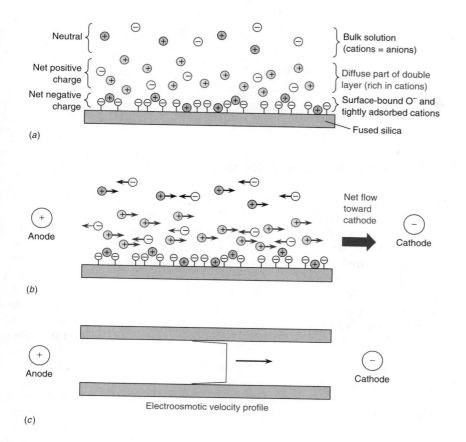

FIGURE 18-12 (a) Electric double layer is created by negatively charged silica surface and excess cations in the diffuse part of the double layer in the solution near the wall. The wall is negative and the diffuse part of the double layer is positive. (b) Predominance of cations in diffuse part of the double layer produces net electroosmotic flow toward the cathode when an external field is applied. (c) Electroosmotic velocity profile is uniform over more than 99.9% of the cross section of the capillary. Capillary tubing is required to maintain constant temperature in the liquid. Temperature variation in larger diameter tubing causes bands to broaden.

double layer is composed of (1) a negative charge at the wall and (2) an equal positive charge in solution near the wall. The thickness of the positive layer, called the *diffuse part of the double layer*, is approximately 1 nm. When an electric field is applied, cations are attracted to the cathode and anions are attracted to the anode. The excess cations in the diffuse part of the double layer drive the entire solution in the capillary toward the cathode (Figure 18-12b,c). The greater the applied electric field, the faster the flow.

Electroosmosis decreases at low pH because the wall loses its negative charge when Si–O⁻ is converted to Si–OH and the number of cations in the double layer diminishes. Electroosmotic velocity is measured by adding an ultraviolet-absorbing neutral solute, such as methanol, to the sample and measuring the time it takes (called the *migration time*) to reach the detector. In one experiment with 30 kV across a 50-cm capillary, the electroosmotic velocity was 4.8 mm/s at pH 9 and 0.8 mm/s at pH 3.

In Figures 18-10 and 18-12, electroosmosis is from left to right because cations in the double layer are attracted to the cathode. Superimposed on electroosmosis of the bulk fluid, electrophoresis transports cations to the right and anions to the left. At neutral or high pH, electroosmosis is faster than electrophoresis and the net flow of anions is to the right. At low pH, electroosmosis is weak and anions may flow to the left and never reach the detector. To separate anions at low pH, you can reverse the polarity of the instrument to make the sample side negative and the detector side positive.

Ions in the diffuse part of the double layer adjacent to the capillary wall are the "pump" that drives electroosmotic flow.

Migration time in electrophoresis is analogous to retention time in chromatography.

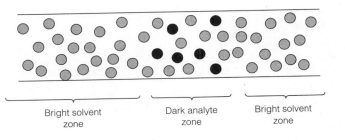

Bright solvent zone	Dark analyte zone	Bright solvent zone

FIGURE 18-13 Principle of indirect detection. When analyte emerges from the capillary, the strong background signal decreases.

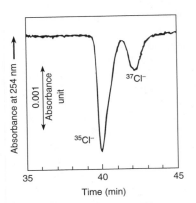

FIGURE 18-14 Separation of natural isotopes of 0.56 mM Cl⁻ by capillary electrophoresis with indirect spectrophotometric detection at 254 nm. The background electrolyte contains 5 mM CrO_4^{2-} to provide absorbance at 254 nm. There are few ways to separate isotopes so cleanly. This impressive separation was done by an undergraduate student at the University of Calgary.

Detecting the Separated Solutes

The most common detector is an ultraviolet absorbance monitor set to a wavelength near 200 nm, where many solutes absorb. It is not possible to use such short wavelengths with larger diameter columns, because the solvent absorbs too much of the radiation. A *fluorescence detector* works for fluorescent analytes or fluorescent derivatives. *Electrochemical detection* is sensitive to analytes that can gain or lose electrons at an electrode. Eluate can also be directed into a *mass spectrometer* (Box 16-2), which provides information on the quantity and molecular structure of analyte. *Conductivity detection* with ion-exchange suppression of the background electrolyte (as in ion chromatography, Figure 18-5) gives 1–10 ppb sensitivity for small ions.

In contrast to the direct detection of analyte discussed so far, **indirect detection** relies on measuring a strong signal from the background electrolyte and a weak signal from the analyte as it passes the detector. Figure 18-13 illustrates indirect fluorescence detection, but the same principle applies to any type of detection. A fluorescent ion with the same sign of charge as the analyte is added to background electrolyte to provide a steady background signal. When the analyte ion emerges, the concentration of background ion necessarily decreases, because electroneutrality must be preserved. If the analyte ion is not fluorescent, the fluorescence level decreases when analyte emerges. What we observe is a *negative* signal.

Figure 18-14 shows indirect ultraviolet detection of Cl⁻ anion in the presence of the ultraviolet-absorbing chromate anion, CrO_4^{2-}. In the absence of analyte, CrO_4^{2-} gives a steady absorbance at 254 nm. When Cl⁻ reaches the detector, there is less CrO_4^{2-} present and Cl⁻ does not absorb; therefore the detector signal *decreases*.

FIGURE 18-15 *Wall effects:* Consecutive injections of proteins in (a) uncoated and (b) covalently coated capillaries at pH 8.5. The coating is a hydrophilic polymer. Most of the protein from the first injection into the uncoated column stuck to the walls and never reached the detector. Adsorption gives irreproducible migration times and peak areas, as well as asymmetric peak shapes. The coated column gives reproducible results with normal peak shapes.

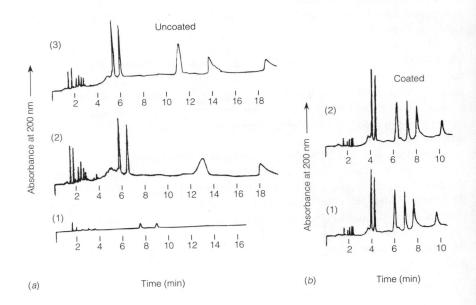

Wall Effects in the Separation of Proteins

Capillary electrophoresis is well suited for the analytical separation of proteins in biochemistry. However, some proteins are tightly adsorbed by the negatively charged capillary wall. Figure 18-15 shows that this problem can be largely overcome by attaching hydrophilic ("water loving") polymers to the silanol groups on the wall.

Ask Yourself

18-E. (a) Capillary electrophoresis was conducted with a solution whose pH was 9, at which the electroosmotic velocity is greater than the electrophoretic velocity. Draw a picture of the capillary, showing the placement of the anode, cathode, injector, and detector. Show the direction of electroosmotic flow and the direction of electrophoretic flow of a cation and an anion. Show the direction of net flow for each ion.

(b) If the pH is reduced to 3, the electroosmotic velocity is less than the electrophoretic velocity. In what directions will cations and anions migrate?

(c) Explain why the detector signal is negative in Figure 18-14.

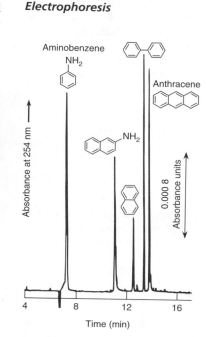

FIGURE 18-16 Separation of neutral molecules by micellar electrokinetic capillary electrophoresis. The average plate count in this experiment is 250 000 plates in 50 cm of capillary length.

18-7 | Types of Capillary Electrophoresis

The type of electrophoresis we have been discussing so far is called **capillary zone electrophoresis,** in which separation is based on different electrophoretic velocities of different ions. Electroosmotic flow of the bulk fluid is toward the cathode (Figure 18-12b). Cations migrate faster than the bulk fluid and anions migrate slower than the bulk fluid. Therefore the order of elution is cations before neutrals before anions. If the electrode polarity is reversed, the order of elution is anions before neutrals before cations. Neither scheme separates neutral molecules from one another.

Micellar Electrokinetic Capillary Chromatography

This mouthful of words describes a form of capillary electrophoresis that separates neutral molecules as well as ions. Figure 18-16 shows an example in which neutral molecules are not all eluted at the same time. The key modification in **micellar electrokinetic capillary chromatography** is that the capillary solution contains *micelles,* which are described in Box 18-1 (you need to read it now).

To understand how neutral molecules are separated, suppose that the background electrolyte contains negatively charged micelles. In Figure 18-17, electroosmotic flow is to the right. Electrophoretic migration of the negatively charged micelles is to the left, but net motion is to the right because the electroosmotic flow is faster than the electrophoretic flow.

In the absence of micelles, all neutral molecules would reach the detector together at a time we designate t_0. Micelles injected with the sample reach the detector at time t_{mc}, which is longer than t_0 because the micelles are anions that migrate upstream.

Order of elution in capillary zone electrophoresis

1. cations (highest mobility first)
2. all neutrals (unseparated)
3. anions (highest mobility last)

FIGURE 18-17 Negatively charged sodium dodecyl sulfate micelles migrate upstream against the electroosmotic flow. Neutral molecules are in dynamic equilibrium between free solution and the inside of the micelle. The more time spent in the micelle, the more the neutral molecule lags behind the electroosmotic flow.

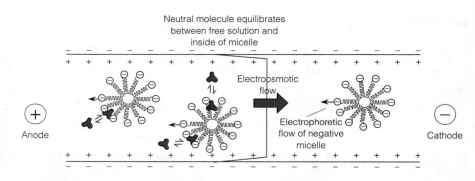

Neutral molecule equilibrates between free solution and inside of micelle

Electroosmotic flow

Electrophoretic flow of negative micelle

Anode (+)   Cathode (−)

Micellar electrokinetic capillary electrophoresis: The more time a solute spends inside the micelle, the longer is its migration time.

If a neutral molecule equilibrates between free solution and the inside of the micelles, its migration time is increased, because it migrates at the slower rate of the micelle part of the time. In this case, the neutral molecule reaches the detector at a time between t_0 and t_{mc}.

The more soluble the neutral molecule is in the micelle, the more time it spends inside the micelle and the longer is its migration time. The nonpolar interior of a sodium dodecyl sulfate micelle dissolves nonpolar solutes best. Polar solutes are not as soluble in the micelles and have a shorter retention time than nonpolar solutes do. Migration times of cations and anions may also be affected by micelles because ions can dissolve in some micelles. Micellar electrokinetic capillary chromatography is truly a form of chromatography because micelles behave like a pseudostationary phase. Solutes partition between the mobile phase (the aqueous solution) and the pseudostationary micelles.

💡 **Box 18-1 *Explanation***

What Is a Micelle?

A **micelle** is an aggregate of molecules with ionic head groups and long, nonpolar tails. Such molecules are called *surfactants,* an example of which is sodium dodecyl sulfate:

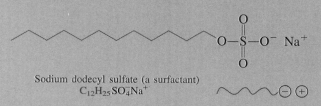

Sodium dodecyl sulfate (a surfactant)
$C_{12}H_{25}SO_4^-Na^+$

The polar head groups of a micelle face outward, where they are surrounded by polar water molecules. The nonpolar tails face inward, where they form a little pocket resembling a nonpolar hydrocarbon solution. *Nonpolar solutes are soluble inside the micelle.*

At low concentrations, surfactant molecules are not associated with one another. When the concentration

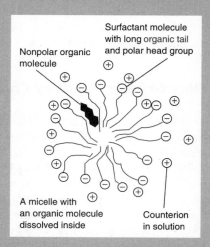

Surfactant molecule with long organic tail and polar head group

Nonpolar organic molecule

A micelle with an organic molecule dissolved inside

Counterion in solution

exceeds the *critical micelle concentration,* spontaneous aggregation into micelles occurs. Isolated surfactant molecules exist in equilibrium with micelles.

Capillary Gel Electrophoresis

In **capillary gel electrophoresis,** macromolecules are separated by *sieving* as they migrate through a gel inside a capillary tube. Small molecules travel faster than large molecules through the gel. Large molecules become entangled in the gel and their motion is slowed. (This behavior is the opposite of that in molecular exclusion chromatography, in which large molecules are excluded from the gel and move faster than small ones.)

Capillary gel electrophoresis is used to sequence DNA, which is composed of four different nucleotides. To learn the order of nucleotides, fragments of DNA are formed with a fluorescent label covalently attached at the end of each fragment. Color Plate 17 shows part of a DNA sequence analysis in which a mixture of fragments with up to 400 nucleotides was separated in a capillary containing a gel. DNA with 30 nucleotides had a migration time of 9 min and DNA with 400 nucleotides required 34 min. The combination of two kinds of fluorescent labels and two wavelengths allows a unique assignment of each of the four possible nucleotides.

Ask Yourself

18-F. (a) Explain why neutral solutes are eluted between times t_0 and t_{mc} in micellar electrokinetic capillary chromatography, where t_0 is the elution time of neutral molecules in the absence of micelles and t_{mc} is the elution time of the micelles.

(b) Micellar electrokinetic capillary chromatography in Figure 18-16 was conducted at pH 10 with anionic micelles and the anode on the sample side, as in Figure 18-10.

  (i) Is aminobenzene ($C_6H_5NH_2$) a cation, an anion, or a neutral molecule in this experiment?
  (ii) Explain how Figure 18-16 allows you to decide which compound, aminobenzene or anthracene, is more soluble in the micelles.

Important Terms

affinity chromatography

anion exchanger

capillary electrophoresis

capillary gel electrophoresis

capillary zone electrophoresis

cation exchanger

deionized water

electroosmosis

electrophoresis

hydrated radius

indirect detection

ion-exchange chromatography

ion chromatography

micellar electrokinetic capillary chromatography

micelle

molecular exclusion chromatography

preconcentration

trace analysis

Problems

18-1. **(a)** Hexanoic acid and 1-aminohexane, adjusted to pH 12 with NaOH, were passed through a cation-exchange column loaded with NaOH at pH 12. State the principal species that will be eluted and the order in which they are expected.

CH$_3$CH$_2$CH$_2$CH$_2$CH$_2$CO$_2$H   CH$_3$CH$_2$CH$_2$CH$_2$CH$_2$CH$_2$NH$_2$
    Hexanoic acid   1-Aminohexane

(b) Hexanoic acid and 1-aminohexane, adjusted to pH 3 with HCl, were passed through a cation-exchange column loaded with HCl at pH 3. State the principal species that will be eluted and the order in which they are expected.

18-2. The exchange capacity of an ion-exchange resin is defined as the number of moles of charged sites per gram of dry resin. Describe how you would measure the exchange capacity of an anion-exchange resin by using standard NaOH, standard HCl, or any other reagent you wish.

18-3. Commercial vanadyl sulfate (VOSO$_4$, FW 163.00) is contaminated with H$_2$SO$_4$ and H$_2$O. A solution was prepared by dissolving 0.244 7 g of impure VOSO$_4$ in 50.0 mL of water. Spectrophotometric analysis indicated that the concentration of the blue VO^{2+} ion was 0.024 3 M. A 5.00-mL sample was passed through a cation-exchange column loaded with H$^+$. VO^{2+} is exchanged for 2H$^+$ by this process. H$_2$SO$_4$ is unchanged by the cation-exchange column.

VOSO$_4$ $\longrightarrow$ | Cation-exchange column loaded with H$^+$ | $\longrightarrow$ H$_2$SO$_4$

H$_2$SO$_4$ $\longrightarrow$ | Cation-exchange column loaded with H$^+$ | $\longrightarrow$ H$_2$SO$_4$

H$^+$ eluted from the column required 13.03 mL of 0.022 74 M NaOH for titration. Find the weight percents of VOSO$_4$, H$_2$SO$_4$, and H$_2$O in the vanadyl sulfate.

18-4. Consider a protein with a net negative charge tightly adsorbed on an anion-exchange gel at pH 8.

(a) How will a gradient from pH 8 to some lower pH be useful for eluting the protein?

(b) How would a gradient of increasing ion concentration (at constant pH) in the eluent be useful for eluting the protein?

18-5. Look up the pK_a values for trimethylamine, dimethylamine, methylamine, and ammonia. Predict the order of elution

of these compounds from a cation-exchange column eluted with a gradient of increasing pH, beginning at pH 7.

18-6. Compounds with $-\overset{OH}{\underset{|}{C}}-\overset{OH}{\underset{|}{C}}-$ or $-\overset{OH}{\underset{|}{C}}-\overset{NH_2}{\underset{|}{C}}-$ linkages can be analyzed by cleavage with periodate. One mole of 1,2-ethanediol consumes one mole of periodate:

$$\begin{matrix} CH_2OH \\ | \\ CH_2OH \end{matrix} \ + \ IO_4^- \ \longrightarrow \ 2CH_2O \ + H_2O + IO_3^-$$

1,2-Ethanediol   Periodate   Formaldehyde   Iodate
(MW 62.068)

To analyze 1,2-ethanediol, oxidation with excess IO$_4^-$ is followed by passage of the whole reaction solution through an anion-exchange resin that binds both IO$_4^-$ and IO$_3^-$. The IO$_3^-$ is then selectively and quantitatively removed from the resin by elution with NH$_4$Cl. The absorbance of eluate is measured at 232 nm to find the quantity of IO$_3^-$ (molar absorptivity (ϵ) = 900 M^{-1} cm^{-1}) produced by the reaction. In one experiment, 0.213 9 g of aqueous 1,2-ethanediol was dissolved in 10.00 mL. Then 1.000 mL of the solution was treated with 3 mL of 0.15 M KIO$_4$ and subjected to ion-exchange separation of IO$_3^-$ from unreacted IO$_4^-$. The eluate (diluted to 250.0 mL) had an absorbance of $A_{232} = 0.521$ in a 1.000-cm cell, and a blank had $A_{232} = 0.049$. Find the weight percent of 1,2-ethanediol in the original sample.

18-7. In *ion-exclusion chromatography,* ions are separated from nonelectrolytes (uncharged molecules) by an ion-exchange column. Nonelectrolytes penetrate the stationary phase, whereas ions with the same sign charge as the stationary phase are repelled by the stationary phase. Because electrolytes have access to less of the column volume, they are eluted before nonelectrolytes. A mixture of trichloroacetic acid (TCA, p$K_a = 0.66$), dichloroacetic acid (DCA, 1.30), and monochloroacetic acid (MCA, 2.86) was separated by passage through a cation-exchange resin eluted with 0.01 M HCl. The order of elution was TCA < DCA < MCA. Explain why the three acids are separated and the order of elution.

18-8. Polystyrene standards of known molecular weight gave the following calibration data in a molecular exclusion column. Prepare a plot of retention time (t_r) versus log (MW) and find the equation of the line by the method of least squares. Find the molecular weight of an unknown with a retention time of 13.00 min.

Molecular weight	Retention time, t_r (min)
8.50×10^6	9.28
3.04×10^6	10.07
1.03×10^6	10.88
3.30×10^5	11.67
1.56×10^5	12.14
6.60×10^4	12.74
2.85×10^4	13.38
9.20×10^3	14.20
3.25×10^3	14.96
5.80×10^2	16.04

18-9. A polystyrene resin molecular exclusion HPLC column has a diameter of 7.8 mm and a length of 30 cm. The solid portion of the particles occupies 20% of the volume, the pores occupy 40%, and the volume between particles occupies 40%.

(a) At what volume would totally excluded molecules be expected to emerge?

(b) At what volume would the smallest molecules be expected?

(c) A mixture of polyethylene glycols of various molecular weights is eluted between 23 and 27 mL. What does this imply about the retention mechanism for these solutes on the column?

18-10. If a capillary is set up as in Figure 18-10 with the injector end positive and the detector end negative, in what order will cations, anions, and neutral molecules be eluted?

18-11. (a) What is electroosmosis?

(b) Why is the electroosmotic flow in a silica capillary five times faster at pH 9 than at pH 3?

(c) When the Si–OH groups on a silica capillary wall are converted to $Si–O(CH_2)_{17}CH_3$ groups, the electroosmotic flow is small and nearly independent of pH. Explain why.

18-12. (a) What is the principal source of zone broadening in ideal capillary electrophoresis?

(b) Why is the detector response *negative* in indirect spectrophotometric detection?

18-13. Explain how neutral molecules can be separated by micellar electrokinetic capillary chromatography. Why is this a form of chromatography?

18-14. (a) Measure the peak width of $^{35}Cl^-$ in Figure 18-14 and calculate the number of theoretical plates.

(b) The distance from injection to the detector is 40 cm in Figure 18-14. From your answer to part **(a)** find the plate height.

(c) Why are the peaks negative in Figure 18-14?

18-15. The water-soluble vitamins niacinamide (a neutral compound), riboflavin (a neutral compound), niacin (an anion), and thiamine (a cation) were separated by micellar electrokinetic capillary chromatography in 15 mM borate buffer (pH 8.0) with 50 mM sodium dodecyl sulfate. The migration times were niacinamide, 8.1 min; riboflavin, 13.0 min; niacin 14.3 min; and thiamine, 21.9 min. What would the order have been in the absence of sodium dodecyl sulfate? Which compound is most soluble in the micelles?

Combustion Analysis Reveals Pollutant Buildup in Trees

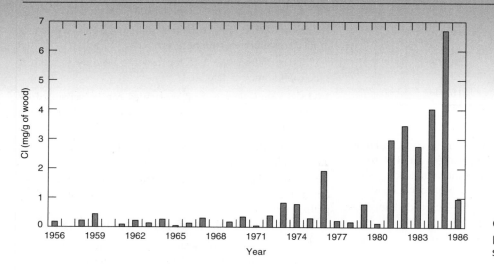

Chlorine content over a 30-year period in rings from a tree in southern Italy.

A combustion procedure was employed to measure Cl, P, and S in pine tree rings from southern Italy. Wood was taken from the tree with a core drill, and rings were sliced apart and dried at 75°C for 3 days. Samples were wrapped in filter paper, leaving an unfolded paper tail as shown in Figure 19-1. The wrapped sample was then placed in the platinum basket of the disassembled *Schöniger flask.*

The 250-mL flask was loaded with 20 mL of 4.4 mM hydrogen peroxide (H_2O_2) and flushed with O_2 gas. The paper tail was ignited and the stopper-sample assembly was immediately inserted into the flask. The flask was inverted to prevent gaseous products from leaking out. Following combustion, 30 min were allowed for absorption of the products by the solution. The elements Cl, P, and S in the wood were converted to chloride (Cl^-), phosphate (PO_4^{3-}), and sulfate (SO_4^{2-}) in the oxidizing solution.

Analysis of the solution by ion chromatography (Chapter 18) showed that the average chlorine uptake by the tree in the period 1971–1986 was 10 times greater than the average uptake in the period 1956–1970. Uptake of phosphorus and sulfur increased by a factor of 2 in the recent 15-year period. The chemical content of the rings presumably mirrors chemical changes in the environment.

Gravimetric and Combustion Analysis

In **gravimetric analysis** the mass of a product is used to calculate the quantity of original analyte. Early in this century, Nobel Prize–winning gravimetric analysis by T. W. Richards and his colleagues measured the atomic weights of Ag, Cl, and N to six-figure accuracy and formed the basis for accurate atomic weight determinations. In **combustion analysis,** a sample is burned in excess oxygen and the products are measured. Combustion is typically used to determine C, H, N, S, and halogens in organic compounds.

Gravimetric procedures were the mainstay of chemical analyses of ores and industrial materials in the eighteenth and nineteenth centuries, long before the chemical basis for the procedures was understood.

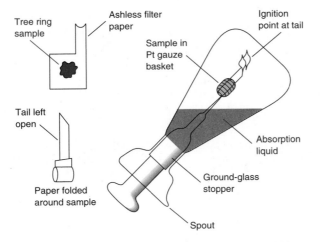

FIGURE 19-1   Schöniger combustion procedure, using 10- to 100-mg tree ring samples. The Pt gauze basket helps catalyze complete combustion of the sample, which is carried out behind a blast shield. *Ashless* filter paper is a special grade that leaves little residue when it is burned. Prior to analysis of tree rings, the filter paper had to be washed several times with distilled water to remove traces of Cl^- and SO_4^{2-}.

19-1 An Example of Gravimetric Analysis

An example of gravimetric analysis is the determination of Cl^- by precipitation with Ag^+:

$$Ag^+ + Cl^- \longrightarrow AgCl(s)$$

The mass of AgCl product tells us how many moles of AgCl were produced. For every mole of AgCl, there must have been one mole of Cl^- in the unknown solution.

EXAMPLE	A Simple Gravimetric Calculation

A 10.00-mL solution containing Cl^- was treated with excess $AgNO_3$ to precipitate 0.436 8 g of AgCl (FW 143.321). What was the molarity of Cl^- in the unknown?

TABLE 19-1 Representative gravimetric analyses

Species analyzed	Precipitated form	Form weighed	Some interfering species
K^+	$KB(C_6H_5)_4$	$KB(C_6H_5)_4$	NH_4^+, Ag^+, Hg^{2+}, Tl^+, Rb^+, Cs^+
Mg^{2+}	$Mg(NH_4)PO_4 \cdot 6H_2O$	$Mg_2P_2O_7$	Many metals except Na^+ and K^+
Ca^{2+}	$CaC_2O_4 \cdot H_2O$	$CaCO_3$ or CaO	Many metals except Mg^{2+}, Na^+, K^+
Ba^{2+}	$BaSO_4$	$BaSO_4$	Na^+, K^+, Li^+, Ca^{2+}, Al^{3+}, Cr^{3+}, Fe^{3+}, Sr^{2+}, Pb^{2+}, NO_3^-
Cr^{3+}	$PbCrO_4$	$PbCrO_4$	Ag^+, NH_4^+
Mn^{2+}	$Mn(NH_4)PO_4 \cdot H_2O$	$Mn_2P_2O_7$	Many metals
Fe^{3+}	$Fe(HCO_2)_3$	Fe_2O_3	Many metals
Co^{2+}	Co(1-nitroso-2-naphtholate)$_3$	$CoSO_4$ (by reaction with H_2SO_4)	Fe^{3+}, Pd^{2+}, Zr^{4+}
Ni^{2+}	Ni(dimethylglyoximate)$_2$	Same	Pd^{2+}, Pt^{2+}, Bi^{3+}, Au^{3+}
Cu^{2+}	CuSCN	CuSCN	NH_4^+, Pb^{2+}, Hg^{2+}, Ag^+
Zn^{2+}	$Zn(NH_4)PO_4 \cdot H_2O$	$Zn_2P_2O_7$	Many metals
Al^{3+}	Al(8-hydroxyquinolate)$_3$	Same	Many metals
Sn^{4+}	Sn(cupferron)$_4$	SnO_2	Cu^{2+}, Pb^{2+}, As(III)
Pb^{2+}	$PbSO_4$	$PbSO_4$	Ca^{2+}, Sr^{2+}, Ba^{2+}, Hg^{2+}, Ag^+, HCl, HNO_3
NH_4^+	$NH_4B(C_6H_5)_4$	$NH_4B(C_6H_5)_4$	K^+, Rb^+, Cs^+
Cl^-	AgCl	AgCl	Br^-, I^-, SCN^-, S^{2-}, $S_2O_3^{2-}$, CN^-
Br^-	AgBr	AgBr	Cl^-, I^-, SCN^-, S^{2-}, $S_2O_3^{2-}$, CN^-
I^-	AgI	AgI	Cl^-, Br^-, SCN^-, S^{2-}, $S_2O_3^{2-}$, CN^-
SCN^-	CuSCN	CuSCN	NH_4^+, Pb^{2+}, Hg^{2+}, Ag^+
CN^-	AgCN	AgCN	Cl^-, Br^-, I^-, SCN^-, S^{2-}, $S_2O_3^{2-}$
F^-	$(C_6H_5)_3SnF$	$(C_6H_5)_3SnF$	Many metals (except alkali metals), SiO_4^{4-}, CO_3^{2-}
ClO_4^-	$KClO_4$	$KClO_4$	
SO_4^{2-}	$BaSO_4$	$BaSO_4$	Na^+, K^+, Li^+, Ca^{2+}, Al^{3+}, Cr^{3+}, Fe^{3+}, Sr^{2+}, Pb^{2+}, NO_3^-
PO_4^{3-}	$Mg(NH_4)PO_4 \cdot 6H_2O$	$Mg_2P_2O_7$	Many metals except Na^+, K^+
NO_3^-	Nitron nitrate	Nitron nitrate	ClO_4^-, I^-, SCN^-, CrO_4^{2-}, ClO_3^-, NO_2^-, Br^-, $C_2O_4^{2-}$

SOLUTION A precipitate weighing 0.463 8 g contains

$$\frac{0.436\ 8\ \text{g AgCl}}{143.321\ \text{g AgCl/mol AgCl}} = 3.048 \times 10^{-3}\ \text{mol AgCl}$$

Because 1 mol of AgCl contains 1 mol of Cl^-, there must have been 3.048×10^{-3} mol of Cl^- in the unknown. The molarity of Cl^- in the unknown is therefore

$$[Cl^-] = \frac{3.048 \times 10^{-3}\ \text{mol}}{0.010\ 00\ \text{L}} = 0.304\ 8\ \text{M}$$

Representative analytical precipitations are listed in Table 19-1. Potentially interfering substances listed in the table may need to be removed prior to analysis. A few common organic **precipitants** (agents that cause precipitation) are listed in Table 19-2.

Ask Yourself

19-A. A 50.00-mL solution containing NaBr was treated with excess $AgNO_3$ to precipitate 0.214 6 g of AgBr (FW 187.772).
    (a) How many moles of AgBr product were isolated?
    (b) What was the molarity of NaBr in the solution?

19-2 *Precipitation*

The ideal gravimetric precipitate should be insoluble, easily filtered, and should possess a known, constant composition. The precipitate should be stable when you heat it to remove the last traces of solvent. Although few substances meet these requirements, techniques described in this section help optimize the properties of precipitates.

Precipitate particles should be large enough to be collected by filtration; they should not be so small that they clog or pass through the filter. Large crystals also have less surface area to which foreign species may become attached. At the other extreme is a *colloid,* whose particles are so small (1–100 nm) that they pass through most filters (Demonstration 19-1).

Crystal Growth

Crystallization occurs in two phases: nucleation and particle growth. During **nucleation,** dissolved molecules come together randomly and form small aggregates. In *particle growth,* more molecules add to the nucleus to form a crystal.

A solution containing more dissolved solute than should be present at equilibrium is said to be **supersaturated.** Nucleation proceeds faster than particle growth in a highly supersaturated solution, creating tiny particles or, worse, a colloid. In a more dilute solution, nucleation is slower, so the nuclei have a chance to grow into larger, more tractable particles.

TABLE 19-2 Common organic precipitating agents

Name	Structure	Some ions precipitated
Dimethylglyoxime		Ni^{2+}, Pd^{2+}, Pt^{2+}
Cupferron		Fe^{3+}, VO_2^+, Ti^{4+}, Zr^{4+}, Ce^{4+}, Ga^{3+}, Sn^{4+}
8-Hydroxyquinoline (oxine)		Mg^{2+}, Zn^{2+}, Cu^{2+}, Cd^{2+}, Pb^{2+}, Al^{3+}, Fe^{3+}, Bi^{3+}, Ga^{3+}, Th^{4+}, Zr^{4+}, UO_2^{2+}, TiO^{2+}
1-Nitroso-2-naphthol		Co^{2+}, Fe^{3+}, Pd^{2+}, Zr^{4+}
Nitron		NO_3^-, ClO_4^-, BF_4^-, WO_4^{2-}
Sodium tetraphenylborate	$Na^+B(C_6H_5)_4^-$	K^+, Rb^+, Cs^+, NH_4^+, Ag^+, organic ammonium ions
Tetraphenylarsonium chloride	$(C_6H_5)_4As^+Cl^-$	$Cr_2O_7^{2-}$, MnO_4^-, ReO_4^-, MoO_4^{2-}, WO_4^{2-}, ClO_4^-, I_3^-

Techniques that promote particle growth include

1. raising the temperature to increase solubility and thereby decrease supersaturation

2. adding precipitant slowly with vigorous mixing, to avoid high local supersaturation where the stream of precipitant first enters the analyte

3. keeping the volume of solution large so that the concentrations of analyte and precipitant are low

Homogeneous Precipitation

In **homogeneous precipitation,** precipitant is generated slowly by a chemical reaction. For example, urea decomposes in boiling water to produce OH^-:

Colloids and Dialysis

Colloids are particles with diameters in the range 1–100 nm. They are larger than molecules, but too small to precipitate. Colloids remain in solution indefinitely, suspended by the Brownian motion (random movement) of solvent molecules.

To make a colloid, heat a beaker containing 200 mL of distilled water to 70°–90°C and leave an identical beaker of water at room temperature. Add 1 mL of 1 M $FeCl_3$ to each beaker and stir. The warm solution turns brown-red in a few seconds, whereas the cold solution remains yellow (Color Plate 18a). The yellow color is characteristic of low molecular weight Fe^{3+} compounds. The red color results from colloidal aggregates of Fe^{3+} ions held together by hydroxide, oxide, and some chloride ions. These particles have a molecular weight of $\sim 10^5$ and a diameter of ~ 10 nm and contain $\sim 10^3$ atoms of Fe.

To demonstrate the size of colloidal particles, we perform a **dialysis** experiment in which two solutions are separated by a *semipermeable membrane*. The membrane has pores through which small molecules, but not large molecules and colloids, can diffuse. Cellulose dialysis tubing (such as catalog number 3787 from A. H. Thomas Co.) has 1–5 nm pores.

Pour some of the brown-red colloidal Fe solution into a dialysis tube knotted at one end; then tie off the other end. Drop the tube into a flask of distilled water to show that the color remains entirely within the bag even after several days (Color Plate 18b,c). For comparison, leave an identical bag containing a dark blue solution of 1 M $CuSO_4 \cdot 5H_2O$ in another flask. Cu^{2+} diffuses out of the bag and the solution in the flask becomes light blue in 24 h. Alternatively, the yellow food coloring, tartrazine, can be used in place of Cu^{2+}. If dialysis is conducted in hot water, it is completed during one class period.

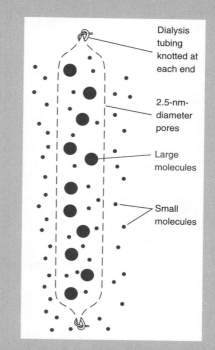

Large molecules remain trapped inside a dialysis bag, whereas small molecules diffuse through the membrane in both directions. The opening of Chapter 17 described how a *microdialysis probe* is used for sampling fluids in living organisms.

Dialysis is used to treat patients suffering from kidney failure. Their blood is run over a dialysis membrane having a very large surface area. Small metabolic waste products in the blood diffuse across the membrane and are diluted into a large volume of liquid going out as waste. Large proteins, which are a necessary part of the blood plasma, cannot cross the membrane and are retained in the blood.

By using urea, we can raise the pH of a solution very gradually, and the slow OH^- formation enhances the particle size of ferric formate precipitate:

$$\text{Formic acid} + OH^- \longrightarrow HCO_2^- + H_2O \qquad\qquad 3HCO_2^- + Fe^{3+} \longrightarrow \underset{\text{Ferric formate}}{Fe(HCO_2)_3 \cdot nH_2O(s)\downarrow}$$

Table 19-3 lists some reagents for homogeneous precipitation.

TABLE 19-3 Common reagents for homogeneous precipitation

Precipitant	Reagent	Reaction	Some elements precipitated
OH^-	Urea	$(H_2N)_2CO + 3H_2O \longrightarrow CO_2 + 2NH_4^+ + 2OH^-$	Al, Ga, Th, Bi, Fe, Sn
S^{2-}	Thioacetamide^a	$\underset{\parallel}{\overset{S}{}}CH_3CNH_2 + H_2O \longrightarrow \underset{\parallel}{\overset{O}{}}CH_3CNH_2 + H_2S$	Sb, Mo, Cu, Cd
SO_4^{2-}	Sulfamic acid	$H_3\overset{+}{N}SO_3^- + H_2O \longrightarrow NH_4^+ + SO_4^{2-} + H^+$	Ba, Ca, Sr, Pb
$C_2O_4^{2-}$	Dimethyl oxalate	$\overset{OO}{\overset{\parallel\parallel}{}}CH_3OCCOCH_3 + 2H_2O \longrightarrow 2CH_3OH + C_2O_4^{2-} + 2H^+$	Ca, Mg, Zn
PO_4^{3-}	Trimethyl phosphate	$(CH_3O)_3P{=}O + 3H_2O \longrightarrow 3CH_3OH + PO_4^{3-} + 3H^+$	Zr, Hf

a. Hydrogen sulfide is volatile and toxic; it should be handled only in a well-vented hood. Thioacetamide is a carcinogen that should be handled with gloves. If thioacetamide contacts your skin, wash yourself thoroughly immediately. Leftover reagent is destroyed by heating at 50°C with 5 mol of NaOCl per mole of thioacetamide, and then washing the products down the drain.

Precipitation in the Presence of Electrolyte

An *electrolyte* is a compound that dissociates into ions when it dissolves. We say that the electrolyte *ionizes* when it dissolves.

Ionic compounds are usually precipitated in the presence of added electrolyte. To understand why, consider how tiny crystallites *coagulate* (come together) into larger crystals. We will illustrate the case of AgCl, which is commonly formed in the presence of 0.1 M HNO$_3$.

FIGURE 19-2 Schematic diagram showing a colloidal particle of AgCl in a solution containing excess Ag$^+$, H$^+$, and NO$_3^-$. The particle has a net positive charge because of adsorbed Ag$^+$ ions. The region of solution surrounding the particle is called the *ionic atmosphere.* It has a net negative charge, because the particle attracts anions and repels cations.

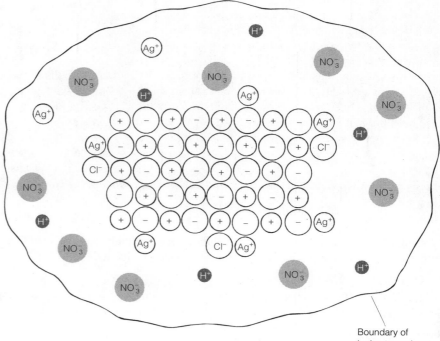

Boundary of ionic atmosphere

Figure 19-2 shows a colloidal particle of AgCl growing in a solution containing excess Ag^+, H^+, and NO_3^-. The particle has an excess positive charge due to **adsorption** of extra silver ions on exposed chloride ions. (To be adsorbed means to be attached to the surface. In contrast, **absorption** involves penetration beyond the surface, to the inside.) The positively charged surface of the solid attracts anions and repels cations from the *ionic atmosphere* in the liquid surrounding the particle.

Colloidal particles must collide with each other to coalesce. However, the negatively charged ionic atmospheres of the particles repel one another. The particles, therefore, must have enough kinetic energy to overcome electrostatic repulsion before they can coalesce. Heating promotes coalescence by increasing the particles' kinetic energy.

Increasing electrolyte concentration (HNO_3 for AgCl) decreases the volume of the ionic atmosphere and allows particles to approach closer together before repulsion becomes significant. Therefore, most gravimetric precipitations are done in the presence of electrolyte.

Although it is common to find the excess common ion adsorbed on the crystal surface, it is also possible to find other ions selectively adsorbed. In the presence of citrate and sulfate, there is more citrate than sulfate adsorbed on a particle of $BaSO_4$.

Digestion

Digestion is the process of allowing a precipitate to stand in contact with the *mother liquor* for some period of time, usually with heating. Digestion promotes slow recrystallization of the precipitate. Particle size increases and impurities tend to be expelled from the crystal.

Mother liquor is the solution from which a substance crystallized.

Purity

Adsorbed impurities are bound to the surface of a crystal. *Absorbed* impurities (within the crystal) are classified as *inclusions* or *occlusions*. Inclusions are impurity ions that randomly occupy sites in the crystal lattic normally occupied by ions that belong in the crystal. Inclusions are more likely when the impurity ion has a size and charge similar to those of one of the ions that belongs to the product. Occlusions are pockets of impurity that are literally trapped inside the growing crystal.

Adsorbed, occluded, and included impurities are said to be **coprecipitated.** That is, the impurity is precipitated along with the desired product, even though the solubility of the impurity has not been exceeded. Coprecipitation tends to be worst in colloidal precipitates (which have a large surface area), such as $BaSO_4$, $Al(OH)_3$, and $Fe(OH)_3$. Some procedures call for washing away the mother liquor, redissolving the precipitate, and *reprecipitating* the product. During the second precipitation, the concentration of impurities in the solution is lower than during the first precipitation, and the degree of coprecipitation therefore tends to be lower. Box 19-1 illustrates effects of coprecipitation and digestion.

Reprecipitation improves the purity of some precipitates.

Occasionally, a trace component that is too dilute to be measured is intentionally concentrated by coprecipitation with a major component of the solution. The procedure is called **gathering,** and the precipitate used to collect the trace component is said to be a *gathering agent.* When the precipitate is dissolved in a small volume of solvent, the concentration of the trace component is high enough for accurate analysis.

Some impurities can be treated with a **masking agent,** which prevents them from reacting with the precipitant. In the gravimetric analysis of Be^{2+}, Mg^{2+}, Ca^{2+}, or Ba^{2+} with the reagent *N-p*-chlorophenylcinnamohydroxamic acid (designated

Example of *gathering:* Se(IV) at concentrations as low as 25 ng/L is gathered by coprecipitation with $Fe(OH)_3$. The precipitate is then dissolved in a small volume of concentrated acid to obtain a more concentrated solution of Se(IV).
Question How many ppb is 25 ng/L?

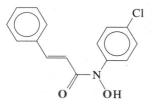

N-p-Chlorophenylcinnamo-
hydroxamic acid (RH)
(ligand donor atoms are **bold**)

RH), impurities such as Ag^+, Mn^{2+}, Zn^{2+}, Cd^{2+}, Hg^{2+}, Fe^{2+}, and Ga^{3+} are kept in solution by excess KCN.

$$Ca^{2+} + 2RH \longrightarrow CaR_2(s)\downarrow + 2H^+$$
Analyte   Precipitate

$$Mn^{2+} + 6CN^- \longrightarrow Mn(CN)_6^{4-}$$
Impurity   Masking agent   Stays in solution

Impurities might collect on the product while it is standing in the mother liquor. This process is called *postprecipitation* and usually involves a supersaturated impurity that does not readily crystallize. An example is the crystallization of magnesium oxalate (MgC_2O_4) on calcium oxalate (CaC_2O_4).

Washing precipitate on a filter helps remove droplets of liquid containing excess solute. Some precipitates can be washed with water, but many require electrolyte to maintain coherence. For these precipitates, the ionic atmosphere is required to neutralize the surface charge of the tiny particles. If electrolyte is washed away with water, the charged solid particles repel one another and the product breaks up. This breaking up, called **peptization,** results in loss of product through the filter. Silver

---

Box 19-1 *Explanation*

A Case Study in Coprecipitation and Digestion

This box requires some explanations from you instead of from Dan. Students at the University of Tulsa experimented with procedural variations to learn the effects of coprecipitation and digestion in the gravimetric analysis of barium by precipitation of $BaSO_4$. The basic procedure below was carried out in duplicate:

1. Pipet 50.0 mL of standard (~0.05 M) $BaCl_2$ into a 400-mL beaker and add 100 mL of H_2O.

2. Add 1 mL of 16 M HCl, which increases the solubility of $BaSO_4$ and thereby decreases supersaturation during precipitation. Solubility is increased because the reaction $SO_4^{2-} + H^+ \rightleftharpoons HSO_4^-$ consumes some dissolved sulfate.

3. Heat to 90°C to increase the solubility of $BaSO_4$ and promote crystallization during precipitation.

4. Add 50 mL of 0.15 M Na_2SO_4 all at once.

5. Digest the precipitate at 90°C for 1 h in the mother liquor.

6. Decant the mother liquor (pour it off), wash the precipitate with hot water, and decant the wash water.

7. Collect the precipitate by filtration through ashless filter paper and wash it with hot water

until no more Cl^- is detected in the filtrate (by precipitation with Ag^+).

8. Transfer the paper and precipitate to a porcelain crucible and ignite it over a Bunsen burner to burn away the paper, leaving pure $BaSO_4$.

9. Cool the crucible in a desiccator and weigh it. Reheat and reweigh until successive weighings agree within ±0.3 mg.

Students A and B each carried out the basic procedure twice and obtained an average of 0.497 g of $BaSO_4$ product (range = 0.495 to 0.499).

Coprecipitation of Ca^{2+}: Student C followed the same procedure, but the initial solution contained 0.5 mmol of $CaCl_2$ in addition to $BaCl_2$. The average mass of his product was 0.505 g (0.503 and 0.507 g). How would you explain this?

Prolonged digestion: Student D carried out the same procedure as student C (with $CaCl_2$ present) but digested her product for 20 h, instead of 1 h, at 90°C. She obtained an average mass of 0.497 g (0.495 and 0.499 g) of product. Why did her results agree with those of students A and B even though Ca^{2+} was present in her initial sample?

chloride peptizes if washed with water, so it is washed with dilute HNO_3 instead. Volatile electrolytes including HNO_3, HCl, NH_4NO_3, NH_4Cl, and $(NH_4)_2CO_3$ are used for washing because they evaporate during drying.

Product Composition

The final product must have a known, stable composition. A **hygroscopic substance** is one that picks up water from the air and is therefore difficult to weigh accurately. Many precipitates contain a variable quantity of water and must be dried under conditions that give a known (possibly zero) stoichiometry of H_2O.

Ignition (strong heating) is used to change the chemical form of some precipitates that do not have a constant composition after drying at moderate temperatures. For example, $Fe(HCO_2)_3 \cdot nH_2O$ is ignited at 850°C for 1 h to give Fe_2O_3, and $Mg(NH_4)PO_4 \cdot 6H_2O$ is ignited at 1 100°C to give $Mg_2P_2O_7$.

In **thermogravimetric analysis,** a sample is heated, and its mass is measured as a function of temperature. Figure 19-3 shows how the composition of calcium salicylate changes in four stages:

Ammonium chloride, for example, decomposes as follows when it is heated:

$$NH_4Cl(s) \xrightarrow{\text{heat}} NH_3(g) + HCl(g)$$

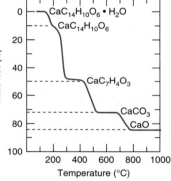

FIGURE 19-3 Thermogravimetric curve for calcium salicylate.

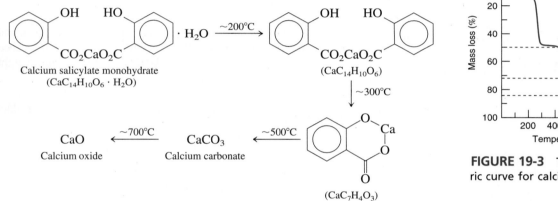

Calcium salicylate monohydrate
$(CaC_{14}H_{10}O_6 \cdot H_2O)$

$(CaC_{14}H_{10}O_6)$

CaO
Calcium oxide

CaCO$_3$
Calcium carbonate

$(CaC_7H_4O_3)$

The composition of the product depends on the temperature and duration of heating.

19-3 *Examples of Gravimetric Calculations*

We now illustrate how to relate the mass of a gravimetric precipitate to the quantity of original analyte. *The general approach is to relate the moles of product to the moles of reactant.*

EXAMPLE **Relating Mass of Product to Mass of Reactant**

If you were performing this analysis, it would be important to determine that impurities in the piperazine are not also precipitated.

The piperazine content of an impure commercial material can be determined by precipitating and weighing piperazine diacetate:

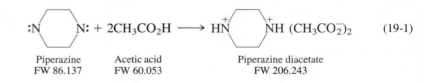

$$\text{:N} \overset{}{\bigcirc} \text{N:} + 2CH_3CO_2H \longrightarrow \text{HN}^+ \overset{}{\bigcirc} \text{NH}^+ (CH_3CO_2^-)_2 \qquad (19\text{-}1)$$

| Piperazine | Acetic acid | Piperazine diacetate |
| FW 86.137 | FW 60.053 | FW 206.243 |

In one experiment, 0.312 6 g of the sample was dissolved in 25 mL of acetone, and 1 mL of acetic acid was added. After 5 min, the precipitate was filtered, washed with acetone, dried at 110°C, and found to weigh 0.712 1 g. What is the weight percent of piperazine in the commercial material?

Reminder:

$$\text{wt \%} = \frac{\text{mass of analyte}}{\text{mass of unknown}} \times 100$$

SOLUTION For each mole of piperazine in the impure material, 1 mol of product is formed:

$$\text{moles of piperazine} = \text{moles of product}$$

$$= \frac{0.712\ 1\ \text{g product}}{206.243\ \dfrac{\text{g product}}{\text{mol product}}} = 3.453 \times 10^{-3}\ \text{mol}$$

This many moles of piperazine corresponds to

$$\text{grams of piperazine} = (3.453 \times 10^{-3}\ \text{mol piperazine})\left(86.137\ \frac{\text{g piperazine}}{\text{mol piperazine}}\right)$$

$$= 0.297\ 4\ \text{g}$$

which gives

$$\text{wt \% piperazine in analyte} = \frac{0.297\ 4\ \text{g piperazine}}{0.312\ 6\ \text{g unknown}} \times 100 = 95.14\%$$

EXAMPLE **When the Stoichiometry Is Not 1:1**

Solid residue weighing 8.444 8 g from an aluminum refining process was dissolved in acid, treated with 8-hydroxyquinoline, and ignited to give Al_2O_3 weighing 0.855 4 g. Find the weight percent of Al in the original mixture.

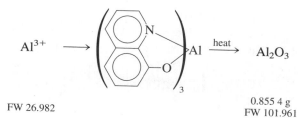

Al^{3+} $\longrightarrow$ $\xrightarrow{\text{heat}}$ Al$_2$O$_3$

FW 26.982

0.855 4 g
FW 101.961

SOLUTION Each mole of product (Al$_2$O$_3$) contains two moles of Al. The mass of product tells us the moles of product, and from this we can find the moles of Al. The moles of product are (0.855 4 g)/(101.961 g/mol) = 0.008 389$_5$ mol Al$_2$O$_3$. Because each mole of product contains two moles of Al, there must have been

$$\text{moles of Al in unknown} = \frac{2 \text{ mol Al}}{\text{mol Al}_2\text{O}_3} \times 0.008\ 389_5 \ \text{mol Al}_2\text{O}_3$$

$$= 0.016\ 77_9 \ \text{mol Al}$$

The mass of Al is (0.016 77$_9$ mol)(26.982 g/mol) = 0.452 7$_3$ g Al. The weight percent of Al in the unknown is

$$\text{wt \% Al} = \frac{0.452\ 7_3 \ \text{g Al}}{8.444\ 8 \ \text{g unknown}} \times 100 = 5.361 \ \%$$

A note from Dan: Whether or not I show it, I keep at least one extra, insignificant figure in my calculations and do not round off until the final answer. Usually, I keep all the digits in my calculator.

EXAMPLE **Calculating How Much Precipitant to Use**

(a) To measure the nickel content in steel, the steel is dissolved in 12 M HCl and neutralized in the presence of citrate ion, which binds iron and keeps it in solution. The slightly basic solution is warmed, and dimethylglyoxime (DMG) is added to precipitate the red DMG-nickel complex. The product is filtered, washed with cold water, and dried at 110°C.

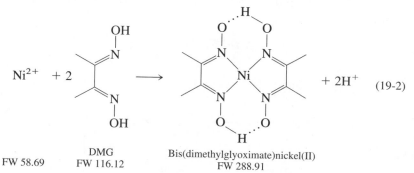

Ni^{2+} + 2 $\longrightarrow$ + 2H$^+$ (19-2)

FW 58.69

DMG
FW 116.12

Bis(dimethylglyoximate)nickel(II)
FW 288.91

If the nickel content is known to be near 3 wt % and you wish to analyze 1.0 g of the steel, what volume of 1.0 wt % DMG in alcohol solution should be used to give a 50% excess of DMG for the analysis? Assume that the density of the alcohol solution is 0.79 g/mL.

1.0 wt % DMG means

$$\frac{1.0 \text{ g DMG}}{100 \text{ g solution}}$$

Density means

$$\frac{\text{grams of solution}}{\text{mL of solution}}$$

SOLUTION Our strategy is to estimate the moles of Ni in 1.0 g of steel. Equation 19-2 tells us that two moles of DMG are required for each mole of Ni. After finding the required number of moles of DMG, we will multiply it by 1.5 to get a 50% excess, to be sure we have enough.

The Ni content of the steel is around 3%, so 1.0 g of steel contains about $(0.03)(1.0 \text{ g}) = 0.03$ g of Ni, which corresponds to $(0.03 \text{ g Ni})/(58.69 \text{ g/mol Ni}) = 5.1 \times 10^{-4}$ mol Ni. This amount of Ni requires

$$2\left(\frac{\text{mol DMG}}{\text{mol Ni}}\right)(5.1 \times 10^{-4} \text{ mol Ni})\left(116.12 \frac{\text{g DMG}}{\text{mol DMG}}\right) = 0.12 \text{ g DMG}$$

A 50% excess of DMG would be $(1.5)(0.12 \text{ g}) = 0.18$ g.

The DMG solution is 1.0 wt %, which means that there are 0.010 g of DMG per gram of solution. The required mass of solution is

$$\left(\frac{0.18 \text{ g DMG}}{0.010 \text{ g DMG/g solution}}\right) = 18 \text{ g solution}$$

The volume of solution is found from the mass of solution and the density:

$$\text{volume} = \frac{\text{mass}}{\text{density}} = \frac{18 \text{ g solution}}{0.79 \text{ g solution/mL}} = 23 \text{ mL}$$

$$\text{density} = \frac{\text{mass}}{\text{volume}}$$

(b) If 1.163 4 g of steel gave 0.179 5 g of $Ni(DMG)_2$ precipitate, what is the weight percent of Ni in the steel?

SOLUTION Here is the strategy. From the mass of precipitate, we can find the moles of precipitate. We know that one mole of precipitate comes from one mole of Ni in Equation 19-2. From the moles of Ni, we can compute the mass of Ni and its weight percent in the steel:

First, find the moles of precipitate in 0.179 5 g of precipitate:

$$\frac{0.179 \text{ 5 g } Ni(DMG)_2}{288.91 \text{ g } Ni(DMG)_2/\text{mol } Ni(DMG)_2} = 6.213 \times 10^{-4} \text{ mol } Ni(DMG)_2$$

There must have been 6.213×10^{-4} mol of Ni in the steel. The mass of Ni in the steel is

$$(6.213 \times 10^{-4} \text{ mol Ni})\left(58.69 \frac{\text{g}}{\text{mol Ni}}\right) = 0.036 \text{ 46 g}$$

and the weight percent of Ni in steel is

$$\text{wt \% Ni} = \frac{0.036 \text{ 46 g Ni}}{1.163 \text{ 4 g steel}} \times 100 = 3.134\%$$

FIGURE 19-4 Gravimetric combustion analysis for carbon and hydrogen.

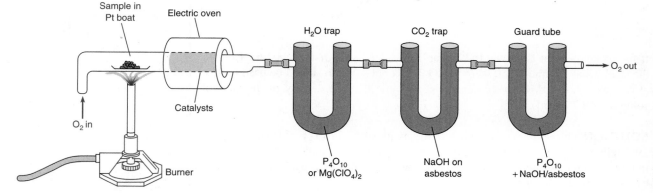

Sample in Pt boat   Electric oven H_2O trap CO_2 trap   Guard tube O_2 out

Catalysts O_2 in   Burner P_4O_{10} or $Mg(ClO_4)_2$ NaOH on asbestos P_4O_{10} + NaOH/asbestos

Ask Yourself

19-C. The lanthanide element cerium was discovered in 1839 and named after the asteroid Ceres. Cerium is a major component of flint lighters. To find the Ce^{4+} content of a solid, an analyst dissolved 4.37 g of the solid and treated it with excess iodate to precipitate $Ce(IO_3)_4$. The precipitate was collected, washed well, dried, and ignited to produce 0.104 g of CeO_2.

$$Ce^{4+} + 4IO_3^- \longrightarrow Ce(IO_3)_4(s) \xrightarrow{\text{heat}} CeO_2(s)$$
$$\text{MW } 172.114$$

(a) How much cerium is contained in 0.104 g of CeO_2?

(b) What was the weight percent of Ce in the original solid?

19-4 Combustion Analysis

A historically important form of gravimetric analysis is *combustion analysis,* used to determine the carbon and hydrogen content of organic compounds burned in excess O_2. Modern combustion analyzers use thermal conductivity, infrared absorption, or electrochemical methods to measure the products.

Gravimetric Combustion Analysis

In gravimetric combustion analysis (Figure 19-4), partially combusted product is passed through catalysts such as Pt gauze, CuO, PbO_2, or MnO_2 at elevated temperature to complete the oxidation to CO_2 and H_2O. The products are flushed through a chamber containing P_4O_{10} ("phosphorus pentoxide"), which absorbs water, and then through a chamber of Ascarite (NaOH on asbestos), which absorbs CO_2. The increase in mass of each chamber tells how much hydrogen and carbon, respectively, were initially present. A guard tube prevents atmospheric H_2O or CO_2 from entering the chambers from the exit.

EXAMPLE Combustion Analysis Calculations

A compound weighing 5.714 mg produced 14.414 mg of CO_2 and 2.529 mg of H_2O upon combustion. Find the weight percent of C and H in the sample.

SOLUTION One mole of CO_2 contains one mole of carbon. Therefore,

moles of C in sample = moles of CO_2 produced

$$= \frac{14.414 \times 10^{-3} \text{ g } CO_2}{44.010 \text{ g } CO_2/\text{mol}} = 3.275 \times 10^{-4} \text{ mol}$$

mass of C in sample $= (3.275 \times 10^{-4} \text{ mol C})\left(12.011 \frac{\text{g}}{\text{mol C}}\right) = 3.934 \text{ mg}$

$$\text{wt \% C} = \frac{3.934 \text{ mg C}}{5.714 \text{ mg sample}} \times 100 = 68.84\%$$

One mole of H_2O contains two moles of H. Therefore,

moles of H in sample $= 2$(moles of H_2O produced)

$$= 2\left(\frac{2.529 \times 10^{-3} \text{ g } H_2O}{18.015\ 2 \text{ g } H_2O/\text{mol}}\right) = 2.808 \times 10^{-4} \text{ mol}$$

$$\text{mass of H in sample} = (2.808 \times 10^{-4} \text{ mol H})\left(1.007\ 9 \frac{\text{g}}{\text{mol H}}\right) = 2.830 \times 10^{-4} \text{ g}$$

$$\text{wt \% H} = \frac{0.283\ 0 \text{ mg H}}{5.714 \text{ mg sample}} \times 100 = 4.952\%$$

Combustion Analysis Today

Figure 19-5 shows how C, H, N, and S are measured in a single operation. An accurately weighed 2-mg sample is sealed in a tin or silver capsule. The analyzer is swept with He gas that has been treated to remove traces of O_2, H_2O, and CO_2. At the start of a run, a measured excess volume of O_2 is added to the He stream. Then the sample capsule is dropped into a preheated ceramic crucible, where the capsule melts and the sample is rapidly oxidized.

$$\text{C, H, N, S} \xrightarrow[O_2]{1\ 050°C} CO_2(g) + H_2O(g) + N_2(g) + \underbrace{SO_2(g) + SO_3(g)}_{95\% \ SO_2}$$

Elemental analyzers use an *oxidation catalyst* to complete the oxidation of sample and a *reduction catalyst* to carry out any required reduction and to remove excess O_2.

The products pass through a hot WO_3 catalyst to complete the combustion of carbon to CO_2. In the next zone, metallic Cu at 850°C reduces SO_3 to SO_2 and removes excess O_2:

$$\text{Cu} + SO_3 \xrightarrow{850°C} SO_2 + CuO(s)$$

$$\text{Cu} + \tfrac{1}{2}O_2 \xrightarrow{850°C} CuO(s)$$

The mixture of CO_2, H_2O, N_2, and SO_2 is separated by gas chromatography, and each component is measured with a thermal conductivity detector described in

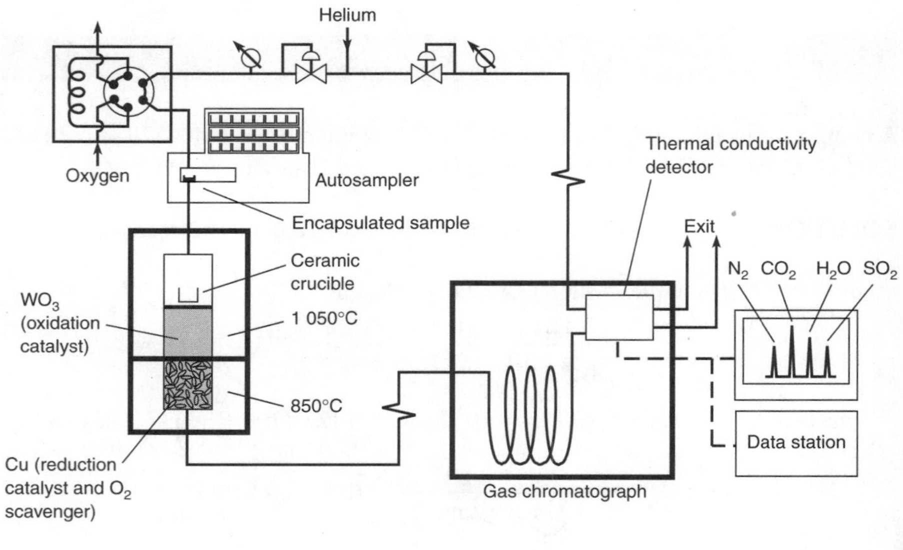

FIGURE 19-5 Schematic diagram of C,H,N,S elemental analyzer that uses gas chromatographic separation and thermal conductivity detection.

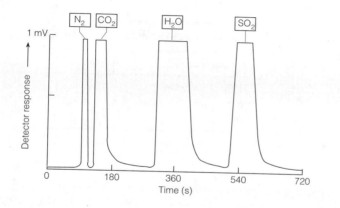

FIGURE 19-6 Gas chromatographic trace from elemental analyzer, showing substantially complete separation of combustion products. The area of each peak (when they are not off-scale) is proportional to the mass of each product.

Section 17-1 (Figure 19-6). Figure 19-7 shows a different C,H,N,S analyzer, which uses infrared absorbance to measure CO_2, H_2O, and SO_2 and thermal conductivity to measure N_2.

A key to elemental analysis is *dynamic flash combustion,* which creates a short burst of gaseous products instead of slowly bleeding products out over several minutes. This feature is important because chromatographic analysis requires that the whole sample be injected at once. Otherwise, the injection zone is so broad that the products cannot be separated.

In dynamic flash combustion, the sample is encapsulated in tin and dropped into the preheated furnace slowly after the flow of a 50 vol % O_2/50 vol % He mixture is started (Figure 19-8). The Sn capsule melts at 235°C and is instantly oxidized to SnO_2, a process that liberates 594 kJ/mol and heats the sample to 1 700°–1 800°C. Because the sample is dropped in before very much O_2 is present, decomposition of the sample occurs prior to oxidation, which minimizes the formation of nitrogen oxides.

The Sn capsule is oxidized to SnO_2, which

1. liberates heat to vaporize and crack sample
2. uses available oxygen immediately
3. ensures that sample oxidation occurs in gas phase
4. acts as an oxidation catalyst

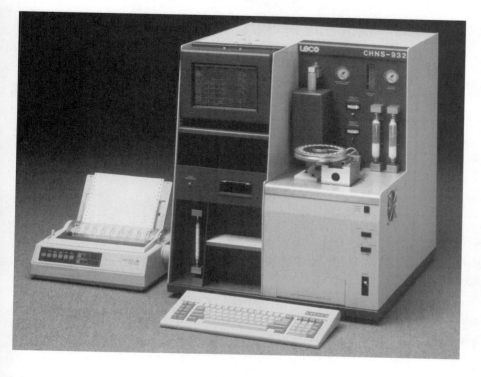

FIGURE 19-7 Combustion analyzer that uses infrared absorbance to measure CO_2, H_2O, and SO_2 and thermal conductivity to measure N_2. Three separate infrared cells in series are equipped with filters to isolate wavelengths absorbed by one of the products. Absorbance is integrated over time as the combustion product mixture is swept through each cell.

Element	Theoretical value (wt %)	Instrument 1	Instrument 2

TABLE 19-4 C, H, and N in acetanilide: $C_6H_5NHCCH_3$ (with O double-bonded structure)

Element	Theoretical value (wt %)	Instrument 1	Instrument 2
C	71.09	71.17 ± 0.41	71.22 ± 1.1
H	6.71	6.76 ± 0.12	6.84 ± 0.10
N	10.36	10.34 ± 0.08	10.33 ± 0.13

Uncertainties are standard deviations from five replicate determinations.

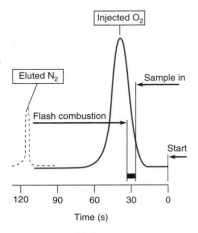

FIGURE 19-8 Sequence of events in dynamic flash combustion.

Oxygen analysis requires a different strategy. The sample is thermally decomposed (a process called **pyrolysis**) without adding oxygen. The gaseous products are passed through nickel-coated carbon at 1 075°C to convert oxygen from the compound into CO (not CO_2). Other products include N_2, H_2, CH_4, and hydrogen halides. Acidic products are absorbed by NaOH/asbestos, and the remaining gases are separated and measured by gas chromatography with a thermal conductivity detector.

For halogen analysis, the combustion product contains HX (X = Cl, Br, I). HX is trapped in aqueous solution and titrated with Ag^+ ions (Section 14-1) by an automated electrochemical process.

Table 19-4 shows analytical results for pure acetanilide obtained with two different commercial instruments. Chemists consider a result within ±0.3 of the theoretical percentage of an element to be good evidence that the compound has the expected formula. For N in acetanilide, ±0.3 corresponds to a relative error of $0.3/10.36 = 3\%$, which is not hard to achieve. For C, ±0.3 corresponds to a relative error of $0.3/71.09 = 0.4\%$, which is not so easy. The standard deviation for C in Instrument 1 is $0.41/71.17 = 0.6\%$, and for Instrument 2 it is $1.1/71.22 = 1.5\%$.

Ask Yourself

19-D. **(a)** What is the difference between combustion and pyrolysis?
 (b) What is the purpose of the WO_3 and Cu in Figure 19-5?
 (c) Why is tin used to encapsulate a sample for combustion analysis?
 (d) Why is sample dropped into the preheated furnace before the oxygen concentration reaches its peak in Figure 19-8?
 (e) What is the balanced equation for the combustion of $C_8H_7NO_2SBrCl$ in a C,H,N,S elemental analyzer?

Important Terms

absorption	dialysis	hygroscopic substance	precipitant
adsorption	digestion	ignition	pyrolysis
colloid	gathering	masking agent	supersaturated solution
combustion analysis	gravimetric analysis	nucleation	thermogravimetric analysis
coprecipitation	homogeneous precipitation	peptization	

Problems

19-1. An organic compound with a molecular weight of 417 was analyzed for ethoxyl (CH_3CH_2O-) groups by the reactions

$$ROCH_2CH_3 + HI \longrightarrow ROH + CH_3CH_2I$$
$$(R = \text{remainder of molecule})$$
$$CH_3CH_2I + Ag^+ + OH^- \longrightarrow AgI(s) + CH_3CH_2OH$$

A 25.42-mg sample produced 29.03 mg of AgI (FW 234.77). How many ethoxyl groups are there in each molecule?

19-2. A 0.050 02-g sample of impure piperazine contained 71.29 wt % piperazine. How many grams of product will be formed if this sample is analyzed by Reaction 19-1?

19-3. A 1.000-g sample of unknown analyzed by Reaction 19-2 gave 2.500 g of bis(dimethylglyoximate)nickel(II). Find the wt % of Ni in the unknown.

19-4. How many milliliters of 2.15 wt % dimethylglyoxime solution should be used to provide a 50.0% excess for Reaction 19-2 with 0.998 4 g of steel containing 2.07 wt % Ni? The density of dimethylglyoxime solution is 0.790 g/mL.

19-5. A solution containing 1.263 g of unknown potassium compound was dissolved in water and treated with excess sodium tetraphenylborate, $Na^+B(C_6H_5)_4^-$ solution to precipitate 1.003 g of insoluble $K^+B(C_6H_5)_4^-$ (FW 358.33). Find the wt % of K in the unknown.

19-6. Twenty dietary iron tablets with a total mass of 22.131 g were ground and mixed thoroughly. Then 2.998 g of the powder was dissolved in HNO_3 and heated to convert all the iron to Fe^{3+}. Addition of NH_3 caused quantitative precipitation of $Fe_2O_3 \cdot xH_2O$, which was ignited to give 0.264 g of Fe_2O_3 (FW 159.69). What is the average mass of $FeSO_4 \cdot 7H_2O$ (FW 278.01) in each tablet?

19-7. *The man in the vat problem.* Long ago a workman at a dye factory fell into a vat containing a hot concentrated mixture of sulfuric and nitric acids. He dissolved completely! Because nobody witnessed the accident, it was necessary to prove that he fell in so that the man's wife could collect his insurance money. The man weighed 70 kg, and a human body contains about 6.3 parts per thousand phosphorus. The acid in the vat was analyzed for phosphorus to see if it contained a dissolved human.

    (a) The vat had 8.00×10^3 L of liquid, and 100.0 mL was analyzed. If the man did fall into the vat, what is the expected quantity of phosphorus in 100.0 mL?

    (b) The 100.0 mL was treated with a molybdate reagent that caused ammonium phosphomolybdate, $(NH_4)_3[P(Mo_{12}O_{40})] \cdot 12H_2O$, to precipitate. This substance was dried at 110°C to remove waters of hydration and heated to 400°C until it reached a constant composition corresponding to the formula $P_2O_5 \cdot 24MoO_3$, which weighed 0.371 8 g. When a fresh mixture of the same acids (not from the vat) was treated in the same manner, 0.033 1 g of $P_2O_5 \cdot 24MoO_3$ (FW 3 596.46) was produced. This *blank determination* gives the amount of phosphorus in the starting reagents. The $P_2O_5 \cdot 24MoO_3$ that could have come from the dissolved man is therefore $0.371\ 8 - 0.033\ 1 = 0.338\ 7$ g. How much phosphorus was present in the 100.0-mL sample? Is this quantity consistent with a dissolved man?

19-8. Consider a mixture of the two solids $BaCl_2 \cdot 2H_2O$ (FW 244.26) and KCl (FW 74.551), in an unknown ratio. When the unknown is heated to 160°C for 1 h, the water of crystallization is driven off:

$$BaCl_2 \cdot 2H_2O(s) \xrightarrow{160°C} BaCl_2(s) + 2H_2O(g)$$

A sample originally weighing 1.783 9 g weighed 1.562 3 g after heating. Calculate the weight percent of Ba, K, and Cl in the original sample. (*Hint:* The mass loss tells how much water was lost, which tells how much $BaCl_2 \cdot 2H_2O$ was present. The remainder of the sample is KCl.)

19-9. Write a balanced equation for the combustion of benzoic acid, $C_6H_5CO_2H$, to give CO_2 and H_2O. How many milligrams of CO_2 and of H_2O will be produced by the combustion of 4.635 mg of benzoic acid?

19-10. Combustion of 8.732 mg of an unknown organic compound gave 16.432 mg of CO_2 and 2.840 mg of H_2O.

    (a) Find the wt % of C and H in the substance.

    (b) Find the smallest reasonable integer mole ratio of C:H in the compound.

19-11. Combustion analysis of a compound known to contain just C, H, N, and O demonstrated that it contains 46.21 wt % C, 9.02 wt % H, 13.74 wt % N, and, by difference, $100 - 46.21 - 9.02 - 13.74 = 31.03$ wt % O. This means that 100 g of unknown would contain 46.21 g of C, 9.02 g of H, etc. Find the atomic ratio C:H:N:O. Then divide each stoichiometry coefficient by the smallest one and express the atomic composition in the lowest reasonable integer ratio ($C_xH_yN_zO_w$, where x, y, z, and w are integers and one of them is 1).

19-12. A method for measuring organic carbon in seawater involves oxidation of the organic materials to CO_2 with $K_2S_2O_8$, followed by gravimetric determination of the CO_2 trapped by a column of NaOH-coated asbestos. A water sample weighing 6.234 g produced 2.378 mg of CO_2 (FW 44.010). Calculate the ppm carbon in the seawater.

19-13. Use the uncertainties from Instrument 1 in Table 19-4 to estimate the uncertainties in the stoichiometry coefficients in the formula $C_8H_{h\pm x}N_{n\pm y}$.

Measuring Calcium in Single Cells

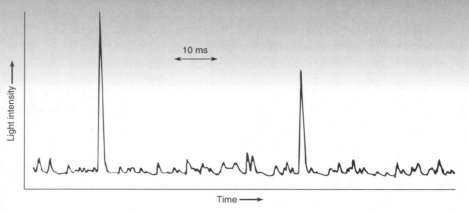

Left: Inductively coupled plasma torch decomposes molecules into atoms that emit specific wavelengths of light. *Right:* Bursts of light recorded from calcium when two individual cells enter the flame.

An inductively coupled plasma is an extremely hot flame that decomposes molecules into excited atoms, which then emit characteristic frequencies of light. Emission intensity is proportional to the quantity of analyte atoms over many orders of magnitude. The photograph shows an inductively coupled plasma consisting of argon ions and electrons whose energy is derived by absorption of radio frequency energy from the water-cooled electrical coil at the base of the torch. When a suspension of animal cells is injected into the flame, the calcium from each cell is vaporized all at once and a single burst of light is observed from each cell. The area of each spike in the graph is proportional to the mass of calcium in each cell. Representative results for three different cell types are given in the table.

Cell type	Ca content (pg/cell)	Cell diameter (μm)	[Ca] in cell (mM)
Mouse fibroblast	0.06 ±0.03	10–15	0.8–2.8
Human pancreas	0.16 ±0.04	15–20	0.9–2.2
Human endothelium	0.27 ±0.04	15–20	1.6–3.7

20

Atomic Spectroscopy

Atomic spectroscopy is a principal tool for measuring metallic elements at parts per million (and lower) levels in industrial and environmental laboratories. This workhorse technique is automated with mechanical sample-changing devices that allow each instrument to turn out hundreds of analyses per day.

20-1 What Is Atomic Spectroscopy?

In atomic spectroscopy (Figure 20-1), a liquid sample is *aspirated* (sucked) through a plastic tube into a flame that is hot enough to break molecules apart into atoms. The concentration of an element in the flame is measured either by its absorption or emission of radiation. For **atomic absorption spectroscopy,** radiation of the correct frequency is shined through the flame (Figure 20-2) and the transmitted radiation is measured. For **atomic emission spectroscopy,** no lamp is required. Radiation is emitted by hot atoms whose electrons have been promoted to excited states in the flame. For both experiments in Figure 20-2, a monochromator is used to select the wavelength of radiation that will be viewed by the detector. It is routine to measure analyte concentrations at the parts per million level with a precision of 2%. To analyze major constituents of an unknown, the sample must be diluted to reduce concentrations to the parts per million level.

   A typical spectrum of a molecule in solution, such as Figure 8-1, has bands that are ~100 nm in width. In contrast, the spectrum of gaseous atoms in a flame has extremely sharp lines with widths of 10^{-3} to 10^{-2} nm (Figure 20-3). Because the lines are so sharp, there is usually little overlap between the spectra of different elements in the same sample. The lack of overlap allows some instruments to measure over 60 elements in a sample simultaneously.

Atomic spectroscopy:

- absorption (requires a lamp with light whose frequency is absorbed by atoms)
- emission (luminescence from excited atoms—no lamp required)

ppm (parts per million) = micrograms of solute per gram of solution

Because the density of dilute aqueous solutions is close to 1 g/mL, ppm usually refers to μg/mL. A concentration of 1.00 ppm of Fe corresponds to 1.00×10^{-6} g Fe/mL = 1.79×10^{-5} M.

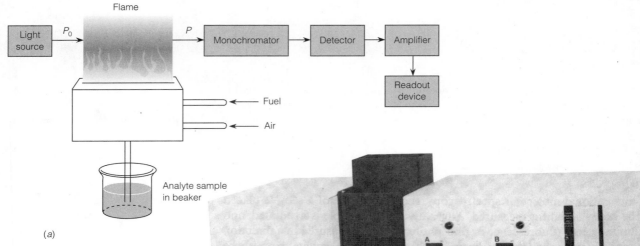

(a)

FIGURE 20-1 (a) Outline of an atomic absorption experiment. (b) An instrument for atomic absorption and emission. The sample in the flask is aspirated into the burner, which is behind the black screen at the upper center. Valves on the left control the flow of fuel and oxidizer. Dials on the right select wavelength and monochromator bandwidth.

(b)

FIGURE 20-2 Absorption and emission of light by atoms in a flame. In atomic absorption, atoms absorb light from the lamp and unabsorbed light reaches the detector. In atomic emission, light is emitted by excited atoms in the flame.

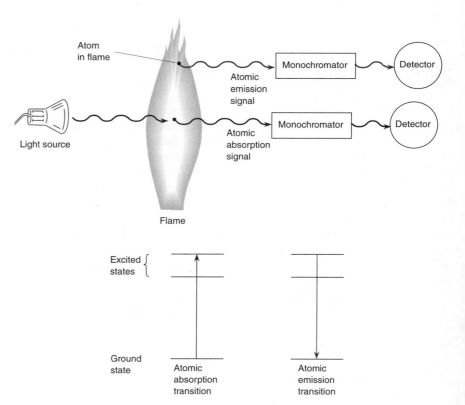

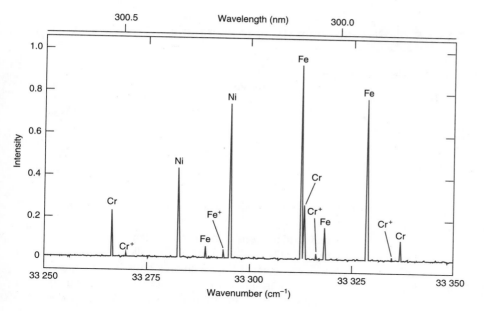

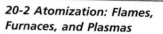

FIGURE 20-3 A tiny portion of the spectrum of a steel hollow-cathode lamp, showing sharp lines characteristic of gaseous Fe, Ni, and Cr atoms and weak lines from Cr^+ and Fe^+ ions. The resolution is 0.001 nm, which is about half of the true linewidths of the signals.

Ask Yourself

20-A. What is the difference between atomic absorption and atomic emission spectroscopy?

20-2 *Atomization: Flames, Furnaces, and Plasmas*

The essential feature of atomic spectroscopy is a flame, an electrically heated graphite tube, or a radio-frequency plasma in which analyte is **atomized** (broken into atoms).

Flames

Most flame spectrometers use a *premix burner,* such as that in Figure 20-4, in which the sample, oxidant, and fuel are mixed before being introduced into the flame. Sample solution is drawn in by the rapid flow of oxidant and breaks into a fine mist when it leaves the tip of the *nebulizer* and strikes a glass bead. The formation of small droplets is termed *nebulization.* The mist flows past a series of baffles to promote further mixing and block large droplets of liquid (which flow out to the drain). A fine mist containing about 5% of the initial sample reaches the flame.

After solvent evaporates in the flame, the remaining sample vaporizes and decomposes to atoms. Many metal atoms (M) form oxides (MO) and hydroxides (MOH) as they rise through the flame. Molecules do not have the same spectra as atoms, so the atomic signal is lowered. If the flame is relatively rich in fuel (a "rich" flame), excess carbon species tend to reduce MO and MOH back to M and thereby increase sensitivity. The opposite of a rich flame is a "lean" flame, which has excess

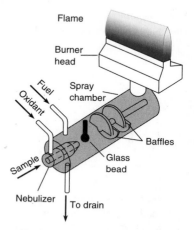

FIGURE 20-4 Premix burner. The slot in the burner head is typically 10 cm long and 0.5 mm wide.

Hotter flames are needed for refractory elements or to decompose metal oxides.

TABLE 20-1	Maximum flame temperatures	
Fuel	*Oxidant*	*Temperature (K)*
Acetylene	Air	2 400–2 700
Acetylene	Nitrous oxide	2 900–3 100
Acetylene	Oxygen	3 300–3 400
Hydrogen	Air	2 300–2 400
Hydrogen	Oxygen	2 800–3 000
Cyanogen	Oxygen	4 800

FIGURE 20-5 Electrically heated graphite-rod furnace for atomic spectroscopy. Light travels the length of the furnace (~38 mm in this case), and sample is injected through the hole at the top.

Furnaces offer increased sensitivity and require less sample than a flame.

The operator must determine reasonable time and temperature for each stage of the analysis. Once a program is established, it can be applied to similar samples.

oxidant and is hotter. Whether a lean flame or a rich flame is used depends on the elements being analyzed.

The most common fuel-oxidant combination is acetylene and air, which produces a flame temperature of 2 400–2 700 K (Table 20-1). When a hotter flame is required to vaporize *refractory* elements (those with high boiling points), the acetylene–nitrous oxide combination is usually used. The height above the burner head at which maximum atomic absorption or emission is observed depends on the element being measured, as well as the flow rates of sample, fuel, and oxidant. Each of these parameters must be optimized for a given analysis.

Furnaces

An electrically heated **graphite furnace** (Figure 20-5) provides greater sensitivity than a flame and requires less sample. Sample with a volume of 1 to 100 μL is injected into the oven through the hole at the center. The light beam travels through windows at each end of the tube. The maximum recommended temperature for a graphite furnace is 2 550°C for not more than 7 s. A surrounding atmosphere of Ar helps prevent graphite oxidation.

The graphite furnace has a high sensitivity because it confines atoms in the optical path for a *residence time* of several seconds. In flame spectroscopy, the sample is diluted during nebulization, and its residence time in the optical path is only a fraction of a second. Flames require a sample volume of at least 1–2 mL, because sample is constantly flowing into the flame. The graphite furnace requires much less sample. In an extreme case, when only nanoliters of kidney tubular fluid were available, a method was devised to reproducibly deliver 0.1 nL to a furnace for analysis of Na and K. Precision with a furnace is rarely better than 5–10% with manual sample injection, but automated injection improves reproducibility.

A skilled operator must determine heating conditions for three or more steps to properly atomize a sample. To analyze Fe in the iron-storage protein ferritin, 10 μL of sample containing ~0.1 ppm Fe is injected into the cold oven. The furnace is programmed to *dry* the sample at 125°C for 20 s to remove solvent. Drying is followed by 60 s of *charring* (also called *pyrolysis*) at 1 400°C to destroy organic matter, which creates smoke that would interfere with the optical measurement. *Atomization* is then carried out at 2 100°C for 10 s, during which the absorbance reaches a maximum and then decreases as Fe evaporates from the oven. The time-integrated absorbance is taken as the analytical signal. Finally, the furnace is heated to 2 500°C for 3 s to vaporize any residue.

Improved performance is obtained with a *L'vov platform* (Figure 20-6a). Without a platform, sample is injected directly onto the inside wall of the furnace. Atomization occurs while the wall is still heating up at a rate of 2 000 K/s (Figure

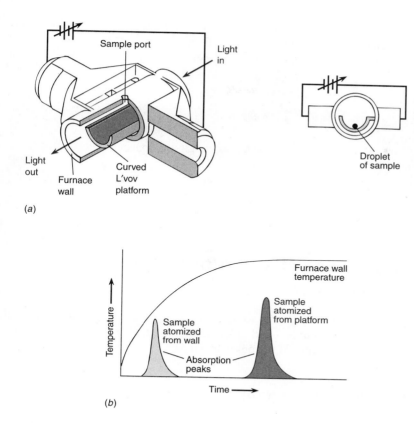

(a)

(b)

FIGURE 20-6 (a) L'vov platform in graphite furnace is uniformly heated by radiation from the resistively heated outer wall. The platform is attached to the wall by one small connection hidden from view. (b) Heating profile comparing analyte evaporation from wall and from platform.

20-6b). If sample is injected onto a platform inside the furnace, the temperature of the sample on the platform lags behind the rising wall temperature. Analyte does not vaporize until the wall reaches constant temperature (Figure 20-6b), and a more reliable measurement results.

The temperature needed to char the sample **matrix** (the medium containing the analyte) may vaporize the analyte. *Matrix modifiers* can retard evaporation of analyte until the matrix has charred away. For example, $Mg(NO_3)_2$ matrix modifier prevents premature evaporation of Al analyte. At high temperature, $Mg(NO_3)_2$ produces gaseous MgO. Aluminum is converted to $Al_2O_3(s)$ during heating. At elevated temperature, Al_2O_3 decomposes to Al and O, and the Al evaporates. However, evaporation of Al is retarded as long as MgO is present, by virtue of the reaction

$$3MgO(g) + 2Al(s) \rightleftharpoons 3Mg(g) + Al_2O_3(s)$$

When all the MgO is gone, Al_2O_3 finally decomposes and evaporates. The $Mg(NO_3)_2$ matrix modifier prevents Al from evaporating until a high temperature is reached.

Inductively Coupled Plasmas

The **inductively coupled plasma** discussed at the beginning of this chapter reaches a much higher temperature than combustion flames. Its high temperature and stability eliminate many problems encountered with conventional flames. The plasma's disadvantage is its expense to purchase and to operate.

The inductively coupled plasma in Figure 20-7 has a radio-frequency induction coil wrapped around the quartz burner head. High-purity argon gas is fed into the

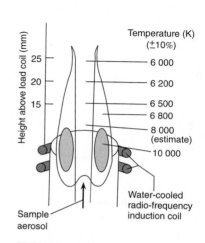

FIGURE 20-7 Temperature profile of a typical inductively coupled plasma used in analytic spectroscopy.

plasma from below. After a spark from a Tesla coil ionizes the Ar gas, free electrons are accelerated by the radio-frequency field and transfer energy to the entire gas by colliding with atoms to produce a temperature of 6 000 to 10 000 K.

Ask Yourself

20-B. Why can we detect smaller samples with lower concentrations with a furnace than with a flame or a plasma?

20-3 How Temperature Affects Atomic Spectroscopy

Temperature determines the degree to which a sample breaks down to atoms and the extent to which an atom is found in its ground, excited, or ionized states. Each effect influences the strength of the observed signal.

The Boltzmann Distribution

Consider a molecule with two energy levels (Figure 20-8) separated by energy ΔE. Call the lower level E_0 and the upper level E^*. In general, an atom (or molecule) may have more than one state available at a given energy level. In Figure 20-8, we show three states at E^* and two at E_0. The number of states at each energy level is called the *degeneracy* of the level. We will call the degeneracies g_0 and g^*.

The **Boltzmann distribution** describes the relative populations of different states at thermal equilibrium. If equilibrium exists (which is not true in all parts of a flame), the relative population (N^*/N_0) of any two states is

Boltzmann distribution:
$$\frac{N^*}{N_0} = \left(\frac{g^*}{g_0}\right)e^{-\Delta E/kT} \tag{20-1}$$

where T is temperature (K) and k is Boltzmann's constant (1.381×10^{-23} J/K).

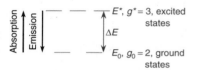

FIGURE 20-8 Two energy levels with different degeneracies. Ground-state atoms can absorb light to be promoted to the excited state. Excited-state atoms can emit light to return to the ground state.

The Effect of Temperature on the Excited-State Population

The lowest excited state of a sodium atom lies 3.371×10^{-19} J/atom above the ground state. The degeneracy of the excited state is 2, whereas that of the ground state is 1. Let's calculate the fraction of sodium atoms in the excited state in an acetylene-air flame at 2 600 K:

$$\frac{N^*}{N_0} = \left(\frac{2}{1}\right)e^{-(3.371\,\times\,10^{-19}\,\text{J})/[(1.381\,\times\,10^{-23}\,\text{J/K})(2\,600\text{K})]} = 0.000\ 167$$

Fewer than 0.02% of the atoms are in the excited state.

How would the fraction of atoms in the excited state change if the temperature were 2 610 K instead?

$$\frac{N^*}{N_0} = \left(\frac{2}{1}\right)e^{-(3.371 \times 10^{-19} \text{ J})/[(1.381 \times 10^{-23} \text{ J/K})(2\ 610 \text{ K})]} = 0.000\ 174$$

The fraction of atoms in the excited state is still less than 0.02%, but that fraction has increased by $[(1.74 - 1.67)/1.67] \times 100 = 4\%$.

A 10-K temperature rise changes the excited-state population by 4% in this example.

The Effect of Temperature on Absorption and Emission

We see that 99.98% of the sodium atoms are in their ground state at 2 600 K. *Varying the temperature by 10 K hardly affects the ground-state population and would not noticeably affect the signal in an atomic absorption experiment.*

How would emission intensity be affected by a 10-K rise in temperature? In Figure 20-8, we see that absorption arises from ground-state atoms, but emission arises from excited-state atoms. Emission intensity is proportional to the population of the excited state. *Because the excited-state population changes by 4% when the temperature rises 10 K, the emission intensity rises by 4%.* It is critical in atomic *emission* spectroscopy that the flame be very stable, or the emission intensity will vary significantly. In atomic *absorption* spectroscopy, flame temperature variation is not as critical.

The inductively coupled plasma is almost always used for emission, not absorption, because it is so hot that a substantial fraction of atoms and ions are excited. Table 20-2 compares excited-state populations for a flame at 2 500 K and a plasma at 6 000 K. Although the fraction of excited atoms is small, each atom emits many photons per second because it is rapidly promoted back to the excited state by collisions.

The magnitude of atomic absorption is not as sensitive to temperature as the intensity of atomic emission, which is exponentially sensitive to temperature.

Ask Yourself

20-C. (a) A ground-state atom absorbs light of wavelength 400 nm to be promoted to an excited state. Find the energy difference (in joules) between the two states.

   (b) If both states have a degeneracy of $g_0 = g^* = 1$, find the fraction of excited-state atoms (N^*/N_0) at thermal equilibrium at 2 500 K.

TABLE 20-2 **Effect of energy separation and temperature on population of excited states**

Wavelength separation of states (nm)	Energy separation of states (J)	Excited-state fraction $(N^*/N_0)^a$	
		2 500 K	*6 000 K*
250	7.95×10^{-19}	1.0×10^{-10}	6.8×10^{-5}
500	3.97×10^{-19}	1.0×10^{-5}	8.3×10^{-3}
750	2.65×10^{-19}	4.6×10^{-4}	4.1×10^{-2}

a. Based on the equation $N^*/N_0 = (g^*/g_0)e^{-\Delta E/KT}$ in which $g^* = g_0 = 1$.

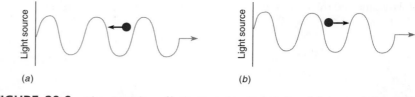

FIGURE 20-9 The Doppler effect. A molecule moving (*a*) toward the radiation source "feels" the electromagnetic field oscillate more often than one moving (*b*) away from the source.

20-4 *Instrumentation*

Requirements for atomic absorption were shown in Figure 20-1. Principal differences between atomic and solution spectroscopy lie in the light source, the sample container (the flame or furnace), and the need to subtract background emission from the observed signal.

The Linewidth Problem

The linewidth of the source must be narrower than the linewidth of the atomic vapor for Beer's law (Section 7-2) to be obeyed. "Linewidth" and "bandwidth" are used interchangeably, but "lines" are narrower than "bands."

In order for absorbance to be proportional to analyte concentration, the linewidth of radiation being measured must be substantially narrower than the linewidth of the absorbing atoms. Atomic absorption lines are very sharp, with an inherent width of $\sim 10^{-4}$ nm.

Two mechanisms broaden the lines in atomic spectroscopy. One is the *Doppler effect* in which an atom moving toward the lamp samples the oscillating electromagnetic wave more frequently than one moving away from the lamp (Figure 20-9). That is, an atom moving toward the source "sees" higher frequency light than that encountered by one moving away. Linewidth is also affected by *pressure broadening* from collisions between atoms. A colliding atom does not absorb exactly the same frequency of radiation as an isolated atom. Broadening is proportional to pressure. The Doppler effect and pressure broadening are similar in magnitude and yield linewidths of 10^{-3} to 10^{-2} nm in atomic spectroscopy.

Hollow-Cathode Lamps

Doppler and pressure effects broaden the atomic lines by one to two orders of magnitude relative to their inherent linewidths.

To produce narrow lines of the correct frequency, we use a **hollow-cathode lamp** containing a vapor of the same element as that being analyzed. The lamp in Figure

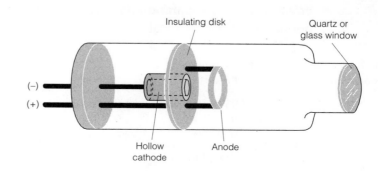

FIGURE 20-10 A hollow-cathode lamp.

FIGURE 20-11 Relative line-widths of hollow-cathode emission, atomic absorption, and a monochromator. The linewidth from the hollow cathode is relatively narrow because the gas temperature in the lamp is lower than a flame temperature (so there is less Doppler broadening) and the pressure in the lamp is lower than the pressure in a flame (so there is less pressure broadening).

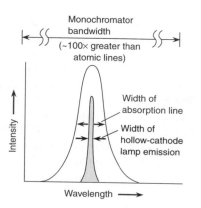

20-10 is filled with Ne or Ar at a pressure of 130–700 Pa. A high voltage between the anode and cathode ionizes the gas and accelerates cations toward the cathode. Ions striking the cathode "sputter" atoms from the metallic cathode into the gas phase. Free metal atoms are excited by collisions with high-energy electrons and then emit photons to return to the ground state. This atomic radiation shown in Figure 20-3 has the same frequency as that absorbed by analyte atoms in the flame or furnace. The linewidth in Figure 20-11 is sufficiently narrow for Beer's law to hold. A lamp with different cathode material is required for each element.

Background Correction

Atomic spectroscopy requires **background correction** to distinguish analyte signal from absorption, emission, and optical scattering by the sample matrix, the flame, plasma, or white-hot graphite furnace. For example, Figure 20-12 shows the absorption spectrum of Fe, Cu, and Pb in a graphite furnace. The sharp atomic signals with a maximum absorbance near 1.0 are superimposed on a flat background with an absorbance of 0.3. If we did not measure the background absorbance, significant errors would result. Background correction is most critical for graphite furnaces that tend to be filled with smoke from the charring step. Optical scatter from smoke must somehow be distinguished from optical absorption by analyte.

For atomic absorption, either *beam chopping* or electrical *modulation* of the hollow-cathode lamp is used to distinguish the signal of the flame from the desired

Background signal arises from absorption, emission, or scatter by everything in the sample besides analyte (the *matrix*) and from absorption, emission, or scatter by the flame, the plasma, or the furnace.

Modulation means applying an oscillating voltage to the lamp to pulse it on and off.

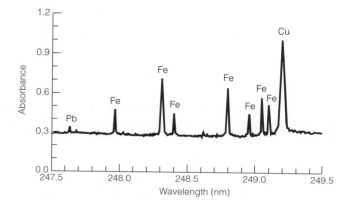

FIGURE 20-12 Graphite furnace absorption spectrum of bronze dissolved in HNO_3.

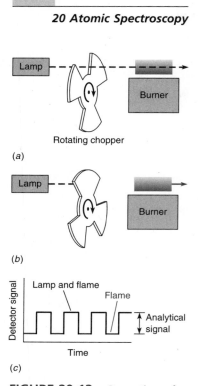

FIGURE 20-13 Operation of a beam chopper for subtracting the signal due to flame background emission. (a) Lamp and flame emission reach detector. (b) Only flame emission reaches detector. (c) Resulting square-wave signal.

Detection limits vary among instruments. An *ultrasonic nebulizer* improves the efficiency of nebulization and lowers the detection limit for some elements by more than a factor of 10.

atomic line at the same wavelength. Figure 20-13 shows light from the lamp being periodically blocked by a rotating chopper. Signal reaching the detector while the beam is blocked must be from flame emission. Signal reaching the detector when the beam is not blocked is from the lamp and the flame. The difference between these two signals is the desired analytical signal.

Beam chopping corrects for flame emission but not for scattering. Most spectrometers provide an additional means to correct for scattering and broad background absorption. Deuterium lamp, Smith-Hieftje (pronounced HEEF-yeh), and Zeeman correction systems are most common. We will discuss only the Smith-Hieftje method.

When a hollow-cathode lamp is run at high electric current, the output is broadened and a dip develops at the central wavelength as a result of absorption by free ground-state atoms located between the window of the lamp and the hot atoms near the cathode (Figure 20-14). For Smith-Hieftje background correction, the lamp is first run at low current to measure absorbance due to analyte plus background. Then the lamp is pulsed with a high current. During the pulse, analyte absorbance is reduced—but not to zero, because the two-humped output in Figure 20-14 still has significant intensity at the analytical wavelength. Most of the light reaching the detector during the pulse is due to broad background emission and scattering, because most of the lamp intensity no longer coincides with the sharp atomic absorption line of analyte. The difference in signals measured before and during the pulse is due to analyte. Smith-Hieftje background correction results in a 10–75% loss of sensitivity because analyte absorbs some fraction of the light during the high-current pulse.

Detection Limits

The **detection limit** is the concentration of an element that gives a signal equal to twice the peak-to-peak noise level of the baseline (Figure 20-15). The baseline noise level is measured while a blank sample is aspirated into the flame.

Detection limits for a furnace are typically 100 times lower than those of a flame (Figure 20-16), because the sample is confined in a small volume for a relatively long time in the furnace. Detection limits for the plasma are intermediate. New instruments improve detection limits for an inductively coupled plasma by viewing emission along the axis of the plasma, instead of perpendicular to the axis. The axial view increases the effective pathlength. Alkali metals have especially strong emission and are commonly measured in clinical analysis with relatively simple equipment (Box 20-1).

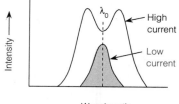

FIGURE 20-14 Distortion of hollow-cathode lamp emission at elevated current. Analyte atomic absorption is centered at wavelength λ_0.

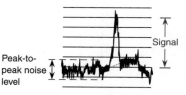

FIGURE 20-15 Measurement of peak-to-peak noise level and signal level. The signal is measured from its base at the midpoint of the noise component along the slightly slanted baseline. This sample exhibits a signal-to-noise ratio of 2.4.

Detection limits (ng/mL)

Example (Fe): 0.7 — Inductively coupled plasma emission; 5 — Flame atomic absorption; 0.02 — Graphite furnace atomic absorption

Element	ICP emission	Flame AA	Graphite furnace AA
Li	0.7	2	0.1
Be	0.07	1	0.02
B	1	500	15
C	10	—	
N			
O			
F			
Ne			
Na	3	0.2	0.005
Mg	0.08	0.3	0.004
Al	2	30	0.01
Si	5	100	0.1
P	7	40 000	30
S	3	—	
Cl			
Ar			
K	20	3	0.1
Ca	0.07	0.5	0.01
Sc	0.3	40	—
Ti	0.4	70	0.5
V	0.7	50	0.2
Cr	2	3	0.01
Mn	0.2	2	0.01
Fe	0.7	5	0.02
Co	1	4	0.02
Ni	3	90	0.1
Cu	0.9	1	0.02
Zn	0.6	0.5	0.001
Ga	10	60	0.5
Ge	20	200	—
As	7	200	0.2
Se	10	250	0.5
Br			
Kr			
Rb	1	7	0.05
Sr	0.2	2	0.1
Y	0.6	200	—
Zr	2	1000	—
Nb	5	2000	—
Mo	3	20	0.02
Tc			
Ru	10	60	1
Rh	20	4	—
Pd	4	10	0.3
Ag	0.8	2	0.005
Cd	0.5	0.4	0.003
In	20	40	0.2
Sn	9	30	0.2
Sb	9	40	0.15
Te	4	30	0.1
I			
Xe			
Cs	40 000	4	0.2
Ba	0.6	10	0.04
La	1	2000	—
Hf	4	2000	—
Ta	10	2000	—
W	8	1000	—
Re	3	600	—
Os	0.2	100	—
Ir	7	400	—
Pt	7	100	0.2
Au	2	10	0.1
Hg	7	150	2
Tl	10	20	0.1
Pb	10	10	0.05
Bi	7	40	0.1
Po			
At			
Rn			
Ce	2	—	—
Pr	9	6000	—
Nd	10	1000	—
Pm			
Sm	10	1000	—
Eu	0.9	20	0.5
Gd	5	2000	—
Tb	6	500	0.1
Dy	2	30	1
Ho	2	40	—
Er	0.7	30	2
Tm	2	900	—
Yb	0.3	4	—
Lu	0.3	300	—
Th	7	—	—
Pa			
U	60	40 000	—
Np			
Pu			
Am			
Cm			
Bk			
Cf			
Es			
Fm			
Md			
No			
Lr			

Legend: Shaded — Requires N_2O/C_2H_2 flame and is therefore better analyzed by inductively coupled plasma. Dark shaded — Best analyzed by emission.

FIGURE 20-16 Flame, furnace, and inductively coupled plasma detection limits (ng/mL = ppb) with instruments from one manufacturer. Reliable quantitative analysis usually requires concentrations 10–100 times greater than the detection limit.

Box 20-1 *Informed Citizen*

Atomic Emission Spectroscopy in the Clinical Laboratory

Before the advent of atomic spectroscopy, laborious gravimetric procedures were required to measure Na^+ and K^+ in serum and urine. Today these measurements are routine in clinical analysis.

The typical flame photometer shown here operates on air and natural gas. Optical filters isolate the strong emission lines of Na, K, Li, Ca, and Ba. Full-scale sensitivity for Na and K corresponds to 3 μg/mL; 50 μg/mL for Ca. The digital readout is calibrated by aspirating standard samples through the narrow tube immersed in the beaker. Lithium is often used as an *internal standard* (Section 16-5) for sodium and potassium.

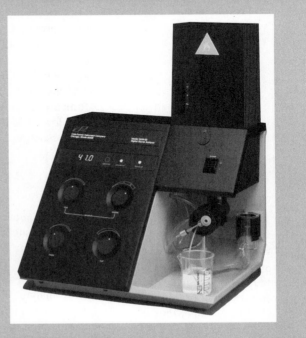

Digital flame photometer.

20-5 Interference

Interference is any effect that changes the signal when analyte concentration remains unchanged. Interference can be corrected by counteracting the source of interference or by preparing standards that exhibit the same interference.

Types of Interference

Types of interference:

spectral: unwanted signals overlap analyte signal

chemical: chemical reactions decrease the concentration of analyte atoms

ionization: ionization of analyte atoms decreases the concentration of neutral atoms

Spectral interference occurs when analyte signal overlaps with signals from other species in the sample or with signals due to the flame or furnace. Interference from the flame can be subtracted by background correction. The best means of dealing with overlap between lines of different elements in the sample is to choose another wavelength for analysis. The spectrum of a molecule is much broader than that of an atom, so spectral interference can occur at many wavelengths. Figure 20-17 shows an example of a plasma containing Y and Ba atoms in addition to YO molecules. Elements that form stable oxides in the flame are a common source of spectral interference.

    Chemical interference is caused by any substance that decreases the extent of atomization of analyte. For example, SO_4^{2-} and PO_4^{3-} hinder the atomization of Ca^{2+}, perhaps by forming nonvolatile salts. *Releasing agents* are chemicals that can be added to a sample to decrease chemical interference. EDTA and 8-hydroxyquinoline protect Ca^{2+} from the interfering effects of SO_4^{2-} and PO_4^{3-}. La^{3+} can also be used as a releasing agent, apparently because it preferentially reacts with PO_4^{3-} and frees the Ca^{2+}. A fuel-rich flame reduces oxidized analyte species that would otherwise hinder atomization. Higher flame temperatures eliminate many kinds of chemical interference.

    Ionization interference is a problem in the analysis of alkali metals, which have the lowest ionization potentials. For any element, we can write a gas-phase ionization reaction:

$$M(g) \rightleftharpoons M^+(g) + e^-(g) \qquad K = \frac{[M^+][e^-]}{[M]} \qquad (20\text{-}2)$$

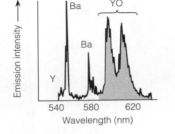

FIGURE 20-17 Emission from a plasma produced by laser irradiation of the high-temperature superconductor $YBa_2Cu_3O_7$. The solid is vaporized by the laser, and excited atoms and molecules in the gas phase emit light at their characteristic wavelengths.

At 2 450 K and a pressure of 0.1 Pa, sodium is 5% ionized. With its lower ionization potential, potassium is 33% ionized under the same conditions. Because ionized atoms have energy levels different from those of neutral atoms, the desired signal is decreased.

    An *ionization suppresser* is an element added to a sample to decrease the ionization of analyte. For example, 1 mg/mL of CsCl is added to the sample for the analysis of potassium, because cesium is more easily ionized than potassium. By producing a high concentration of electrons in the flame, ionization of Cs reverses Reaction 20-2 for K. This reversal is an example of Le Châtelier's principle.

An Ar plasma eliminates common interferences. The plasma is twice as hot as a conventional flame, and the residence time of analyte in the plasma is about twice as long. Atomization is more complete and the signal is correspondingly enhanced. Formation of analyte oxides and hydroxides is negligible. The plasma is relatively free of background radiation. Figure 20-18 shows plasma emission calibration curves that are linear over nearly five orders of magnitude. In flames and furnaces, the linear range spans only two orders of magnitude.

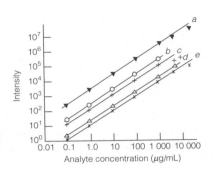

FIGURE 20-18 Analytical calibration curves for emission from (a) Ba^+, (b) Cu, (c) Na, (d) Fe, and (e) Ba in an inductively coupled plasma. Note that both axes are logarithmic.

Ask Yourself

20-E. What is meant by (a) spectral, (b) chemical, and (c) ionization interference?

20-6 Quantitative Analysis by Standard Addition

In **standard addition,** known quantities of analyte are added to the unknown, and the increased signal allows us to tell how much analyte was in the original unknown. This method requires a linear response to analyte. We illustrate standard addition for atomic spectroscopy, but the method applies to most types of chemical analysis.

Consider an inductively coupled plasma analysis in which the intensity of light emitted by analyte atoms in the plasma is proportional to their concentration in the unknown solution. A sample with unknown initial concentration $[X]_i$ gives an emission intensity I_X. Then a known concentration of standard S (a known concentration of analyte) is added to the sample and an emission intensity I_{S+X} is observed. Because emission is proportional to analyte concentration, we can say that

$$\frac{\text{concentration of analyte in unknown}}{\text{concentration of analyte plus standard in mixture}} = \frac{\text{signal from unknown}}{\text{signal from mixture}}$$

Standard addition equation:
$$\frac{[X]_i}{[X]_f + [S]_f} = \frac{I_X}{I_{S+X}} \qquad (20\text{-}3)$$

where $[X]_f$ is the final concentration of unknown analyte after adding the standard and $[S]_f$ is the final concentration of standard after addition to the unknown.

Bear in mind that the chemical species X and S are the same.

EXAMPLE   Standard Addition

A blood serum sample containing Na^+ gave an emission signal of 4.27 mV in a flame photometer. A small volume of concentrated standard Na^+ was then added to increase the Na^+ concentration by 0.104 M, without significantly diluting the sample. This "spiked" serum sample gave a signal of 7.98 mV in atomic emission. Find the original concentration of Na^+ in the serum.

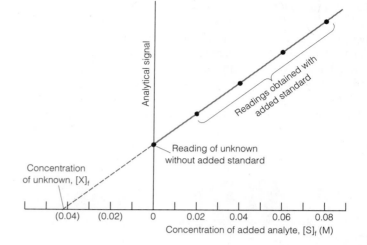

FIGURE 20-19 Graphical treatment of the method of standard addition.

SOLUTION Because the serum was not significantly diluted, $[Na^+]_f = [Na^+]_i$. Equation 20-3 allows us to solve for $[Na^+]_i$ in the original serum:

$$\frac{[X]_i}{[X]_f + [S]_f} = \frac{[Na^+]_i}{[Na^+]_i + 0.104 \text{ M}} = \frac{4.27 \text{ mV}}{7.98 \text{ mV}} \Rightarrow [Na^+]_i = 0.120 \text{ M}$$

An alternative procedure is to make a series of standard additions and to plot the results, as shown in Figure 20-19. The *x*-axis is the concentration of added analyte *after* it has been mixed with sample. The *x*-intercept of the extrapolated line is equal to the concentration of unknown *after* it has been diluted to the final volume. In Figure 20-19, this concentration is near 0.042 M. The most useful range of standard additions should increase the analytical signal to between 1.5 and 3 times its original value.

Standard addition is especially appropriate when the sample matrix is complex or unknown. *Matrix* refers to everything in the unknown, other than analyte. The matrix may have unknown constituents that you could not incorporate into standard solutions to make a calibration curve. When we add a small volume of concentrated standard, we do not change the concentration of the matrix very much. The assumption in standard addition is that the matrix has the same effect on added standard as it has on the original analyte in the unknown.

Ask Yourself

20-F. Suppose that 5.00 mL of serum containing an unknown concentration of potassium, $[K^+]_i$, gave an atomic emission signal of 3.00 mV. After adding 1.00 mL of 30.0 mM K^+ standard and diluting the mixture to 10.00 mL, the emission signal increased to 4.00 mV.

(a) Find the concentration of added standard, $[S]_f$, in the mixture.

(b) The initial 5.00-mL sample was diluted to 10.00 mL. Therefore the serum potassium concentration decreased from $[K^+]_i$ to $[K^+]_f = \frac{1}{2}[K^+]_i$. Use Equation 20-3 to find the original K^+ content of the serum, $[K^+]_i$.

Key Equations

Boltzmann distribution

$$\frac{N^*}{N_0} = \left(\frac{g^*}{g_0}\right)e^{-\Delta E/kT}$$

$N^* =$ population of excited state

$N_0 =$ population of ground state

$\Delta E =$ energy difference between excited and ground states

$g^* =$ number of states with energy E^*

$g_0 =$ number of states with energy E_0

Standard addition

$$\frac{[X]_i}{[X]_f + [S]_f} = \frac{I_X}{I_{S+X}}$$

$[X]_i =$ concentration of analyte in initial unknown

$[X]_f =$ concentration of analyte after standard addition

$[S]_f =$ concentration of standard after addition to unknown

Important Terms

atomic absorption spectroscopy

atomic emission spectroscopy

atomization

background correction

Boltzmann distribution

chemical interference

detection limit

graphite furnace

hollow-cathode lamp

inductively coupled plasma

ionization interference

matrix

spectral interference

standard addition

Problems

20-1. Compare the advantages and disadvantages of furnaces and flames in atomic absorption spectroscopy.

20-2. Compare the advantages and disadvantages of the inductively coupled plasma and flames in atomic spectroscopy.

20-3. In which technique, atomic absorption or atomic emission, is flame temperature stability more critical? Why?

20-4. Explain how Smith-Hieftje background correction works.

20-5. What is the purpose of a matrix modifier in atomic spectroscopy?

20-6. The first excited state of Ca is reached by absorption of 422.7-nm light.

(a) What is the energy difference (J) between the ground and excited states? (*Hint:* See Section 7-1.)

(b) The degeneracies are $g^*/g_0 = 3$ for Ca. What is the ratio N^*/N_0 at 2 500 K?

(c) By what percentage will the fraction in (b) be changed by a 15-K rise in temperature?

(d) Find the ratio N^*/N_0 at 6 000 K.

20-7. The first excited state of Cu is reached by absorption of 327-nm radiation.

(a) What is the energy difference (J) between the ground and excited states?

(b) The ratio of degeneracies is $g^*/g_0 = 3$ for Cu. What is the ratio N^*/N_0 at 2 400 K?

(c) By what percentage will the fraction in part (b) be changed by a 15-K rise in temperature?

(d) What will the ratio N^*/N_0 be at 6 000 K?

20-8. The atomic absorption signal shown on page 282 was obtained with 0.048 5 μg Fe/mL in a graphite furnace. Measure the signal height and the peak-to-peak noise level. The detection limit is the concentration of Fe that gives a signal equal to twice the peak-to-peak noise level. Estimate the detection limit for Fe.

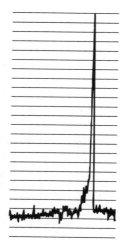

20-9. *Standard curve.* (Review Section 8-1 on standard curves.) A series of potassium standards gave the following emission intensities at 404.3 nm. Construct a standard curve and find the concentration of potassium in the unknown.

Sample (μg K/mL)	Relative emission
Blank	0
5.00	124
10.0	243
20.0	486
30.0	712
Unknown	417

20-10. *Standard addition.* An unknown sample of Cu^{2+} gave an absorbance of 0.262 in an atomic absorption analysis. Then 1.00 mL of solution containing 100.0 ppm (= μg/mL) Cu^{2+} was mixed with 95.0 mL of unknown, and the mixture was diluted to 100.0 mL in a volumetric flask. The absorbance of the new solution was 0.500.

(a) Denoting the initial, unknown concentration as $[Cu^{2+}]_i$, write an expression for the final concentration, $[Cu^{2+}]_f$, after dilution. Your answer will be an expression containing $[Cu^{2+}]_i$.

(b) Find the final concentration (ppm) of added standard Cu^{2+}.

(c) Use Equation 20-3 to find $[Cu^{2+}]_i$ in the unknown.

20-11. *Standard addition.* Li concentration was determined by atomic emission, using standard additions. From the data in the following table, prepare a graph similar to Figure 20-19 to find the concentration of Li in pure unknown. The Li standard contained 1.62 μg Li/mL.

Unknown (mL)	Standard (mL)	Final volume (mL)	Emission intensity (arbitrary units)
10.00	0.00	100.0	309
10.00	5.00	100.0	452
10.00	10.00	100.0	600
10.00	15.00	100.0	765
10.00	20.00	100.0	906

20-12. *Internal standard.* (Review Section 16-5 on internal standards.) Mn was used as an internal standard for measuring Fe by atomic absorption. A standard mixture containing 2.00 μg Mn/mL and 2.50 μg Fe/mL gave a quotient (Fe signal/Mn signal) = 1.05. A mixture with a volume of 6.00 mL was prepared by mixing 5.00 mL of unknown Fe solution with 1.00 mL containing 13.5 μg Mn/mL. The absorbance of this mixture at the Mn wavelength was 0.128, and the absorbance at the Fe wavelength was 0.185. Find the molarity of the unknown Fe solution.

20-13. *Standard addition.* An unknown containing element X was mixed with aliquots of a standard solution of element X for atomic absorption spectroscopy. The standard solution contained 1 000.0 μg of X per milliliter.

Volume of unknown (mL)	Volume of standard (mL)	Total volume (mL)	Absorbance
10.00	0	100.0	0.163
10.00	1.00	100.0	0.240
10.00	2.00	100.0	0.319
10.00	3.00	100.0	0.402
10.00	4.00	100.0	0.478

(a) Calculate the concentration (μg X/mL) of added standard in each solution.

(b) Prepare a graph similar to Figure 20-19 to determine the concentration of X in the unknown.

20-14. *Internal standard.* (Review Section 16-5 on internal standards.) A solution was prepared by mixing 10.00 mL of unknown (X) with 5.00 mL of standard (S) containing 8.24 μg S/mL and diluting to 50.0 mL. The measured signal quotient was (signal due to X/signal due to S) = 1.69.

(a) In a separate experiment it was found that for equal concentrations of X and S, the signal due to X was 0.93 times as intense as the signal due to S. Find the concentration of X in the unknown.

(b) Answer the same question if in a separate experiment it was found that for the concentration of X equal to 3.42 times the concentration of S, the signal due to X was 0.93 times as intense as the signal due to S.

Experiments

Experiments in this chapter illustrate major analytical techniques described in the text.[1] Although the procedures are safe when carried out with reasonable care, *all chemical experiments are potentially hazardous.* Any solution that fumes (such as concentrated HCl) and all nonaqueous solvents should be handled in a fume hood. Pipetting should never be done by mouth. Spills on your body should immediately be flooded with water and your instructor should be notified for possible further action. Spills on the benchtop should be cleaned immediately. Toxic chemicals should not be flushed down the drain. Your instructor should establish a safe procedure for disposing of each chemical that you use (see Box 2-1).

When equipment permits, it is environmentally friendly and (in the long term) economical to reduce the scale of an experiment. For example, 50-mL burets can be replaced by 10-mL burets at some loss in analytical precision, but little educational loss. Micropipets and small volumetric flasks can replace large glass pipets and large volumetric flasks (again at a sacrifice of precision). With properly chosen electrodes, potentiometric titrations can be performed on a scale of a few milliliters, instead of tens of milliliters, by using a syringe or a micropipet to deliver titrant. Instructions in this chapter were written for the commonly available, large-size lab equipment, but your modification to a smaller scale is encouraged.

21-1 Penny Statistics

U.S. pennies minted after 1982 have a Zn core with a Cu overlayer. Prior to 1982, pennies were made of brass, with a uniform composition (95 wt % Cu/5 wt % Zn). In 1982, both the heavier brass coins and the lighter zinc coins were made. In this experiment,[2] your class will weigh many coins and pool the data to answer the following questions: (1) Do pennies from different years have the same mass? (2) Do pennies from different mints have the same mass? Read Section 2-3 on the analytical balance and Chapter 4 on statistics prior to doing this experiment.

Gathering Data

Each student should collect and weigh enough pennies to the nearest milligram to provide a total set that contains 300 to 500 brass coins and a similar number of zinc coins. Instructions are given for a spreadsheet, but the same operations can be carried out with a calculator. Compile all the class data in a spreadsheet. Each column should list the masses of pennies from only one calendar year. Use the spreadsheet "sort" function to sort each column so that the lightest mass is at the top of the column and the heaviest is at the bottom. There will be two columns for 1982, in which both types of coins were made. Select a year other than 1982 for which you have many coins and divide the coins into those made in Denver (with a "D" beneath the year) and those minted in Philadelphia (with no mark beneath the year).

Discrepant Data

At the bottom of each column, list the mean (Equation 4-1) and standard deviation (Equation 4-2). Retain at least one extra digit beyond the milligram place to avoid round-off errors in your calculations.

[1]For a computerized list of experiments published in the *Journal of Chemical Education,* see *J. Chem. Ed.* **1994,** *71,* 450.

[2]T. H. Richardson, *J. Chem. Ed.* **1991,** *68,* 310.

Damaged or corroded coins may have masses different from the general population. Discard grossly discrepant masses lying ≥4 standard deviations from the mean in any one year. (For example, if one column has an average of 3.000 g and a standard deviation of 0.030 g, the 4 standard deviation limit is 3.000 ±(4 × 0.030) = 3.000 ±0.120 g. A mass that is ≤2.880 or ≥3.120 g should be discarded.) After rejecting discrepant data, recompute the average and standard deviation for each column.

Confidence Intervals and t-Test

Select the two years (≥1982) in which the zinc coins have the highest and lowest average masses. For each of the two years, compute the 95% confidence interval by using Equation 4-3. Use the *t*-test (Equation 4-4) to compare the two mean values at the 95% confidence level. Are the two average masses significantly different? Try the same for two years of brass coins (≤1982). Try the same for the one year whose coins you segregated into those from Philadelphia and those from Denver. Do the two mints produce coins with the same mass?

Gaussian Distribution of Masses

List the masses of all pennies made in or after 1983 in a single column, sorted from lowest to highest mass. There should be at least 300 masses listed. Divide the data into 0.01-g intervals (e.g., 2.480 to 2.489 g) and prepare a bar graph, like that shown in Figure 21-1. Find the mean ($\bar{x}$), median, and standard deviation (*s*) for all coins in the graph. For random (Gaussian) data, only 3 out of 1 000 measurements should lie outside of $\bar{x} \pm 3s$. Indicate which bars (if any) lie beyond ±3*s*. In Figure 21-1 two bars at the right are outside of $\bar{x} \pm 3s$.

Least-Squares Analysis: Do Pennies Have the Same Mass Each Year?

Prepare a graph like that in Figure 21-2 in which the ordinate (*y*-axis) is the mass of zinc pennies minted each year since 1982 and the abscissa (*x*-axis) is the year. For simplicity, let 1982 be year 1, 1983 be year 2, and so on. If the mass of a penny increases systematically from year to year, then the least-squares line through the data will have a positive slope. If the mass decreases, the slope will be negative. If the mass is constant, the slope will be zero. Even if the mass is really constant, your selection of coins is random and the slope is not exactly zero.

We want to know if the slope is *significantly* different from zero. Suppose that you have data for 14 years. Enter all of the data into two columns of the least-squares spreadsheet in Figure 4-8. Column B (x_i) is the year (1 to 14) and column C (y_i) is the mass of each penny. Your table will have several hundred entries. (If you are not using a spread-

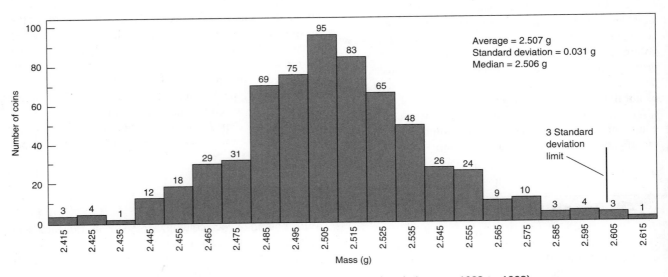

FIGURE 21-1 Distribution of 613 penny masses (made in years 1982 to 1992) measured in Dan's house by Jimmy Kusznir and Doug Harris in December 1992.

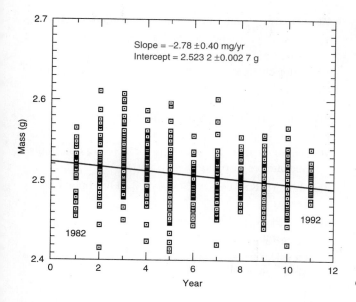

FIGURE 21-2 Penny mass versus year for 440 coins. Because the slope of the least-squares line is *significantly* less than zero, we conclude that the average mass of older pennies is greater than the average mass of newer pennies.

3. One year should be divided into one column from Denver and one from Philadelphia.

Discrepant Data

1. List the mean ($\bar{x}$) and standard deviation (s) for each column.

2. Discard data that lie outside of $\bar{x} \pm 4s$ and recompute $\bar{x}$ and s.

Confidence Intervals and t-Test

1. For the year ≥1982 with highest average mass:
95% confidence interval ($= \bar{x} \pm ts/\sqrt{n}$) =
<u>2.5190 - 2.5477</u>

For the year ≥1982 with lowest average mass:
95% confidence interval = <u>2.4909 - 2.5068</u>
Comparison of means with *t*-test:
$t_{calculated}$ (Equation 4-4) = <u>4.819</u>
t_{table} (Table 4-2) = <u>2.021</u>
Is the difference significant? <u>yes</u>

2. For the year ≤1982 with highest average mass:
95% confidence interval = <u>3.0834 - 3.1019</u>
For the year ≤1982 with lowest average mass:
95% confidence interval = <u>3.0579 - 3.0758</u>
Comparison of means with *t*-test:
$t_{calculated}$ (Eq. 4-4) = <u>3.212</u>
t_{table} (Table 4-2) = <u>2.000</u>
Is the difference significant? <u>yes</u>

3. For Philadelphia versus Denver coins in one year:

Philadelphia 95% confidence interval =
<u>2.5147 - 2.5425</u>
Denver 95% confidence interval =
<u>2.4835 - 2.5466</u>
Comparison of means with *t*-test:
$t_{calculated}$ (Eq. 4-4) = <u>1.035</u>
t_{table} (Table 4-2) = <u>2.086</u>
Is the difference significant? <u>no</u>

sheet, just tabulate the average mass for each year. Your table will have only 14 entries.) Calculate the slope (m) and intercept (b) of the best straight line through all points and find the uncertainties in slope (σ_m) and intercept (σ_b).

Use Student's *t* to find the 95% confidence interval for the slope:

$$\text{confidence interval for slope} = m \pm t\sigma_m \quad (21\text{-}1)$$

where *t* is from Table 4-2 for $n - 2$ degrees of freedom. For example, if you have $n = 300$ pennies, $n - 2 = 298$, and it would be reasonable to use the value of t ($= 1.960$) at the bottom of the table for $n = \infty$. If the least-squares slope is $m \pm \sigma_m = -2.78 \pm 0.40$ mg/year, then the 95% confidence interval is $m \pm t\sigma_m = -2.78 \pm (1.960)(0.40) = -2.78 \pm 0.78$ mg/year.

The 95% confidence interval is $-2.78 \pm 0.78 = -3.56$ to -2.00 mg/year. We are 95% confident that the true slope is in this range and is, therefore, not zero. We conclude that older zinc pennies are heavier than newer zinc pennies.

Reporting Your Results

Gathering Data

1. Attach a table of masses, with one column sorted by mass for each year.

2. Divide 1982 into two columns, one for light (zinc) and one for heavy (brass) pennies.

Gaussian Distribution of Masses

Prepare a graph analogous to Figure 21-1 with labels showing the $\pm 3s$ limits.

Least-Squares Analysis

Prepare a graph analogous to Figure 21-2 and construct a least-squares table or spreadsheet analogous to Table 4-5 or Figure 4-8.

$m \pm \sigma_m$ = _____ $t_{95\% \text{ confidence}}$ = _____

$t_{99\% \text{ confidence}}$ = _____

95% confidence: $m \pm t\sigma_m$ = _____

   Does interval include zero? _____

99% confidence: $m \pm t\sigma_m$ = _____

   Does interval include zero? _____

Is there a systematic increase or decrease of penny mass with year? _____

21-2 Statistical Evaluation of Acid-Base Indicators

This experiment introduces you to the use of indicators and to the statistical concepts of mean, standard deviation, Q test, and t-test.[3] You will compare the accuracy of different indicators in locating the end point in the titration of the base "tris" with hydrochloric acid:

$$(HOCH_2)_3CNH_2 \;+\; H^+ \longrightarrow (HOCH_2)_3CNH_3^+$$

Tris(hydroxymethyl)aminomethane   (21-2)
       "tris"

Study Section 2-3 on using a balance and Section 2-4 on the buret before this experiment.

Reagents

~0.1 M HCl: Each student needs ~500 mL of unstandardized solution, all from a single batch that will be analyzed by the whole class.

Tris: Solid, primary standard powder should be available (~4 g/student).

Indicators: Bromothymol blue (BB), methyl red (MR), bromocresol green (BG), methyl orange (MO), and erythrosine (E) should be available in dropper bottles.

BB: Dissolve 100 mg in 16.0 mL of 0.01 M NaOH and dilute to 250 mL with distilled water.

MR: Dissolve 20 mg in 60 mL of ethanol and add 40 mL of distilled water.

BG: Dissolve 100 mg in 14.3 mL of 0.01 M NaOH and dilute to 250 mL with distilled water.

[3]D. T. Harvey, *J. Chem. Ed.* **1991,** *68,* 329.

MO: Dissolve 25 mg in 250 mL of distilled water.

E: Dissolve 25 mg in 250 mL of distilled water.

Color changes to use for the titration of tris with HCl are

BB: blue (pH 7.6) → yellow (pH 6.0) (end point is disappearance of green)

MR: yellow (pH 6.0) → red (pH 4.8) (end point is disappearance of orange)

BG: blue (pH 5.4) → yellow (pH 3.8) (end point is green)

MO: yellow (pH 4.4) → red (pH 3.1) (end point is first appearance of orange)

E: red (pH 3.6) → orange (pH 2.2) (end point is first appearance of orange)

Procedure

1.   Calculate the molecular weight of tris and the mass required to react with 35 mL of 0.10 M HCl. Weigh this much tris into a 125-mL flask.

2.   It is good practice to rinse a buret with a new solution to wash away traces of previous reagents. Wash your 50-mL buret with three 10-mL portions of 0.1 M HCl and discard the washings. Tilt and rotate the buret so that the liquid washes the walls, and drain the liquid through the stopcock. Then fill the buret with 0.1 M HCl to near the 0-mL mark, allow a minute for the liquid to settle, and record the reading to the nearest 0.01 mL.

3.   The first titration will be rapid, to allow you to find the approximate end point of the titration. Add ~20 mL of HCl from the buret to the flask and swirl to dissolve the tris. Add 2–4 drops of indicator and titrate with ~1-mL aliquots of HCl to find the end point.

4.   Based on the first titration, calculate how much tris is required to cause each succeeding titration to require 35–40 mL of HCl. Weigh this much tris into a clean flask (it need not be dry). Refill your buret to near 0 mL and record the reading. Repeat the titration in step 3, but use 1 drop at a time near the end point. When you are very near the end point, use less than a drop at a time. To do this, carefully suspend a fraction of a drop from the buret tip and touch it to the inside wall of the flask. Carefully tilt the flask so that the bulk solution overtakes the droplet and then swirl the flask to mix the solution.

TABLE 21-1 Pooled data

Indicator	Number of measurements (n)	Number of students (S)	Mean HCl molarity (M)[a] ($\bar{x}$)	s_x (M)[b]	Relative standard deviation (%) $100 \times s_x/\bar{x}$	s_p (M)[c]
BB MR	28	5	0.095 65	0.002 25	2.35	0.001 09
BG MO E	29	4	0.086 41	0.001 13	1.31	0.001 09

a. Computed from all values that were not discarded with the Q test.
b. s_x = standard deviation of all n measurements (degrees of freedom = $n - 1$).
c. s_p = pooled standard deviation for S students (degrees of freedom = $n - S$). Computed with a modification of Equation 4-5:

$$s_p = \sqrt{\frac{s_1^2(n_1 - 1) + s_2^2(n_2 - 1) + s_3^2(n_3 - 1) + \ldots}{n - S}}$$

where there is one term in the numerator for each student using that indicator.

Record the total volume of HCl required to reach the end point to the nearest 0.01 mL. Calculate the true mass of tris with the buoyancy equation 2-1 (density of tris = 1.327 g/mL). Calculate the molarity of HCl.

5. Repeat the titration to obtain at least six accurate measurements of the HCl molarity.

6. Use the Q test (Section 4-3) to decide whether any results should be discarded. Report your retained values, their mean (Equation 4-1), their standard deviation (Equation 4-2), and the percent relative standard deviation (= 100 × standard deviation/mean). Learn to compute standard deviation with the function built into your calculator.

Analyzing Class Data

Pool the data from your class to fill in Table 21-1, which shows two possible results. The quantity s_x is the standard deviation of all results reported by many students. The pooled standard deviation, s_p, is derived from the standard deviations reported by each student. If two students see the end point differently, each result might be very reproducible, but their reported molarities will be different. Together, they will generate a large value of s_x (because their results are so different), but a small value of s_p (because each one was reproducible).

Select the pair of indicators giving average HCl molarities that are farthest apart. Use the t-test (Equation 4-4) to decide whether the average molarities are significantly dif-

ferent from each other at the 95% confidence level.[4] When you calculate the pooled standard deviation for Equation 4-4, the values of s_1 and s_2 in Equation 4-5 are the values of s_x (not s_p) in Table 21-1. Select the pair of indicators giving the second most different molarities and use the t-test again to see whether or not this second pair of results is significantly different.

Reporting Your Results

Your Individual Data

Trial	Mass of tris from balance (g)	True mass of tris (Eq. 2-1)	HCl volume (mL)	Calculated HCl molarity (M)
1				
2				
3				
4				
5				
6				

[4]A condition for using the t-test with Equations 4-4 and 4-5 is that the standard deviations for the two sets of measurements should not be significantly different from each other. Comparison of standard deviations for Table 21-1 requires the F test, which is beyond the scope of this book. For the correct procedure, see D. C. Harris, *Quantitative Chemical Analysis* (New York: Freeman, 1995), pp. 66–68 and D. C. Harris, *Solutions Manual for Quantitative Chemical Analysis* (New York: Freeman, 1995), p. 18. Alternatively, see J. C. Miller and J. N. Miller, *Statistics for Analytical Chemistry* (New York: Wiley, 1984), pp. 53–59. Instructors may find it enlightening to go through a comparison of standard deviations in class, as discussed in reference 3.

On the basis of the Q test, should any molarity be discarded? If so, which one? _____

Mean value of retained results: _____

Standard deviation: _____

Relative standard deviation (%): _____

Pooled Class Data

1. Attach your copy of Table 21-1 with all entries filled in.
2. Compare the two most different molarities in Table 21-1. (Show your *t*-test and state your conclusion.)
3. Compare the two second most different molarities in Table 21-1.

21-3 *Calibration of Volumetric Glassware*

For most accurate results in future experiments, you should calibrate your own volumetric glassware (burets, pipets, flasks, etc.) to measure the exact volumes delivered or contained. This experiment develops basic skills in using a balance and handling volumetric glassware. Before beginning, you should study Sections 2-3, 2-4, and 2-9.

Calibration of a 50-mL Buret

In this procedure, you will construct a graph (like that in Figure 3-2) needed to convert the measured volume delivered by your buret to the true volume delivered at 20°C.

1. Fill the buret with distilled water and force any air bubbles out the tip. See that the buret drains without leaving drops on the walls. If drops are left, clean the buret with soap and water or soak it with peroxydisulfate–sulfuric acid cleaning solution.[5] Ad-

---

[5]Cleaning solution is prepared by your instructor by dissolving 36 g of ammonium peroxydisulfate, $(NH_4)_2S_2O_8$, in a loosely stoppered 2.2-L ("one-gallon") bottle of 98 wt % sulfuric acid. Add more peroxydisulfate every few weeks, as necessary, to maintain the oxidizing capacity of the solution. *Caution:* Cleaning solution eats dirt, grease, clothing, and people. Wear gloves and pour it very carefully. Place used reagent back into the original bottle. It can be reused many times.

just the meniscus to be at or slightly below 0.00 mL and touch the buret tip to a beaker to remove the suspended drop of water. Allow the buret to stand for 5 min while you weigh a 125-mL flask fitted with a rubber stopper. (Hold the flask with a tissue or paper towel, not with your hands, to avoid changing its mass with fingerprint residue.) If the level of the liquid in the buret has changed, tighten the stopcock and repeat the procedure.

2. Drain ~10 mL of water (at a rate of <20 mL/min) into the weighed flask and cap it tightly to prevent evaporation. Allow about 30 s for the film of liquid on the walls to descend before you read the buret. Estimate all readings to the nearest 0.01 mL. Weigh the flask again to determine the mass of water delivered.

3. Now drain the buret from 10 to ~20 mL and measure the mass of water delivered. Repeat the procedure for 30, 40, and 50 mL. Then do the entire procedure (10, 20, 30, 40, 50 mL) a second time.

4. Use Table 2-4 to convert the mass of water to the volume delivered. Repeat any set of duplicate buret corrections that do not agree to within 0.04 mL. Prepare a calibration graph like that in Figure 3-2, showing the correction factor at each 10-mL interval.

EXAMPLE   Buret Calibration

When draining the buret at 24°C, you observe the following values:

Final reading	10.01	10.08 mL
Initial reading	0.03	0.04
Difference	9.98	10.04 mL
Mass	9.984	10.056 g
Actual volume delivered	10.02	10.09 mL
Correction	+0.04	+0.05 mL
Average correction		+0.045 mL

To calculate the actual volume delivered when 9.984 g of water is delivered at 24°C, use the correction factor in Table 2-4: volume = (9.984 g)(1.003 8 mL/g) = 10.02 mL. The average correction for both sets of data is +0.045 mL.

To obtain the correction for a volume greater than 10 mL, add successive masses of water collected in the flask. Suppose that the following masses were measured:

Volume interval (mL)	Mass delivered (g)
0.03 to 10.01	9.984
10.01 to 19.90	9.835
19.90 to 30.06	10.071
Sum 30.03 mL	29.890 g

The actual volume of water delivered is $(29.890 \text{ g})(1.003\ 8$ mL/g$) = 30.00$ mL. Because the indicated volume is 30.03 mL, the buret correction at 30 mL is -0.03 mL.

What does this mean? Suppose that Figure 3-2 applies to your buret. If you begin a titration at 0.04 mL and end at 29.00 mL, you would deliver 28.96 mL if the buret were perfect. In fact, Figure 3-2 tells you that the buret delivers 0.03 mL less than the indicated amount, so only 28.93 mL was actually delivered. To use the calibration curve, you should begin all titrations near 0.00 mL. Use the calibration curve whenever you use your buret.

Other Calibrations

Pipets are calibrated by weighing water delivered from them. A volumetric flask is calibrated by weighing it empty and then weighing it filled to the mark with distilled water. Perform each procedure at least twice. Compare your results with the tolerances in Tables 2-1, 2-2, and 2-3.

21-4 *Gravimetric Determination of Calcium as $CaC_2O_4 \cdot H_2O$*

Calcium ion can be analyzed by precipitation with oxalate in basic solution to form $CaC_2O_4 \cdot H_2O$.[6] The precipitate is soluble in acidic solution because oxalate reacts with H^+ to make $HC_2O_4^-$. Large, easily filtered, pure crystals of product are obtained if the precipitation is carried out slowly. This goal can be achieved by dissolving Ca^{2+} and $C_2O_4^{2-}$ in acidic solution and gradually raising the pH by thermal decomposition of urea, as described in Section 19-2. Read Sections 19-1 to 19-3 for this experiment.

Reagents

Ammonium oxalate solution: Make 1 L of solution containing 40 g of $(NH_4)_2C_2O_4$ plus 25 mL of 12 M HCl. Each student will need 80 mL of this solution.
Methyl red indicator: Dissolve 20 mg in 60 mL of ethanol and add 40 mL of distilled water.
0.1 M HCl: (225 mL/student) Dilute 8.3 mL of 37 wt % HCl up to 1 L.
Urea: 45 g/student.
Unknowns: Prepare 1 L of solution containing 15–18 g of $CaCO_3$ plus 38 mL of 12 M HCl. Each student will need 100 mL of this solution. Alternatively, solid unknowns are available from Thorn Smith.[7]

Procedure

1.   Dry three medium-porosity, sintered-glass funnels for 1–2 h at 105°C; cool them in a desiccator for 30 min and weigh them. Repeat the procedure with 30-min heating periods until successive weighings agree to within 0.3 mg. Use a paper towel or tongs, not your fingers, to handle the funnels. Alternatively, use a 900-W kitchen microwave oven to dry your crucibles to constant mass in two heating periods of 4 min each, followed by heating periods of 2 min each (with 15 min allowed for cooldown after each cycle).

2.   Use a few small portions of unknown to rinse a 25-mL transfer pipet and discard the washings. Use a rubber bulb or other mechanical device, not your mouth, to provide suction. Transfer exactly 25 mL of unknown to each of three 250- or 400-mL beakers and dilute each with ~75 mL of 0.1 M HCl. Add 5 drops of methyl red indicator solution to each beaker. This indicator is red below pH 4.8 and yellow above pH 6.0.

3.   Add ~25 mL of ammonium oxalate solution to each beaker while stirring with a glass rod. Remove the rod and rinse it into the beaker. Add ~15 g of solid urea to each sample, cover it with a watchglass, and boil gently for ~30 min until the indicator turns yellow.

4.   Filter each hot solution through a weighed funnel, using suction (Figure 2-13). Add ~3 mL of ice-cold distilled water to the beaker and use a rubber policeman to help transfer the remaining solid to the

[6]C. H. Hendrickson and P. R. Robinson, *J. Chem. Ed.* **1979**, *56*, 341.

[7]Thorn Smith, Inc., 7755 Narrow Gauge Road, Beulah, MI 49617.

funnel. Repeat this procedure with small portions of ice-cold water until all of the precipitate has been transferred. Finally, use two 10-mL portions of ice-cold water to rinse each beaker and pour the washings over the precipitate.

5. Dry the precipitate, first with aspirator suction for 1 min, then in an oven at 105°C for 1–2 h. Bring each filter to constant weight. The product is somewhat hygroscopic, so only one filter at a time should be removed from the desiccator and weighed rapidly. Alternatively, dry the precipitate in a microwave oven once for 4 min, followed by several 2-min periods, with cooling for 15 min before weighing. The water of crystallization is not lost.

6. Calculate the average molarity ($\bar{x}$) of Ca^{2+} in the unknown solution or the average wt % of Ca in the unknown solid. Report the standard deviation (s) and percent relative standard deviation ($= 100 \times s/\bar{x}$).

21-5 Gravimetric Measurement of Phosphorus in Plant Food

The nutrients in plant foods—nitrogen, phosphorus, and potassium—are indicated on the label by three numbers, such as 15-30-20. This notation means that the fertilizer contains 15 wt % N, 30 wt % P_2O_5, and 20 wt % K_2O. (This is a fictitious but customary notation. Phosphorus is not in the form P_2O_5 and potassium is not in the form K_2O in the fertilizer.)

This experiment[8] measures the phosphorus content of plant food by the reaction

$$\underset{\text{From plant food}}{HPO_4^{2-}} + NH_3 + Mg^{2+} + 6H_2O \rightleftharpoons$$

$$MgNH_4PO_4 \cdot 6H_2O(s) \qquad (21\text{-}3)$$
$$\text{FW 245.41}$$

The experiment is not as accurate as others in this book, but it can be carried out with tap water and chemicals from the grocery store. The product is dried at room temperature to avoid loss of water of crystallization above 40°C. In more accurate determinations of phosphate, the solid

[8]S. Solomon, A. Lee, and D. Bates, *J. Chem. Ed.* **1993**, *70*, 410.

$MgNH_4PO_4 \cdot 6H_2O$ is *ignited* (heated strongly) and decomposes to $Mg_2P_2O_7$. Read Sections 19-1 to 19-3 for this experiment.

Reagents

Epsom salts: $MgSO_4 \cdot 7H_2O$ (FW 246.52), 5–15 g/student.
Ammonia: (200 mL/student) Use aqueous ammonia from the grocery (~4 wt % NH_3) or prepare 4.0 wt % NH_3 (= 2.3 M) by diluting 160 mL of concentrated (28 wt %) NH_3 up to 1.0 L.
Isopropyl alcohol: Each student needs 100 mL of rubbing alcohol.

Procedure

1. Weigh out 10 to 11 g of commercial plant food to the nearest 0.01 g and dissolve it in 140 mL of water in a 250-mL beaker. Stir the mixture well and filter it (Figures 2-14 and 2-15) into a 1-L beaker to remove insoluble residue. A coffee basket filter can be used instead of laboratory filter paper. Rinse the 250-mL beaker twice with 5-mL portions of water and pour the washings through the filter into the 1-L beaker.

2. From the plant food package label, estimate how many moles of phosphorus are in the solid that you weighed out. (The middle number on the package gives the hypothetical wt % of P_2O_5, FW 141.94.) Calculate how many grams of Epsom salts ($MgSO_4 \cdot 7H_2O$, FW 246.52) are required for a 50% molar excess of Mg^{2+} for Reaction 21-3. Dissolve the required amount of Epsom salts in water, using 10 mL of H_2O per gram of $MgSO_4 \cdot 7H_2O$, and pour the resulting solution into the filtered plant food solution.

3. In a fume hood or a well-ventilated area, gradually add 200 mL of ammonia solution (measured with a beaker) to the plant food solution while stirring well with a glass rod. Allow the white precipitate to stand for at least 15 min.

4. Collect the precipitate by filtering through two layers of fluted filter paper or two layers of coffee filters. Then add 50 mL of isopropyl alcohol to the 1-L beaker and swirl to dislodge particles of solid. Pour the mixture over the precipitate in the filter and allow the liquid to drain. Repeat the process once more with another 50 mL of alcohol.

5. Remove the filter paper from the funnel and spread it on several layers of paper towel to dry

overnight, protected from dust. The next day, use a spatula to break up lumps. Then allow an additional day of drying, so that the solid is a free-flowing powder.

6. Weigh an empty 250-mL beaker. Scrape the solid from the filter into the beaker and weigh the beaker again to find the mass of solid precipitate. To be certain that the product is dry, allow the beaker to stand for at least another day protected from dust and weigh it again. If successive weighings agree to within ± 0.1 g, the product is dry enough.

7. From the mass of dry product, calculate the moles of product in Reaction 21-3 and the moles of phosphorus analyzed. From the moles of phosphorus, compute the moles of P_2O_5 and grams of P_2O_5 in the solid plant food. Report the wt % P_2O_5 in the plant food. Pool the class results to find the mean and standard deviation for the wt % P_2O_5 in each type of plant food analyzed. Compare the class results to the package labels.

21-6 Preparing Standard Acid and Base

Hydrochloric acid and sodium hydroxide are the most common strong acids and bases used in the laboratory. Both reagents need to be standardized to learn their exact concentration. Section 11-5 provides background information for the procedures described below.

Reagents

50 wt % NaOH: (3 mL/student) Dissolve 50 g of reagent grade NaOH in 50 mL of distilled water and allow the suspension to settle overnight. Na_2CO_3 is insoluble in the solution and precipitates. Store the solution in a tightly sealed polyethylene bottle and handle it gently to avoid stirring the precipitate when liquid is withdrawn.
Phenolphthalein indicator: Dissolve 50 mg in 50 mL of ethanol and add 50 mL of distilled water.
Bromocresol green indicator: Dissolve 100 mg in 14.3 mL of 0.01 M NaOH and dilute to 250 mL with distilled water.
Concentrated (37 wt %) HCl: 10 mL/student.
Primary standards: Potassium acid phthalate (~ 2.5 g/student) and sodium carbonate (~ 1.0 g/student).
0.05 M NaCl: 50 mL/student.

Standardizing NaOH

1. Dry primary standard grade potassium hydrogen phthalate for 1 hr at 105°C and store it in a capped bottle in a desiccator.

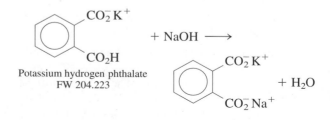

Potassium hydrogen phthalate
FW 204.223

2. Boil 1 L of distilled water for 5 min to expel CO_2. Pour the water into a polyethylene bottle, which should be tightly capped whenever possible. Calculate the volume of 50 wt % NaOH needed to prepare 1 L of 0.1 M NaOH. (The density of 50 wt % NaOH is 1.50 g per milliliter of solution.) Use a graduated cylinder to transfer this much NaOH to the bottle of water. (*Caution:* 50 wt % NaOH eats people. Flood any spills on your skin with water.) Mix well and cool the solution to room temperature (preferably overnight).

3. Weigh four samples of solid potassium hydrogen phthalate and dissolve each in ~ 25 mL of distilled water in a 125-mL flask. Each sample should contain enough solid to react with ~ 25 mL of 0.1 M NaOH. Add 3 drops of phenolphthalein to each flask and titrate one rapidly to find the end point. The buret should have a loosely fitted cap to minimize entry of CO_2 from the air.

4. Calculate the volume of NaOH required for each of the other three samples and titrate them carefully. During each titration, periodically tilt and rotate the flask to wash all liquid from the walls into the bulk solution. Near the end, deliver less than 1 drop of titrant at a time. To do this, carefully suspend a fraction of a drop from the buret tip, touch it to the inside wall of the flask, wash it into the bulk solution by careful tilting, and swirl the solution. The end point is the first appearance of faint pink color that persists for 15 s. (The color will slowly fade as CO_2 from the air dissolves in the solution.)

5. Calculate the average molarity ($\bar{x}$), the standard deviation (s), and the percent relative standard deviation ($= 100 \times s/\bar{x}$). If you were careful, the relative standard deviation should be <0.2%.

Standardizing HCl

1. Use the table on the inside cover of this book to calculate the volume of ~37 wt % HCl that should be added to 1 L of distilled water to produce 0.1 M HCl and prepare this solution.

2. Dry primary standard grade sodium carbonate for 1 h at 105°C and cool it in a desiccator.

3. Weigh four samples, each containing enough Na_2CO_3 to react with ~25 mL of 0.1 M HCl, and place each in a 125-mL flask. When you are ready to titrate each one, dissolve it in ~25 mL of distilled water. Add 3 drops of bromocresol green indicator and titrate one rapidly to a green color to find the approximate end point.

$$2HCl + Na_2CO_3 \longrightarrow CO_2 + 2NaCl + H_2O$$
$$\text{FW 105.99}$$

4. Carefully titrate each sample until it just turns from blue to green. Then boil the solution to expel CO_2. The solution should return to a blue color. Carefully add HCl from the buret until the solution turns green again and report the volume of acid at this point.

5. Perform one blank titration of 50 mL of 0.05 M NaCl containing 3 drops of indicator. Subtract the volume of HCl needed for the blank from that required to titrate Na_2CO_3.

6. Calculate the mean HCl molarity, standard deviation, and percent relative standard deviation.

21-7 Analysis of a Mixture of Carbonate and Bicarbonate

This procedure involves two titrations. First, total alkalinity ($= [HCO_3^-] + 2[CO_3^{2-}]$) is measured by titrating the mixture with standard HCl to a bromocresol green end point:

$$HCO_3^- + H^+ \longrightarrow H_2CO_3$$
$$CO_3^{2-} + 2H^+ \longrightarrow H_2CO_3$$

A separate aliquot of unknown is treated with excess standard NaOH to convert HCO_3^- to CO_3^{2-}:

$$HCO_3^- + OH^- \longrightarrow CO_3^{2-} + H_2O$$

Then all the carbonate is precipitated with $BaCl_2$:

$$Ba^{2+} + CO_3^{2-} \longrightarrow BaCO_3(s)$$

The excess NaOH is immediately titrated with standard HC to determine how much HCO_3^- was present. From the tota alkalinity and bicarbonate concentration, you can calculate the original carbonate concentration.

Reagents

Standard NaOH and standard HCl: From Experimen 21-6.

CO_2-free water: Boil 500 mL of distilled water to expe carbon dioxide and pour the water into a 500-mL plasti bottle. Screw the cap on tightly and allow the water to coo to room temperature. Keep tightly capped when not in use

10 wt % aqueous $BaCl_2$: 35 mL/student.

Bromocresol green and phenolphthalein indicators: Se Experiment 21-6.

Unknowns: Solid unknowns (2.5 g/student) can be prepared from reagent-grade sodium or potassium carbonate and bicarbonate. Unknowns should be stored in a desiccato but should not be heated. Even mild heating at 50°–100°C converts $NaHCO_3$ to Na_2CO_3.

Procedure

1. Accurately weigh 2.0–2.5 g of unknown into a 250-mL volumetric flask by weighing the sample in a capped weighing bottle, delivering some to a funnel in the volumetric flask, and reweighing the bottle. Continue this process until the desired mass of reagent has been transferred to the funnel. Rinse the funnel repeatedly with small portions of CO_2-free water to dissolve the sample. Remove the funnel, dilute to the mark, and mix well.

2. Pipet a 25.00-mL aliquot of unknown solution into a 250-mL flask and titrate with standard 0.1 M HCl, using bromocresol green indicator as described in the previous experiment for standardizing HCl. Repeat this procedure with two more 25.00-mL aliquots.

3. Pipet 25.00 mL of unknown and 50.00 mL of standard 0.1 M NaOH into a 250-mL flask. Swirl and add 10 mL of 10 wt % $BaCl_2$, using a graduated cylinder. Swirl again to precipitate $BaCO_3$, add 2 drops of phenolphthalein indicator, and immediately

titrate with standard 0.1 M HCl. Repeat this procedure with two more 25.00-mL samples of unknown.

4. From the results of step 2, calculate the total alkalinity and its standard deviation. From the results of step 3, calculate the bicarbonate concentration and its standard deviation. Using the standard deviations as estimates of uncertainty, calculate the concentration (and uncertainty) of carbonate in the sample. Express the composition of the solid unknown in a form such as 63.4 (± 0.5) wt % K_2CO_3 and 36.6 (± 0.2) wt % $NaHCO_3$.

<hr>

21-8 Kjeldahl Nitrogen Analysis

The Kjeldahl nitrogen analysis described in Section 11-6 is widely used to measure the nitrogen content of pure organic compounds and complex substances such as milk, cereal, and flour. Digestion in boiling H_2SO_4 with a catalyst converts organic nitrogen into NH_4^+. The solution is then made basic, and the liberated NH_3 is distilled into a known amount of HCl (Reactions 11-5 to 11-7). Unreacted HCl is titrated with NaOH to determine how much HCl was consumed by NH_3. Because the solution to be titrated contains both HCl and NH_4^+, we choose an indicator that permits titration of HCl without beginning to titrate NH_4^+. Problem 11-22 shows that bromocresol green (transition range 3.8–5.4) fulfills this purpose.

Reagents

Standard NaOH and standard HCl: From Experiment 21-6.
Bromocresol green and phenolphthalein indicators: See Experiment 21-6.
Potassium sulfate: 10 g/student.
Selenium-coated boiling chips: Hengar selenium-coated granules are convenient. Alternative catalysts for Kjeldahl digestion are 0.1 g of Se, 0.2 g of $CuSeO_3$, or a crystal of $CuSO_4$.
Concentrated (98 wt %) H_2SO_4: 25 mL/student.
Unknowns: Unknowns are pure acetanilide, N-2-hydroxyethylpiperazine-N'-2-ethanesulfonic acid (HEPES buffer, Table 10-2), tris(hydroxymethyl)aminomethane (tris buffer, Table 10-2), or the p-toluenesulfonic acid salts of ammonia, glycine, or nicotinic acid. Each student needs enough unknown to produce 2–3 mmol of NH_3.

Digestion

1. Dry your unknown at 105°C for 45 min and accurately weigh an amount that will produce 2–3 mmol of NH_3. Place the sample in a *dry* 500-mL Kjeldahl flask (Figure 11-7), so that as little as possible sticks to the walls. Add 10 g of K_2SO_4 (to raise the boiling temperature) and three selenium-coated boiling chips. Pour in 25 mL of 98 wt % H_2SO_4, washing down any solid from the walls. (*Caution:* Concentrated H_2SO_4 eats people. If you get any on your skin, flood it immediately with water, followed by soap and water.)

2. In a fume hood, clamp the flask at a 30° angle away from you. Heat gently with a burner until foaming ceases and the solution becomes homogeneous. Continue boiling gently for an additional 30 min.

3. Cool the flask for 30 min *in the air,* and then in an ice bath for 15 min. Slowly, and with constant stirring, add 50 mL of ice-cold distilled water. Dissolve any solids that crystallize. Transfer the liquid to the 500-mL distillation flask in Figure 21-3. Wash the Kjeldahl flask five times with 10-mL portions of distilled water and pour the washings into the distillation flask.

Distillation

1. Set up the apparatus in Figure 21-3 and tighten the connections well. Pipet 50.00 mL of standard 0.1 M HCl into the receiving beaker and clamp the funnel in place below the liquid level.

2. Add 5–10 drops of phenolphthalein indicator to the three-neck flask shown in Figure 21-3 and secure the stoppers. Pour 60 mL of 50 wt % NaOH into the adding funnel and drip this into the distillation flask over a period of 1 min until the indicator turns pink. (*Caution:* 50 wt % NaOH eats people. Flood any spills on your skin with water.) Do not let the last 1 mL through the stopcock, so that gas cannot escape from the flask. Close the stopcock and heat the flask gently until two-thirds of the liquid has distilled.

3. Remove the funnel from the receiving beaker *before* removing the burner from the flask (to avoid sucking distillate back into the condenser). Rinse the funnel well with distilled water and catch the rinses in the beaker. Add 6 drops of bromocresol green indicator solution to the beaker and carefully titrate to

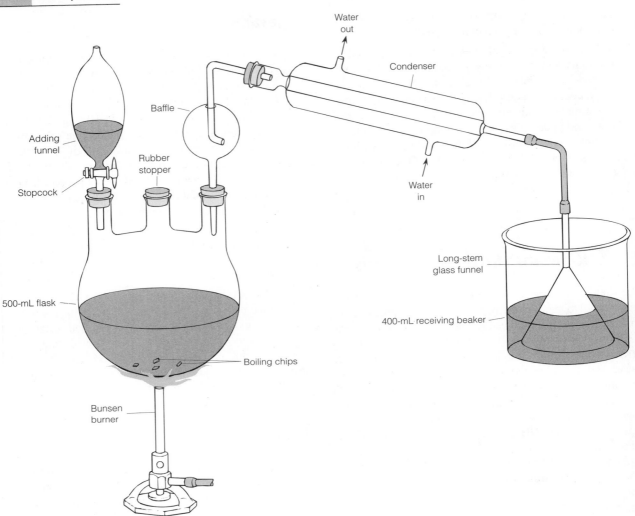

FIGURE 21-3 Apparatus for Kjeldahl distillation.

the blue end point with standard 0.1 M NaOH. You are looking for the first appearance of light blue color. (Practice titrations with HCl and NaOH beforehand to familiarize yourself with the end point.)

4. Calculate the wt % of nitrogen in the unknown.

21-9 Using a pH Electrode for an Acid-Base Titration

This experiment uses a pH electrode to follow the course of an acid-base titration. You will observe how pH changes slowly during most of the reaction and rapidly near the equivalence point. You will compute the first and second derivatives of the titration curve to locate the end point. From the mass of unknown acid or base and the moles of titrant, you can calculate the molecular weight of the unknown. Section 11-4 provides background for this experiment.

Reagents

Standard NaOH and standard HCl: From Experiment 21-6.

Bromocresol green and phenolphthalein indicators: See Experiment 21-6.

pH calibration buffers: pH 7 and pH 4. Use commercial standards.

Unknowns: Unknowns should be stored in a desiccator by your instructor.

Suggested acid unknowns: potassium hydrogen phthalate (MW 204.22), 2-(*N*-morpholino)ethanesulfonic acid (MES, Table 10-2, MW 195.24), imidazole hydrochloride (Table 10-2, MW 104.54, hygroscopic), potassium hydrogen iodate (Table 11-4, FW 389.91).

Suggested base unknowns: tris (Table 11-4, MW 121.14), imidazole (MW 68.08), disodium hydrogen phosphate (Na_2HPO_4, FW 141.96), sodium glycinate (may be found in chemical catalogs as glycine, sodium salt hydrate, $H_2NCH_2CO_2Na \cdot xH_2O$, FW 97.05 + x(18.015)). For sodium glycinate, one objective of the titration is to find the number of waters of hydration from the molecular weight.

Procedure

1. Your instructor will recommend a mass of unknown (5–8 mmol) for you to weigh accurately and dissolve in distilled water in a 250-mL volumetric flask. Dilute to the mark and mix well.

2. Following instructions for your particular pH meter, calibrate a meter and glass electrode, using buffers with pH values near 7 and 4. Rinse the electrodes well with distilled water and blot them dry with a tissue before immersing in any new solution.

3. The first titration is intended to be rough, so you will know the approximate end point in the next titration. For the rough titration, pipet 25.0 mL of unknown into a 125-mL flask. If you are titrating an unknown acid, add 3 drops of phenolphthalein indicator and titrate with standard 0.1 M NaOH to the pink end point, using a 50-mL buret. If you are titrating an unknown base, add 3 drops of bromocresol green indicator and titrate with standard 0.1 M HCl to the green end point. Add 0.5 mL of titrant at a time so that you can estimate the equivalence volume to within 0.5 mL. Near the end point, the indicator temporarily changes color as titrant is added. If you recognize this, you can slow down the rate of addition and estimate the end point to within a few drops.

4. Now comes the careful titration. Pipet 100.0 mL of unknown solution into a 250-mL beaker containing a magnetic stirring bar. Position the electrode(s) in the liquid so that the stirring bar will not strike the electrode. If a combination electrode is used, the small hole near the bottom on the side must be immersed in the solution. This hole is the salt bridge to the reference electrode. Allow the electrode to equilibrate for 1 min with stirring and record the pH.

5. Add 1 drop of indicator and begin the titration. The equivalence volume will be four times greater than it was in step 3. Add ~1.5-mL aliquots of titrant and record the exact volume, the pH, and the color 30 s after each addition. When you are within 2 mL of the equivalence point, add titrant in 2-drop increments. When you are within 1 mL, add titrant in 1-drop increments. Continue with 1-drop increments until you are 0.5 mL past the equivalence point. The equivalence point has the most rapid change in pH. Add five more 1.5-mL aliquots of titrant and record the pH after each.

Analysis of the Data

1. Construct a graph of pH versus titrant volume. Mark on your graph where the indicator color change(s) was (were) observed.

2. Following the example in Table 11-3 and Figure 11-4, compute the first derivative (the slope, $\Delta pH/\Delta V$), for each data point within ± 1 mL of the equivalence volume. From your graph, estimate the equivalence volume as accurately as you can, as shown in Figure 21-4.

3. Following the example in Table 11-3, compute the second derivative (the slope of the slope, $\Delta(slope)/\Delta V$). Prepare a graph like Figure 11-5 and locate the equivalence volume as accurately as you can.

4. Go back to your graph from step 1 and mark where the indicator color changes were observed. Compare the indicator end point to the end point estimated from the first and second derivatives.

5. From the equivalence volume and the mass of unknown, calculate the molecular weight of the unknown.

21-10 EDTA Titration of Ca^{2+} and Mg^{2+} in Natural Waters

The most common multivalent metal ions in streams, lakes, and oceans are Ca^{2+} and Mg^{2+}. In this experiment you will find the total concentration of metal ions that can react with EDTA (Chapter 15), and you will assume that this equals the concentration of Ca^{2+} and Mg^{2+}. In a second

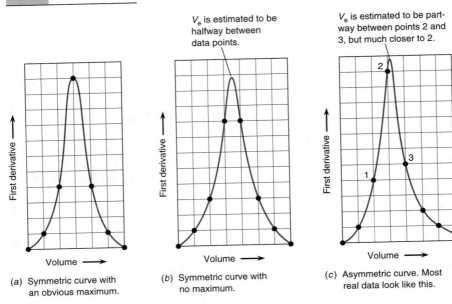

FIGURE 21-4 Three examples of locating the maximum position of the first derivative of a titration curve. Circles are observed data points and the line is drawn by a person interpreting the data.

experiment, you will analyze Ca^{2+} separately after precipitating $Mg(OH)_2$ with strong base. This experiment (and the last part of Experiment 21-13) could be the basis for a class project to study the composition of natural waters collected in different seasons and different locations.

Reagents

EDTA: $Na_2H_2EDTA \cdot 2H_2O$ (FW 372.25), 0.6 g/student.
Buffer (pH 10): Add 142 mL of 28 wt % aqueous NH_3 to 17.5 g of NH_4Cl and dilute to 250 mL with distilled water.
Eriochrome black T indicator: Dissolve 0.2 g of the solid indicator in 15 mL of triethanolamine plus 5 mL of absolute ethanol.
Hydroxynaphthol blue indicator: 0.5 g/student.
Unknowns: Collect water from streams or lakes or from the ocean. To minimize bacterial growth, plastic jugs should be filled to the top and tightly sealed. Refrigeration is recommended.

Procedure

1. Dry $Na_2H_2EDTA \cdot 2H_2O$ (FW 372.25) at 80°C for 1 h and cool in a desiccator. Accurately weigh out ~0.6 g and dissolve it with heating in 400 mL of distilled water in a 500-mL volumetric flask. Cool to room temperature, dilute to the mark, and mix well.

2. Pipet a sample of unknown into a 250-mL flask. A 1.000-mL sample of seawater or a 50.00-mL

sample of tap water is usually reasonable. If you use 1.000 mL of seawater, add 50 mL of distilled water. To each sample, add 3 mL of pH 10 buffer and 6 drops of Eriochrome black T indicator. Titrate with EDTA from a 50-mL buret and note when the color changes from wine red to blue. You may need to practice finding the end point several times by adding a little tap water and titrating with more EDTA. Save a solution at the end point to use as a color comparison for other titrations.

3. Repeat the titration with three samples to accurately measure the total Ca^{2+} and Mg^{2+} concentration. Perform a blank titration with 50 mL of distilled water and subtract the value of the blank from each result.

4. For the determination of Ca^{2+}, pipet four samples of unknown into clean flasks (adding 50 mL of distilled water if you use 1.000 mL of seawater). Add 30 drops of 50 wt % NaOH to each solution and swirl for 2 min to precipitate $Mg(OH)_2$ (which may not be visible). Add ~0.1 g of solid hydroxynaphthol blue to each flask. (This indicator is used because it remains blue at higher pH than Eriochrome black T does.) Titrate one sample rapidly to find the end point; practice finding it several times, if necessary.

5. Titrate the other three samples carefully. After reaching the blue end point, allow each sample to sit for 5 min with occasional swirling so that any $Ca(OH)_2$ precipitate can redissolve. Then titrate back to the blue end point. (Repeat this procedure if

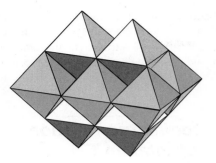

FIGURE 21-5 Structure of $V_{10}O_{28}^{6-}$ anion, consisting of 10 VO_6 octahedra sharing edges with one another. Each octahedron has a vanadium atom at the center and an oxygen atom at each vertex.

the blue color turns to red upon standing.) Perform a blank titration with 50 mL of distilled water.

6. Calculate the total concentration of Ca^{2+} and Mg^{2+}, as well as the individual concentrations of each ion. Calculate the relative standard deviation of replicate titrations.

21-11 *Synthesis and Analysis of Ammonium Decavanadate*

The species in a solution of V^{5+} are a delicate function of both pH and concentration.

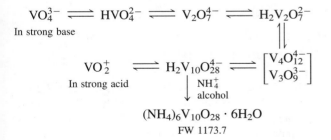

$$VO_4^{3-} \rightleftharpoons HVO_4^{2-} \rightleftharpoons V_2O_7^{4-} \rightleftharpoons H_2V_2O_7^{2-}$$

In strong base

$$VO_2^+ \rightleftharpoons H_2V_{10}O_{28}^{4-} \rightleftharpoons \begin{bmatrix} V_4O_{12}^{4-} \\ V_3O_9^{3-} \end{bmatrix}$$

In strong acid

$$\downarrow NH_4^+ \text{ alcohol}$$

$$(NH_4)_6V_{10}O_{28} \cdot 6H_2O$$

FW 1173.7

The decavanadate ion ($V_{10}O_{28}^{6-}$), whose ammonium salt you will isolate, consists of 10 VO_6 octahedra sharing edges with one another (Figure 21-5).

After preparing this salt, you will determine the vanadium content by a redox titration and NH_4^+ by the Kjeldahl method.[9] For the redox titration, V^{5+} is first reduced to

[9]G. G. Long, R. L. Stanfield, and C. F. Hentz, Jr., *J. Chem. Ed.* **1979**, *56*, 195.

V^{4+} with sulfurous acid and then titrated with standard permanganate (Color Plate 19).

$$V_{10}O_{28}^{6-} + 5H_2SO_3 + 26H^+ \longrightarrow$$

V^{5+} Sulfurous acid

$$10VO^{2+} + 5H_2SO_4 + 13H_2O$$

V^{4+}

$$5VO^{2+} + MnO_4^- + H_2O \longrightarrow 5VO_2^+ + Mn^{2+} + 2H^+$$

Blue   Purple   Yellow   Colorless

Reagents

Ammonium metavanadate: 3 g NH_4VO_3/student.
50 vol % aqueous acetic acid: 4 mL/student.
95 wt % aqueous ethanol: 180 mL/student.
Potassium permanganate: 1.6 g $KMnO_4$/student.
Sodium oxalate: 1 g $Na_2C_2O_4$/student.
0.9 M H_2SO_4: 1 L/student.
1.5 M H_2SO_4: 80 mL/student.
Sodium hydrogen sulfite: 2 g $NaHSO_3$/student.
For Kjeldahl analysis: Standard HCl, standard NaOH, and bromocresol green indicator, as in Experiment 21-6.

Synthesis of $(NH_4)_6V_{10}O_{28} \cdot 6H_2O$

1. Heat 3.0 g of ammonium metavanadate (NH_4VO_3) in 100 mL of distilled water with constant stirring (but not boiling) until most or all of the solid has dissolved. Filter the solution and add 4 mL of 50 vol % aqueous acetic acid with stirring.

2. Add 150 mL of 95 wt % ethanol with stirring and then cool the solution in a refrigerator or ice bath.

3. After maintaining a temperature of 0°–10°C for 15 min, filter the orange product with suction and wash with two 15-mL portions of ice-cold 95 wt % ethanol.

4. Dry the product in the air (protected from dust) for 2 days. Typical yield is 2.0–2.5 g.

Preparation and Standardization of $KMnO_4$

1. Prepare a 0.02 M permanganate solution by dissolving 1.6 g of $KMnO_4$ in 500 mL of distilled water. Boil gently for 1 h, cover, and allow the solution to cool overnight. Filter through a clean, fine sintered-glass funnel, discarding the first 20 mL of filtrate. Store the solution in a clean glass amber bottle. Do not let the solution touch the cap.

2. Dry sodium oxalate ($Na_2C_2O_4$) at 105°C for 1 h, cool in a desiccator, and weigh three ~0.25-g samples into 500-mL flasks or 400-mL beakers. To each, add 250 mL of 0.9 M H_2SO_4 that has been recently boiled and cooled to room temperature. Stir with a thermometer to dissolve the sample and add 90–95% of the theoretical amount of $KMnO_4$ solution needed for the titration. (This can be calculated from the mass of $KMnO_4$ used to prepare the permanganate solution. The chemistry is Reaction 6-1.)

3. Leave the solution at room temperature until it is colorless. Then heat it to 55°–60°C and complete the titration by adding $KMnO_4$ until the first pale pink color persists. Proceed slowly near the end, allowing 30 s for each drop to lose its color.

4. As a blank, titrate 250 mL of 0.9 M H_2SO_4 to the same pale pink color. Subtract the blank volume from each titration volume. Computing the average molarity of $KMnO_4$.

Vanadium Analysis

1. Accurately weigh two 0.3-g samples of ammonium decavanadate into 250-mL flasks and dissolve each in 40 mL of 1.5 M H_2SO_4 (with warming, if necessary).

2. In a fume hood, add 50 mL of distilled water and 1 g of $NaHSO_3$ to each sample and dissolve with swirling. After 5 min, boil gently for 15 min to remove excess H_2SO_3 by the reaction $H_2SO_3(aq) \rightarrow SO_2(g)\uparrow + H_2O(l)$.

3. Titrate the warm solution with standard 0.02 M $KMnO_4$ from a 50-mL buret. The end point is when the yellow color of VO_2^+ takes on a dark shade (from excess MnO_4^-) that persists for 15 s.

4. Calculate the average wt % of vanadium in the ammonium decavanadate and compare your result to the theoretical value.

Ammonium Ion Analysis

Ammonium ion is analyzed by a modification of the Kjeldahl procedure. Transfer 0.6 g of accurately weighed ammonium decavanadate to the distillation flask in Figure

21-3 and add 200 mL of distilled water. Then proceed as described in the section "Distillation" in Experiment 21-8. Calculate the wt % of nitrogen in the ammonium decavanadate and compare your result to the theoretical value.

21-12 *Iodimetric Titration of Vitamin C*

You should read Section 15-5 before you do this experiment. Ascorbic acid (vitamin C) is a mild reducing agent that reacts rapidly with triiodide (Reaction 15-7). In this experiment[10] you will generate a known excess of I_3^- by the reaction of iodate with iodide (Reaction 15-4), allow the reaction with ascorbic acid to proceed, and then back-titrate the excess I_3^- with thiosulfate (Reaction 15-5 and Color Plate 14).

Reagents

Starch indicator: Make a paste of 5 g of soluble starch and 5 mg of Hg_2I_2 in 50 mL of distilled water. Pour the paste into 500 mL of boiling distilled water and boil until it is clear.
Sodium thiosulfate: 9 g $Na_2S_2O_3 \cdot 5H_2O$/student.
Sodium carbonate: 50 mg Na_2CO_3/student.
Potassium iodate: 1 g KIO_3/student.
Potassium iodide: 12 g KI/student.
0.5 M H_2SO_4: 30 mL/student.
Vitamin C: Dietary supplement containing 100 mg of vitamin C per tablet is suitable. Each student needs six tablets.
0.3 M H_2SO_4: 180 mL/student.

Preparation and Standardization of Thiosulfate Solution

1. Prepare 0.07 M $Na_2S_2O_3$ by dissolving 8.7 g of $Na_2S_2O_3 \cdot 5H_2O$ in 500 mL of freshly boiled distilled water containing 0.05 g of Na_2CO_3. Store in a tightly capped amber bottle. Sodium thiosulfate is not a primary standard, so the molarity of the solution is not exactly known.

2. Prepare ~0.01 M KIO_3 by accurately weighing ~1 g of solid reagent and dissolving it in a 500-mL volumetric flask. From the mass of KIO_3 (FW 214.00), compute the molarity of the solution.

[10]D. N. Bailey, *J. Chem. Ed.* **1974,** *51,* 488.

3. Standardize the thiosulfate solution as follows: Pipet 50.00 mL of KIO_3 solution into a 250-mL flask. Add 2 g of solid KI and 10 mL of 0.5 M H_2SO_4. Immediately titrate with thiosulfate until the solution has lost almost all color (pale yellow). Then add 2 mL of starch indicator and complete the titration. Repeat the titration with two additional 50.00-mL volumes of KIO_3 solution. From the stoichiometries of Reactions 15-4 and 15-5, compute the average molarity of thiosulfate and the relative standard deviation.

Analysis of Vitamin C

1. Dissolve two tablets in 60 mL of 0.3 M H_2SO_4, using a glass rod to help break the solid. (Some solid binding material will not dissolve.)

2. Add 2 g of solid KI and 50.00 mL of standard KIO_3. Then titrate with standard thiosulfate as above. Add 2 mL of starch indicator just before the end point.

3. Repeat the analysis two more times. Find the mean and relative standard deviation for the number of milligrams of vitamin C per tablet.

21-13 Potentiometric Halide Titration with Ag^+

Mixtures of halides can be titrated with Ag^+ by using the apparatus in Figure 6-3 to monitor the course of the reaction as described in Sections 6-3 and 14-1. In this experiment, you will measure the Cl^- and I^- content of a carefully prepared solid unknown. The second procedure can be used to measure Cl^- in natural waters, which could be the basis for a class project.

Reagents

Unknowns: Each student receives a vial containing 0.22–0.44 g of KCl plus 0.50–1.00 g of KI (weighed accurately). The object is to determine the quantity of each salt in the mixture.
Buffer (pH 2): (6 mL/student) Titrate 1 M H_2SO_4 with 1 M NaOH to a pH near 2.0.
Silver nitrate: 1.2 g $AgNO_3$/student.

Procedure

1. Pour your unknown into a 50- or 100-mL beaker. Dissolve the solid in ~20 mL of distilled water and pour it into a 100-mL volumetric flask. Rinse the vial and beaker five times with small portions of H_2O and transfer the washings to the flask. Dilute to the mark and mix well.

2. Dry 1.2 g of $AgNO_3$ (FW 169.87) at 105°C for 1 h and cool in a desiccator for 30 min with minimal exposure to light. Accurately weigh 1.2 g and dissolve it in a 100-mL volumetric flask.

3. Set up the apparatus shown in Figure 6-3. The Ag electrode is a 3-cm length of silver wire soldered to copper wire. The copper wire is fitted with a jack that goes to the reference socket of a pH meter. The reference electrode for this titration is a glass pH electrode connected to its usual socket on the meter. If a combination pH electrode is employed, the reference jack of the combination electrode is not used. The silver electrode should be taped to the inside of the 100-mL beaker so that the Ag/Cu junction remains dry for the entire titration. The stirring bar should not hit either electrode.

4. Pipet 25.00 mL of unknown into the beaker, add 3 mL of pH 2 buffer, and begin magnetic stirring. Record the initial level of $AgNO_3$ in a 50-mL buret and add ~1 mL of titrant to the beaker. Turn the pH meter to the millivolt scale and record the volume and voltage. It is convenient (but not essential) to set the initial reading to +800 mV by adjusting the meter.

5. Titrate the solution with ~1-mL aliquots until 50 mL of titrant has been added or until you can see two abrupt voltage changes. You need not allow more than 15–30 s for each point. Record the volume and voltage at each point. Make a graph of millivolts versus milliliters to find approximate positions (±1 mL) of the two end points.

6. Turn the pH meter to standby, remove the beaker, rinse the electrodes well with distilled water, and blot them dry with a tissue.[11] Clean the beaker and set up the titration apparatus again. (The beaker need not be dry.)

---

[11] Silver halide adhering to the glass electrode can be removed by soaking the electrode in concentrated sodium thiosulfate solution after the experiment is finished. Thorough cleaning is not necessary between steps 6 and 7. Silver halides can be saved and converted back to pure $AgNO_3$ by the procedure of E. Thall, *J. Chem. Ed.* **1981**, *58*, 561.

7. Now perform an accurate titration, using 1-drop aliquots near the end points (and 1-mL aliquots elsewhere). You need not allow more than 30 s per point for equilibration.

8. Prepare a graph of millivolts versus milliliters and locate the end points, as illustrated in Figure 6-4. The I^- end point is taken as the intersection of the two dashed lines in the inset of Figure 6-4. The Cl^- end point is the steepest part of the curve at the second break. You can use the first derivative (see Figure 21-4) to locate the Cl^- end point. Calculate the wt % of KI and wt % of KCl in your solid unknown.

Analyzing Cl^- in Streams, Lakes, or Salt Water

A variation of the preceding procedure could make a class project in environmental analysis. For example, you could study changes in streams as a function of the season or recent rainfall. You could study stratification (layering) of water in lakes. The measurement responds to all ions that precipitate with Ag^+, of which Cl^- is, by far, the dominant ion in natural waters.

You can use the electrodes in Figure 6-3, or you can construct the rugged combination electrode shown in Figure 21-6.[12] The indicator electrode in Figure 21-6 is a bare silver wire in contact with analyte solution. The inner (reference) chamber of the combination electrode in Figure 21-6 contains a copper wire dipped into $CuSO_4$ solution. The inner electrode maintains a constant potential because the concentration of Cu^{2+} in the solution is constant. The solution makes electrical contact at the septum, where a piece of thread leaves enough space for solution to slowly drain from the electrode into the external sample solution.

1. Collect water from streams or lakes or from the ocean. To minimize bacterial growth, plastic jugs should be filled to the top and tightly sealed. Refrigeration is recommended.

2. Prepare 4 mM $AgNO_3$ solution with an accurately known concentration, as in step 2 at the beginning of this experiment. One student can prepare enough reagent for five people by using 0.68 g of $AgNO_3$ in a 1-L volumetric flask.

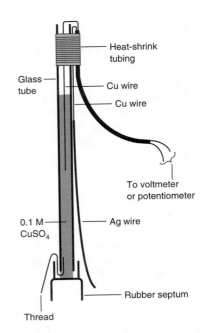

FIGURE 21-6 Combination electrode constructed from 8-mm-outer-diameter glass tubing. Strip the insulation off the ends of a two-wire copper cable. One bare Cu wire extends into the glass tube. A silver wire soldered to the other Cu wire is left outside the glass tube. The assembly is held together at the top with heat-shrink tubing. When ready for use, a cotton thread is inserted into the bottom of the tube and a 14/20 serum cap moistened with 0.1 M $CuSO_4$ is fitted over the end so that part of the thread is inside and part is outside. After adding 0.1 M $CuSO_4$ to the inside of the tube, the electrode is ready for use.

3. Measure with a graduated cylinder 100 mL of a natural water sample and pour it into a 250-mL beaker. Position the combination electrode from Figure 21-6 or the pair of electrodes from Figure 6-3 in the beaker so that a magnetic stirring bar will not hit the electrode.

4. Carry out a rough titration by adding 1.5-mL increments of titrant to the unknown and reading the voltage after 30 s to the nearest millivolt. Prepare a graph of voltage versus volume of titrant to locate the end point, which will not be as abrupt as those in Figure 6-4. If necessary, adjust the volume of unknown in future steps so that the end point comes at 20–40 mL. If you need less unknown, make up the difference with distilled water.

5. Carry out a more careful titration with fresh unknown. Add three-quarters of the titrant volume required to reach the equivalence point all at once.

[12]G. Lisensky and K. Reynolds, *J. Chem. Ed.* **1991,** *68,* 334; R. Ramette, *Chemical Equilibrium* (Reading, MA: Addison-Wesley, 1981), p. 649.

Then add ~0.4-mL increments (8 drops from a 50-mL buret) of titrant until you are 5 mL past the equivalence point. Allow 30 s (or longer, if necessary) for the voltage to stabilize after each addition. The end point is the steepest part of the curve, which can be estimated by the method shown in Figure 21-4.

6. Calculate the molarity and parts per million (μg/mL) of Cl^- in the unknown. Use data from several studies with the same sample to find the mean and standard deviation of ppm Cl^-.

21-14 *Spectrophotometric Nitrite Determination of Aquarium Water*

Background discussion for this experiment[13,14] is found in Sections 6-4 and 8-1.

Reagents

Sodium nitrite: 1 g $NaNO_2$ (FW 68.995)/student.
Sodium oxalate: 4 g $Na_2C_2O_4$ (FW 134.00)/student.
Potassium permanganate: 1 g $KMnO_4$ (FW 158.03)/student.
0.9 M H_2SO_4: 1 L/student.
Concentrated (96 wt %) H_2SO_4: 20 mL/student.
Color-forming reagent: (Reaction 8-1, 15 mL/student) Mix 1.0 g of sulfanilamide, 0.10 g of N-(1-naphthyl)-ethylenediamine dihydrochloride, and 10 mL of 85 wt % H_3PO_4 and dilute with distilled water to 100 mL. Store in a dark bottle in the refrigerator.

Procedure

1. Prepare and standardize 0.01 M $KMnO_4$ solution as described in Experiment 21-11. Reduce the quantities of $KMnO_4$ and $Na_2C_2O_4$ by a factor of 2.

2. Prepare 0.018 M $NaNO_2$ by dissolving 0.62 g of $NaNO_2$ in 500 mL of distilled water.

3. Follow steps 1–4 in Section 6-4 (page 108) to standardize the $NaNO_2$.

[13]K. D. Hughes, *Anal. Chem.* **1993**, *65*, 883A.

[14]*Standard Methods for the Examination of Water and Wastewater,* 18th ed. (Washington, DC: American Public Health Association, 1992).

4. Withdraw ~30 mL of aquarium water and filter it to remove suspended solids. Analyze duplicate 10.00-mL aliquots of freshly drawn aquarium water by the procedure in Section 8-1 (page 135). Use three or four standard points for the calibration curve such that the unknown lies within the range of the standards (Figure 8-2).

21-15 *Measuring Ammonia in a Fishtank with an Ion-Selective Electrode*

Reagents

Standard ammonia solution: Dissolve ~0.382 g of NH_4Cl (FW 53.491) in a 1-L volumetric flask to obtain a solution containing ~100 μg of nitrogen per milliliter. From the mass of NH_4Cl that you weighed, calculate the nitrogen concentration in μg/mL.

Procedure[13,14]

1. Using appropriate dilutions of your standard ammonia solution, prepare standards containing 3, 1, 0.3, and 0.1 μg N/mL in 100-mL volumetric flasks.

2. Pour the 100-mL standard solution containing 0.1 μg N/mL into a clean, dry 150-mL beaker. Immerse an ammonia ion-selective electrode and a reference electrode in the solution and begin magnetic stirring. Do not stir the solution so rapidly that air bubbles are drawn into the liquid. Add 1.0 mL of 10 M NaOH (to convert NH_4^+ into NH_3) and record the voltage to the nearest millivolt when the reading stabilizes. Do not allow more than 5 min before reading the voltage, because the temperature of the solution will increase from the stirring and the reading will change.

3. Repeat the same procedure for the other three standards. Prepare a graph of mV versus log[N], where [N] is the concentration of nitrogen in μg/mL. You should observe a reasonably straight line.

4. Measure 100 mL of freshly collected aquarium water in a graduated cylinder. Pour it into a clean, dry 150-mL beaker and repeat the process outlined in step 2.

5. From the least-squares slope and intercept of the calibration curve from step 3, and the potential measured in step 4, compute the concentration of ammonia nitrogen in the aquarium water.

6. Repeat steps 4 and 5 once more with a fresh sample to obtain a duplicate measurement.

21-16 Spectrophotometric Determination of Iron in Dietary Tablets

In this experiment,[15] iron from dietary supplement tablets is dissolved in acid, reduced to Fe^{2+} with hydroquinone, and complexed with o-phenanthroline to form an intensely colored complex (Color Plate 3).

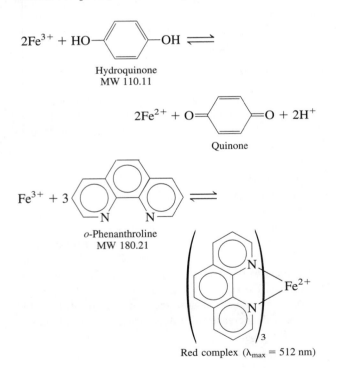

$$2Fe^{3+} + HO—\langle\rangle—OH \rightleftharpoons$$

Hydroquinone
MW 110.11

$$2Fe^{2+} + O=\langle\rangle=O + 2H^+$$

Quinone

$$Fe^{3+} + 3 \; \langle\text{o-Phenanthroline}\rangle \rightleftharpoons$$

o-Phenanthroline
MW 180.21

$$\left(\text{N} \atop \text{N}\right)_3 Fe^{2+}$$

Red complex (λ_{max} = 512 nm)

Reagents

Hydroquinone: (20 mL/student) Freshly prepared solution containing 10 g/L in distilled water. Store in an amber bottle.
Trisodium citrate: (20 mL/student) 25 g/L of Na_3(citrate) · $2H_2O$ (FW 294.10) in distilled water.

[15]R. C. Atkins, *J. Chem. Ed.* **1975**, *52*, 550.

o-Phenanthroline: (25 mL/student) Dissolve 2.5 g in 100 mL of ethanol and add 900 mL of distilled water. Store in an amber bottle.
6 M HCl: (25 mL/student) Dilute 124 mL of 37 wt % HCl up to 250 mL with distilled water.
Standard Fe (40 µg Fe/mL): (35 mL/student) Dissolve 0.281 g of reagent-grade $Fe(NH_4)_2(SO_4)_2 \cdot 6H_2O$ (FW 392.14) in distilled water in a 1-L volumetric flask containing 1 mL of 98 wt % H_2SO_4.

Procedure

1. Place one tablet in a 125-mL flask or 100-mL beaker and boil gently (*in a fume hood*) with 25 mL of 6 M HCl for 15 min. Filter the solution into a 100-mL volumetric flask. Wash the beaker and filter several times with small portions of distilled water to complete a quantitative transfer. Allow the solution to cool, dilute to the mark, and mix well. Dilute 5.00 mL of this solution to 100.0 mL in a fresh volumetric flask. If the label indicates that the tablet contains <15 mg of Fe, use 10.00 mL instead of 5.00 mL.

2. Pipet 10.00 mL of standard Fe solution into a beaker and measure the pH (with pH paper or a glass electrode). Add sodium citrate solution 1 drop at a time until a pH of ~3.5 is reached. Count the drops needed. (It should require ~30 drops.)

3. Pipet a fresh 10.00-mL aliquot of Fe standard into a 100-mL volumetric flask and add the same number of drops of citrate required in step 2. Add 2.00 mL of hydroquinone solution and 3.00 mL of o-phenanthroline solution, dilute to the mark with distilled water, and mix well.

4. Prepare three more solutions from 5.00, 2.00, and 1.00 mL of Fe standard and prepare a blank containing no Fe. Use sodium citrate solution in proportion to the volume of Fe solution. (If 10 mL of Fe requires 30 drops of citrate solution, then 5 mL of Fe requires 15 drops of citrate solution.)

5. Determine how many drops of citrate solution are needed to bring 10.00 mL of the iron tablet solution to pH 3.5. This will require about 3.5 or 7 mL of citrate, depending on whether 5 or 10 mL of unknown was diluted in the second part of step 1.

6. Transfer 10.00 mL of the solution containing the dissolved tablet to a 100-mL volumetric flask. Add the required amount of citrate solution, 2.00 mL of hydroquinone solution, and 3.00 mL of o-phenanthroline solution. Dilute to the mark and mix well.

7. Allow the solutions to stand for at least 15 min. Then measure the absorbance of each at 512 nm in a 1-cm cell. (The color is stable, so all solutions should be prepared and all the absorbances measured at once.) Use distilled water in the reference cuvet and subtract the absorbance of the blank from the absorbance of the Fe standards.

8. Make a graph of absorbance versus micrograms of Fe in the standards. Find the slope and intercept (and standard deviations) by the method of least squares, as described in Section 4-4. Calculate the molarity of $Fe(o\text{-phenanthroline})_3^{2+}$ in each solution and find the average molar absorptivity (ϵ in Beer's law) from the four absorbances. (Remember that all the iron has been converted to the phenanthroline complex.) If a Spectronic 20 is used for absorbance measurements, assume that the pathlength is 1.1 cm for the sake of this calculation.

9. From the calibration curve, find the number of milligrams of Fe in the tablet. Use Equation 4-16 to find the uncertainty in the number of milligrams of Fe.

21-17 *Microscale Spectrophoto-metric Measurement of Iron in Foods by Standard Addition*

This microscale experiment[16] uses the same chemistry as that of Experiment 21-16 to measure iron in foods such as broccoli, peas, cauliflower, spinach, beans, and nuts. A possible project is to compare processed (canned or frozen) vegetables to fresh vegetables. Instructions are provided for small volumes, but the experiment can be scaled up to fit available equipment. Section 20-6 describes the method of standard addition. You do not need to read the rest of Chapter 20 to understand Section 20-6.

Reagents

2.0 M HCl: (15 mL/student) Dilute 165 mL of concentrated (37 wt %) HCl up to 1 L.
Hydroquinone: 4 mL/student; prepared as in Experiment 21-16.
Trisodium citrate dihydrate: 4 g/student.

[16]Idea based on P. E. Adams, *J. Chem. Ed.* **1995**, *72*, 649.

o-Phenanthroline: 4 mL/student; prepared as in Experiment 21-16.
Standard Fe (40 μg Fe/mL): 4 mL/student; prepared as in Experiment 21-16.
6 M HCl: Dilute 500 mL of 37 wt % HCl up to 1 L with distilled water. Store in a bottle and reuse many times for soaking crucibles.

Procedure

1. Fill a clean porcelain crucible with 6 M HCl in the hood and allow it to stand for 1 h to remove traces of iron from previous uses. Rinse well with distilled water and dry. After weighing the empty crucible, add 5–6 g of finely chopped food sample and weigh again to obtain the mass of food. (Some foods, like frozen peas, should not be chopped because they will lose their normal liquid content.)

2. This step could require 3 h, during which you can be doing other lab work. Carefully heat the crucible with a Bunsen burner in a hood (Figure 21-7). Use a low flame to *dry* the sample, being careful to avoid spattering. Increase the flame temperature to *char* the sample. Keep the crucible lid and tongs nearby. If the sample bursts into flames, use tongs to place the lid on the crucible to smother the flame. After charring, use the hottest possible flame (bottom of crucible should be red hot) to *ignite* the black solid, converting it to white ash. Continue ignition until all traces of black disappear.

3. After cooling the crucible to room temperature, add 10.00 mL of 2.0 M HCl by pipet and swirl gently for 5 min to dissolve the ash. Filter the mixture through a small filter and collect the filtrate in a vial or small flask. You need to recover >8 mL for the analysis.

4. Weigh 0.71 g of trisodium citrate dihydrate into each of four 10-mL volumetric flasks. Using a 2-mL volumetric pipet or a 1-mL micropipet, add 2.00 mL

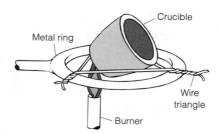

FIGURE 21-7 Positioning a crucible above a burner.

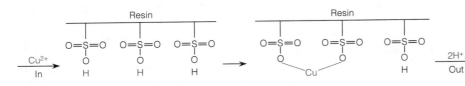

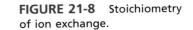

FIGURE 21-8 Stoichiometry of ion exchange.

of ash solution to each flask. Add 4 mL of distilled water and swirl to dissolve the citrate. The solution will have a pH near 3.6. Using a micropipet, add 0.20 mL of hydroquinone solution and 0.30 mL of phenanthroline solution to each flask.

5. Label the volumetric flasks 0 through 3. Add no Fe standard to flask 0. Using a micropipet, add 0.250 mL of Fe standard to flask 1. Add 0.500 mL of Fe standard to flask 2 and 0.750 mL to flask 3. The four flasks now contain 0, 1, 2, and 3 μg Fe/mL, in addition to Fe from the food. Dilute each to the mark with distilled water, mix well, and allow 15 min to develop full color.

6. Prepare a blank by mixing 0.71 g of trisodium citrate dihydrate, 2.00 mL of 2.0 M HCl, 0.20 mL of hydroquinone solution, 0.30 mL of phenanthroline solution and diluting to 10 mL. The blank does not require a volumetric flask.

7. Measure the absorbance of each solution at 512 nm in a 1-cm cell with distilled water in the reference cell. Before each measurement, remove all liquid from the cuvet with a Pasteur pipet. Then use ~1 mL of your next solution (delivered with a clean, dry Pasteur pipet) to wash the cuvet. Remove and discard the washing. Repeat the washing once more with fresh solution and discard the washing. Finally, add your new solution to the cuvet for measuring absorbance.

8. Subtract the absorbance of the blank from each reading and make a graph like that in Figure 20-19 to find the Fe content of the unknown solution. Calculate the wt % of Fe in the food.

21-18 *Properties of an Ion-Exchange Resin*

This experiment[17] explores properties of a cation-exchange resin, which is an organic polymer containing sulfonic acid

[17]Part of this experiment is from M. W. Olson and J. M. Crawford, *J. Chem. Ed.* **1975**, *52*, 546.

groups ($-SO_3H$). When a cation, such as Cu^{2+}, flows into the resin, the cation is tightly bound by sulfonate groups, releasing one H^+ for each positive charge bound to the resin (Figure 21-8). Cu^{2+} can be displaced from the resin by a large excess of H^+ or by an excess of any other cation for which the resin has some affinity.

First, known quantities of NaCl, $Fe(NO_3)_3$, and NaOH will be passed through the resin in the H^+ form. The H^+ released by each cation will be measured by titration with NaOH.

In the second part of the experiment, you will analyze a sample of impure vanadyl sulfate ($VOSO_4 \cdot 2H_2O$). As supplied commercially, this salt contains $VOSO_4$, H_2SO_4, and H_2O. A solution will be prepared from a known mass of reagent. The VO^{2+} content can be assayed spectrophotometrically, and the total cation (VO^{2+} and H^+) content can be assayed by ion exchange. Together, these measurements enable you to establish the quantities of $VOSO_4$, H_2SO_4, and H_2O in the sample.

Reagents

Bio-Rad Dowex 50W-X2 (100/200 mesh) cation-exchange resin: 1.1 g/student.

0.3 M NaCl: (5–10 mL/student) Accurately weigh ~17.5 g of NaCl (FW 58.44) and dissolve it in a 1-L volumetric flask.

0.1 M Fe(NO₃)₃ · 6H₂O: (5–10 mL/student) Accurately weigh ~35 g of $Fe(NO_3)_3 \cdot 6H_2O$ (FW 349.95) and dissolve it in a 1-L volumetric flask.

VOSO₄: The common grade (usually designated "purified") is used for this experiment. Students can make their own solutions and measure the absorbance at 750 nm, or a bottle of stock solution (25 mL per student) can be supplied. The stock should contain 8 g/L (accurately weighed) and be labeled with the absorbance.

0.02 M NaOH: Each student should prepare an accurate 1/5 dilution of standard 0.1 M NaOH. For example, 50.0 mL of 0.1 M NaOH can be pipetted into a 250-mL volumetric flask and diluted to volume with distilled water.

0.1 M HCl: 50 mL/student.

Phenolphthalein indicator: See Experiment 21-6.

Procedure

1. Prepare a chromatography column from glass tubing that has a 0.7-cm diameter and is 15 cm long, by fitting the tubing at the bottom with a cork that has a small hole to serve as the outlet. Place a small ball of glass wool above the cork to retain the resin. Use a glass rod to plug the outlet temporarily. (Alternatively, an inexpensive column such as 0.7 × 15 cm Econo-Column from Bio-Rad Laboratories[18] works well in this experiment.) Fill the column with distilled water, close it off, and test for leaks. Then drain the water until 2 cm remains and close the column again.

2. Make a slurry of 1.1 g of Bio-Rad Dowex 50W-X2 (100/200 mesh) cation-exchange resin in 5 mL of distilled water and pour it into the column. If the resin cannot be poured into the column all at once, allow some to settle, remove the supernatant liquid with a pipet, and pour in the rest of the resin. If the column is stored between laboratory periods, it should be upright and capped, and should contain distilled water above the level of the resin.[19]

3. The general procedure for analysis of a sample is as follows:

(a) Generate the H^+-saturated resin by passing $\sim$10 mL of 1 M HCl through the column. Apply the liquid sample to the glass wall so as not to disturb the resin.

(b) Wash the column with $\sim$15 mL of distilled water. Use the first few milliliters to wash the glass walls and allow the water to soak into the resin before continuing the washing.[20]

(c) Place a clean 125-mL flask under the outlet and pipet the sample onto the column.

(d) After reagent has soaked in, wash it through with 10 mL of H_2O, collecting all eluate.

(e) Add 3 drops of phenolphthalein to the flask and titrate with standard 0.02 M NaOH.

4. Analyze 2.000-mL aliquots of 0.3 M NaCl and 0.1 M $Fe(NO_3)_3$, following the procedure in step 3.

Calculate the theoretical volume of NaOH needed for each titration. If you do not come within 2% of this volume, repeat the analysis.

5. Pass 10.0 mL of your 0.02 M NaOH through the column as directed in step 3 and titrate the eluate. Explain what you observe.

6. Analyze 10.00 mL of $VOSO_4$ solution as described in step 3.

7. Step 6 gives the total cation content $(= VO^2 + 2H_2SO_4)$ of 10.00 mL of $VOSO_4$ solution. From the absorbance at 750 nm and the molar absorptivity of $VO^{2+}(\epsilon = 18.0 \ M^{-1} \ cm^{-1})$, calculate the concentration of VO^{2+} in the solution. By difference, calculate the concentration of H_2SO_4 in $VOSO_4$ reagent. From the difference between the mass of $VOSO_4$ reagent that was weighed out and the content of $VOSO_4$ and H_2SO_4, calculate the mass of H_2O in the $VOSO_4$ reagent. Express the composition of the vanadyl sulfate in the form $(VOSO_4)_{1.00}(H_2SO_4)_x(H_2O)_y$.

21-19 Using an Internal Standard in Gas Chromatography or High-Performance Liquid Chromatography

This experiment uses an internal standard (Section 16-5) for quantitative analysis. You will analyze several mixtures to establish the validity of Equation 16-8. Directions are purposely general, because the specific mixture and analytical conditions depend on the equipment and columns employed. You will receive an unknown solution containing two or three compounds from a specified list of possibilities.[21] Pure samples of each possible component of the unknown should also be available.

Procedure

1. Identify the components of your mixture. Run a chromatogram of the pure unknown. Then mix pure

---

[18]Bio-Rad Laboratories, 2000 Alfred Nobel Drive, Hercules, CA 94547 (Phone: 510-741-1000).

[19]When the experiment is finished, the resin can be collected, washed with 1 M HCl and water, and reused.

[20]Unlike most other chromatography resins, the one used in this experiment retains water when allowed to run "dry." Ordinarily, you must not let liquid fall below the top of the solid phase in a chromatography column.

[21]We have done this experiment with packed gas chromatography columns with Carbowax or dinonyl phthalate stationary phases. We used dichloromethane, chloroform, ethyl acetate, 1-propanol, and toluene as unknowns. We have also done this experiment by HPLC, using a C_{18} bonded phase eluted with 65 vol % methanol in water. The unknowns included toluene, biphenyl, naphthalene, 2-naphthol, and 9-fluorenone.

samples of suspected components with samples of the unknown and run a new chromatogram of each. By observing which peaks grow or whether new peaks appear, you should be able to identify each species in your unknown.

2. Perform a quantitative analysis of one component (designated X) of the unknown, which is well separated from all other components. Select an internal standard (designated S) that is not a component of the unknown and is separated from other peaks in the chromatogram. By trial and error, prepare a mixture of unknown plus S such that the peak areas (or heights, if the peaks are sharp) of X and S are within a factor of 2 of each other. The mass (or volume) of S and the mass of unknown (not X, but total unknown) in the mixture must be accurately known.

3. Prepare known mixtures of pure X and pure S to verify Equation 16-8. If a solvent is used, the mixtures should have the same concentration of S required in step 2, plus variable quantities of unknown. If pure liquids are used without a solvent, the concentrations of both S and X will necessarily vary. Prepare a graph in which you plot the signal ratio (peak area of X/peak area of S) versus the known concentration ratio (concentration of X/concentration of S). The graph should span the range including the signal ratio measured for the unknown mixture plus standard.

4. From your graph, calculate the concentration of species X in the unknown.

21-20 *Analysis of Sulfur in Coal by Ion Chromatography*

When coal is burned, sulfur in the coal is converted to $SO_2(g)$, which is further oxidized to H_2SO_4 in the atmosphere. Rainfall laden with H_2SO_4 is harmful to plant life. Measuring the sulfur content of coal is therefore important to efforts to limit man-made sources of acid rain. This experiment[22] measures sulfur in coal by first heating the coal in the presence of air in a flux (Section 2-10) containing Na_2CO_3 and MgO, which converts the sulfur to Na_2SO_4. The product is dissolved in water, and sulfate ion is mea-

[22]E. Koubek and A. E. Stewart, *J. Chem. Ed.* **1992,** *69,* A146.

sured by ion chromatography. Although you will be given a sample of coal to analyze, consider how you would obtain a representative sample from an entire trainload of coal being delivered to a utility company.

Reagents

Coal: (1 g/student) Coal can be obtained from electric power companies and some heavy industries. Coal is also available as a Standard Reference Material.[23]
Flux: (4 g/student) 67 wt % MgO/33 wt % Na_2CO_3.
6 M HCl: 25 mL/student.
Phenolphthalein indicator: See Experiment 21-6.
Ammonium sulfate: Solid reagent for preparing standards.

Procedure

1. Grind the coal to a fine powder with a mortar and pestle. Mix 1 g of coal (accurately weighed) with ~3 g of flux in a porcelain crucible. Mix thoroughly with a spatula and tap the crucible to pack the powder. Gently cover the mixture with ~1 g of additional flux. Cover the crucible and place it in a muffle furnace. Then turn on the furnace and set the temperature to 800°C and leave the sample in the furnace at 800°C overnight. The reaction is over when all black particles have disappeared. A burner can be used in place of the furnace, but the burner should not be left unattended overnight.

2. After cooling to room temperature, place the crucible into a 150-mL beaker and add 100 mL of distilled water. Heat the beaker on a hot plate to just below boiling for 20 min to dissolve as much solid as possible. Pour the liquid through filter paper in a conical funnel directly into a 250-mL volumetric flask. Wash the crucible and beaker three times with 25-mL portions of distilled water and pour the washings through the filter. Add 5 drops of phenolphthalein indicator to the flask and neutralize with 6 M HCl until the pink color disappears. Dilute to 250 mL and mix well.

3. Pipet 25.00 mL of sulfate solution from the 250-mL volumetric flask in step 2 into a 100-mL volumetric flask and dilute to volume to prepare a fourfold dilution for ion chromatography. Inject a

[23]Standards containing 0.5–5 wt % S are available from NIST Standard Reference Materials Program, Room 204, Building 202, Gaithersburg, MD 20899-0001 (Phone: 301-975-6776; E-mail: SRMINFO@enh.nist.gov).

sample of this solution into an ion chromatograph[24] to be sure that the concentration is in a reasonable range for analysis. More or less dilution may be necessary. Prepare the correct dilution for your equipment.

4. Assuming that the coal contains 3 wt % sulfur, calculate the concentration of SO_4^{2-} in the solution in step 3. Using ammonium sulfate and appropriate volumetric glassware, prepare five standards containing 0.1, 0.5, 1.0, 1.5, and 2.0 times the calculated concentration of SO_4^{2-} in the unknown.

5. Analyze all solutions by ion chromatography. Prepare a calibration curve from the standards, plotting peak area versus SO_4^{2-} concentration. Use the least-squares fit to find the concentration of SO_4^{2-} in the unknown. Calculate the wt % of S in the coal.

21-21 *Measuring Carbon Monoxide in Automobile Exhaust by Gas Chromatography*

Carbon monoxide is a colorless, odorless, poisonous gas emitted from automobile engines because of incomplete combustion of fuel to CO_2. A well-tuned car with a catalytic converter might emit 0.01 vol % CO, whereas an old "clunker" could emit as much as 15 vol % CO! In this experiment you will collect samples of auto exhaust and measure the CO content by gas chromatography.[25] A possible class project is to compare different types of cars and different states of maintenance of vehicles. CO emission is greatest within the first few minutes after starting a cold engine. After warm-up, a well-tuned vehicle may emit too little CO to detect with an inexpensive gas chromatograph. CO can be measured as a function of time after start-up.

Chromatography is performed at 50°–60°C with a 2-m-long packed column containing 5A molecular sieves with He

FIGURE 21-9 Collecting a sample of standard gas mixture from a lecture bottle. Remove the vent needle when withdrawing gas into the syringe. Do not open the tank valve so much that the connections pop open. (*Caution: Handle CO only in a hood.*)

carrier gas and thermal conductivity detection. The column should be flushed periodically by disconnecting it from the detector and flowing He through for 30 h. Flushing after 50–100 injections desorbs H_2O and CO_2 from the sieves.

Reagents

CO gas standard: Lecture bottle containing 1 vol % CO in N_2.[26]

Procedure

1. Attach with heavy tape a heat-resistant hose to the exhaust pipe of a car. (*Caution: Avoid breathing the exhaust.*) Use the free end of the hose to collect exhaust in a heavy-walled, 0.5-L plastic zipper-type bag from the grocery. Flush the bag well with exhaust before sealing it tightly. Allow the contents to come to room temperature for analysis.

2. Inject a 1-mL sample of air into the gas chromatograph by using a gas-tight syringe and adjust the temperature and/or flow rate if necessary so that N_2 is eluted within 2 min. You should see peaks for O_2 and N_2.

3. Inject 1.00-mL of standard 1 vol % CO in N_2. One way to obtain gas from a lecture bottle is to attach a hose to the tank (*in a hood*) with a serum stopper on a glass tube at the end of the hose (Figure 21-9). Place a needle in the serum stopper and slowly bleed gas from the tank to flush the hose. Insert a gas-tight syringe into the serum stopper, remove the vent needle, and slowly withdraw gas into the syringe.

[24]For example, chromatography can be done with 25–50 μL of sample on a 4-mm-diameter × 250-mm-long Dionex IonPac AS5 analytical column and an AG5 guard column using 2.2 mM Na_2CO_3/2.8 mM $NaHCO_3$ eluent at 2.0 mL/min with ion suppression. SO_4^{2-} is eluted near 6 min and is detected by its conductivity on a full-scale setting of 30 microsiemens. Many combinations of column and eluent are suitable for this analysis.
[25]D. Jaffe and S. Herndon, *J. Chem. Ed.* **1995,** *72,* 364.
[26]Available, for example, from Scott Specialty Gases, Route 611, Plumsteadville, PA 18949 (Phone: 215-766-8861).

Then close the tank to prevent pressure buildup in the tubing. When you inject the standard into the chromatograph, you should see a peak for CO with about three times the retention time of N_2. Adjust the detector attenuation so the CO peak is near full scale. Reinject the standard twice and measure the peak area each time.

4. Inject two 1.00-mL samples of auto exhaust and measure the peak area each time. Compute the vol % of CO in the unknown from its average peak area:

$$\frac{\text{vol \% CO in the unknown}}{\text{vol \% CO in the standard}} =$$
$$\frac{\text{peak area of unknown/detector attenuation}}{\text{peak area of standard/detector attenuation}}$$

21-22 Amino Acid Analysis by Capillary Electrophoresis

In this experiment[27] you will hydrolyze a protein with HCl to break the protein into its component amino acids (see Table 12-1 and page 222). After adding an internal standard, the amino acids and the internal standard will be derivatized (chemically converted) to a form that absorbs light strongly at 420 nm.

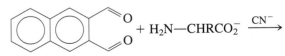

Naphthalene-
2,3-dicarboxaldehyde
(NDA, MW 184.19)   Amino acid

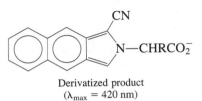

Derivatized product
(λ_{max} = 420 nm)

The mixture will then be separated by capillary electrophoresis, and the quantity of each component will be measured relative to that of the internal standard. The goal is to find the relative number of each amino acid in the protein.

[27]P. L. Weber and D. R. Buck, *J. Chem. Ed.* **1994,** *71,* 609.

Reagents

Protein: (6 mg/student) Use a pure protein such as lysozyme or cytochrome *c*. The amino acid content should be available in the literature for you to compare with your results.[28]

6 M HCl: Dilute 124 mL of concentrated (37 wt %) HCl up to 250 mL with distilled water.

0.05 M NaOH: Dissolve 0.50 g of NaOH (FW 40.00) in 250 mL of distilled water.

1.5 M NH₃: Dilute 26 mL of 28 wt % NH_3 up to 250 mL with distilled water.

10 mM KCN: Dissolve 16 mg of KCN (FW 65.12) in 25 mL of distilled water.

Borate buffer (pH 9.0): To prepare 20 mM buffer, dissolve 0.76 g of sodium tetraborate ($Na_2B_4O_7 \cdot 10H_2O$, FW 381.37) in 70 mL of distilled water. Using a pH electrode, adjust the pH to 9.0 with 0.3 M HCl (a 1:20 dilution of 6 M HCl with distilled water) and dilute to 100 mL with distilled water.

Run buffer (20 mM borate–50 mM sodium dodecyl sulfate, pH 9.0): Prepare this as you prepared borate buffer but add 1.44 g of $CH_3(CH_2)_{11}OSO_3Na$ (FW 288.38) prior to adjusting the pH with HCl.

Amino acids: Prepare 100 mL of standard solution containing all 15 of the amino acids in Figure 21-10, each at a concentration near 2.5 mM in a solvent of 0.05 M NaOH. Table 12-1 gives molecular weights of amino acids.

α-Aminoadipic acid internal standard: Prepare a 5-mM solution by dissolving 40 mg of $HO_2C(CH_2)_3CH(NH_2)CO_2H$ (MW 161.16) in 50 mL of distilled water.

Naphthalene-2,3-dicarboxaldehyde (NDA): Prepare a 10-mM solution by dissolving 18 mg of NDA in 10 mL of acetonitrile.

Hydrolysis of the Protein

1. Fit a glass tube (17 × 55 mm) with a rubber septum and evacuate it through a needle. Dissolve 6 mg of protein in 0.5 mL of 6 M HCl in a small vial. Purge the solution with N_2 for 1 min to remove O_2 and

[28]For hen egg white lysozyme, the amino acid content is $S_{10}T_7H_1G_{12}A_{12}(E_2Q_3)Y_3(D_8N_{13})V_6M_2I_6L_8F_3R_{11}K_6C_8W_8P_2$. (R. E. Canfield and A. K. Liu, *J. Biol. Chem.* **1965,** *240,* 2000; D. C. Philips, *Scientific American,* May 1966.) For horse cytochrome *c,* the composition is $S_0T_{10}H_3G_{12}A_6(E_9Q_3)Y_4(D_3N_5)V_3M_2I_6L_6F_4R_2K_{19}C_2W_1P_4$. (E. Margoliash and A. Schejter, *Adv. Protein Chem.* **1966,** *21,* 114.) Both proteins are available from Sigma Chemical Co., P.O. Box 14508, St. Louis, MO 63178 (Phone: 314-771-5750).

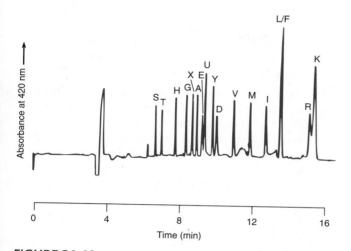

FIGURE 21-10 Electropherogram of standard mixture of NDA-derivatized amino acids plus internal standard, all at equal concentrations. Abbreviations for amino acids are given in Table 12-1. The internal standard, designated X, is α-aminoadipic acid. U is an unidentified peak. Acid hydrolysis converts Q into E and converts N into D, so Q and N are not observed. The analysis of lysine (K) is not reliable because its NDA derivative is unstable. Cysteine (C) and tryptophan (W) are degraded during acid hydrolysis and are not observed. Proline (P) is not a primary amine, so it does not react with NDA to form a detectable product.

immediately transfer the liquid by syringe into the evacuated tube. Add 0.5 mL of fresh HCl to the vial, purge, and transfer again into the tube. Heat the part of the glass tube containing the liquid at 100°–110°C in an oil bath for 18–24 h, preferably behind a shield.

2.   After cooling, remove the septum and transfer the liquid to a 25-mL round-bottom flask. Evaporate the solution to dryness with gentle heat and suction from a water aspirator. Use a trap like that in Figure 2-13 whenever you use an aspirator. Rinse the hydrolysis tube with 1 mL of distilled water, add it to the flask, and evaporate to dryness again. Dissolve the residue in 1.0 mL of 0.05 M NaOH and filter it through a 0.45-μm pore size syringe filter. The total concentration of all amino acids in this solution is ~50 mM.

Derivatization

3.   Using a micropipet, place 345 μL of 20 mM borate buffer into a small screw-cap vial. Add 10 μL

of the standard amino acid solution. Then add 10 μL of 5 mM α-aminoadipic acid (the internal standard), 90 μL of 10 mM KCN, and 75 μL of 10 mM naphthalene-2,3-dicarboxaldehyde. The fluorescent yellow-green color of the amino acid–NDA product should appear within minutes. After 25 min, add 25 μL of 1.5 M NH₃ to react with excess NDA. Wait 15 min before electrophoresis.

4.   Place 345 μL of 20 mM borate buffer (pH 9.0) into a small screw-cap vial. Add 10 μL of the hydrolyzed protein solution in 0.05 M NaOH from step 2. Then add 10 μL of 5 mM α-aminoadipic acid (the internal standard), 60 μL of 10 mM KCN, and 50 μL of 10 mM naphthalene-2,3-dicarboxaldehyde. After 25 min, add 25 μL of 1.5 M NH₃ to react with excess NDA. After 15 min, begin electrophoresis. (Precise timing reduces variations between the standard and the unknown.)

Electrophoresis and Analysis of Results

5.   (*Caution: Electrophoresis uses a dangerously high voltage of 20–24 kV. Be sure to follow all safety procedures.*) Electrophoresis is conducted with a 50-μm-inner-diameter uncoated silica capillary and spectrophotometric detection at 420 nm. Precondition the column by injecting 30 μL of 0.1 M NaOH and flushing 15 min later with 30 μL each of distilled water and then run buffer. The column should be reconditioned in the same manner after every 3–4 runs.

6.   Inject 3 nL of standard amino acid mixture from step 3. The electropherogram should look similar to Figure 21-10. Measure the area of each peak (or the height, if you do not have a computer for measurement of area). Repeat the injection and measure the areas again.

7.   Find the quotient

$$\frac{\text{area of amino acid peak}}{\text{area of internal standard peak}}$$

for each amino acid in the standard mixture. (Use peak heights if area is not available.) Prepare a table showing the relative areas for each peak in each injection of standard and find the average quotient for each amino acid from both runs.

8.   Inject 3 nL of derivatized, hydrolyzed protein from step 4 and measure the same quotient as mea-

sured in step 7. Repeat the injection and find the average quotient from both injections.

9. Find the concentration of each amino acid in the protein hydrolysate by using Equation 16-8. The numerator on the right side of Equation 16-8 was found in step 8, and the denominator on the right side was found in step 7. (Use the average values from the two sets of runs.) The denominator on the left side of Equation 16-8 is known from the masses of amino acids and internal standard weighed into the standard solution. For each amino acid, you can now solve Equation 16-8 to find the numerator on the left side, which is the quotient

$$\frac{\text{concentration of amino acid in protein hydrolysate}}{\text{concentration of internal standard in protein hydrolysate}}$$

But you know the concentration of internal standard in the protein hydrolysate from the volume and concentration of internal standard used in step 4. Therefore, you can calculate the concentration of each amino acid in the hydrolysate.

10. Find the mole ratio of amino acids in the protein. If there were no experimental error, you could divide all concentrations by the lowest one. Because the lowest concentration has a large relative error, pick an amino acid with two or three times the concentration of the least concentrated amino acid. Define this concentration to be exactly 2 or 3. Then compute the molarities of other amino acids relative to the chosen amino acid. Your result is a formula such as $S_{10.6}T_{6.6}H_{0.82}G_{12.5}A_{11.2}E_{5.3}Y_{=3}D_{19.8}V_{6.1}M_{2.2}I_{5.7}(L + F)_{11.5}$.

Glossary

abscissa The horizontal (x) axis of a graph.

absolute uncertainty The uncertainty associated with a measurement.

absorbance, A Defined as $A = \log(P_0/P)$, where P_0 is the radiant power of light striking a sample on one side and P is the radiant power emerging from the other side.

absorption Occurs when a substance is taken up *inside* another. See also **adsorption.**

absorption spectrum A graph of absorbance or transmittance of light as a function of wavelength, frequency, or wavenumber.

accuracy How close the measured value is to the "true" value.

acid A substance that increases the concentration of H^+ when added to water.

acid dissociation constant, K_a The equilibrium constant for the reaction of an acid, HA, with H_2O: $HA + H_2O \rightleftharpoons A^- + H_3O^+$

acid wash Procedure in which glassware is soaked in 3–6 M HCl for >1 h to remove traces of cations adsorbed on the surface of the glass and to replace them with H^+. The soaking in acid is followed by rinsing well with distilled water and soaking in distilled water to remove the acid.

acidic solution One in which the concentration of H^+ (actually the activity of H^+) is greater than the concentration (activity) of OH^-.

adsorption Occurs when a substance becomes attached to the *surface* of another substance. See also **absorption.**

adsorption chromatography A technique in which the solute equilibrates between the mobile phase and adsorption sites on the stationary phase.

adsorption indicator Used for precipitation titrations, it becomes attached to a precipitate and changes color when the surface charge of the precipitate changes sign at the equivalence point.

aerosol A suspension of very fine liquid or solid particles in air or gas. Examples include fog and smoke.

affinity chromatography A technique in which a particular solute is retained on a column by a specific interaction with a molecule bound to the stationary phase.

aliquot Portion.

amine A compound with the general formula RNH_2, R_2NH, or R_3N, where R is any group of atoms.

amino acid One of 20 building blocks of proteins, having the general structure

$$\overset{\displaystyle R}{\underset{\displaystyle {}^+H_3NCHCO_2^-}{|}}$$

where R is a different substituent for each acid.

ammonium ion *The* ammonium ion is NH_4^+. *An* ammonium ion is any ion of the type RNH_3^+, $R_2NH_2^+$, R_3NH^+, or R_4N^+, where R is an organic substitutent.

ampere, A One ampere is the current that will produce a force of exactly 2×10^{-7} N/m when that current flows through two "infinitely" long, parallel conductors of negligible cross section, with a spacing of 1 m, in a vacuum.

amphiprotic molecule One that can act as both a proton donor and a proton acceptor. The intermediate species of polyprotic acids are amphiprotic.

analyte The substance being analyzed.

anion A negatively charged ion.

anion exchanger An ion exchanger with positively charged groups covalently attached to the support. It reversibly binds anions.

anode The electrode at which oxidation occurs. In electrophoresis, it is the positively charged electrode.

antibody A protein manufactured by an organism for sequestering and marking foreign molecules for destruction.

antigen A molecule that is foreign to an organism and stimulates the production of antibodies.

antilogarithm The antilogarithm of a is b if $10^a = b$.

aqueous solution One in which water is the solvent.

ashless filter paper Specially treated paper that leaves a negligible residue after ignition. It is used for gravimetric analysis.

atomic absorption spectroscopy A technique in which the absorption of light by gaseous atoms in a flame or furnace is used to measure the concentration of atoms.

atomic emission spectroscopy A technique is which the emission of light by thermally excited atoms in a flame or furnace is used to measure the concentration of atoms.

atomic weight The number of grams of an element containing Avogadro's number of atoms.

atomization The process in which a compound is decomposed into its atoms at high temperature.

autoprotolysis The reaction of a solvent in which two molecules of the same species transfer a proton from one to the other; e.g., $CH_3OH + CH_3OH \rightleftharpoons CH_3OH_2^+ + CH_3O^-$.

auxiliary complexing agent A species, such as ammonia, that is added to a solution to stabilize another species and keep that other species in solution. It binds loosely enough to be displaced by a titrant.

average The sum of measured values divided by the number of values.

back titration One in which an excess of standard reagent is added to react with analyte. Then the excess reagent is titrated with a second standard reagent or with a standard solution of analyte.

background correction In atomic spectroscopy, a means of distinguishing signal due to analyte from signal due to absorption, emission, or scattering by the flame, furnace, plasma, or sample matrix.

base A substance that decreases the concentration of H^+ when added to water.

base hydrolysis constant or **base "dissociation" constant, K_b** The equilibrium constant for the reaction of a base, B, with H_2O: $B + H_2O \rightleftharpoons BH^+ + OH^-$.

basic solution One in which the concentration (actually, the activity) of OH^- is greater than the concentration (activity) of H^+.

Beer's law Relates the absorbance (A) of a sample to the concentration (c) of the absorbing species, the pathlength (b), and the molar absorptivity (ϵ) of the absorbing species: $A = \epsilon bc$.

blank A solution containing all the reagents for an analysis, but no deliberately added analyte.

blank titration One in which a solution containing all the reagents except analyte is titrated. The volume of titrant used in the blank titration should be subtracted from the volume used to titrate unknown.

blocking Occurs when metal binds tightly to a metal ion indicator and is not readily released to the titrant (such as EDTA). A blocked indicator is unsuitable for a titration because no color change is observed at the end point.

Boltzmann distribution The relative population of two states at thermal equilibrium:

$$\frac{N_2}{N_1} = \left(\frac{g_2}{g_1}\right) e^{-(E_2-E_1)/kT}$$

where N_i is the population of the state, g_i is the degeneracy of the state, E_i is the energy of the state, k is Boltzmann's constant, and T is temperature in kelvins. Degeneracy refers to the number of states with the same energy.

bomb Sealed vessel for conducting high-temperature, high-pressure reactions.

bonded stationary phase In chromatography, a stationary liquid phase covalently attached to the solid support.

buffer A mixture of an acid and its conjugate base. A buffered solution is one that resists changes in pH when acid or base is added.

buoyancy Occurs when an object is weighed in air and the observed mass is less than the true mass because the object has displaced an equal volume of air from the balance pan.

buret A calibrated glass tube with a stopcock at the bottom. Used to deliver known volumes of liquid.

calibration curve A graph showing the value of some property versus concentration of analyte. When the corresponding property of an unknown is measured, its concentration can be determined from the graph.

capillary electrophoresis Separation of a mixture into its components by using a strong electric field imposed between the two ends of a narrow capillary tube filled with electrolyte solution.

capillary gel electrophoresis Same as capillary electrophoresis, but the tube is filled with a gel through which solutes can migrate.

capillary zone electrophoresis A form of capillary electrophoresis in which ionic solutes are separated because of differences in their electrophoretic mobility.

carboxylate anion The conjugate base (RCO_2^-) of a carboxylic acid.

carboxylic acid A molecule with the general structure RCO_2H, where R is any group of atoms.

cathode The electrode at which reduction occurs. In electrophoresis, it is the negatively charged electrode.

cation A positively charged ion.

cation exchanger An ion exchanger with negatively charged groups covalently attached to the support. It reversibly binds cations.

character The part of a logarithm to the left of the decimal point.

charge balance A statement that the sum of all positive charges in solution equals the magnitude of the sum of all negative charges in solution.

chelating ligand A ligand that binds to a metal through more than one atom.

chemical interference In atomic spectroscopy, any chemical reaction that decreases the efficiency of atomization.

chromatogram A graph showing the concentration of solutes emerging from a chromatography column as a function of elution time or volume.

chromatography A technique in which molecules in the mobile phase are separated because of their different affinities for a stationary phase. The greater the affinity for the stationary phase, the longer the molecule is retained.

co-chromatography Simultaneous chromatography of a known compound with an unknown. If a known and an unknown have the same retention time on several different columns, they are probably identical.

colloid A dissolved particle with a diameter in the range 1–100 nm.

combination electrode Consists of a glass pH electrode with a concentric reference electrode built on the same body.

combustion analysis A technique in which a sample is heated in an atmosphere of O_2 to oxidize it to CO_2 and H_2O, which are collected and measured. Modifications permit the simultaneous analysis of N, S, and halogens.

common ion effect Occurs when a salt is dissolved in a solution already containing one of the ions of the salt. The salt is less soluble than it would be in a solution without that ion. An application of Le Châtelier's principle.

complexometric titration One in which the reaction between analyte and titrant involves complex formation.

composite sample A representative sample prepared from a heterogeneous material. If the material consists of distinct regions, the composite is made by taking portions from each region, with relative amounts proportional to the size of each region.

compound electrode An ion-selective electrode consisting of a conventional electrode surrounded by a barrier that is selectively permeable to the analyte of interest. Alternatively, the barrier region might convert external analyte into a different species, to which the inner electrode is sensitive.

concentration An expression of the quantity per unit volume or unit mass of a substance. Common measures of concentration are molarity (mol/L) and molality (mol/kg of solvent).

confidence interval The range of values within which there is a specified probability that the true value will occur.

conjugate acid-base pair An acid and a base that differ through the gain or loss of a single proton.

coprecipitation Occurs when a substance whose solubility is not exceeded precipitates along with one whose solubility is exceeded.

coulomb, C The amount of charge per second that flows past any point in a circuit when the current is one ampere.

current, I The amount of charge flowing through a circuit per unit time.

cuvet A cell used to hold samples for spectrophotometric measurements.

deionized water Water that has been passed through a cation exchanger in the H^+ form and an anion exchanger in the OH^- form to remove ions from the solution.

density The mass per unit volume of a substance.

derivatization Chemical alteration of an analyte so that it can be detected conveniently or separated conveniently.

desiccant A drying agent.

desiccator A sealed chamber in which samples can be dried in the presence of a desiccant and/or by vacuum pumping.

detection limit That concentration of an element that gives a signal equal to twice the peak-to-peak noise level of the baseline.

determinate error See **systematic error.**

dialysate Liquid retained by the membrane in dialysis. For a microdialysis probe, dialysate is the liquid exiting the outlet of the probe.

dialysis A technique in which solutions are placed on either side of a semipermeable membrane that allows small molecules, but not large molecules, to cross. The small molecules in the two solutions diffuse across and equilibrate with each other. The large molecules are retained on their original side.

digestion The process in which a precipitate is left (usually warm) in the presence of mother liquor to promote recrystallization and particle growth. Purer, more easily filterable crystals result. Also, any chemical treatment in which a substance is decomposed into a form suitable for analysis.

direct titration One in which the analyte is treated with titrant, and the volume of titrant required for complete reaction is measured.

displacement titration An EDTA titration procedure in which analyte is treated with excess $MgEDTA^{2-}$ to displace Mg^{2+}: $M^{n+} + MgEDTA^{2-} \rightleftharpoons MEDTA^{n-4} + Mg^{2+}$. The liberated Mg^{2+} is then titrated with EDTA. This procedure is useful if there is no suitable indicator for direct titration of M^{n+}.

disproportionation An oxidation reaction in which an element in one oxidation state produces the same element in higher and lower oxidation states (e.g., $Cu^+ \rightleftharpoons Cu^{2+} + Cu(s)$).

$E^{\circ\prime}$ The effective standard reduction potential at pH 7 (or at some other specified conditions).

electric field The potential difference (volts) between two points divided by the distance (meters) separating the points.

electric potential The potential difference (volts) between two points is the energy (joules) needed to transport one coulomb of positive charge from the negative point to the positive point.

electroactive species Any species that can be oxidized or reduced at an electrode.

electrochemical detector Chromatography detector that measures current when an electroactive solute emerges from the column and passes over a working electrode held at a fixed potential with respect to a reference electrode. Also called *amperometric detector.*

electrode A device at which or through which electrons flow into or out of chemical species involved in a redox reaction.

electrokinetic injection In capillary electrophoresis, the use of an electric field to inject sample into the capillary. Because different species have different mobilities, the injected sample does not have the same composition as the original sample.

electrolysis Process in which passage of an electric current causes a chemical reaction to occur.

electromagnetic spectrum The spectrum of all electromagnetic radiation (visible light, radio waves, X-rays, etc.).

electronic balance A balance that uses an electromagnet to balance the load on the pan. The mass of the load is proportional to the current needed to balance it.

electroosmosis Bulk flow of fluid in a capillary tube caused by an electric field. Ions in the diffuse part of the double layer at the wall of the capillary serve as the "pump."

electrophoresis Migration of ions in solution in an electric field. Cations move toward the cathode and anions move toward the anode.

eluate or **effluent** What comes out of a chromatography column.

eluent The solvent applied to the beginning of a chromatography column.

eluent strength Also called *solvent strength*. A measure of the absorption energy of a solvent on the stationary phase in chromatography. The greater the eluent strength, the more rapidly will the solvent displace solutes from the column.

elution The process of passing a liquid or a gas through a chromatography column.

end point The point in a titration at which there is a sudden change in a physical property, such as indicator color, pH, conductivity, or absorbance. Used as a measure of the equivalence point.

equilibrium constant, K For the reaction $aA + bB \rightleftharpoons cC + dD$, $K = [C]^c[D]^d/[A]^a[B]^b$, where [X] is the concentration of species X. The equilibrium constant is more correctly written in terms of activities, instead of concentrations.

equivalence point The point in a titration at which the quantity of titrant is exactly sufficient for stoichiometric reaction with the analyte.

excited state Any state of an atom or a molecule having more than its minimum possible energy.

extraction The process in which a solute is allowed to equilibrate between two phases, usually for the purpose of separating solutes from one another.

extrapolation The estimation of a value that lies beyond the range of measured data.

Fajans titration A precipitation titration in which the end point is signaled by adsorption of a colored indicator on the precipitate.

Faraday constant $9.648\ 530\ 9 \times 10^4$ C/mol of charge.

filtrate Liquid that passes through a filter.

flame ionization detector A gas chromatography detector in which solute is burned in a H_2-air flame to produce CHO^+ ions. The current carried through the flame by these ions is proportional to the concentration of susceptible species in the eluate.

fluorescence The process in which a molecule emits a photon 10^{-8} to 10^{-4} s after absorbing a photon.

flux The medium that is melted in sample preparation to decompose and dissolve the sample.

formal concentration or **analytical concentration** The molarity of a substance were it not to change its chemical form upon being dissolved. This measure represents the total number of moles of substance dissolved in a liter of solution, regardless of any reactions that do take place when the solute is dissolved.

formal potential The potential of a half-reaction (relative to a standard hydrogen electrode) when the formal concentrations of reactants and products are unity. Any other conditions (such as pH, ionic strength, and concentrations of ligands) must also be specified.

formation constant or **stability constant** The equilibrium constant for the reaction of a metal with a ligand to form a metal-ligand complex.

formula weight, FW The mass containing one mole of the indicated chemical formula of a substance. For example, the formula weight of $CuSO_4 \cdot 5H_2O$ is the sum of the masses of copper, sulfate, and five water molecules.

frequency The number of oscillations of a wave per second.

fusion Process in which an otherwise insoluble substance is dissolved in a molten salt such as Na_2CO_3, Na_2O_2, or KOH. Once the substance has dissolved, the melt is cooled, dissolved in aqueous solution, and analyzed.

galvanic cell One that produces electricity by means of a spontaneous chemical reaction.

gas chromatography A form of chromatography in which the mobile phase is a gas.

gathering A process in which a trace constituent of a solution is intentionally coprecipitated with a major constituent.

Gaussian distribution Theoretical bell-shaped distribution of measurements when all error is random. The center of the curve is the mean (μ) and the width is characterized by the standard deviation (σ). A normalized Gaussian distribution, also called the *normal error curve*, has an area of unity and is given by

$$y = \left(\frac{1}{\sigma\sqrt{2\pi}}\right)e^{-(x-\mu)^2/2\sigma^2}$$

glass electrode One that has a thin glass membrane across which a pH-dependent voltage develops. The voltage (and hence pH) is measured by a pair of reference electrodes on either side of the membrane.

gradient elution Chromatography in which the composition of the mobile phase is progressively changed to increase the eluent strength of the solvent.

graphite furnace A hollow graphite rod that can be heated electrically to about 2 500 K and is used to decompose and atomize a sample for atomic spectroscopy.

grating Either a reflective or a transmitting surface with closely spaced lines to disperse light into its component wavelengths.

gravimetric analysis Any analytical method that relies on measuring the mass of a substance (such as a precipitate) to complete the analysis.

ground state The state of an atom or a molecule with the minimum possible energy.

guard column In HPLC, a short column that is packed with the same material as the main column and is placed between the injector and the main column. The guard column removes impurities that might irreversibly bind to and degrade the main column. Also called *precolumn*.

Henderson-Hasselbalch equation A logarithmic, rearranged form of the acid dissociation equilibrium equation: $pH = pK_a + \log([A^-]/[HA])$.

hertz, Hz The unit of frequency, s^{-1}.

heterogeneous sample A sample that is not uniform throughout.

high-performance liquid chromatography (HPLC) A chromatographic technique using very small stationary-phase particles and high pressure to force solvent through the column.

hollow-cathode lamp One that emits sharp atomic lines characteristic of the element from which the cathode is made.

homogeneous precipitation A technique in which a precipitating agent is generated slowly by a reaction in homogeneous solution, effecting a slow crystallization instead of a rapid precipitation of product.

homogeneous sample A sample that has the same composition everywhere.

hydrated radius The effective radius of an ion or a molecule plus its associated water molecules in solution.

hydrodynamic injection In capillary electrophoresis, the use of a pressure difference between the two ends of the capillary to inject sample into the capillary.

hydrolysis Reaction with water. The reaction $B + H_2O \rightleftharpoons BH^+ + OH^-$ is called hydrolysis of a base.

hydronium ion, H_3O^+ What we really mean when we write $H^+(aq)$.

hydrophilic substance One that is soluble in water or attracts water to its surface.

hydrophobic substance One that is insoluble in water or repels water from its surface.

hygroscopic substance One that readily picks up water from the atmosphere.

ignition The heating to high temperature of a gravimetric precipitate to convert it to a known, constant composition that can be weighed.

immunoassay An analytical measurement based on the use of antibodies.

indeterminate error See **random error.**

indicator A compound with a physical property (usually color) that changes abruptly near the equivalence point of a chemical reaction.

indicator electrode One whose potential depends on the concentration (actually the activity) of one or more species in contact with the electrode.

indicator error The difference between the indicator end point of a titration and the true equivalence point.

indirect detection Chromatographic detection based on the *absence* of signal from a background species. For example, in ion chromatography a light-absorbing ionic species can be added to the eluent. Analyte that does not absorb light replaces an equivalent amount of light-absorbing eluent when the analyte emerges from the column, thereby decreasing the observed absorbance.

indirect titration One that is used when the analyte cannot be directly titrated. For example, analyte A may be precipitated with excess reagent R to make AR. The product is filtered and the excess R washed away. Then AR is dissolved in a new solution, and R is titrated.

inductively coupled plasma A high-temperature plasma that derives its energy from an oscillating radio-frequency field. It is used to atomize a sample for atomic emission spectroscopy.

intercept For a straight line whose equation is $y = mx + b$, b is the intercept. It is the value of y when $x = 0$.

interference The effect when the presence of one substance changes the signal in the analysis of another substance.

internal standard A known quantity of a compound added to a solution containing an unknown quantity of analyte. The concentration of analyte is then measured relative to that of the internal standard.

ion chromatography A high-performance version of ion-exchange chromatography, with a key modification that removes eluent ions before detecting analyte ions.

ion-exchange chromatography A technique in which solute ions are retained by oppositely charged sites in the stationary phase.

ion-selective electrode One whose potential is selectively dependent on the concentration of one particular ion in solution.

ionization interference In atomic spectroscopy, a lowering of signal intensity as a result of ionization of analyte atoms.

isocratic elution Chromatography using a single solvent for the mobile phase.

isoelectric focusing A technique in which a sample containing polyprotic molecules is subjected to a strong electric field in a medium with a pH gradient. Each species migrates until it reaches the region of its isoelectric pH. In that region, the molecule has no net charge, ceases to migrate, and remains focused in a narrow band.

joule, J SI unit of energy. One joule is expended when a force of 1 N acts over a distance of 1 m. This energy is equivalent to that required to raise 102 g (about $\frac{1}{4}$ pound) by 1 m.

junction potential An electric potential difference that exists at the junction between two different electrolyte solutions or substances. It arises in solutions as a result of unequal rates of diffusion of different ions.

Kjeldahl nitrogen analysis Procedure for the analysis of nitrogen in organic compounds. The compound is digested with boiling H_2SO_4 to convert nitrogen to NH_4^+, which is treated with base and distilled as NH_3 into a standard acid solution. The number of moles of acid consumed equals the number of moles of NH_3 liberated from the compound.

Le Châtelier's principle States that if a system at equilibrium is disturbed, the direction in which it proceeds back to equilibrium is such that the disturbance is partially offset.

least squares See **method of least squares.**

Lewis acid One that can form a chemical bond by sharing a pair of electrons donated by another species.

Lewis base One that can form a chemical bond by sharing a pair of its electrons with another species.

ligand An atom or a group attached to a central atom in a molecule. The term is often used to mean any group attached to anything else of interest.

liquid-based ion-selective electrode One that has a hydrophobic membrane separating an inner reference electrode from the analyte solution. The membrane contains a liquid ion exchanger that transports analyte across the membrane to produce an electric potential difference between the two sides of the membrane.

liquid chromatography Chromatography with a liquid mobile phase.

liter, L A volume that is exactly 1 000 cm^3.

logarithm The base 10 logarithm of n is a if $10^a = n$ (which means log $n = a$).

luminescence Any emission of light by a molecule.

mantissa The part of a logarithm to the right of the decimal point.

masking The process of adding a chemical substance (a masking agent) to a sample to prevent one or more components from interfering in a chemical analysis.

masking agent A reagent that selectively reacts with one (or more) component(s) of a solution to prevent the component(s) from interfering in a chemical analysis.

mass spectrometer A device that ionizes (and often fragments) gaseous molecules and then separates them by mass and measures the quantity of each ion.

matrix The medium containing analyte. For many analyses, it is important that standards be prepared in the same matrix as the unknown.

matrix modifier Substance added to sample for atomic spectroscopy to retard analyte evaporation until the matrix is fully charred.

mean The average of a set of all results.

mechanical balance A balance having a beam that pivots on a fulcrum. Standard masses are used to measure the mass of an unknown.

meniscus The curved surface of a liquid.

metal ion indicator A compound whose color changes when it binds to a metal ion.

method of least squares Process of fitting a mathematical function to a set of measured points by minimizing the sum of the squares of the distances from the points to the curve.

micellar electrokinetic capillary chromatography A form of capillary electrophoresis in which a micelle-forming surfactant is present. Migration times of solutes depend on the fraction of time spent in the micelles.

micelle An aggregate of molecules with ionic head groups and long, nonpolar tails. The inside of the micelle resembles hydrocarbon solvent, whereas the outside interacts strongly with aqueous solution.

microdialysis probe A needle with a semipermeable (dialysis) membrane near the tip. When inserted into a living organism, small molecules from the organism enter the probe. Fluid from the probe is analyzed by chromatography, electrochemistry, etc.

microporous particles Chromatographic stationary phase consisting of porous particles 3–10 μm in diameter, with high efficiency and high capacity for solute.

mobile phase In chromatography, the phase that travels through the column.

Mohr titration Argentometric titration in which the end point is signaled by the formation of red $Ag_2CrO_4(s)$.

molality The number of moles of solute per kilogram of solvent.

molar absorptivity, ϵ, or **extinction coefficient** The constant of proportionality in Beer's law: $A = \epsilon bc$, where A is absorbance, b is pathlength, and c is the molarity of the absorbing species.

molarity, M The number of moles of solute per liter of solution.

molecular exclusion chromatography or **gel filtration** or **gel permeation chromatography** or **molecular sieve chromatography** A technique in which the stationary phase has a porous structure into which small molecules can enter but large molecules cannot. Molecules are separated by size, with larger molecules moving faster than smaller ones.

molecular sieve A solid particle, with pores the size of small molecules, used to separate small molecules in gas chromatography and used as a desiccant.

molecular weight, MW The number of grams of a substance that contains Avogadro's number of molecules.

monochromator A device (usually a prism, grating, or filter) for selecting a single wavelength of light.

monodentate ligand One that binds to a metal ion through only one atom.

mortar and pestle A mortar is a hard ceramic or steel vessel in which a solid sample is ground with a hard pestle.

mother liquor The solution from which a substance has crystallized.

multidentate ligand One that binds to a metal ion through more than one atom.

nebulization The process of breaking a liquid into a mist of fine droplets.

Nernst equation For the half-reaction $a A + n e^- \rightleftharpoons b B$, the Nernst equation giving the half-cell potential, E, is

$$E = E^\circ - \left(\frac{0.059\ 16}{n} \right) \log \left(\frac{[B]^b}{[A]^a} \right) \quad \text{(at 25°C)}$$

where E° is the standard reduction potential that applies when $[A] = [B] = 1$ M (actually when the activities of A and B are unity).

neutralization The process in which a stoichiometric equivalent of acid (or base) is added to a base (or acid).

nonpolar compound A compound with little or no charge separation (little or no dipole moment), such as a compound containing only hydrogen and carbon.

normal-phase chromatography A chromatographic separation utilizing a polar stationary phase and a less polar mobile phase.

nucleation The process whereby molecules in solution come together randomly to form small aggregates.

open tubular column In chromatography, a capillary column whose walls are coated with stationary phase.

ordinate The vertical (y) axis of a graph.

oxidant See **oxidizing agent.**

oxidation A loss of electrons or an increase of the oxidation state.

oxidizing agent or **oxidant** A substance that takes electrons in a chemical reaction.

packed column A chromatography column filled with stationary-phase particles.

parallax error The apparent displacement of an object when the observer changes position. Occurs when the scale of an instrument is viewed from a position that is not perpendicular to the scale; consequently, the apparent reading is not the true reading.

partition chromatography A technique in which separation is achieved by equilibration of solute between two phases.

parts per billion, ppb An expression of concentration that refers to nanograms (10^{-9} g) of solute per gram of solution.

parts per million, ppm An expression of concentration that refers to micrograms (10^{-6} g) of solute per gram of solution.

peptization Occurs when washing some ionic precipitates with distilled water washes away the ions that neutralize the charges of individual particles and help to hold the particles together. The washed particles then disintegrate and pass through the filter with the wash liquid.

pH Defined as pH $= -\log \mathcal{A}_{H^+}$, where $\mathcal{A}_{H^+}$ is the activity of H^+. For all applications in this book, we use the approximation pH $= -\log [H^+]$.

phosphorescence Relatively long-lived emission of light occurring 10^{-4} to 10^2 s after absorption of a photon.

photon A "particle" of light with energy $h\nu$, where h is Planck's constant and ν is the frequency of the light.

pipet A glass tube calibrated to deliver a fixed or variable volume of liquid.

pK The negative logarithm of an equilibrium constant: p$K = -\log K$.

plate height The length of a chromatography column divided by the number of theoretical plates in the column.

polar compound A compound with a significant amount of charge separated on different atoms. Polar compounds tend to attract other polar compounds fairly strongly by electrostatic forces and, sometimes, hydrogen bonding.

polyprotic acids and bases Compounds that can donate or accept more than one proton.

potentiometry An analytical method in which an electric potential difference (a voltage) of a cell is measured.

power The amount of energy per unit time (J/s) being expended.

precipitant A reagent that causes a precipitate to form.

precision The reproducibility of a measured result.

preconcentration The process of concentrating trace components of a mixture prior to their analysis.

primary standard A reagent that is pure enough and stable enough to be used directly after weighing. The entire mass is considered to be pure reagent.

pyrolysis Thermal decomposition of a substance.

Q **test** A statistical test used to decide whether a discrepant datum should be rejected or retained from a set of measurements.

qualitative analysis The process of determining the identity of the constituents of a substance.

quantitative analysis The process of measuring how much of a constituent is present in a substance.

random error or **indeterminate error** A type of error that can be either positive or negative and cannot be eliminated. It arises from the ultimate limitations on a physical measurement.

random sample Bulk sample constructed by taking portions of the entire lot at random.

reagent Refers to any substance used in a chemical reaction.

redox reaction A chemical reaction involving transfer of electrons from one species to another.

reducing agent or **reductant** A substance that donates electrons in a chemical reaction.

reductant See **reducing agent.**

reduction A gain of electrons or a lowering of the oxidation state.

reference electrode One that maintains a constant potential against which the potential of another half-cell can be measured.

refractive index detector Liquid chromatography detector that measures the change in the refractive index of a solution as solutes emerge from a column.

relative uncertainty The uncertainty of a quantity divided by the value of the quantity. It is usually expressed as a percentage of the measured quantity.

resistance A measure of the retarding force opposing the flow of electric current.

resolution How close two bands in a spectrum or a chromatogram can be to each other and still be seen as two peaks. In chromatography, it is defined as the difference in retention times of adjacent peaks divided by their width at the base.

retention time The time, measured from injection, needed for a solute to be eluted from a chromatography column.

reverse-phase chromatography A technique in which the stationary phase is less polar than the mobile phase.

rubber policeman A glass rod with a flattened piece of rubber on the tip. The rubber is used to scrape solid particles from glass surfaces in gravimetric analysis.

salt An ionic solid.

salt bridge A conducting ionic medium in contact with two electrolyte solutions. It allows ions to flow without immediately contaminating one electrolyte solution with the other.

sample cleanup Removal of portions of the sample that do not contain analyte and may interfere with analysis.

sample preparation The process of converting a heterogeneous bulk sample into a homogeneous laboratory sample prior to analysis. Sample preparation also refers to concentrating a very dilute sample or eliminating species that interfere with the analysis.

sampling The process of obtaining a small bulk sample whose composition is representative of the lot to be analyzed.

saturated calomel electrode (S.C.E.) A calomel electrode saturated with KCl. The electrode half-reaction is $Hg_2Cl_2(s) + 2e^- \rightleftharpoons 2Hg(l) + 2Cl^-$.

saturated solution One containing the maximum amount of a compound that can dissolve at equilibrium.

segregated material A material containing distinct regions of different composition.

selectivity coefficient With respect to an ion-selective electrode, a measure of the relative response of the electrode to two different ions. In ion-exchange chromatography, this is the equilibrium constant for displacement of one ion by another from the resin.

semipermeable membrane A barrier across which some molecules, but not others, can cross.

septum A disk, usually made of silicone rubber, covering the injection port of a gas chromatograph. The sample is injected by syringe through the septum.

SI units The units of an international system of measurement based on the meter, kilogram, second, ampere, kelvin, candela, mole, radian, and steradian.

significant figure The number of significant figures in a quantity is the minimum number of figures needed to express the quantity in scientific notation. In experimental data, the first uncertain figure is the last significant figure.

silver-silver chloride electrode A common reference electrode containing a silver wire coated with AgCl paste and dipped in a solution saturated with AgCl and (usually) KCl. The half-reaction is $AgCl(s) + e^- \rightleftharpoons Ag(s) + Cl^-$.

slope For a straight line whose equation is $y = mx + b$, the value of m is the slope. It is the ratio $\Delta y/\Delta x$ for any segment of the line.

solid-phase extraction A procedure that uses a chromatographic solid phase in a short column to separate one or more analytes from a mixture. Solutes adsorbed on the column can be eluted with a small volume of solvent.

solid-state ion-selective electrode An ion-selective electrode with a membrane made of an inorganic crystal. Ion-exchange equilibrium between the solution and the crystal surface creates a voltage across the crystal that depends on the concentration of analyte on each side.

solubility product, K_{sp} The equilibrium constant for the dissolution of a solid salt to give its ions in solution. For the reaction $M_mN_n(s) \rightleftharpoons mM^{n+} + nN^{m-}$, $K_{sp} = [M^{n+}]^m[N^{m-}]^n$.

solute A minor component of a solution.

solvent The major constituent of a solution.

species Chemists refer to any molecule or ion as a "species." The term is both singular and plural.

spectral interference In atomic spectroscopy, any physical process that affects the intensity of radiation at the analytical wavelength. Created by substances that absorb, scatter, or emit radiation of the analytical wavelength.

spectrophotometer A device that measures absorption of electromagnetic radiation. It includes a source, a wavelength selector (monochromator), and a detector.

spectrophotometric titration One in which absorption of electromagnetic radiation is used to monitor the progress of the chemical reaction.

spectrophotometry Any method using electromagnetic radiation to measure chemical concentrations.

stability constant See **formation constant.**

standard addition Analytical procedure in which known quantities of analyte are added to the unknown; the accompanying increase in the signal allows the operator to determine how much analyte was in the original unknown.

standard curve A graph showing the response of an analytical technique to known quantities of analyte.

standard deviation Measures how closely data are clustered about the mean value. For a finite set of data, the standard deviation, s, is computed from the formula

$$s = \sqrt{\frac{\Sigma(x_i - \bar{x})^2}{n - 1}}$$

where n is the number of results, x_i is an individual result, and $\bar{x}$ is the mean result.

standard hydrogen electrode (S.H.E.) or **normal hydrogen electrode (N.H.E.)** One in which $H_2(g)$ bubbles over a catalytic Pt surface in contact with aqueous H^+. The activities of H_2 and H^+ are both unity in the hypothetical standard electrode. The concentrations are approximately 1 atm for H_2 and 1 M for H^+. The cell reaction is $H^+ + e^- \rightleftharpoons \frac{1}{2}H_2(g)$.

standard reduction potential, $E°$ The voltage that would be measured when the hypothetical cell containing the desired half-reaction (with all species at unit activity—approximately 1 M, 1 atm, or pure solid or liquid) is connected to a standard hydrogen electrode.

standard solution A solution whose composition is known by virtue of the way it was made from a reagent of known purity or by virtue of its reaction with a known quantity of a standard reagent.

standardization The process whereby the concentration of a reagent is determined by reaction with a known quantity of a second reagent.

stationary phase In chromatography, the phase that does not move.

strong acids and bases Those that are completely dissociated (to H^+ or OH^-) in water.

strong electrolyte One that dissociates completely into its ions when dissolved.

Student's t A statistical tool used to express confidence intervals and to compare results from different experiments.

supersaturated solution One that contains more dissolved solute than would be present at equilibrium.

suppressed-ion chromatography Separation of ions by using an ion-exchange column and subsequently removing the ionic eluent.

surfactant A molecule, such as soap, with an ionic or polar headgroup and a long, nonpolar tail. Surfactants aggregate in aqueous solution to form micelles.

systematic error or **determinate error** A type of error due to procedural or instrumental factors that cause a measurement to be systematically too large or too small. The error can, in principle, be discovered and corrected.

t-test Used to decide whether the results of two experiments are within experimental uncertainty of each other.

tare The mass of an empty vessel used for weighing. When a balance is "tared," the empty mass is set to zero.

temperature programming Technique in gas chromatography in which the temperature of the column is raised during the separation of a mixture.

theoretical plate An imaginary segment of a chromatography column in which one equilibration of solute occurs between the stationary and mobile phases. The number of theoretical plates on a column with Gaussian bandshapes is $N = 5.55\ t_r^2/w_{1/2}^2 = t_r^2/\sigma^2$, where t_r is the retention time of a band, $w_{1/2}^2$ is the width at half-height, and σ is the standard deviation of the band.

thermal conductivity detector A device that detects bands eluted from a gas chromatography column by measuring changes in the thermal conductivity of the gas stream.

thermogravimetric analysis A technique in which the mass of a substance is measured as the substance is heated. Changes in mass reflect decomposition of the substance, often to well-defined products.

titrant The substance added to the analyte in a titration.

titration A procedure in which one substance (titrant) is carefully added to another (analyte) until complete reaction has occurred. The quantity of titrant required for complete reaction tells how much analyte is present.

titration error The difference between the observed end point and the true equivalence point in a titration.

trace analysis The measurement of extremely low levels of analyte.

transmittance, *T* Defined as $T = P/P_0$, where P_0 is the radiant power of light striking the sample on one side and P is the radiant power of light emerging from the other side of the sample.

ultraviolet detector Liquid chromatography detector that measures ultraviolet absorbance of solutes emerging from the column.

volatile Easily vaporized.

Volhard titration Titration of Ag^+ with SCN^-, in which the formation of the red complex $Fe(SCN)^{2+}$ marks the end point.

volt, V Unit of electric potential. If the potential difference between two points is one volt, it requires one joule of energy to move one coulomb of charge between the two points.

volume percent Defined as (volume of solute/volume of solution) $\times$ 100.

volumetric analysis A technique in which the volume of material needed to react with analyte is measured.

volumetric flask One having a tall, thin neck with a calibration mark. When the liquid level is at the calibration mark, the flask contains its specified volume of liquid.

watt, W The SI unit of power, equal to an energy flow of one joule per second. When an electric current of one ampere flows through a potential difference of one volt, the power is one watt.

wavelength, λ The distance between consecutive crests of a wave.

wavenumber, $\tilde{\nu}$ The reciprocal of the wavelength, λ.

weak acids and bases Those whose dissociation constants are not large.

weak electrolyte One that partially dissociates into ions when it dissolves.

weight percent Defined as (mass of solute/mass of solution) $\times$ 100.

y-intercept Same as *intercept.*

zwitterion A molecule with a positive charge localized in one position and a negative charge localized at another position.

Appendix A

Solubility Products†

Formula	K_{sp}
Azides: L = N_3^-	
CuL	4.9×10^{-9}
AgL	2.8×10^{-9}
Hg_2L_2	7.1×10^{-10}
TlL	2.2×10^{-4}
$PdL_2(\alpha)$	2.7×10^{-9}
Bromates: L = BrO_3^-	
$BaL \cdot H_2O$	7.8×10^{-6}
AgL	5.5×10^{-5}
TlL	1.7×10^{-4}
PbL_2	7.9×10^{-6}
Bromides: L = Br^-	
CuL	5×10^{-9}
AgL	5.0×10^{-13}
Hg_2L_2	5.6×10^{-23}
TlL	3.6×10^{-6}
HgL_2	1.3×10^{-19}
PbL_2	2.1×10^{-6}
Carbonates: L = CO_3^{2-}	
MgL	3.5×10^{-8}
CaL (calcite)	4.5×10^{-9}
CaL (aragonite)	6.0×10^{-9}
SrL	9.3×10^{-10}
BaL	5.0×10^{-9}
Y_2L_3	2.5×10^{-31}
La_2L_3	4.0×10^{-34}
MnL	5.0×10^{-10}
FeL	2.1×10^{-11}
CoL	1.0×10^{-10}
NiL	1.3×10^{-7}
CuL	2.3×10^{-10}
Ag_2L	8.1×10^{-12}
Hg_2L	8.9×10^{-17}
ZnL	1.0×10^{-10}
CdL	1.8×10^{-14}
PbL	7.4×10^{-14}

Formula	K_{sp}
Chlorides: L = Cl^-	
CuL	1.9×10^{-7}
AgL	1.8×10^{-10}
Hg_2L_2	1.2×10^{-18}
TlL	1.8×10^{-4}
PbL_2	1.7×10^{-5}
Chromates: L = CrO_4^{2-}	
BaL	2.1×10^{-10}
CuL	3.6×10^{-6}
Ag_2L	1.2×10^{-12}
Hg_2L	2.0×10^{-9}
Tl_2L	9.8×10^{-13}
Cobalticyanides: L = $Co(CN)_6^{3-}$	
Ag_3L	3.9×10^{-26}
$(Hg_2)_3L_2$	1.9×10^{-37}
Cyanides: L = CN^-	
AgL	2.2×10^{-16}
Hg_2L_2	5×10^{-40}
ZnL_2	3×10^{-16}
Ferrocyanides: L = $Fe(CN)_6^{4-}$	
Ag_4L	8.5×10^{-45}
Zn_2L	2.1×10^{-16}
Cd_2L	4.2×10^{-18}
Pb_2L	9.5×10^{-19}
Fluorides: L = F^-	
LiL	1.7×10^{-3}
MgL_2	6.6×10^{-9}
CaL_2	3.9×10^{-11}
SrL_2	2.9×10^{-9}
BaL_2	1.7×10^{-6}
ThL_4	5×10^{-29}
PbL_2	3.6×10^{-8}
Hydroxides: L = OH^-	
MgL_2	7.1×10^{-12}
CaL_2	6.5×10^{-6}

(continued)

†Solubility products generally apply at 25°C and zero ionic strength. The designations α, β, or γ after some formulas refer to particular crystalline forms.

421

Formula	K_{sp}
$BaL_2 \cdot 8H_2O$	3×10^{-4}
YL_3	6×10^{-24}
LaL_3	2×10^{-21}
CeL_3	6×10^{-22}
$UO_2 (\rightleftharpoons U^{4+} + 4OH^-)$	6×10^{-57}
$UO_2L_2 (\rightleftharpoons UO_2^{2+} + 2OH^-)$	4×10^{-23}
MnL_2	1.6×10^{-13}
FeL_2	7.9×10^{-16}
CoL_2	1.3×10^{-15}
NiL_2	6×10^{-16}
CuL_2	4.8×10^{-20}
VL_3	4.0×10^{-35}
CrL_3	1.6×10^{-30}
FeL_3	1.6×10^{-39}
CoL_3	3×10^{-45}
$VOL_2 (\rightleftharpoons VO^{2+} + 2OH^-)$	3×10^{-24}
PdL_2	3×10^{-29}
ZnL_2 (amorphous)	3.0×10^{-16}
$CdL_2 (\beta)$	4.5×10^{-15}
HgO (red) $(\rightleftharpoons Hg^{2+} + 2OH^-)$	3.6×10^{-26}
$Cu_2O (\rightleftharpoons 2Cu^+ + 2OH^-)$	4×10^{-30}
$Ag_2O (\rightleftharpoons 2Ag^+ + 2OH^-)$	3.8×10^{-16}
AuL_3	3×10^{-6}
$AlL_3 (\alpha)$	3×10^{-34}
GaL_3 (amorphous)	10^{-37}
InL_3	1.3×10^{-37}
$SnO (\rightleftharpoons Sn^{2+} + 2OH^-)$	6×10^{-27}
PbO (yellow) $(\rightleftharpoons Pb^{2+} + 2OH^-)$	8×10^{-16}
PbO (red) $(\rightleftharpoons Pb^{2+} + 2OH^-)$	5×10^{-16}
Iodates: $L = IO_3^-$	
CaL_2	7.1×10^{-7}
SrL_2	3.3×10^{-7}
BaL_2	1.5×10^{-9}
YL_3	7.1×10^{-11}
LaL_3	1.0×10^{-11}
CeL_3	1.4×10^{-11}
ThL_4	2.4×10^{-15}
$UO_2L_2 (\rightleftharpoons UO_2^{2+} + 2IO_3^-)$	9.8×10^{-8}
CrL_3	5×10^{-6}
AgL	3.1×10^{-8}
Hg_2L_2	1.3×10^{-18}
TlL	3.1×10^{-6}
ZnL_2	3.9×10^{-6}
CdL_2	2.3×10^{-8}
PbL_2	2.5×10^{-13}
Iodides: $L = I^-$	
CuL	1×10^{-12}
AgL	8.3×10^{-17}
$CH_3HgL (\rightleftharpoons CH_3Hg^+ + I^-)$	3.5×10^{-12}
$CH_3CH_2HgL (\rightleftharpoons CH_3CH_2Hg^+ + I^-)$	7.8×10^{-5}
TlL	5.9×10^{-8}
Hg_2L_2	1.1×10^{-28}
SnL_2	8.3×10^{-6}
PbL_2	7.9×10^{-9}
Oxalates: $L = C_2O_4^{2-}$	
CaL	1.3×10^{-8}

Formula	K_{sp}
SrL	4×10^{-7}
BaL	1×10^{-6}
La_2L_3	1×10^{-25}
ThL_2	4.2×10^{-22}
$UO_2L (\rightleftharpoons UO_2^{2+} + C_2O_4^{2-})$	2.2×10^{-9}
Phosphates: $L = PO_4^{3-}$	
$MgHL \cdot 3H_2O (\rightleftharpoons Mg^{2+} + HL^{2-})$	1.7×10^{-6}
$CaHL \cdot 2H_2O (\rightleftharpoons Ca^{2+} + HL^{2-})$	2.6×10^{-7}
$SrHL (\rightleftharpoons Sr^{2+} + HL^{2-})$	1.2×10^{-7}
$BaHL (\rightleftharpoons Ba^{2+} + HL^{2-})$	4.0×10^{-8}
LaL	3.7×10^{-23}
$Fe_3L_2 \cdot 8H_2O$	1×10^{-36}
$FeL \cdot 2H_2O$	4×10^{-27}
$(VO)_3L_2 (\rightleftharpoons 3VO^{2+} + 2L^{3-})$	8×10^{-26}
Ag_3L	2.8×10^{-18}
$Hg_2HL (\rightleftharpoons Hg_2^{2+} + HL^{2-})$	4.0×10^{-13}
$Zn_3L_2 \cdot 4H_2O$	5×10^{-36}
Pb_3L_2	3.0×10^{-44}
GaL	1×10^{-21}
InL	2.3×10^{-22}
Sulfates: $L = SO_4^{2-}$	
CaL	2.4×10^{-5}
SrL	3.2×10^{-7}
BaL	1.1×10^{-10}
RaL	4.3×10^{-11}
Ag_2L	1.5×10^{-5}
Hg_2L	7.4×10^{-7}
PbL	6.3×10^{-7}
Sulfides: $L = S^{2-}$	
MnL (pink)	3×10^{-11}
MnL (green)	3×10^{-14}
FeL	8×10^{-19}
$CoL (\alpha)$	5×10^{-22}
$CoL (\beta)$	3×10^{-26}
$NiL (\alpha)$	4×10^{-20}
$NiL (\beta)$	1.3×10^{-25}
$NiL (\gamma)$	3×10^{-27}
CuL	8×10^{-37}
Cu_2L	3×10^{-49}
Ag_2L	8×10^{-51}
Tl_2L	6×10^{-22}
$ZnL (\alpha)$	2×10^{-25}
$ZnL (\beta)$	3×10^{-23}
CdL	1×10^{-27}
HgL (black)	2×10^{-53}
HgL (red)	5×10^{-54}
SnL	1.3×10^{-26}
PbL	3×10^{-28}
In_2L_3	4×10^{-70}
Thiocyanates: $L = SCN^-$	
CuL	4.0×10^{-14}
AgL	1.1×10^{-12}
Hg_2L_2	3.0×10^{-20}
TlL	1.6×10^{-4}
HgL_2	2.8×10^{-20}

Acid Dissociation Constants

Name	Structure[†]	pK_a[‡]	K_a
Acetic acid (ethanoic acid)	CH_3CO_2H	4.757	1.75×10^{-5}
Alanine	NH_3^+ $\mid$ $CHCH_3$ $\mid$ CO_2H	2.348 (CO_2H) 9.867 (NH_3)	4.49×10^{-3} 1.36×10^{-10}
Aminobenzene (aniline)	⬡—NH_3^+	4.601	2.51×10^{-5}
2-Aminobenzoic acid (anthranilic acid)	⬡ NH_3^+ CO_2H	2.08 (CO_2H) 4.96 (NH_3)	8.3×10^{-3} 1.10×10^{-5}
2-Aminoethanol (ethanolamine)	$HOCH_2CH_2NH_3^+$	9.498	3.18×10^{-10}
2-Aminophenol	⬡ OH NH_3^+	4.78 (NH_3) (20°) 9.97 (OH) (20°)	1.66×10^{-5} 1.05×10^{-10}
Ammonia	NH_4^+	9.244	5.70×10^{-10}
Arginine	NH_3^+ $\mid$ $CHCH_2CH_2CH_2NHC$ $\mid$ CO_2H NH_2^+ NH_2	1.823 (CO_2H) 8.991 (NH_3) (12.48) (NH_2)	1.50×10^{-2} 1.02×10^{-9} 3.3×10^{-13}
Arsenic acid (hydrogen arsenate)	O $\parallel$ HO—As—OH $\mid$ OH	2.24 6.96 11.50	5.8×10^{-3} 1.10×10^{-7} 3.2×10^{-12}
Arsenious acid (hydrogen arsenite)	$As(OH)_3$	9.29	5.1×10^{-10}

(continued)

[†]Each acid is written in its protonated form. The acidic protons are indicated in bold type.
[‡]pK_a values refer to 25°C unless otherwise indicated. Values in parentheses are considered to be less reliable.

Name	Structure	pK_a	K_a
Asparagine	(structure: NH_3^+, $CHCH_2CNH_2$ (with O), CO_2H)	2.14 (CO_2H) 8.72 (NH_3)	7.2×10^{-3} 1.9×10^{-9}
Aspartic acid	(structure: NH_3^+, $CHCH_2CO_2H$ (β), $\alpha \rightarrow CO_2H$)	1.990 (α-CO_2H) 3.900 (β-CO_2H) 10.002 (NH_3)	1.02×10^{-2} 1.26×10^{-4} 9.95×10^{-11}
Benzene-1,2,3-tricarboxylic acid (hemimellitic acid)	(benzene ring with three CO_2H)	2.88 4.75 7.13	1.32×10^{-3} 1.78×10^{-5} 7.4×10^{-8}
Benzoic acid	(benzene ring)—CO_2H	4.202	6.28×10^{-5}
2,2'-Bipyridine	(two pyridine rings, N, N, H^+)	4.35	4.5×10^{-5}
Boric acid (hydrogen borate)	$B(OH)_3$	9.236 (12.74) (20°) (13.80) (20°)	5.81×10^{-10} 1.82×10^{-13} 1.58×10^{-14}
Bromoacetic acid	$BrCH_2CO_2H$	2.902	1.25×10^{-3}
Butane-2,3-dione dioxime (dimethylglyoxime)	(structure: HON, NOH, CH_3, CH_3)	10.66 12.0	2.2×10^{-11} 1×10^{-12}
Butanoic acid	$CH_3CH_2CH_2CO_2H$	4.819	1.52×10^{-5}
cis-Butenedioic acid (maleic acid)	(structure: CO_2H, CO_2H)	1.910 6.332	1.23×10^{-2} 4.66×10^{-7}
trans-Butenedioic acid (fumaric acid)	(structure: CO_2H, HO_2C)	3.053 4.494	8.85×10^{-4} 3.21×10^{-5}
Butylamine	$CH_3CH_2CH_2CH_2NH_3^+$	10.640	2.29×10^{-11}
Carbonic acid (hydrogen carbonate)	$HO-C-OH$ (with O)	6.352 10.329	4.45×10^{-7} 4.69×10^{-11}
Chloroacetic acid	$ClCH_2CO_2H$	2.865	1.36×10^{-3}
Chlorous acid (hydrogen chlorite)	$HOCl=O$	1.95	1.12×10^{-2}
Chromic acid (hydrogen chromate)	$HO-Cr-OH$ (with O, O)	−0.2 (20°) 6.51	1.6 3.1×10^{-7}
Citric acid (2-hydroxypropane-1,2,3-tricarboxylic acid)	$HO_2CCH_2CCH_2CO_2H$ (with CO_2H, OH)	3.128 4.761 6.396	7.44×10^{-4} 1.73×10^{-5} 4.02×10^{-7}

Name	Structure	pK_a	K_a
Cyanoacetic acid	$NCCH_2CO_2H$	2.472	3.37×10^{-3}
Cyclohexylamine	⬡$-NH_3^+$	10.64	2.3×10^{-11}
Cysteine	$\overset{NH_3^+}{\underset{CO_2H}{CHCH_2SH}}$	(1.71) (CO_2H) 8.36 (SH) 10.77 (NH_3)	1.95×10^{-2} 4.4×10^{-9} 1.70×10^{-11}
Dichloroacetic acid	Cl_2CHCO_2H	1.30	5.0×10^{-2}
Diethylamine	$(CH_3CH_2)_2NH_2^+$	10.933	1.17×10^{-11}
1,2-Dihydroxybenzene (catechol)	(benzene with two OH)	9.40 12.8	4.0×10^{-10} 1.6×10^{-13}
1,3-Dihydroxybenzene (resorcinol)	(benzene with two OH)	9.30 11.06	5.0×10^{-10} 8.7×10^{-12}
D-2,3-Dihydroxybutanedioc acid (D-tartaric acid)	$HO_2C\overset{OH}{C}H\overset{OH}{C}HCO_2H$	3.036 4.366	9.20×10^{-4} 4.31×10^{-5}
Dimethylamine	$(CH_3)_2NH_2^+$	10.774	1.68×10^{-11}
2,4-Dinitrophenol	O_2N-(benzene with NO_2)-OH	4.11	7.8×10^{-5}
Ethane-1,2-dithiol	$HSCH_2CH_2SH$	8.85 (30°) 10.43 (30°)	1.4×10^{-9} 3.7×10^{-11}
Ethylamine	$CH_3CH_2NH_3^+$	10.636	2.31×10^{-11}
Ethylenediamine (1,2-diaminoethane)	$H_3\overset{+}{N}CH_2CH_2\overset{+}{N}H_3$	6.848 9.928	1.42×10^{-7} 1.18×10^{-10}
Ethylenedinitrilotetraacetic acid (EDTA)	$(HO_2CCH_2)_2\overset{+}{N}HCH_2CH_2\overset{+}{N}H(CH_2CO_2H)_2$	0.0 (CO_2H) 1.5 (CO_2H) 2.0 (CO_2H) 2.66 (CO_2H) 6.16 (NH) 10.24 (NH)	1.0 0.032 0.010 0.002 2 6.9×10^{-7} 5.8×10^{-11}
Formic acid (methanoic acid)	HCO_2H	3.745	1.80×10^{-4}
Glutamic acid	$\overset{NH_3^+}{\underset{\alpha \to CO_2H}{CHCH_2CH_2CO_2H}}$ γ	2.23 (α-CO_2H) 4.42 (γ-CO_2H) 9.95 (NH_3)	5.9×10^{-3} 3.8×10^{-5} 1.12×10^{-10}
Glutamine	$\overset{NH_3^+}{\underset{CO_2H}{CHCH_2CH_2\overset{O}{C}NH_2}}$	2.17 (CO_2H) 9.01 (NH_3)	6.8×10^{-3} 9.8×10^{-10}

(continued)

Name	Structure	pK_a	K_a
Glycine (aminoacetic acid)	$\overset{NH_3^+}{\underset{CO_2H}{\overset{\mid}{\underset{\mid}{CH_2}}}}$	2.350 (CO_2H) 9.778 (NH_3)	4.47×10^{-3} 1.67×10^{-10}
Guanidine	$\overset{{}^+NH_2}{\underset{}{H_2N-\overset{\mid}{C}-NH_2}}$	13.54 (27°)	2.9×10^{-14}
1,6-Hexanedioic acid (adipic acid)	$HO_2CCH_2CH_2CH_2CH_2CO_2H$	4.42 5.42	3.8×10^{-5} 3.8×10^{-6}
Histidine	$\overset{NH_3^+}{\underset{CO_2H}{\overset{\mid}{\underset{\mid}{CHCH_2}}}}$ (imidazole ring)	1.7 (CO_2H) 6.02 (NH) 9.08 (NH_3)	2×10^{-2} 9.5×10^{-7} 8.3×10^{-10}
Hydrazoic acid (hydrogen azide)	$H\overset{+}{N}=N=\overset{-}{N}$	4.65	2.2×10^{-5}
Hydrogen cyanate	$HOC\equiv N$	3.48	3.3×10^{-4}
Hydrogen cyanide	$HC\equiv N$	9.21	6.2×10^{-10}
Hydrogen fluoride	HF	3.17	6.8×10^{-4}
Hydrogen peroxide	$HOOH$	11.65	2.2×10^{-12}
Hydrogen sulfide	H_2S	7.02 ~18	9.5×10^{-8} $\sim 10^{-18}$
Hydrogen thiocyanate	$HSC\equiv N$	0.9	0.13
Hydroxyacetic acid (glycolic acid)	$HOCH_2CO_2H$	3.831	1.48×10^{-4}
Hydroxybenzene (phenol)	(benzene ring)—OH	9.98	1.05×10^{-10}
2-Hydroxybenzoic acid (salicylic acid)	(benzene ring with CO_2H and OH)	2.97 (CO_2H) 13.74 (OH)	1.07×10^{-3} 1.82×10^{-14}
Hydroxylamine	$HO\overset{+}{N}H_3$	5.96	1.10×10^{-6}
8-Hydroxyquinoline (oxine)	(quinoline ring with HO and $\overset{+}{N}H$)	4.91 (NH) 9.81 (OH)	1.23×10^{-5} 1.55×10^{-10}
Hypochlorous acid (hydrogen hypochlorite)	$HOCl$	7.53	3.0×10^{-8}
Hypophosphorous acid (hydrogen hypophosphite)	$\overset{O}{\underset{}{\overset{\parallel}{H_2POH}}}$	1.23	5.9×10^{-2}
Imidazole (1,3-diazole)	(imidazole ring with NH^+ and NH)	6.993	1.02×10^{-7}
Iminodiacetic acid	$H_2\overset{+}{N}(CH_2CO_2H)_2$	1.82 (CO_2H) 2.84 (CO_2H) 9.79 (NH_2)	1.51×10^{-2} 1.45×10^{-3} 1.62×10^{-10}
Iodic acid (hydrogen iodate)	$\overset{O}{\underset{}{\overset{\parallel}{HO-I=O}}}$	0.77	0.17

Name	Structure	pK_a	K_a
Iodoacetic acid	ICH_2CO_2H	3.175	6.68×10^{-4}
Isoleucine	$\overset{\overset{+}{N}H_3}{\underset{CO_2H}{CHCH(CH_3)CH_2CH_3}}$	2.319 (CO_2H) 9.754 (NH_3)	4.80×10^{-3} 1.76×10^{-10}
Leucine	$\overset{\overset{+}{N}H_3}{\underset{CO_2H}{CHCH_2CH(CH_3)_2}}$	2.329 (CO_2) 9.747 (NH_3)	4.69×10^{-3} 1.79×10^{-10}
Lysine	$\alpha \rightarrow \overset{\overset{+}{N}H_3}{\underset{CO_2H}{CHCH_2CH_2CH_2CH_2\overset{+}{N}H_3}} \quad \epsilon$	2.04 (CO_2H) 9.08 (α-NH_3) 10.69 (ϵ-NH_3)	9.1×10^{-3} 8.3×10^{-10} 2.0×10^{-11}
Malonic acid (propanedioic acid)	$HO_2CCH_2CO_2H$	2.847 5.696	1.42×10^{-3} 2.01×10^{-6}
Mercaptoacetic acid (thioglycolic acid)	$HSCH_2CO_2H$	(3.60) (CO_2H) 10.55 (SH)	2.5×10^{-4} 2.82×10^{-11}
2-Mercaptoethanol	$HSCH_2CH_2OH$	9.72	1.91×10^{-10}
Methionine	$\overset{\overset{+}{N}H_3}{\underset{CO_2H}{CHCH_2CH_2SCH_3}}$	2.20 (CO_2H) 9.05 (NH_3)	6.3×10^{-3} 8.9×10^{-10}
Methylamine	$CH_3\overset{+}{N}H_3$	10.64	2.3×10^{-11}
4-Methylaniline (*p*-toluidine)	$CH_3-\!\!\left\langle\!\bigcirc\!\right\rangle\!\!-\overset{+}{N}H_3$	5.084	8.24×10^{-6}
2-Methylphenol (*o*-cresol)	2-methylphenol structure	10.09	8.1×10^{-11}
4-Methylphenol (*p*-cresol)	$CH_3-\!\!\left\langle\!\bigcirc\!\right\rangle\!\!-OH$	10.26	5.5×10^{-11}
Morpholine (perhydro-1,4-oxazine)	morpholine structure $O\!\!\left\langle\;\right\rangle\!\!\overset{+}{N}H_2$	8.492	3.22×10^{-9}
1-Naphthoic acid	1-naphthoic acid structure (CO_2H)	3.70	2.0×10^{-4}
2-Naphthoic acid	2-naphthoic acid structure (CO_2H)	4.16	6.9×10^{-5}
1-Naphthol	1-naphthol structure (OH)	9.34	4.6×10^{-10}

(continued)

Name	Structure	pK_a	K_a
2-Naphthol		9.51	3.1×10^{-10}
Nitrilotriacetic acid	$H\overset{+}{N}(CH_2CO_2H)_3$	1.1 (CO₂H) (20°) 1.650 (CO₂H) (20°) 2.940 (CO₂H) (20°) 10.334 (NH) (20°)	8×10^{-2} 2.24×10^{-2} 1.15×10^{-3} 4.63×10^{-11}
4-Nitrobenzoic acid	$O_2N-\!\!\langle\ \rangle\!\!-CO_2H$	3.442	3.61×10^{-4}
Nitroethane	$CH_3CH_2NO_2$	8.57	2.7×10^{-9}
4-Nitrophenol	$O_2N-\!\!\langle\ \rangle\!\!-OH$	7.15	7.1×10^{-8}
N-Nitrosophenylhydroxylamine (cupferron)		4.16	6.9×10^{-5}
Nitrous acid	$HON{=}O$	3.15	7.1×10^{-4}
Oxalic acid (ethanedioic acid)	HO_2CCO_2H	1.252 4.266	5.60×10^{-2} 5.42×10^{-5}
Oxoacetic acid (glyoxylic acid)	$\overset{O}{\overset{\|}{H}}CCO_2H$	3.46	3.5×10^{-4}
Oxobutanedioic acid (oxaloacetic acid)	$HO_2CCH_2\overset{O}{\overset{\|}{C}}CO_2H$	2.56 4.37	2.8×10^{-3} 4.3×10^{-5}
2-Oxopentanedioic (α-ketoglutaric acid)	$HO_2CCH_2CH_2\overset{O}{\overset{\|}{C}}CO_2H$	1.85 4.44	1.41×10^{-2} 3.6×10^{-5}
2-Oxopropanoic acid (pyruvic acid)	$CH_3\overset{O}{\overset{\|}{C}}CO_2H$	2.55	2.8×10^{-3}
1,5-Pentanedioic acid (glutaric acid)	$HO_2CCH_2CH_2CH_2CO_2H$	4.34 5.43	4.6×10^{-5} 3.7×10^{-6}
1,10-Phenanthroline		4.86	1.38×10^{-5}
Phenylacetic acid	$\langle\ \rangle\!\!-CH_2CO_2H$	4.310	4.90×10^{-5}
Phenylalanine	$\overset{\overset{+}{N}H_3}{\underset{CO_2H}{\mid}}CHCH_2-\!\!\langle\ \rangle$	2.20 (CO₂H) 9.31 (NH₃)	6.3×10^{-3} 4.9×10^{-10}
Phosphoric acid (hydrogen phosphate)	$HO\!-\!\overset{O}{\overset{\|}{\underset{\underset{OH}{\mid}}{P}}}\!-\!OH$	2.148 7.199 12.15	7.11×10^{-3} 6.32×10^{-8} 7.1×10^{-13}

Name	Structure	pK_a	K_a
Phosphorous acid (hydrogen phosphite)		1.5 6.79	3×10^{-2} 1.62×10^{-7}
Phthalic acid (benzene-1,2-dicarboxylic acid)		2.950 5.408	1.12×10^{-3} 3.90×10^{-6}
Piperazine (perhydro-1,4-diazine)		5.333 9.731	4.65×10^{-6} 1.86×10^{-10}
Piperidine		11.123	7.53×10^{-12}
Proline		1.952 (CO_2H) 10.640 (NH_2)	1.12×10^{-2} 2.29×10^{-11}
Propanoic acid	$CH_3CH_2CO_2H$	4.874	1.34×10^{-5}
Propenoic acid (acrylic acid)	$H_2C{=}CHCO_2H$	4.258	5.52×10^{-5}
Propylamine	$CH_3CH_2CH_2NH_3^+$	10.566	2.72×10^{-11}
Pyridine (azine)		5.229	5.90×10^{-6}
Pyridine-2-carboxylic acid (picolinic acid)		1.01 (CO_2H) 5.39 (NH)	9.8×10^{-2} 4.1×10^{-6}
Pyridine-3-carboxylic acid (nicotinic acid)		2.05 (CO_2H) 4.81 (NH)	8.9×10^{-3} 1.55×10^{-5}
Pyridoxal-5-phosphate		1.4 (POH) 3.44 (OH) 6.01 (POH) 8.45 (NH)	0.04 3.6×10^{-4} 9.8×10^{-7} 3.5×10^{-9}
Pyrophosphoric acid (hydrogen diphosphate)	$(HO)_2POP(OH)_2$	0.8 2.2 6.70 9.40	0.16 6×10^{-3} 2.0×10^{-7} 4.0×10^{-10}
Serine		2.187 (CO_2H) 9.209 (NH_3)	6.50×10^{-3} 6.18×10^{-10}
Succinic acid (butanedioic acid)	$HO_2CCH_2CH_2CO_2H$	4.207 5.636	6.21×10^{-5} 2.31×10^{-6}

(continued)

Name	Structure	pK_a	K_a
Sulfuric acid (hydrogen sulfate)	$HO-\overset{\displaystyle O}{\underset{\displaystyle O}{S}}-OH$	1.99 (pK_2)	1.02×10^{-2}
Sulfurous acid (hydrogen sulfite)	$HO\overset{\displaystyle O}{S}OH$	1.91 7.18	1.23×10^{-2} 6.6×10^{-8}
Thiosulfuric acid (hydrogen thiosulfate)	$HO\overset{\displaystyle O}{\underset{\displaystyle O}{S}}SH$	0.6 1.6	0.3 0.03
Threonine	$\overset{\displaystyle NH_3^+}{\underset{\displaystyle CO_2H}{CHCHOHCH_3}}$	2.088 (CO_2H) 9.100 (NH_3)	8.17×10^{-3} 7.94×10^{-10}
Trichloroacetic acid	Cl_3CCO_2H	0.66	0.22
Triethanolamine	$(HOCH_2CH_2)_3NH^+$	7.762	1.73×10^{-8}
Triethylamine	$(CH_3CH_2)_3NH^+$	10.715	1.93×10^{-11}
1,2,3-Trihydroxybenzene (pyrogallol)	(OH, OH, OH substituted benzene)	8.94 11.08 (14)	1.15×10^{-9} 8.3×10^{-12} 10^{-14}
Trimethylamine	$(CH_3)_3NH^+$	9.800	1.58×10^{-10}
Tris(hydroxymethyl)aminomethane (tris or tham)	$(HOCH_2)_3CNH_3^+$	8.075	8.41×10^{-9}
Tryptophan	$\overset{\displaystyle NH_3^+}{\underset{\displaystyle CO_2H}{CHCH_2}}$—(indole)	2.35 (CO_2H) 9.33 (NH_3)	4.5×10^{-3} 4.7×10^{-10}
Tyrosine	$\overset{\displaystyle NH_3^+}{\underset{\displaystyle CO_2H}{CHCH_2}}$—(benzene)—OH	2.17 (CO_2H) 9.19 (NH_3) 10.47 (OH)	6.8×10^{-3} 6.5×10^{-10} 3.4×10^{-11}
Valine	$\overset{\displaystyle NH_3^+}{\underset{\displaystyle CO_2H}{CHCH(CH_3)_2}}$	2.286 (CO_2H) 9.718 (NH_3)	5.18×10^{-3} 1.91×10^{-10}

Standard Reduction Potentials

Reaction[†]	$E°$ (volts)
Aluminum	
$Al^{3+} + 3e^- \rightleftharpoons Al(s)$	−1.677
$Al(OH)_4^- + 3e^- \rightleftharpoons Al(s) + 4OH^-$	−2.328
Arsenic	
$H_3AsO_4 + 2H^+ + 2e^- \rightleftharpoons H_3AsO_3 + H_2O$	0.575
$H_3AsO_3 + 3H^+ + 3e^- \rightleftharpoons As(s) + 3H_2O$	0.247\ 5
$As(s) + 3H^+ + 3e^- \rightleftharpoons AsH_3(g)$	−0.238
Barium	
$Ba^{2+} + 2e^- \rightleftharpoons Ba(s)$	−2.906
Beryllium	
$Be^{2+} + 2e^- \rightleftharpoons Be(s)$	−1.968
Boron	
$2B(s) + 6H^+ + 6e^- \rightleftharpoons B_2H_6(g)$	−0.150
$B_4O_7^{2-} + 14H^+ + 12e^- \rightleftharpoons 4B(s) + 7H_2O$	−0.792
$B(OH)_3 + 3H^+ + 3e^- \rightleftharpoons B(s) + 3H_2O$	−0.889
Bromine	
$BrO_4^- + 2H^+ + 2e^- \rightleftharpoons BrO_3^- + H_2O$	1.745
$HOBr + H^+ + e^- \rightleftharpoons \frac{1}{2}Br_2(l) + H_2O$	1.584
$BrO_3^- + 6H^+ + 5e^- \rightleftharpoons \frac{1}{2}Br_2(l) + 3H_2O$	1.513
$Br_2(aq) + 2e^- \rightleftharpoons 2Br^-$	1.098
$Br_2(l) + 2e^- \rightleftharpoons 2Br^-$	1.078
$Br_3^- + 2e^- \rightleftharpoons 3Br^-$	1.062
$BrO^- + H_2O + 2e^- \rightleftharpoons Br^- + 2OH^-$	0.766
$BrO_3^- + 3H_2O + 6e^- \rightleftharpoons Br^- + 6OH^-$	0.613
Cadmium	
$Cd^{2+} + 2e^- \rightleftharpoons Cd(s)$	−0.402
$Cd(NH_3)_4^{2+} + 2e^- \rightleftharpoons Cd(s) + 4NH_3$	−0.613
Calcium	
$Ca(s) + 2H^+ + 2e^- \rightleftharpoons CaH_2(s)$	0.776
$Ca^{2+} + 2e^- \rightleftharpoons Ca(s)$	−2.868
$Ca(acetate)^+ + 2e^- \rightleftharpoons Ca(s) + acetate^-$	−2.891
$CaSO_4(s) + 2e^- \rightleftharpoons Ca(s) + SO_4^{2-}$	−2.936

[†]All species are aqueous unless otherwise indicated.

Reaction[†]	$E°$ (volts)
Carbon	
$C_2H_2(g) + 2H^+ + 2e^- \rightleftharpoons C_2H_4(g)$	0.731
$O{=}\langle\ \rangle{=}O + 2H^+ + 2e^- \rightleftharpoons HO{-}\langle\ \rangle{-}OH$	0.700
$CH_3OH + 2H^+ + 2e^- \rightleftharpoons CH_4(g) + H_2O$	0.583
dehydroascorbic acid $+ 2H^+ + 2e^- \rightleftharpoons$ ascorbic acid $+ H_2O$	0.390
$(CN)_2(g) + 2H^+ + 2e^- \rightleftharpoons 2HCN(aq)$	0.373
$H_2CO + 2H^+ + 2e^- \rightleftharpoons CH_3OH$	0.237
$C(s) + 4H^+ + 4e^- \rightleftharpoons CH_4(g)$	0.131\ 5
$HCO_2H + 2H^+ + 2e^- \rightleftharpoons H_2CO + H_2O$	−0.029
$CO_2(g) + 2H^+ + 2e^- \rightleftharpoons CO(g) + H_2O$	−0.103\ 8
$CO_2(g) + 2H^+ + 2e^- \rightleftharpoons HCO_2H$	−0.114
$2CO_2(g) + 2H^+ + 2e^- \rightleftharpoons H_2C_2O_4$	−0.432
Cerium	
$Ce^{4+} + e^- \rightleftharpoons Ce^{3+}$	1.72 1.70 1 F $HClO_4$ 1.44 1 F H_2SO_4 1.61 1 F HNO_3 1.47 1 F HCl
$Ce^{3+} + 3e^- \rightleftharpoons Ce(s)$	−2.336
Cesium	
$Cs^+ + e^- \rightleftharpoons Cs(s)$	−3.026
Chlorine	
$HClO_2 + 2H^+ + 2e^- \rightleftharpoons HOCl + H_2O$	1.674
$HClO + H^+ + e^- \rightleftharpoons \frac{1}{2}Cl_2(g) + H_2O$	1.630
$ClO_3^- + 6H^+ + 5e^- \rightleftharpoons \frac{1}{2}Cl_2(g) + 3H_2O$	1.458
$Cl_2(aq) + 2e^- \rightleftharpoons 2Cl^-$	1.396

(continued)

Reaction	$E°$ (volts)
$Cl_2(g) + 2e^- \rightleftharpoons 2Cl^-$	1.360 4
$ClO_4^- + 2H^+ + 2e^- \rightleftharpoons ClO_3^- + H_2O$	1.226
$ClO_3^- + 3H^+ + 2e^- \rightleftharpoons HClO_2 + H_2O$	1.157
$ClO_3^- + 2H^+ + e^- \rightleftharpoons ClO_2 + H_2O$	1.130
$ClO_2 + e^- \rightleftharpoons ClO_2^-$	1.068
Chromium	
$Cr_2O_7^{2-} + 14H^+ + 6e^- \rightleftharpoons 2Cr^{3+} + 7H_2O$	1.36
$CrO_4^{2-} + 4H_2O + 3e^-$	−0.12
$\quad \rightleftharpoons Cr(OH)_3\ (s, \text{hydrated}) + 5OH^-$	
$Cr^{3+} + e^- \rightleftharpoons Cr^{2+}$	−0.42
$Cr^{3+} + 3e^- \rightleftharpoons Cr(s)$	−0.74
$Cr^{2+} + 2e^- \rightleftharpoons Cr(s)$	−0.89
Cobalt	
$Co^{3+} + e^- \rightleftharpoons Co^{2+}$	1.92
	1.817 8 F H_2SO_4
	1.850 4 F HNO_3
$Co(NH_3)_6^{3+} + e^- \rightleftharpoons Co(NH_3)_6^{2+}$	0.1
$CoOH^+ + H^+ + 2e^- \rightleftharpoons Co(s) + H_2O$	0.003
$Co^{2+} + 2e^- \rightleftharpoons Co(s)$	−0.282
$Co(OH)_2(s) + 2e^- \rightleftharpoons Co(s) + 2OH^-$	−0.746
Copper	
$Cu^+ + e^- \rightleftharpoons Cu(s)$	0.518
$Cu^{2+} + 2e^- \rightleftharpoons Cu(s)$	0.339
$Cu^{2+} + e^- \rightleftharpoons Cu^+$	0.161
$CuCl(s) + e^- \rightleftharpoons Cu(s) + Cl^-$	0.137
$Cu(IO_3)_2(s) + 2e^- \rightleftharpoons Cu(s) + 2IO_3^-$	−0.079
$Cu(\text{ethylenediamine})_2^+ + e^-$	−0.119
$\quad \rightleftharpoons Cu(s) + 2\text{ ethylenediamine}$	
$CuI(s) + e^- \rightleftharpoons Cu(s) + I^-$	−0.185
$Cu(EDTA)^{2-} + 2e^- \rightleftharpoons Cu(s) + EDTA^{4-}$	−0.216
$Cu(OH)_2(s) + 2e^- \rightleftharpoons Cu(s) + 2OH^-$	−0.222
$Cu(CN)_2^- + e^- \rightleftharpoons Cu(s) + 2CN^-$	−0.429
$CuCN(s) + e^- \rightleftharpoons Cu(s) + CN^-$	−0.639
Fluorine	
$F_2(g) + 2e^- \rightleftharpoons 2F^-$	2.890
$F_2O(g) + 2H^+ + 4e^- \rightleftharpoons 2F^- + H_2O$	2.168
Gallium	
$Ga^{3+} + 3e^- \rightleftharpoons Ga(s)$	−0.549
$GaOOH(s) + 3H_2O + 3e^- \rightleftharpoons Ga(s) + 3OH^-$	−1.320
Germanium	
$Ge^{2+} + 2e^- \rightleftharpoons Ge(s)$	0.1
$H_4GeO_4 + 4H^+ + 4e^- \rightleftharpoons Ge(s) + 4H_2O$	−0.039
Gold	
$Au^+ + e^- \rightleftharpoons Au(s)$	1.69
$Au^{3+} + 2e^- \rightleftharpoons Au^+$	1.41
$AuCl_2^- + e^- \rightleftharpoons Au(s) + 2Cl^-$	1.154
$AuCl_4^- + 2e^- \rightleftharpoons AuCl_2^- + 2Cl^-$	0.926
Hydrogen	
$2H^+ + 2e^- \rightleftharpoons H_2(g)$	0.000 0
$H_2O + e^- \rightleftharpoons \frac{1}{2}H_2(g) + OH^-$	−0.828 0
Indium	
$In^{3+} + 3e^- \rightleftharpoons In(s)$	−0.338
$In^{3+} + 2e^- \rightleftharpoons In^+$	−0.444
$In(OH)_3(s) + 3e^- \rightleftharpoons In(s) + 3OH^-$	−0.99
Iodine	
$IO_4^- + 2H^+ + 2e^- \rightleftharpoons IO_3^- + H_2O$	1.589
$H_5IO_6 + 2H^+ + 2e^- \rightleftharpoons HIO_3 + 3H_2O$	1.567
$HOI + H^+ + e^- \rightleftharpoons \frac{1}{2}I_2(s) + H_2O$	1.430

Reaction	$E°$ (volts)
$ICl_3(s) + 3e^- \rightleftharpoons \frac{1}{2}I_2(s) + 3Cl^-$	1.28
$ICl(s) + e^- \rightleftharpoons \frac{1}{2}I_2(s) + Cl^-$	1.22
$IO_3^- + 6H^+ + 5e^- \rightleftharpoons \frac{1}{2}I_2(s) + 3H_2O$	1.210
$IO_3^- + 5H^+ + 4e^- \rightleftharpoons HOI + 2H_2O$	1.154
$I_2(aq) + 2e^- \rightleftharpoons 2I^-$	0.620
$I_2(s) + 2e^- \rightleftharpoons 2I^-$	0.535
$I_3^- + 2e^- \rightleftharpoons 3I^-$	0.535
$IO_3^- + 3H_2O + 6e^- \rightleftharpoons I^- + 6OH^-$	0.269
Iron	
$Fe(\text{phenanthroline})_3^{3+} + e^-$	1.147
$\quad \rightleftharpoons Fe(\text{phenanthroline})_3^{2+}$	
$Fe(\text{bipyridyl})_3^{3+} + e^- \rightleftharpoons Fe(\text{bipyridyl})_3^{2+}$	1.120
$FeOH^{2+} + H^+ + e^- \rightleftharpoons Fe^{2+} + H_2O$	0.900
$FeO_4^{2-} + 3H_2O + 3e^- \rightleftharpoons FeOOH(s) + 5OH^-$	0.80
$Fe^{3+} + e^- \rightleftharpoons Fe^{2+}$	0.771
	0.732 1 F HCl
	0.767 1 F $HClO_4$
	0.746 1 F HNO_3
$FeOOH(s) + 3H^+ + e^- \rightleftharpoons Fe^{2+} + 2H_2O$	0.74
$\text{ferricinium}^+ + e^- \rightleftharpoons \text{ferrocene}$	0.400
$Fe(CN)_6^{3-} + e^- \rightleftharpoons Fe(CN)_6^{4-}$	0.356
$FeOH^+ + H^+ + 2e^- \rightleftharpoons Fe(s) + H_2O$	−0.16
$Fe^{2+} + 2e^- \rightleftharpoons Fe(s)$	−0.44
$FeCO_3(s) + 2e^- \rightleftharpoons Fe(s) + CO_3^{2-}$	−0.756
Lanthanum	
$La^{3+} + 3e^- \rightleftharpoons La(s)$	−2.379
Lead	
$Pb^{4+} + 2e^- \rightleftharpoons Pb^{2+}$	1.69 1 F HNO_3
$PbO_2(s) + 4H^+ + SO_4^{2-} + 2e^-$	1.685
$\quad \rightleftharpoons PbSO_4(s) + 2H_2O$	
$PbO_2(s) + 4H^+ + 2e^- \rightleftharpoons Pb^{2+} + 2H_2O$	1.458
$3PbO_2(s) + 2H_2O + 4e^- \rightleftharpoons Pb_3O_4(s) + 4OH^-$	0.269
$Pb_3O_4(s) + H_2O + 2e^-$	0.224
$\quad \rightleftharpoons 3PbO(s, \text{red}) + 2OH^-$	
$Pb_3O_4(s) + H_2O + 2e^-$	0.207
$\quad \rightleftharpoons 3PbO(s, \text{yellow}) + 2OH^-$	
$Pb^{2+} + 2e^- \rightleftharpoons Pb(s)$	−0.126
$PbF_2(s) + 2e^- \rightleftharpoons Pb(s) + 2F^-$	−0.350
$PbSO_4(s) + 2e^- \rightleftharpoons Pb(s) + SO_4^{2-}$	−0.355
Lithium	
$Li^+ + e^- \rightleftharpoons Li(s)$	−3.040
Magnesium	
$Mg(OH)^+ + H^+ + 2e^- \rightleftharpoons Mg(s) + H_2O$	−2.022
$Mg^{2+} + 2e^- \rightleftharpoons Mg(s)$	−2.360
$Mg(C_2O_4)(s) + 2e^- \rightleftharpoons Mg(s) + C_2O_4^{2-}$	−2.493
$Mg(OH)_2(s) + 2e^- \rightleftharpoons Mg(s) + 2OH^-$	−2.690
Manganese	
$MnO_4^- + 4H^+ + 3e^- \rightleftharpoons MnO_2(s) + 2H_2O$	1.692
$Mn^{3+} + e^- \rightleftharpoons Mn^{2+}$	1.56
$MnO_4^- + 8H^+ + 5e^- \rightleftharpoons Mn^{2+} + 4H_2O$	1.507
$Mn_2O_3(s) + 6H^+ + 2e^- \rightleftharpoons 2Mn^{2+} + 3H_2O$	1.485
$MnO_2(s) + 4H^+ + 2e^- \rightleftharpoons Mn^{2+} + 2H_2O$	1.230
$Mn(EDTA)^- + e^- \rightleftharpoons Mn(EDTA)^{2-}$	0.825
$MnO_4^- + e^- \rightleftharpoons MnO_4^{2-}$	0.56
$3Mn_2O_3(s) + H_2O + 2e^-$	0.002
$\quad \rightleftharpoons 2Mn_3O_4(s) + 2OH^-$	
$Mn_3O_4(s) + 4H_2O + 2e^-$	−0.352
$\quad \rightleftharpoons 3Mn(OH)_2(s) + 2OH^-$	

Reaction	$E°$ (volts)
$Mn^{2+} + 2e^- \rightleftharpoons Mn(s)$	-1.182
$Mn(OH)_2(s) + 2e^- \rightleftharpoons Mn(s) + 2OH^-$	-1.565
Mercury	
$2Hg^{2+} + 2e^- \rightleftharpoons Hg_2^{2+}$	0.908
$Hg^{2+} + 2e^- \rightleftharpoons Hg(l)$	0.852
$Hg_2^{2+} + 2e^- \rightleftharpoons 2Hg(l)$	0.796
$Hg_2SO_4(s) + 2e^- \rightleftharpoons 2Hg(l) + SO_4^{2-}$	0.614
$Hg_2Cl_2(s) + 2e^- \rightleftharpoons 2Hg(l) + 2Cl^-$	$\begin{cases} 0.268 \\ 0.241 \end{cases}$ (saturated calomel electrode)
$Hg(OH)_3^- + 2e^- \rightleftharpoons Hg(l) + 3OH^-$	0.231
$Hg(OH)_2 + 2e^- \rightleftharpoons Hg(l) + 2OH^-$	0.206
$Hg_2Br_2(s) + 2e^- \rightleftharpoons 2Hg(l) + 2Br^-$	0.140
$HgO(s, \text{yellow}) + H_2O + 2e^-$ $\rightleftharpoons Hg(l) + 2OH^-$	$0.098\,3$
$HgO(s, \text{red}) + H_2O + 2e^- \rightleftharpoons Hg(l) + 2OH^-$	$0.097\,7$
Molybdenum	
$MoO_4^{2-} + 2H_2O + 2e^- \rightleftharpoons MoO_2(s) + 4OH^-$	-0.818
$MoO_4^{2-} + 4H_2O + 6e^- \rightleftharpoons Mo(s) + 8OH^-$	-0.926
$MoO_2(s) + 2H_2O + 4e^- \rightleftharpoons Mo(s) + 4OH^-$	-0.980
Nickel	
$NiOOH(s) + 3H^+ + e^- \rightleftharpoons Ni^{2+} + 2H_2O$	2.05
$Ni^{2+} + 2e^- \rightleftharpoons Ni(s)$	-0.236
$Ni(CN)_4^{2-} + e^- \rightleftharpoons Ni(CN)_3^{2-} + CN^-$	-0.401
$Ni(OH)_2(s) + 2e^- \rightleftharpoons Ni(s) + 2OH^-$	-0.714
Nitrogen	
$HN_3 + 3H^+ + 2e^- \rightleftharpoons N_2(g) + NH_4^+$	2.079
$N_2O(g) + 2H^+ + 2e^- \rightleftharpoons N_2(g) + H_2O$	1.769
$2NO(g) + 2H^+ + 2e^- \rightleftharpoons N_2O(g) + H_2O$	1.587
$NO^+ + e^- \rightleftharpoons NO(g)$	1.46
$2NH_3OH^+ + H^+ + 2e^- \rightleftharpoons N_2H_5^+ + 2H_2O$	1.40
$NH_3OH^+ + 2H^+ + 2e^- \rightleftharpoons NH_4^+ + H_2O$	1.33
$N_2H_5^+ + 3H^+ + 2e^- \rightleftharpoons 2NH_4^+$	1.250
$HNO_2 + H^+ + e^- \rightleftharpoons NO(g) + H_2O$	0.984
$NO_3^- + 4H^+ + 3e^- \rightleftharpoons NO(g) + 2H_2O$	0.955
$NO_3^- + 3H^+ + 2e^- \rightleftharpoons HNO_2 + H_2O$	0.940
$NO_3^- + 2H^+ + e^- \rightleftharpoons \frac{1}{2}N_2O_4(g) + H_2O$	0.798
$N_2(g) + 8H^+ + 6e^- \rightleftharpoons 2NH_4^+$	0.274
$N_2(g) + 5H^+ + 4e^- \rightleftharpoons N_2H_5^+$	-0.214
$N_2(g) + 2H_2O + 4H^+ + 2e^- \rightleftharpoons 2NH_3OH^+$	-1.83
$\frac{3}{2}N_2(g) + H^+ + e^- \rightleftharpoons HN_3$	-3.334
Oxygen	
$OH + H^+ + e^- \rightleftharpoons H_2O$	2.56
$O(g) + 2H^+ + 2e^- \rightleftharpoons H_2O$	$2.430\,1$
$O_3(g) + 2H^+ + 2e^- \rightleftharpoons O_2(g) + H_2O$	2.075
$H_2O_2 + 2H^+ + 2e^- \rightleftharpoons 2H_2O$	1.763
$HO_2 + H^+ + e^- \rightleftharpoons H_2O_2$	1.44
$\frac{1}{2}O_2(g) + 2H^+ + 2e^- \rightleftharpoons H_2O$	$1.229\,1$
$O_2(g) + 2H^+ + 2e^- \rightleftharpoons H_2O_2$	0.695
$O_2(g) + H^+ + e^- \rightleftharpoons HO_2$	-0.05
Palladium	
$Pd^{2+} + 2e^- \rightleftharpoons Pd(s)$	0.915
$PdO(s) + 2H^+ + 2e^- \rightleftharpoons Pd(s) + H_2O$	0.79
$PdCl_6^{4-} + 2e^- \rightleftharpoons Pd(s) + 6Cl^-$	0.615
$PdO_2(s) + H_2O + 2e^- \rightleftharpoons PdO(s) + 2OH^-$	0.64
Phosphorus	
$\frac{1}{4}P_4(s, \text{white}) + 3H^+ + 3e^- \rightleftharpoons PH_3(g)$	-0.046
$\frac{1}{4}P_4(s, \text{red}) + 3H^+ + 3e^- \rightleftharpoons PH_3(g)$	-0.088
$H_3PO_4 + 2H^+ + 2e^- \rightleftharpoons H_3PO_3 + H_2O$	-0.30

Reaction	$E°$ (volts)
$H_3PO_4 + 5H^+ + 5e^- \rightleftharpoons \frac{1}{4}P_4(s, \text{white}) + 4H_2O$	-0.402
$H_3PO_3 + 2H^+ + 2e^- \rightleftharpoons H_3PO_2 + H_2O$	-0.48
$H_3PO_2 + H^+ + e^- \rightleftharpoons \frac{1}{4}P_4(s) + 2H_2O$	-0.51
Platinum	
$Pt^{2+} + 2e^- \rightleftharpoons Pt(s)$	1.18
$PtO_2(s) + 4H^+ + 4e^- \rightleftharpoons Pt(s) + 2H_2O$	0.92
$PtCl_4^{2-} + 2e^- \rightleftharpoons Pt(s) + 4Cl^-$	0.755
$PtCl_6^{2-} + 2e^- \rightleftharpoons PtCl_4^{2-} + 2Cl^-$	0.68
Potassium	
$K^+ + e^- \rightleftharpoons K(s)$	-2.936
Rubidium	
$Rb^+ + e^- \rightleftharpoons Rb(s)$	-2.943
Scandium	
$Sc^{3+} + 3e^- \rightleftharpoons Sc(s)$	-2.09
Selenium	
$SeO_4^{2-} + 4H^+ + 2e^- \rightleftharpoons H_2SeO_3 + H_2O$	1.150
$H_2SeO_3 + 4H^+ + 4e^- \rightleftharpoons Se(s) + 3H_2O$	0.739
$Se(s) + 2H^+ + 2e^- \rightleftharpoons H_2Se(g)$	-0.082
$Se(s) + 2e^- \rightleftharpoons Se^{2-}$	-0.67
Silicon	
$Si(s) + 4H^+ + 4e^- \rightleftharpoons SiH_4(g)$	-0.147
$SiO_2(s, \text{quartz}) + 4H^+ + 4e^-$ $\rightleftharpoons Si(s) + 2H_2O$	-0.990
$SiF_6^{2-} + 4e^- \rightleftharpoons Si(s) + 6F^-$	-1.24
Silver	
$Ag^{2+} + e^- \rightleftharpoons Ag^+$	1.989
$Ag^{3+} + 2e^- \rightleftharpoons Ag^+$	1.9
$AgO(s) + H^+ + e^- \rightleftharpoons \frac{1}{2}Ag_2O(s) + \frac{1}{2}H_2O$	1.40
$Ag^+ + e^- \rightleftharpoons Ag(s)$	$0.799\,3$
$Ag_2C_2O_4(s) + 2e^- \rightleftharpoons 2Ag(s) + C_2O_4^{2-}$	0.465
$AgN_3(s) + e^- \rightleftharpoons Ag(s) + N_3^-$	0.293
$AgCl(s) + e^- \rightleftharpoons Ag(s) + Cl^-$	$\begin{cases} 0.222 \\ 0.197 \end{cases}$ (saturated KCl)
$AgBr(s) + e^- \rightleftharpoons Ag(s) + Br^-$	0.071
$Ag(S_2O_3)_2^{3-} + e^- \rightleftharpoons Ag(s) + 2S_2O_3^{2-}$	0.017
$AgI(s) + e^- \rightleftharpoons Ag(s) + I^-$	-0.152
$Ag_2S(s) + H^+ + 2e^- \rightleftharpoons 2Ag(s) + SH^-$	-0.272
Sodium	
$Na^+ + \frac{1}{2}H_2(g) + e^- \rightleftharpoons NaH(s)$	-2.367
$Na^+ + e^- \rightleftharpoons Na(s)$	$-2.714\,3$
Strontium	
$Sr^{2+} + 2e^- \rightleftharpoons Sr(s)$	-2.889
Sulfur	
$S_2O_8^{2-} + 2e^- \rightleftharpoons 2SO_4^{2-}$	2.01
$S_2O_6^{2-} + 4H^+ + 2e^- \rightleftharpoons 2H_2SO_3$	0.57
$4SO_2 + 4H^+ + 6e^- \rightleftharpoons S_4O_6^{2-} + 2H_2O$	0.539
$SO_2 + 4H^+ + 4e^- \rightleftharpoons S(s) + 2H_2O$	0.450
$2H_2SO_3 + 2H^+ + 4e^- \rightleftharpoons S_2O_3^{2-} + 3H_2O$	0.40
$S(s) + 2H^+ + 2e^- \rightleftharpoons H_2S(g)$	0.174
$S(s) + 2H^+ + 2e^- \rightleftharpoons H_2S(aq)$	0.144
$S_4O_6^{2-} + 2H^+ + 2e^- \rightleftharpoons 2HS_2O_3^-$	0.10
$5S(s) + 2e^- \rightleftharpoons S_5^{2-}$	-0.340
$2S(s) + 2e^- \rightleftharpoons S_2^{2-}$	-0.50
$2SO_3^{2-} + 3H_2O + 4e^- \rightleftharpoons S_2O_3^{2-} + 6OH^-$	-0.566
$SO_3^{2-} + 3H_2O + 4e^- \rightleftharpoons S(s) + 6OH^-$	-0.659
$SO_4^{2-} + 4H_2O + 6e^- \rightleftharpoons S(s) + 8OH^-$	-0.751
$SO_4^{2-} + H_2O + 2e^- \rightleftharpoons SO_3^{2-} + 2OH^-$	-0.936

(continued)

Reaction	$E°$ (volts)
$2SO_3^{2-} + 2H_2O + 2e^- \rightleftharpoons S_2O_4^{2-} + 4OH^-$	-1.130
$2SO_4^{2-} + 2H_2O + 2e^- \rightleftharpoons S_2O_6^{2-} + 4OH^-$	-1.71

Thallium

Reaction	$E°$ (volts)
$Tl^{3+} + 2e^- \rightleftharpoons Tl^+$	1.280 0.77 1 F HCl 1.22 1 F H$_2$SO$_4$ 1.23 1 F HNO$_3$ 1.26 1 F HClO$_4$
$Tl^+ + e^- \rightleftharpoons Tl(s)$	-0.336

Tin

Reaction	$E°$ (volts)
$Sn(OH)_3^+ + 3H^+ + 2e^- \rightleftharpoons Sn^{2+} + 6H_2O$	0.142
$Sn^{4+} + 2e^- \rightleftharpoons Sn^{2+}$	0.139 1 F HCl
$SnO_2(s) + 4H^+ \rightleftharpoons Sn^{2+} + 2H_2O$	-0.094
$Sn^{2+} + 2e^- \rightleftharpoons Sn(s)$	-0.141
$SnF_6^{2-} + 4e^- \rightleftharpoons Sn(s) + 6F^-$	-0.25
$Sn(OH)_6^{2-} + 2e^- \rightleftharpoons Sn(OH)_3^- + 3OH^-$	-0.93
$Sn(s) + 4H_2O + 4e^- \rightleftharpoons SnH_4(g) + 4OH^-$	-1.316
$SnO_2(s) + H_2O + 2e^- \rightleftharpoons SnO(s) + 2OH^-$	-0.961

Titanium

Reaction	$E°$ (volts)
$TiO^{2+} + 2H^+ + e^- \rightleftharpoons Ti^{3+} + H_2O$	0.1
$Ti^{3+} + e^- \rightleftharpoons Ti^{2+}$	-0.9
$TiO_2(s) + 4H^+ + 4e^- \rightleftharpoons Ti(s) + 2H_2O$	-1.076
$TiF_6^{2-} + 4e^- \rightleftharpoons Ti(s) + 6F^-$	-1.191
$Ti^{2+} + 2e^- \rightleftharpoons Ti(s)$	-1.60

Tungsten

Reaction	$E°$ (volts)
$W(CN)_8^{3-} + e^- \rightleftharpoons W(CN)_8^{4-}$	0.457
$W^{6+} + e^- \rightleftharpoons W^{5+}$	0.26 12 F HCl
$WO_3(s) + 6H^+ + 6e^- \rightleftharpoons W(s) + 3H_2O$	-0.091
$W^{5+} + e^- \rightleftharpoons W^{4+}$	-0.3 12 F HCl

Reaction	$E°$ (volts)
$WO_2(s) + 2H_2O + 4e^- \rightleftharpoons W(s) + 4OH^-$	-0.982
$WO_4^{2-} + 4H_2O + 6e^- \rightleftharpoons W(s) + 8OH^-$	-1.060

Uranium

Reaction	$E°$ (volts)
$UO_2^+ + 4H^+ + e^- \rightleftharpoons U^{4+} + 2H_2O$	0.39
$UO_2^{2+} + 4H^+ + 2e^- \rightleftharpoons U^{4+} + 2H_2O$	0.273
$UO_2^{2+} + e^- \rightleftharpoons UO_2^+$	0.16
$U^{4+} + e^- \rightleftharpoons U^{3+}$	-0.577
$U^{3+} + 3e^- \rightleftharpoons U(s)$	-1.642

Vanadium

Reaction	$E°$ (volts)
$VO_2^+ + 2H^+ + e^- \rightleftharpoons VO^{2+} + H_2O$	1.001
$VO^{2+} + 2H^+ + e^- \rightleftharpoons V^{3+} + H_2O$	0.337
$V^{3+} + e^- \rightleftharpoons V^{2+}$	-0.255
$V^{2+} + 2e^- \rightleftharpoons V(s)$	-1.125

Xenon

Reaction	$E°$ (volts)
$H_4XeO_6 + 2H^+ + 2e^- \rightleftharpoons XeO_3 + 3H_2O$	2.38
$XeF_2 + 2H^+ + 2e^- \rightleftharpoons Xe(g) + 2HF$	2.2
$XeO_3 + 6H^+ + 6e^- \rightleftharpoons Xe(g) + 3H_2O$	2.1

Yttrium

Reaction	$E°$ (volts)
$Y^{3+} + 3e^- \rightleftharpoons Y(s)$	-2.38

Zinc

Reaction	$E°$ (volts)
$ZnOH^+ + H^+ + 2e^- \rightleftharpoons Zn(s) + 2H_2O$	-0.497
$Zn^{2+} + 2e^- \rightleftharpoons Zn(s)$	-0.762
$Zn(NH_3)_4^{2+} + 2e^- \rightleftharpoons Zn(s) + 4NH_3$	-1.04
$ZnCO_3(s) + 2e^- \rightleftharpoons Zn(s) + CO_3^{2-}$	-1.06
$Zn(OH)_3^- + 2e^- \rightleftharpoons Zn(s) + 3OH^-$	-1.183
$Zn(OH)_4^{2-} + 2e^- \rightleftharpoons Zn(s) + 4OH^-$	-1.199
$Zn(OH)_2(s) + 2e^- \rightleftharpoons Zn(s) + 2OH^-$	-1.249
$ZnO(s) + H_2O + 2e^- \rightleftharpoons Zn(s) + 2OH^-$	-1.260
$ZnS(s) + 2e^- \rightleftharpoons Zn(s) + S^{2-}$	-1.405

Oxidation Numbers and Balancing Redox Equations

The *oxidation number*, or *oxidation state*, is a bookkeeping device used to keep track of the number of electrons formally associated with a particular element. The oxidation number is meant to tell how many electrons have been lost or gained by a neutral atom when it forms a compound. Because oxidation numbers have no real physical meaning, they are somewhat arbitrary, and not all chemists will assign the same oxidation number to a given element in an unusual compound. However, there are some ground rules that provide a useful start.

1. The oxidation number of an element by itself—e.g., $Cu(s)$ or $Cl_2(g)$—is zero.

2. The oxidation number of H is almost always $+1$, except in metal hydrides—e.g., NaH—in which H is -1.

3. The oxidation number of oxygen is almost always -2. The only common exceptions are peroxides, in which two oxygen atoms are connected and each has an oxidation number of -1. Two examples are hydrogen peroxide ($H-O-O-H$) and its anion ($H-O-O^-$). The oxidation number of oxygen in gaseous O_2 is, of course, zero.

4. The alkali metals (Li, Na, K, Rb, Cs, Fr) almost always have an oxidation number of $+1$. The alkaline earth metals (Be, Mg, Ca, Sr, Ba, Ra) are almost always in the $+2$ oxidation state.

5. The halogens (F, Cl, Br, I) are usually in the -1 oxidation state. Exceptions occur when two different halogens are bound to each other or when a halogen is bound to more than one atom. When different halogens are bound to each other, we assign the oxidation number -1 to the more electronegative halogen.

The sum of the oxidation numbers of each atom in a molecule must equal the charge of the molecule. In H_2O, for example, we have

$$
\begin{aligned}
\text{2 hydrogen} &= 2(+1) = +2 \\
\text{oxygen} & = -2 \\
\hline
\text{net charge} & = 0
\end{aligned}
$$

In SO_4^{2-}, sulfur must have an oxidation number of -6 so that the sum of the oxidation numbers will be -2:

$$
\begin{aligned}
\text{oxygen} &= 4(-2) = -8 \\
\text{sulfur} & = +6 \\
\hline
\text{net charge} & = -2
\end{aligned}
$$

In benzene (C_6H_6), the oxidation number of each carbon must be -1 if hydrogen is assigned the number $+1$. In cyclohexane (C_6H_{12}), the oxidation number of each carbon must be -2, for the same reason. The carbons in benzene are in a higher oxidation state than those in cyclohexane.

The oxidation number of iodine in ICl_2^- is $+1$. This is unusual, because halogens are usually -1. However, because chlorine is more electronegative than iodine, we assign Cl as -1, thereby forcing I to be $+1$.

The oxidation number of As in As_2S_3 is $+3$, and the value for S is -2. This is arbitrary, but reasonable. Because S is more electronegative than As, we make S negative and As positive; and because S is in the same family as oxygen, which is usually -2, we assign S as -2, thus leaving As as $+3$.

The oxidation number of S in $S_4O_6^{2-}$ (tetrathionate) is $+2.5$. The *fractional oxidation state* comes about because six O atoms contribute -12. Because the charge is -2, the four S atoms must contribute $+10$. The average oxidation number of S must be $+\frac{10}{4} = 2.5$.

The oxidation number of Fe in $K_3Fe(CN)_6$ is $+3$. To make this assignment, we first recognize cyanide (CN^-) as a common ion that carries a charge of -1. Six cyanide ions give -6, and

three potassium ions (K^+) give $+3$. Therefore, Fe should have an oxidation number of $+3$ for the whole formula to be neutral. In this approach, it is not necessary to assign individual oxidation numbers to carbon and nitrogen, as long as we recognize that the charge of CN is -1.

Problems

Answers are given at the end of this appendix.

1. Write the oxidation state of the boldface atom in each of the following species.

(a) **Ag**Br

(b) **S**$_2$O$_3^{2-}$

(c) **Se**F$_6$

(d) H**S**$_2$O$_3^-$

(e) H**O**$_2$

(f) **N**O

(g) **Cr**$^{3+}$

(h) **Mn**O$_2$

(i) **Pb**(OH)$_3^-$

(j) **Fe**(OH)$_3$

(k) **Cl**O$^-$

(l) K$_4$**Fe**(CN)$_6$

(m) **Cl**O$_2$

(n) **Cl**O$_2^-$

(o) **Mn**(CN)$_6^{4-}$

(p) **N**$_2$

(q) **N**H$_4^+$

(r) **N**$_2$H$_5^+$

(s) H**As**O$_3^{2-}$

(t) (CH$_3$)$_4$**Li**$_4$

(u) **P**$_4$O$_{10}$

(v) **C**$_2$H$_6$O

(w) **V**O(SO$_4$)

(x) **Fe**$_3$O$_4$

2. Identify the oxidizing agent and the reducing agent on the left side of each of the following reactions.

(a) $Cr_2O_7^{2-} + 3Sn^{2+} + 14H^+$
$$\rightarrow 2Cr^{3+} + 3Sn^{4+} + 7H_2O$$

(b) $4I^- + O_2 + 4H^+ \rightarrow 2I_2 + 2H_2O$

(c) $5CH_3\overset{\displaystyle O}{\overset{\displaystyle \|}{C}}H + 2MnO_4^- + 6H^+ \rightarrow$
$$5CH_3\overset{\displaystyle O}{\overset{\displaystyle \|}{C}}OH + 2Mn^{2+} + 3H_2O$$

Balancing Redox Reactions

To balance a reaction involving oxidation and reduction, we must first identify which element is oxidized and which is reduced. We then break the net reaction into two imaginary *half-reactions,* one of which involves only oxidation and the other only reduction. Although free electrons never appear in a balanced net reaction, they do appear in balanced half-reactions. If we are dealing with aqueous solutions, we proceed to balance each half-reaction, using H_2O and either H^+ or OH^-, as necessary. *A reaction is balanced when the number of atoms of each element is the same on both sides, and the net charge is the same on both sides.*

Acidic Solutions

Here are the steps we will follow:

1.   Assign oxidation numbers to the elements that are oxidized or reduced.

2.   Break the reaction into two half-reactions, one involving oxidation and the other reduction.

3.   For each half-reaction, balance the number of atoms that are oxidized or reduced.

4.   Balance the electrons to account for the change in oxidation number by adding electrons to one side of each half-reaction.

5.   Balance oxygen atoms by adding H_2O to one side of each half-reaction.

6.   Balance the H atoms by adding H^+ to one side of each half-reaction.

7.   Multiply each half-reaction by the number of electrons in the other half-reaction so that the number of electrons on each side of the total reaction will cancel. Then add

the two half-reactions and simplify to the smallest integral coefficients.

EXAMPLE	Balancing a Redox Equation

Balance the following equation using H^+, but not OH^-:

$$\underset{+2}{Fe^{2+}} + \underset{+7}{\underset{\text{Permanganate}}{MnO_4^-}} \rightleftharpoons \underset{+3}{Fe^{3+}} + \underset{+2}{Mn^{2+}}$$

SOLUTION

1. *Assign oxidation numbers.* These are assigned for Fe and Mn in each species in the above reaction.

2. *Break the reaction into two half-reactions.*

oxidation half-reaction: $\underset{+2}{Fe^{2+}} \rightleftharpoons \underset{+3}{Fe^{3+}}$

reduction half-reaction: $\underset{+7}{MnO_4^-} \rightleftharpoons \underset{+2}{Mn^{2+}}$

3. *Balance the atoms that are oxidized or reduced.* Because there is only one Fe or Mn in each species on each side of the equation, the atoms of Fe and Mn are already balanced.

4. *Balance electrons.* Electrons are added to account for the change in each oxidation state.

$$Fe^{2+} \rightleftharpoons Fe^{3+} + e^-$$
$$MnO_4^- + 5e^- \rightleftharpoons Mn^{2+}$$

In the second case, we need $5e^-$ on the left side to take Mn from +7 to +2.

5. *Balance oxygen atoms.* There are no oxygen atoms in the Fe half-reactions. There are four oxygen atoms on the left side of the Mn reaction, so we add four molecules of H_2O to the right side:

$$MnO_4^- + 5e^- \rightleftharpoons Mn^{2+} + 4H_2O$$

6. *Balance hydrogen atoms.* The Fe equation is already balanced. The Mn equation needs $8H^+$ on the left.

$$MnO_4^- + 5e^- + 8H^+ \rightarrow Mn^{2+} + 4H_2O$$

At this point, each half-reaction must be completely balanced (the same number of atoms and charge on each side), *or you have made a mistake.*

7. *Multiply and add the reactions.* We multiply the Fe equation by 5 and the Mn equation by 1 and add:

$$5Fe^{2+} \rightleftharpoons 5Fe^{3+} + 5e^-$$
$$\underline{MnO_4^- + 5e^- + 8H^+ \rightleftharpoons Mn^{2+} + 4H_2O}$$
$$5Fe^{2+} + MnO_4^- + 8H^+ \rightleftharpoons 5Fe^{3+} + Mn^{2+} + 4H_2O$$

The total charge on each side is +17, and we find the same number of atoms of each element on each side. The equation is balanced.

Basic Solutions

The method many people prefer for basic solutions is to balance the equation first with H^+. The answer can then be converted to one in which OH^- is used instead. This is done by adding to each side of the equation a number of hydroxide ions equal to the number of H^+ ions appearing in the equation. For example, to balance Equation D-1 with OH^- instead of H^+, proceed as follows:

$$2I_2 + IO_3^- + 10Cl^- + 6H^+ \rightleftharpoons 5ICl_2^- + 3H_2O \qquad \text{(D-1)}$$
$$\underline{+6OH^- \qquad\qquad\qquad + 6OH^-}$$
$$2I_2 + IO_3^- + 10Cl^- + \underbrace{6H^+ + 6OH^-}_{\substack{6H_2O \\ \Downarrow \\ 3H_2O}} \rightleftharpoons 5ICl_2^- + 3H_2O + 6OH^-$$

Realizing that $6H^+ + 6OH^- = 6H_2O$, and canceling $3H_2O$ on each side, gives the final result:

$$2I_2 + IO_3^- + 10Cl^- + 3H_2O \rightleftharpoons 5ICl_2^- + 6OH^-$$

Problems

3. Balance the following reactions using H^+, but not OH^-.

(a) $Fe^{3+} + Hg_2^{2+} \rightleftharpoons Fe^{2+} + Hg^{2+}$

(b) $Ag + NO_3^- \rightleftharpoons Ag^+ + NO$

(c) $VO^{2+} + Sn^{2+} \rightleftharpoons V^{3+} + Sn^{4+}$

(d) $SeO_4^{2-} + Hg + Cl^- \rightleftharpoons SeO_3^{2-} + Hg_2Cl_2$

(e) $CuS + NO_3^- \rightleftharpoons Cu^{2+} + SO_4^{2-} + NO$

(f) $S_2O_3^{2-} + I_2 \rightleftharpoons I^- + S_4O_6^{2-}$

(g) $Cr_2O_7^{2-} + CH_3\overset{\overset{O}{\|}}{C}H \rightleftharpoons CH_3\overset{\overset{O}{\|}}{C}OH + Cr^{3+}$

(h) $MnO_4^{2-} \rightleftharpoons MnO_2 + MnO_4^-$

(i) $ClO_3^- \rightleftharpoons Cl_2 + O_2$

4. Balance the following equations by using OH^-, but not H^+.

(a) $PbO_2 + Cl^- \rightleftharpoons ClO^- + Pb(OH)_3^-$

(b) $HNO_2 + SbO^+ \rightleftharpoons NO + Sb_2O_5$

(c) $Ag_2S + CN^- + O_2 \rightleftharpoons S + Ag(CN)_2^- + OH^-$

(d) $HO_2^- + Cr(OH)_3 \rightleftharpoons CrO_4^{2-} + OH^-$

(e) $ClO_2 + OH^- \rightleftharpoons ClO_2^- + ClO_3^-$

(f) $WO_3^- + O_2 \rightleftharpoons HW_6O_{21}^{5-} + OH^-$

Answers

1. (a) +1
(b) +2
(c) +6
(d) +2
(e) −1/2
(f) +2
(g) +3
(h) +4

(i) +2
(j) +3
(k) +1
(l) +2
(m) +4
(n) +3
(o) +2
(p) 0

(q) −3
(r) −2
(s) +3
(t) −4
(u) +5
(v) −2
(w) +4
(x) +8/3

2.

	Oxidizing agent	Reducing agent
(a)	$Cr_2O_7^{2-}$	Sn^{2+}
(b)	O_2	I^-
(c)	MnO_4^-	CH_3CHO

3. (a) $2Fe^{3+} + Hg_2^{2+} \rightleftharpoons 2Fe^{2+} + 2Hg^{2+}$

(b) $3Ag + NO_3^- + 4H^+ \rightleftharpoons 3Ag^+ + NO + 2H_2O$

(c) $4H^+ + 2VO^{2+} + Sn^{2+} \rightleftharpoons 2V^{3+} + Sn^{4+} + 2H_2O$

(d) $2Hg + 2Cl^- + SeO_4^{2-} + 2H^+ \rightleftharpoons$
$\qquad\qquad\qquad Hg_2Cl_2 + SeO_3^{2-} + H_2O$

(e) $3CuS + 8NO_3^- + 8H^+ \rightleftharpoons$
$\qquad\qquad\qquad 3Cu^{2+} + 3SO_4^{2-} + 8NO + 4H_2O$

(f) $2S_2O_3^{2-} + I_2 \rightleftharpoons S_4O_6^{2-} + 2I^-$

(g) $Cr_2O_7^{2-} + 3CH_3CHO + 8H^+ \rightleftharpoons$
$\qquad\qquad\qquad 2Cr^{3+} + 3CH_3CO_2H + 4H_2O$

(h) $4H^+ + 3MnO_4^{2-} \rightleftharpoons MnO_2 + 2MnO_4^- + 2H_2O$

(i) $2H^+ + 2ClO_3^- \rightleftharpoons Cl_2 + \frac{5}{2}O_2 + H_2O$

4. (a) $H_2O + OH^- + PbO_2 + Cl^- \rightleftharpoons Pb(OH)_3^- + ClO^-$

(b) $4HNO_2 + 2SbO^+ + 2OH^- \rightleftharpoons 4NO + Sb_2O_5 + 3H_2O$

(c) $Ag_2S + 4CN^- + \frac{1}{2}O_2 + H_2O \rightleftharpoons$
$\qquad\qquad\qquad S + 2Ag(CN)_2^- + 2OH^-$

(d) $2HO_2^- + Cr(OH)_3 \rightleftharpoons CrO_4^{2-} + OH^- + 2H_2O$

(e) $2ClO_2 + 2OH^- \rightleftharpoons ClO_2^- + ClO_3^- + H_2O$

(f) $12WO_3^- + 3O_2 + 2H_2O \rightleftharpoons 2HW_6O_{21}^{5-} + 2OH^-$

Solutions to "Ask Yourself" Questions

Chapter 1

1-A. **(a)** A heterogeneous material has different compositions in various regions. A homogeneous material is the same everywhere.

(b) The composition of a random heterogeneous material varies randomly from place to place. A segregated heterogeneous material has relatively large regions of distinctly different compositions.

(c) A random sample is selected by taking material at random from the lot. A composite sample is selected deliberately by taking predetermined portions of the lot from selected regions. Random sampling is appropriate for a random heterogeneous lot. Composite sampling is appropriate when the lot is segregated into regions of different composition.

1-B. See Table 1-3.

1-C. **(a)** Office worker: $2.2 \times 10^6 \dfrac{\text{cal}}{\text{day}} \times 4.184 \dfrac{\text{J}}{\text{cal}}$
$$= 9.2 \times 10^6 \text{ J/day}$$

Mountain climber: $3.4 \times 10^6 \dfrac{\text{cal}}{\text{day}} \times 4.184 \dfrac{\text{J}}{\text{cal}}$
$$= 14.2 \times 10^6 \text{ J/day}$$

(b) $60 \dfrac{\text{s}}{\text{min}} \times 60 \dfrac{\text{min}}{\text{h}} \times 24 \dfrac{\text{h}}{\text{day}} = 8.64 \times 10^4 \text{ s/day}$

(c) $\dfrac{9.2 \times 10^6 \text{ J/day}}{8.64 \times 10^4 \text{ s/day}} = 1.1 \times 10^2 \text{ W};$

$$14.2 \times 10^6 \text{ J/day} = 1.6 \times 10^2 \text{ W}$$

(d) The office worker consumes more power than the light bulb.

1-D. **(a)** $\left(1.67 \dfrac{\text{g solution}}{\text{mL}}\right)\left(1\,000 \dfrac{\text{mL}}{\text{L}}\right) = 1\,670 \dfrac{\text{g solution}}{\text{L}}$

(b) $\left(0.705 \dfrac{\text{g HClO}_4}{\text{g solution}}\right)\left(1\,670 \dfrac{\text{g solution}}{\text{L}}\right)$
$$= 1.18 \times 10^3 \dfrac{\text{g HClO}_4}{\text{L}}$$

(c) $\left(1.18 \times 10^3 \dfrac{\text{g}}{\text{L}}\right) \bigg/ \left(100.458 \dfrac{\text{g}}{\text{mol}}\right) = 11.7 \dfrac{\text{mol}}{\text{L}}$

1-E. **(a)** $\left(1.50 \dfrac{\text{g solution}}{\text{mL solution}}\right)\left(1\,000 \dfrac{\text{mL solution}}{\text{L solution}}\right)$
$$= 1.50 \times 10^3 \dfrac{\text{g solution}}{\text{L solution}}$$

(b) $\left(0.480 \dfrac{\text{g HBr}}{\text{g solution}}\right)\left(1.50 \times 10^3 \dfrac{\text{g solution}}{\text{L solution}}\right)$
$$= 7.20 \times 10^2 \dfrac{\text{g HBr}}{\text{L solution}}$$

(c) $\left(7.20 \times 10^2 \dfrac{\text{g HBr}}{\text{L solution}}\right)\bigg/\left(80.912 \dfrac{\text{g HBr}}{\text{mol}}\right) = 8.90 \text{ M}$

(d) $M_{\text{con}} \cdot V_{\text{con}} = M_{\text{dil}} \cdot V_{\text{dil}}$
$(8.90 \text{ M})(x \text{ mL}) = (0.160 \text{ M})(250 \text{ mL}) \Rightarrow x = 4.49 \text{ mL}$

Chapter 2

2-A. The lab notebook must (1) state what was done; (2) state what was observed; and (3) be understandable to a stranger.

2-B. **(a)** $m = \dfrac{(24.913 \text{ g})\left(1 - \dfrac{0.001\,2 \text{ g/mL}}{8.0 \text{ g/mL}}\right)}{\left(1 - \dfrac{0.001\,2 \text{ g/mL}}{1.00 \text{ g/mL}}\right)} = 24.939 \text{ g}$

(b) $(24.939 \text{ g})/(0.998\,00 \text{ g/mL}) = 24.989 \text{ mL}$

2-C. Dissolve $(0.250\,0 \text{ L})(0.150\,0 \text{ mol/L}) = 0.037\,50$ mol of K_2SO_4 in less than 250 mL of water in a 250-mL volumetric flask. Add more water and mix. Dilute to the 250.0-mL mark and invert the flask 20 times for complete mixing.

2-D. Transfer pipet. The last liquid should be blown out of the serological pipet because it is calibrated down to the tip. With the measuring pipet, simply drain liquid from the 0-mL mark to the 1-mL mark.

2-E. $(10.000\,0 \text{ g})(1.002\,0 \text{ mL/g}) = 10.020 \text{ mL}$

2-F. $S^{2-} + 2H^+ \rightarrow H_2S$. $H_2S(g)$ is lost when the solution is boiled to dryness.

Chapter 3

3-A. **(a)** 5   **(b)** 4   **(c)** 3
3-B. **(a)** 3.71   **(b)** 10.7   **(c)** 4.0×10^1
  (d) 2.85×10^{-6} **(e)** 12.625 1   **(f)** 6.0×10^{-4}
  (g) 242
3-C. **(a)** Carmen **(b)** Cynthia **(c)** Chastity **(d)** Cheryl
3-D. **(a)** Percent relative uncertainty in mass
$$= (0.002/4.635) \times 100 = 0.04_3\%$$
  Percent relative uncertainty in volume
$$= (0.05/1.13) \times 100 = 4.4\%$$
  (b) Density $= \dfrac{4.635 \pm 0.002 \text{ g}}{1.13 \pm 0.05 \text{ mL}}$

$$= \dfrac{4.635 \ (\pm 0.04_3\%) \text{ g}}{1.13 \ (\pm 4.4\%) \text{ mL}} = 4.10 \pm? \text{ g/mL}$$
  Uncertainty $= \sqrt{(0.04_3)^2 + (4.4)^2} = 4.4\%$
  4.4% of 4.10 = 0.18. The answer can be written 4.1 $\pm$0.2 g/mL
3-E. $-196° = 77.15 \text{ K} = -320.8°\text{F}$

Chapter 4

4-A. $\bar{x} = \dfrac{821 + 783 + 834 + 855}{4} = 823._2$

$$s = \sqrt{\dfrac{(821 - 823._2)^2 + (783 - 823._2)^2 + (834 - 823._2)^2 + (855 - 823._2)^2}{4 - 1}}$$

$$= 30._3$$
  relative standard deviation $= (30._3/823._2) \times 100 = 3.6_8\%$
  median $= (821 + 834)/2 = 827._5$
  range $= 855 - 783 = 72$
4-B. **(a)** $\bar{x} = \dfrac{117 + 119 + 111 + 115 + 120}{5} = 116._4$

$$s = \sqrt{\dfrac{(117 - 116._4)^2 + (119 - 116._4)^2 + \ldots + (120 - 116._4)^2}{5 - 1}}$$

$$= 3._{58}$$
  (b) $s_{\text{pooled}} = \sqrt{\dfrac{2.8^2(4 - 1) + 3.58^2(5 - 1)}{4 + 5 - 2}} = 3.27$

$$t_{\text{calculated}} = \dfrac{|111.0 - 116.4|}{3.27} \sqrt{\dfrac{4 \cdot 5}{4 + 5}} = 2.46$$
  $t_{\text{table}} = 2.365$ for 95% confidence and $4 + 5 - 2 = 7$ degrees of freedom.
  $t_{\text{calculated}} > t_{\text{table}}$, so the difference is significant at the 95% confidence level.
4-C. $Q = (216 - 204)/(216 - 192) = 0.50 < 0.64$. Retain 216.
4-D.

x_i	y_i	x_iy_i	x_i^2	$d_i(= y_i - mx_i - b)$	d_i^2
1	3	3	1	-0.167	0.027 89
3	2	6	9	0.333	0.110 89
5	0	0	25	-0.167	0.027 89
$\Sigma x_i = 9$	$\Sigma y_i = 5$	$\Sigma x_iy_i = 9$	$\Sigma(x_i^2) = 35$		$\Sigma(d_i^2) = 0.166\ 67$

$D = n\ \Sigma(x_i^2) - (\Sigma x_i)^2 = 3 \cdot 35 - 9^2 = 24$
$m = [n\ \Sigma x_iy_i - \Sigma x_i\Sigma y_i]/D = [3 \cdot 9 - 9 \cdot 5]/24 = -0.750$
$b = [\Sigma(x_i^2)\Sigma y_i - \Sigma(x_iy_i)\Sigma x_i]/D = [35 \cdot 5 - 9 \cdot 9]/24 = 3.917$

$$\sigma_y = \sqrt{\dfrac{\Sigma(d_i^2)}{3 - 2}} = \sqrt{\dfrac{0.166\ 67}{1}} = 0.408\ 2$$

$$\sigma_m = \sigma_y\sqrt{\dfrac{n}{D}} = 0.408\ 2\sqrt{\dfrac{3}{24}} = 0.144$$

$$\sigma_b = \sigma_y\sqrt{\dfrac{\Sigma(x_i^2)}{D}} = 0.408\ 2\sqrt{\dfrac{35}{24}} = 0.493$$

$y\ (\pm 0.4_{08}) = -0.75_0\ (\pm 0.14_4)x + 3.9_{17}\ (\pm 0.4_{93})$

4-E. $x = \dfrac{y - b}{m} = \dfrac{1.00 - 3.9_{17}}{-0.750} = 3.89$

Uncertainty in $x = \dfrac{\sigma_y}{|m|}\sqrt{1 + \dfrac{x^2n}{D} + \dfrac{\Sigma(x_i^2)}{D} - \dfrac{2x\Sigma x_i}{D}}$

$$= \dfrac{0.408}{0.75}\sqrt{1 + \dfrac{(3.89)^2(3)}{24} + \dfrac{35}{24} - \dfrac{2\ (3.89)(9)}{24}}$$

$$= 0.65$$

Chapter 5

5-A. **(a)** and **(b)**

$\begin{array}{lr}\text{HOBr} + \text{OCl}^- \rightleftharpoons \text{HOCl} + \text{OBr}^- & K_1 = 1/15 \\ \text{HOCl} \rightleftharpoons \text{H}^+ + \text{OCl}^- & K_2 = 3.0 \times 10^{-8} \\ \hline \text{HOBr} \rightleftharpoons \text{H}^+ + \text{OBr}^- & K = K_1K_2 = 2 \times 10^{-9}\end{array}$

  (c) Consumption of a product drives the reaction forward (to the right).
5-B. **(a)** $\text{PbBr}_2(s) \overset{K_{sp}}{\rightleftharpoons} \text{Pb}^{2+} + 2\text{Br}^-$
     FW 367.0 x $2x$
  $x(2x)^2 = 2.1 \times 10^{-6} \Rightarrow x = 8.0_7 \times 10^{-3} \text{ M} = 2.9_6 \text{ g/L}$
$$= 1.4_8 \text{ g/0.500 L}$$
  (b) $[\text{Pb}^{2+}](0.063\ 4)^2 = 2.1 \times 10^{-6} \Rightarrow$
  $x = 5.2_2 \times 10^{-4} \text{ M} = 0.19_2 \text{ g/L} = 0.09_6 \text{ g/0.500 L}$
5-C. neutralize . . . conjugate
5-D. **(a)** $[\text{Mg}^{2+}][\text{OH}^-]^2 = K_{sp} = 7.1 \times 10^{-12} = (0.050)[\text{OH}^-]^2$
     0.050 M   ? $\Rightarrow [\text{OH}^-] = 1.1_9 \times 10^{-5} \text{ M}$
  (b) $[\text{H}^+] = K_w/[\text{OH}^-] = 8.3_9 \times 10^{-10} \text{ M}$
$$\Rightarrow \text{pH} = -\log [\text{H}^+] = 9.08$$
5-E. The stronger acid is (A), which has the larger value of K_a.

$$\text{Cl}_2\text{CHCO}_2\text{H} \overset{K_a}{\rightleftharpoons} \text{Cl}_2\text{CHCO}_2^- + \text{H}^+$$
$$\text{ClCH}_2\text{CO}_2\text{H} \overset{K_a}{\rightleftharpoons} \text{ClCH}_2\text{CO}_2^- + \text{H}^+$$

The stronger base is (C), which has the larger value of K_b.

$$\text{H}_2\text{NNH}_2 + \text{H}_2\text{O} \overset{K_b}{\rightleftharpoons} \text{H}_2\text{NNH}_3^+ + \text{OH}^-$$
$$\text{H}_2\text{NCONH}_2 + \text{H}_2\text{O} \overset{K_b}{\rightleftharpoons} \text{H}_2\text{NCONH}_3^+ + \text{OH}^-$$

Chapter 6

6-A. **(a)** The concentrations of reagents used in an analysis are determined either by weighing out pure primary standards or by reaction with such standards. If the standards are not pure, none of the concentrations will be correct.

(b) In a blank titration, the quantity of titrant required to reach the end point in the absence of analyte is measured. This quantity is subtracted from the quantity of titrant needed in the presence of analyte.

(c) In a direct titration, titrant reacts directly with analyte. In a back titration, a known excess of reagent that reacts with analyte is used. The excess is then titrated.

6-B. **(a)** $0.197\ 0$ g of ascorbic acid $= 1.118\ 5 \times 10^{-3}$ mol, which requires $1.118\ 5 \times 10^{-3}$ mol I_3^-. $[I_3^-] = 1.118\ 5 \times 10^{-3}$ mol$/0.029\ 41$ L $= 0.038\ 03$ M

(b) $(0.031\ 63$ L $I_3^-)(0.038\ 03$ M$) = 1.203 \times 10^{-3}$ mol I_3^-
$= 1.203 \times 10^{-3}$ mol ascorbic acid

(c) $(1.203 \times 10^{-3}$ mol ascorbic acid$)(176.126$ g/mol$) =$
$0.211\ 9$ g ascorbic acid

$$\frac{0.211\ 9 \text{ g ascorbic acid}}{0.424\ 2 \text{ g tablet}} \times 100 = 49.94 \text{ wt \%}$$

6-C. **(a)**
$Ag^+ + Br^- \rightleftharpoons AgBr(s)$ $K = 1/K_{sp}(AgBr) = 2.0 \times 10^{12}$
$Ag^+ + Cl^- \rightleftharpoons AgCl(s)$ $K = 1/K_{sp}(AgCl) = 5.6 \times 10^9$
AgBr precipitates first.

(b) mol $Br^- =$ mol Ag^+ at first end point
$= (0.015\ 55$ L$)(0.033\ 33$ M$)$
$= 5.183 \times 10^{-4}$ mol
$[Br^-] = (5.183 \times 10^{-4}$ mol$)/(0.025\ 00$ L$)$
$= 0.020\ 73$ M

(c) mol $Cl^- =$ mol Ag^+ at second end point $-$ mol Ag^+
at first end point
$= (0.042\ 23 - 0.015\ 55$ L$)(0.033\ 33$ M$)$
$= 8.892 \times 10^{-4}$ mol
$[Cl^-] = (8.892 \times 10^{-4}$ mol$)/(0.025\ 00$ L$)$
$= 0.035\ 57$ M

6-D. **(a)** total volume $KMnO_4 = 50.00 + 9.68 - 0.07$
$= 59.61$ mL
total mol $KMnO_4 = (0.009\ 22$ M$)(0.059\ 61$ L$)$
$= 5.496 \times 10^{-4}$ mol

(b) volume $Na_2C_2O_4 = 30.00$ mL
mol $Na_2C_2O_4 = (0.025\ 00$ M$)(0.030\ 00$ L$)$
$= 7.500 \times 10^{-4}$ mol

(c) mol $KMnO_4$ reacting with $Na_2C_2O_4$
$= (7.500 \times 10^{-4}$ mol $Na_2C_2O_4)\left(\dfrac{2 \text{ mol } KMnO_4}{5 \text{ mol } Na_2C_2O_4}\right)$
$= 3.000 \times 10^{-4}$ mol

(d) $KMnO_4$ reacting with $NaNO_2$
$=$ (total mol $KMnO_4 - KMnO_4$ reacting with $Na_2C_2O_4$)
$= 5.496 \times 10^{-4} - 3.00 \times 10^{-4} = 2.496 \times 10^{-4}$ mol

(e) mol $NaNO_2$ reacting with $KMnO_4$
$= (2.496 \times 10^{-4}$ mol $KMnO_4)\left(\dfrac{5 \text{ mol } NaNO_2}{2 \text{ mol } KMnO_4}\right)$
$= 6.240 \times 10^{-4}$ mol

(f) $[NaNO_2] = \dfrac{6.24 \times 10^{-4} \text{ mol}}{0.050\ 00 \text{ L}} = 0.012\ 48$ M

6-E. **(a)** Some $Ag_2CrO_4(s)$ must be present before red color is apparent. The blank titration measures the volume of Ag^+ that must be added for us to see the red color of $Ag_2CrO_4(s)$. This volume is subtracted from subsequent titrations.

(b) AgCl is more soluble than AgSCN and will slowly dissolve in the presence of excess SCN^-. This consumes the SCN^- and causes the red end point color to fade. Nitrobenzene coats the AgCl and inhibits reaction with SCN^-.

(c) Consider the titration of C^+ (in a flask) by A^- (from a buret). Before the equivalence point, there is excess C^+ in solution. Selective adsorption of C^+ on the CA crystal surface gives the crystal a positive charge. After the equivalence point there is excess A^- in solution. Selective adsorption of A^- on the CA crystal surface gives it a negative charge.

(d) Beyond the equivalence point, there is excess $Fe(CN)_6^{4-}$ in solution. Adsorption of this anion on the precipitate will make the particles *negative*.

Chapter 7

7-A. **(a)** $\nu = c/\lambda = (2.998 \times 10^8 \text{ m/s})/(100 \times 10^{-9} \text{ m})$
$= 2.998 \times 10^{15} \text{ s}^{-1} = 2.998 \times 10^{15} \text{ Hz}$
$\tilde{\nu} = 1/\lambda = 1/(100 \times 10^{-9} \text{ m}) = 10^7 \text{ m}^{-1}$
$(10^7 \text{ m}^{-1})\left(\dfrac{1 \text{ m}}{100 \text{ cm}}\right) = 10^5 \text{ cm}^{-1}$
$E = h\nu = (6.626\ 2 \times 10^{-34} \text{ J s})(2.998 \times 10^{15} \text{ s}^{-1})$
$= 1.986 \times 10^{-18} \text{ J}$
$(1.986 \times 10^{-18} \text{ J/photon})(6.022 \times 10^{23} \text{ photons/}$
$\text{mol}) = 1\ 196 \text{ kJ/mol}$

(b), (c), (d) These are done in an analogous manner. The final results for all four parts are

λ	ν (Hz)	$\tilde{\nu}$ (cm^{-1})	E (kJ/mol)	Spectral region	Molecular process
100 nm	2.998×10^{15}	10^5	1.196×10^3	ultraviolet	electronic excitation
500 nm	5.996×10^{14}	2×10^4	239.3	visible	electronic excitation
10 μm	2.998×10^{13}	$1\ 000$	11.96	infrared	molecular vibration
1 cm	2.998×10^{10}	1	$0.011\ 96$	microwave	molecular rotation

7-B. **(a)** $A = \epsilon bc$

$= (1.05 \times 10^3 \text{ M}^{-1} \text{ cm}^{-1})(1.00 \text{ cm})(2.33 \times 10^{-4} \text{ M})$

$= 0.245$

(b) $T = 10^{-A} = 0.569 = 56.9\%$

(c) Doubling b will double $A \Rightarrow A = 0.489 \Rightarrow T = 10^{-A} = 32.4\%$

(d) Doubling c also doubles $A \Rightarrow A = 0.489 \Rightarrow T = 10^{-A} = 32.4\%$

(e) $A = \epsilon bc$

$= (2.10 \times 10^3 \text{ M}^{-1} \text{ cm}^{-1})(1.00 \text{ cm})(2.33 \times 10^{-4} \text{ M})$

$= 0.489$

7-C. **(a)** The single-beam instrument measures light transmitted through one sample at a time. A double-beam instrument uses a beam chopper to direct light alternately through the sample and reference cuvets.

(b) If absorbance is too high, too little light reaches the detector for accurate measurement. If absorbance is too low, there is too little difference between sample and reference for accurate measurement.

(c) Fingerprints scatter light, which appears to be absorption to the spectrophotometer.

Chapter 8

8-A. **(a)** $\epsilon = \dfrac{A}{cb} = \dfrac{0.267 - 0.019}{(3.15 \times 10^{-6} \text{ M})(1.000 \text{ cm})}$

$= 7.87 \times 10^4 \text{ M}^{-1} \text{ cm}^{-1}$

(b) $c = \dfrac{A}{\epsilon b} = \dfrac{0.175 - 0.019}{(7.87 \times 10^4 \text{ M}^{-1} \text{ cm}^{-1})(1.000 \text{ cm})}$

$= 1.98 \times 10^{-6} \text{ M}$

8-B. **(a)** 163×10^{-6} L of 1.43×10^{-3} M Fe(III)

$= 2.33 \times 10^{-7}$ mol Fe(III)

(b) 1.17×10^{-7} mol transferrin in 2.00×10^{-3} L

$\Rightarrow 5.83 \times 10^{-5}$ M transferrin

(c) Prior to the equivalence point, all added Fe(III) binds to the protein to form a red complex whose absorbance is shown in the figure. After the equivalence point, no more protein-binding sites are available. The slight increase in absorbance arises from the color of the iron reagent in the titrant.

8-C. The calibration graph is shown below. The least-squares equation of the straight line is $y = (6.142 \times 10^{-3})x +$

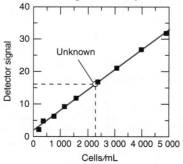

1.804, where y is detector signal and x is cells/mL. For the unknown signal $y = 15.9$, we solve the equation $15.9 = (6.142 \times 10^{-3})x + 1.804$ to find $x = 2\,295$ cells/mL.

8-D.

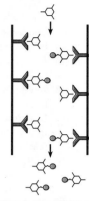

Step 1: Column is loaded with labeled TNT and excess is washed away. There is no fluorescence in liquid leaving the column.

Step 3: Some unlabeled TNT analyte displaces labeled TNT. Fluorescence is now seen in liquid leaving the column.

Chapter 9

9-A. **(a)** **(i)** pH $= -\log (1.0 \times 10^{-3}) = 3.00$

(ii) pH $= -\log (1.0 \times 10^{-14}/1.0 \times 10^{-2}) = 12.00$

(b) **(i)** pH $= -\log (3.2 \times 10^{-5}) = 4.49$

(ii) pH $= -\log (1.0 \times 10^{-14}/0.007\,7) = 11.89$

(c) $[\text{H}^+] = 10^{-4.44} = 3.6 \times 10^{-5}$ M

(d) $[\text{H}^+] = 1.0 \times 10^{-14}/0.007\,7 = 1.3 \times 10^{-12}$ M

It is derived from $\text{H}_2\text{O} \rightleftharpoons \text{H}^+ + \text{OH}^-$

(e) The concentration of the strong base is too low to perturb the pH of pure water. The pH is very close to 7.00.

9-B. **(a)** $\text{HCO}_2\text{H} \rightleftharpoons \text{HCO}_2^- + \text{H}^+$

(b) HCO_2^-

(c) $K_a = \dfrac{[\text{HCO}_2^-][\text{H}^+]}{[\text{HCO}_2\text{H}]} = 1.80 \times 10^{-4}$

(d) $K_b = \dfrac{[\text{HCO}_2\text{H}][\text{OH}^-]}{[\text{HCO}_2^-]}$

(e) $K_b = \dfrac{K_w}{K_a} = 5.6 \times 10^{-11}$

9-C. **(a)** Let $x = [\text{H}^+] = [\text{A}^-]$ and $0.100 - x = [\text{HA}]$.

$\dfrac{x^2}{0.100 - x} = 1.00 \times 10^{-5} \Rightarrow x = 9.95 \times 10^{-4}$ M

$\Rightarrow$ pH $= -\log x = 3.00$

$\dfrac{[\text{A}^-]}{[\text{A}^-] + [\text{HA}]} = \dfrac{9.95 \times 10^{-4}}{0.100} = 9.95 \times 10^{-3}$

(b) $\quad\text{HA} \quad\rightleftharpoons\quad \text{H}^+ \quad+\quad \text{A}^-$

$\quad 0.045\,0 - 10^{-2.78} \quad 10^{-2.78} \quad 10^{-2.78}$

$K_a = \dfrac{(10^{-2.78})^2}{0.045\,0 - 10^{-2.78}} = 6.4 \times 10^{-5} \Rightarrow \text{p}K_a = 4.19$

(c) $HA \rightleftharpoons H^+ + A^-$

$$\underset{F-x}{} \quad \underset{x}{} \quad \underset{x}{}$$

$$\frac{[A^-]}{[HA] + [A^-]} = \frac{x}{F - x + x} = 0.006\,0$$

With $F = 0.045\,0$ M, $x = 2.7 \times 10^{-4}$ M

$$\Rightarrow K_a = \frac{x^2}{F - x} = 1.6 \times 10^{-6} \Rightarrow pK_a = 5.79$$

9-D. (a) Let $x = [OH^-] = [BH^+]$ and $0.100 - x = [B]$.

$$\frac{x^2}{0.100 - x} = 1.00 \times 10^{-5}$$

$$\Rightarrow x = 9.95 \times 10^{-4}\ M$$

$$\Rightarrow [H^+] = \frac{K_w}{x} = 1.005 \times 10^{-11} \Rightarrow pH = 11.00$$

$$\frac{[BH^+]}{[B] + [BH^+]} = \frac{9.95 \times 10^{-4}}{0.100} = 9.95 \times 10^{-3}$$

(b)

$$\underset{0.10 - (K_w/10^{-9.28})}{B} + H_2O \rightleftharpoons \underset{K_w/10^{-9.28}}{BH^+} + \underset{K_w/10^{-9.28}}{OH^-}$$

$$K_b = \frac{(K_w/10^{-9.28})^2}{0.10 - (K_w/10^{-9.28})} = 3.6 \times 10^{-9}$$

(c)

$$\underset{0.10 - x}{B} + H_2O \overset{K_b}{\longrightarrow} \underset{x}{BH^+} + \underset{x}{OH^-}$$

$$\frac{[BH^+]}{[BH^+] + [B]} = 0.020$$

$$\frac{x}{0.10 - x + x} = 0.020 \Rightarrow x = 0.002\,0$$

$$K_b = \frac{x^2}{0.10 - x} = \frac{(0.002\,0)^2}{0.10 - 0.002\,0} = 4.1 \times 10^{-5}$$

9-E. (a) At pH 2.00: $\alpha_{HA} = \dfrac{10^{-2.00}}{10^{-2.00} + 10^{-3.00}} = 0.90_9$

$$\alpha_{A^-} = \frac{10^{-3.00}}{10^{-2.00} + 10^{-3.00}} = 0.090_9$$

$$\frac{[HA]}{[A^-]} = \frac{0.90_9}{0.090_9} = 10._0$$

(b), (c)

	α_{HA}	α_{A^-}	$[HA]/[A^-]$
pH = 3.00	0.50_0	0.50_0	1.0_0
pH = 4.00	0.090_9	0.90_9	0.10_0

Chapter 10

10-A. (a) $pH = pK_a + \log\left(\dfrac{[A^-]}{[HA]}\right) = 5.00 + \log\left(\dfrac{0.050}{0.100}\right) = 4.70$

(b) $pH = 3.745 + \log\left(\dfrac{[HCO_2^-]}{[HCO_2H]}\right)$

pH:	3.000	3.745	4.000
$[HCO_2^-]/[HCO_2H]$:	0.180	1.00	1.80

10-B. (a) $pH = pK_a + \log\left(\dfrac{[tris]}{[trisH^+]}\right)$

$$= 8.075 + \log\left(\frac{(10.0\ g)/(121.136\ g/mol)}{(10.0\ g)/(157.597\ g/mol)}\right)$$

$$= 8.19$$

(b)

	tris	+	H^+	$\rightarrow$	trisH$^+$	+	H_2O
Initial mmol:	82.55		5.25		63.45		
Final mmol:	77.30		—		68.70		

$$pH = 8.075 + \log\left(\frac{77.30}{68.70}\right) = 8.13$$

(c)

	tris H$^+$	+	OH^-	$\rightarrow$	tris	+	H_2O
Initial mmol:	63.45		5.25		82.55		
Final mmol:	58.20		—		87.80		

$$pH = 8.075 + \log\left(\frac{87.80}{58.20}\right) = 8.25$$

10-C. (a) mol tris $= 10.0/g(121.136\ g/mol) = 0.082\,55$ mol

$$= 82.55\ mmol$$

	tris	+	H^+	$\rightarrow$	trisH$^+$	+	H_2O
Initial mmol:	82.5_5		x				—
Final mmol:	$82.5_5 - x$		—		x		

$$pH = pK_a + \log\left(\frac{[tris]}{[trisH^+]}\right)$$

$$7.60 = 8.075 + \log\left(\frac{82.5_5 - x}{x}\right)$$

$$-0.475 = \log\left(\frac{82.5_5 - x}{x}\right)$$

$$\Rightarrow 10^{-0.475} = 10^{\log[(82.5_5 - x)/x]}$$

$$0.335_0 = \frac{82.5_5 - x}{x} \Rightarrow x = 61.8_4\ mmol$$

volume of HCl required

$$= (0.061\,8_4\ mol)/(1.20\ M) = 51.5\ mL$$

(b) I would weigh out 0.020 0 mol of acetic acid (= 1.201 g) and place it in a beaker with ~75 mL of water. While monitoring the pH with a pH electrode, I would add 3 M NaOH (~4 mL is required) until the pH is exactly 5.00. I would then pour the solution into a 100-mL volumetric flask and wash the beaker several times with a few milliliters of distilled water. Each washing would be added to the volumetric flask, to ensure quantitative transfer from the beaker to the flask. After swirling the volumetric flask to mix the solution, I would carefully add water up to the 100-mL mark, insert the cap, and invert 20 times to ensure complete mixing.

10-D. (a)

Compound	pK_a
Hydroxybenzene	9.98
Propanoic acid	4.87
Cyanoacetic acid	2.47 ← Most suitable, because pK_a is closest to pH
Sulfuric acid	1.99

(b) Buffer capacity is based on the ability of buffer to react with added acid or base, without making a large change in the ratio of concentrations $[A^-]/[HA]$. The greater the concentration of each component, the less relative change is brought about by reaction with a small increment of added acid or base.

(c)

Compound	pK_a (for conjugate acid)
Ammonia	9.24 ← Most suitable, because pK_a is closest to pH
Aniline	4.60
Hydrazine	8.48
Pyridine	5.23

10-E. **(a)** The quotient $[HIn]/[In^-]$ changes from 10:1 when $pH = pK_{HIn} - 1$ to 1:10 when $pH = pK_{HIn} + 1$. This change is generally sufficient to cause a complete color change.

(b) red, orange, yellow

Chapter 11

11-A. The titration reaction is $H^+ + OH^- \rightarrow H_2O$ and $V_e = 5.00$ mL because

$$\underbrace{(V_e(\text{mL}))(0.100 \text{ M})}_{\text{mmol HCl at } V_e} = \underbrace{(50.00 \text{ mL})(0.010\,0 \text{ M})}_{\text{Initial mmol NaOH}} \Rightarrow V_e = 5.00 \text{ mL}$$

Representative calculations:

$V_a = 1.00$ mL:

$$OH^- = \underbrace{(50.00 \text{ mL})(0.010\,0 \text{ M})}_{\text{Initial mmol NaOH}} -$$

$$\underbrace{(1.00 \text{ mL})(0.100 \text{ M})}_{\text{Added mmol HCl}} = 0.400 \text{ mmol}$$

$[OH^-] = (0.400 \text{ mmol})/(51.00 \text{ mL}) = 0.007\,84$ M
$[H^+] = K_w/(0.007\,84 \text{ M}) = 1.28 \times 10^{-12}$ M
$pH = -\log(1.28 \times 10^{-12}) = 11.89$

$V_a = V_e = 5.00$ mL:

$$H_2O \rightleftharpoons \underset{x}{H^+} + \underset{x}{OH^-}$$

$K_w = x^2 \Rightarrow x = 1.00 \times 10^{-7}$ M $\Rightarrow pH = 7.00$

$V_a = 5.01$ mL:

excess $H^+ = (0.01 \text{ mL})(0.100 \text{ M}) = 0.001$ mmol
$[H^+] = (0.001 \text{ mmol})/(55.01 \text{ mL}) = 1._8 \times 10^{-5}$ M
$pH = -\log(1._8 \times 10^{-5}) = 4.74$

V_a (mL)	pH	V_a (mL)	pH	V_a (mL)	pH
0.00	12.00	4.50	10.96	5.10	3.74
1.00	11.89	4.90	10.26	5.50	3.05
2.00	11.76	4.99	9.26	6.00	2.75
3.00	11.58	5.00	7.00	8.00	2.29
4.00	11.27	5.01	4.74	10.00	2.08

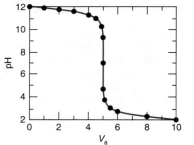

11-B. The titration reaction is $HCO_2H + OH^- \rightarrow HCO_2^- + H_2O$

$$\underbrace{(V_e(\text{mL}))(0.050\,0 \text{ M})}_{\text{mmol KOH at } V_e} = \underbrace{(50.0 \text{ mL})(0.050\,0 \text{ M})}_{\text{Initial mmol } HCO_2H}$$

$$\Rightarrow V_e = 50.0 \text{ mL}$$

Representative calculations:

$V_b = 0$ mL:

$$HA \rightleftharpoons H^+ + A^-$$
$$0.050\,0 - x \qquad x \qquad x$$

$K_a = 1.80 \times 10^{-4} \quad pK_a = 3.745$

$$\frac{x^2}{0.050\,0 - x} = K_a \Rightarrow x = 2.91 \times 10^{-3}$$

$$\Rightarrow pH = 2.54$$

$V_b = 48.0$ mL:

$$HA + OH^- \rightarrow A^- + H_2O$$

Initial mmol:	2.50	2.40	—
Final mmol:	0.10	—	2.40

$$pH = pK_a + \log\left(\frac{[A^-]}{[HA]}\right) = 3.745 + \log\left(\frac{2.40}{0.10}\right)$$
$$= 5.13$$

$V_b = 50.0$ mL: formal conc. of A^-

$$= \frac{(50.0 \text{ mL})(0.050\,0 \text{ M})}{100.0 \text{ mL}} = 0.025\,0 \text{ M}$$

$$A^- + H_2O \rightleftharpoons HA + OH^-$$
$$0.025\,0 - x \qquad x \qquad x$$

$$K_b = K_w/K_a = 5.56 \times 10^{-11}$$

$$\frac{x^2}{0.025\,0 - x} = K_b \Rightarrow x = 1.18 \times 10^{-6}$$

$$\Rightarrow pH = -\log\left(\frac{K_w}{x}\right) = 8.07$$

$V_b = 60.0$ mL: excess $[OH^-]$

$$= \frac{(10.0 \text{ mL})(0.050\,0 \text{ M})}{110.0 \text{ mL}} = 4.55 \times 10^{-3} \text{ M}$$

$$\Rightarrow pH = 11.66$$

V_b (mL)	pH	V_b (mL)	pH	V_b (mL)	pH
0.0	2.54	45.0	4.70	50.5	10.40
10.0	3.14	48.0	5.13	51.0	10.69
20.0	3.57	49.0	5.44	52.0	10.99
25.0	3.74	49.5	5.74	55.0	11.38
30.0	3.92	50.0	8.07	60.0	11.66
40.0	4.35				

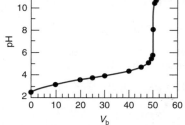

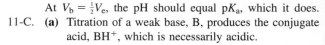

At $V_b = \frac{1}{2}V_e$, the pH should equal pK_a, which it does.

11-C. **(a)** Titration of a weak base, B, produces the conjugate acid, BH^+, which is necessarily acidic.

(b) The titration reaction is $B + H^+ \rightarrow BH^+$.

Find V_e: $(V_e)(0.200 \text{ M}) = (100.0 \text{ mL})(0.100 \text{ M})$

$\Rightarrow V_e = 50.0 \text{ mL}$

Representative calculations:

$V_a = 0.0 \text{ mL}$:

$$\begin{array}{ccccc} B & + H_2O & \rightleftharpoons & BH^+ & + OH^- \\ 0.100 - x & & & x & x \end{array}$$

$$K_b = 2.6 \times 10^{-6}$$

$$\frac{x^2}{0.100 - x} = K_b = 2.6 \times 10^{-6}$$

$$\Rightarrow x = 5.09 \times 10^{-4}$$

$$pH = -\log\left(\frac{K_w}{x}\right) = 10.71$$

$V_a = 20.0 \text{ mL}$:

$$\begin{array}{ccccc} B & + H^+ & \rightarrow & BH^+ \end{array}$$

Initial mmol:	10.00	4.00	—
Final mmol:	6.00	—	4.00

$$pH = \underset{\underset{K_a = K_w/K_b}{\uparrow}}{pK_a} \text{ (for } BH^+) + \log\left(\frac{[B]}{[BH^+]}\right)$$

$$= 8.41 + \log\left(\frac{6.00}{4.00}\right) = 8.59$$

$V_a = V_e = 50 \text{ mL}$: All B has been converted to the conjugate acid, BH^+. The formal concentration of BH^+ is

$$F' = \frac{(100.0 \text{ mL})(0.100 \text{ M})}{150.0 \text{ mL}} = 0.066 \, 7 \text{ M}$$

The pH is determined by the reaction

$$\begin{array}{ccccc} BH^+ & \rightleftharpoons & B & + H^+ \\ 0.066 \, 7 - x & & x & x \end{array}$$

$$\frac{x^2}{0.066 \, 7 - x} = K_a = \frac{K_w}{K_b} \Rightarrow x = 1.60 \times 10^{-5}$$

$$\Rightarrow pH = 4.80$$

$V_a = 51.0 \text{ mL}$: There is excess $[H^+]$.

$$\text{excess } [H^+] = \frac{(1.0 \text{ mL})(0.200 \text{ M})}{(151.0 \text{ mL})}$$

$$= 1.32 \times 10^{-3} \Rightarrow pH = 2.88$$

V_a (mL)	pH	V_a	pH	V_a	pH
0.0	10.71	30.0	8.23	50.0	4.80
10.0	9.01	40.0	7.81	50.1	3.88
20.0	8.59	49.0	6.72	51.0	2.88
25.0	8.41	49.9	5.71	60.0	1.90

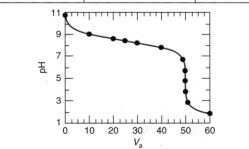

11-D. **(a)** Figure 11-1: bromothymol blue: blue → yellow

Figure 11-2: thymol blue: yellow → blue

Figure 11-10 ($pK_a = 8$): thymolphthalein:

colorless → blue

(b) The derivatives are shown in the spreadsheet below. In the first derivative graph, the maximum is near 119 mL. In Figure 11-5, the second derivative graph gives an end point of 118.9 μL.

	A	B	C	D	E	F
1			First Derivative		Second Derivative	
2	μL NaOH	pH	μL	Derivative	μL	Derivative
3	107	6.921				
4	110	7.117	108.5	6.533E-02		
5	113	7.359	111.5	8.067E-02	110	5.11E-03
6	114	7.457	113.5	9.800E-02	112.5	8.67E-03
7	115	7.569	114.5	1.120E-01	114	1.40E-02
8	116	7.705	115.5	1.360E-01	115	2.40E-02
9	117	7.878	116.5	1.730E-01	116	3.70E-02
10	118	8.090	117.5	2.120E-01	117	3.90E-02
11	119	8.343	118.5	2.530E-01	118	4.10E-02
12	120	8.591	119.5	2.480E-01	119	-5.00E-03
13	121	8.794	120.5	2.030E-01	120	-4.50E-02
14	122	8.952	121.5	1.580E-01	121	-4.50E-02
15						
16	C4 = (A4+A3)/2				E5 = (C5+C4)/2	
17	D4 = (B4-B3)/(A4-A3)				F5 = (D5-D4)/(C5-C4)	

11-E. **(a)** Tris(hydroxymethyl)aminomethane ($H_2NC(CH_2OH)_3$), sodium carbonate (Na_2CO_3), or borax ($Na_2B_4O_7 \cdot 10H_2O$) can be used to standardize HCl. Potassium hydrogen phthalate ($HO_2C-C_6H_4-CO_2^-K^+$) or potassium hydrogen iodate ($KH(IO_3)_2$) can be used to standardize NaOH.

(b) 30 mL of 0.05 M $OH^- = 1.5$ mmol $OH^- = 1.5$ mmol potassium hydrogen phthalate. $(1.5 \times 10^{-3}$ mol)(204.233 g/mol) = 0.30 g of potassium hydrogen phthalate.

11-F. **(a)** 5.00 mL of 0.033 6 M HCl = 0.168_0 mmol

6.34 mL of 0.010 0 M NaOH = 0.063 4 mmol

HCl consumed by $NH_3 = 0.168_0 - 0.063 \, 4$

$= 0.104_6$ mmol

mol NH_3 = mol HCl = 0.104_6 mmol

(b) mol nitrogen = mol $NH_3 = 0.104_6$ mmol

$(0.104_6 \times 10^{-3}$ mol N)(14.006 7 g/mol) = 1.46_5 mg of nitrogen

(c) $(256 \text{ μL})\left(\dfrac{1 \text{ mL}}{1 \, 000 \text{ μL}}\right)(37.9 \text{ mg protein/mL}) =$

9.70_2 mg protein

(d) wt % N = $\dfrac{1.46_5 \text{ mg N}}{9.70_2 \text{ mg protein}} \times 100 = 15.1$ wt %

11-G. **(a)** Your spreadsheet should reproduce the results in Figure 11-10.

(b)

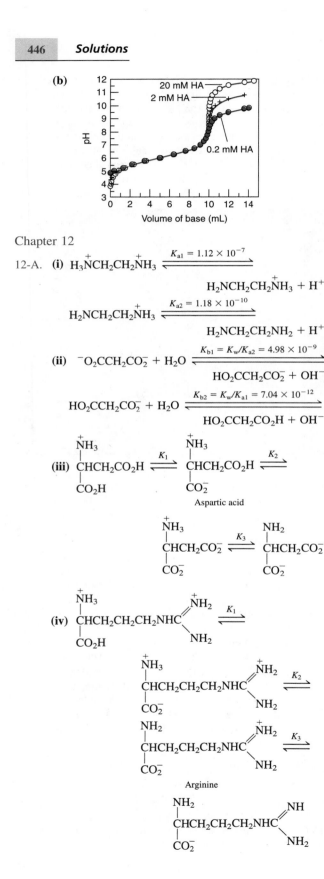

Chapter 12

12-A. **(i)** $H_3\overset{+}{N}CH_2CH_2\overset{+}{N}H_3 \xrightleftharpoons{K_{a1} = 1.12 \times 10^{-7}}$

$$H_2NCH_2CH_2\overset{+}{N}H_3 + H^+$$

$H_2NCH_2CH_2\overset{+}{N}H_3 \xrightleftharpoons{K_{a2} = 1.18 \times 10^{-10}}$

$$H_2NCH_2CH_2NH_2 + H^+$$

(ii) $^-O_2CCH_2CO_2^- + H_2O \xrightleftharpoons{K_{b1} = K_w/K_{a2} = 4.98 \times 10^{-9}}$

$$HO_2CCH_2CO_2^- + OH^-$$

$HO_2CCH_2CO_2^- + H_2O \xrightleftharpoons{K_{b2} = K_w/K_{a1} = 7.04 \times 10^{-12}}$

$$HO_2CCH_2CO_2H + OH^-$$

(iii)
Aspartic acid

(iv)
Arginine

12-B. **(a)** $H_2SO_3 \rightleftharpoons HSO_3^- + H^+$

   $ 0.050 - x \quad x \quad x$

   $\dfrac{x^2}{0.050 - x} = K_1 = 1.23 \times 10^{-2} \Rightarrow x = 1.94 \times 10^{-2}$

   $[HSO_3^-] = [H^+] = 1.94 \times 10^{-2}$ M $\Rightarrow$ pH = 1.71

   $[H_2SO_3] = 0.050 - x = 0.031$ M

   $[SO_3^{2-}] = \dfrac{K_2[HSO_3^-]}{[H^+]} = K_2 = 6.6 \times 10^{-8}$ M

(b) pH $\approx \frac{1}{2}(pK_1 + pK_2) = \frac{1}{2}(1.91 + 7.18) = 4.54$

   $[H^+] = 10^{-pH} = 2.9 \times 10^{-5}$ M; $[HSO_3^-] \approx 0.050$ M

   $[H_2SO_3] = \dfrac{[H^+][HSO_3^-]}{K_1} = \dfrac{(2.9 \times 10^{-5})(0.050)}{1.23 \times 10^{-2}}$

   $= 1.2 \times 10^{-4}$ M

   $[SO_3^{2-}] = \dfrac{K_2[HSO_3^-]}{[H^+]}$

   $= \dfrac{(6.6 \times 10^{-8})(0.050)}{2.9 \times 10^{-5}} = 1.1 \times 10^{-4}$ M

(c) $SO_3^{2-} + H_2O \rightleftharpoons HSO_3^- + OH^-$

   $\phantom{SO_3^{2-}} 0.050 - x \quad\quad x \quad x$

   $\dfrac{x^2}{0.050 - x} = K_{b1} = \dfrac{K_w}{K_{a2}} = 1.52 \times 10^{-7}$

   $\Rightarrow x = 8.7 \times 10^{-5}$

   $[HSO_3^-] = 8.7 \times 10^{-5}$ M; $[H^+] = \dfrac{K_w}{x} =$

   1.15×10^{-10} M $\Rightarrow$ pH = 9.94

   $[SO_3^{2-}] = 0.050 - x = 0.050$ M

   $[H_2SO_3] = \dfrac{[H^+][HSO_3^-]}{K_1} = 8.1 \times 10^{-13}$ M

12-C. **(a)**

	pH 9.00	pH 11.00
Principal species:	OH / OH	O⁻ / OH
Secondary species:	O⁻ / OH	O⁻ / O⁻

(b) HC^-: pH $\approx \frac{1}{2}(pK_2 + pK_3) = \frac{1}{2}(8.36 + 10.77) = 9.56$

12-D. **(a)** $(50.0 \text{ mL})(0.050\,0 \text{ M}) = V_{e1}(0.100 \text{ M})$

   $\Rightarrow V_{e1} = 25.0$ mL

   $V_{e2} = 2V_{e1} = 50.0$ mL

(b) 0 mL: $H_2A \rightleftharpoons H^+ + HA^-$

   $ 0.050\,0 - x \quad x \quad x$

   $\dfrac{x^2}{0.050\,0 - x} = K_1 = 1.42 \times 10^{-3}$

   $\Rightarrow x = 7.75 \times 10^{-3} \Rightarrow$ pH = 2.11

   12.5 mL: pH = pK_1 = 2.85

   25.0 mL: pH $\approx \frac{1}{2}(pK_1 + pK_2)$

   $= \frac{1}{2}(2.847 + 5.696) = 4.27$

   37.5 mL: pH = pK_2 = 5.70

   50.0 mL: H_2A has been converted to A^{2-} at a formal concentration of

$$F = \frac{(50.0 \text{ mL})(0.050\ 0 \text{ M})}{100.0 \text{ mL}} = 0.025\ 0 \text{ M}$$

$$\underset{0.025\ 0 - x}{A^{2-}} + H_2O \rightleftharpoons \underset{x}{HA^-} + \underset{x}{OH^-}$$

$$\frac{x^2}{0.025\ 0 - x} = K_{b1} = \frac{K_w}{K_{a2}} \Rightarrow x = 1.12 \times 10^{-5}$$

$$\Rightarrow pH = -\log\left(\frac{K_w}{x}\right) = 9.05$$

55.0 mL: There is an excess of 5.0 mL of OH^-:

$$[OH^-] = \frac{(5.0 \text{ mL})(0.100 \text{ M})}{105.0 \text{ mL}}$$

$$= 0.004\ 76 \text{ M} \Rightarrow pH = 11.68$$

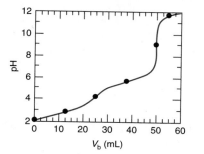

Chapter 13

13-A. **(a)** $\underset{\text{Oxidant}}{I_2 + 2e^- \rightleftharpoons 2I^-}$

(b) $\underset{\text{Reductant}}{2S_2O_3^{2-} \rightleftharpoons S_4O_6^{2-} + 2e^-}$

(c) $1.00 \text{ g } S_2O_3^{2-}/(112.12 \text{ g/mol}) = 8.92 \text{ mmol } S_2O_3^{2-}$
$= 8.92 \text{ mmol } e^-$
$(8.92 \times 10^{-3} \text{ mol})(9.649 \times 10^4 \text{ C/mol}) = 861 \text{ C}$

(d) current (A) = coulombs/s = 861 C/60 s = 14.3 A

(e) work = $E \cdot q$ = (0.200 V)(861 C) = 172 J

13-B. **(a)** $Pt(s) \mid Br_2(l) \mid HBr(aq, 0.10 \text{ M}) \parallel Al(NO_3)_3(aq, 0.010 \text{ M}) \mid Al(s)$

(b)

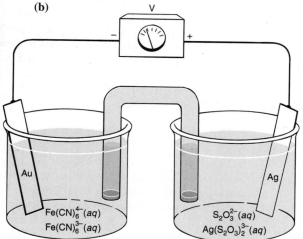

right: $\quad Ag(S_2O_3)_2^{3-} + e^- \rightleftharpoons Ag(s) + 2S_2O_3^{2-}$
left: $\quad\quad\quad\quad Fe(CN)_6^{4-} \rightleftharpoons Fe(CN)_6^{3-} + e^-$
net: $Fe(CN)_6^{4-} + Ag(S_2O_3)_2^{3-} \rightleftharpoons Fe(CN)_6^{3-} + Ag(s) + 2S_2O_3^{2-}$

13-C. $Pt(s) \mid H_2(g, 1 \text{ atm}) \mid H^+(aq, 1 \text{ M}) \parallel Fe^{2+}(aq, 1 \text{ M}), Fe^{3+}(aq, 1 \text{ M}) \mid Pt(s)$
$E°$ for the reaction $Fe^{3+} + e^- \rightleftharpoons Fe^{2+}$ is 0.771 V. Therefore electrons flow from left to right through the meter.

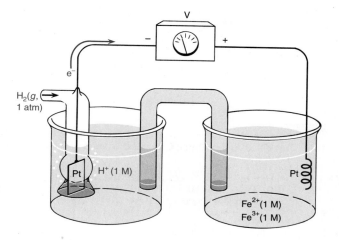

13-D. **(a)** $E = E° - \left(\frac{0.059\ 16}{3}\right)\log\left(\frac{P_{AsH_3}}{[H^+]^3}\right)$

$E = -0.238 - \left(\frac{0.059\ 16}{3}\right)\log\left(\frac{1.00/760}{(10^{-3.00})^3}\right)$

$= -0.359 \text{ V}$

(b) $Zn(s) \mid Zn^{2+}(0.1 \text{ M}) \parallel Cu^{2+}(0.1 \text{ M}) \mid Cu(s)$

right half-cell: $Cu^{2+} + 2e^- \rightleftharpoons Cu(s) \quad E_+° = 0.339 \text{ V}$
left half-cell: $\;\; Zn^{2+} + 2e^- \rightleftharpoons Zn(s) \quad E_-° = -0.762 \text{ V}$

$$E = \left\{0.339 - \left(\frac{0.059\ 16}{2}\right)\log\left(\frac{1}{0.1}\right)\right\} - \left\{-0.762 - \left(\frac{0.059\ 16}{2}\right)\log\left(\frac{1}{0.1}\right)\right\} = 1.101 \text{ V}$$

Because the voltage is positive, electrons are transferred from Zn to Cu. The net reaction is $Cu^{2+} + Zn(s) \rightleftharpoons Cu(s) + Zn^{2+}$.

13-E. **(a)**
right half-cell: $Cu^{2+} + 2e^- \rightleftharpoons Cu(s) \quad E_+° = 0.339 \text{ V}$
left half-cell: $\;Zn^{2+} + 2e^- \rightleftharpoons Zn(s) \quad E_-° = -0.762 \text{ V}$
$E° = E_+° - E_-° = 1.101 \text{ V}$
$K = 10^{nE°/0.059\ 16} = 10^{2(1.101)/0.059\ 16} = 1.7 \times 10^{37}$

(b)
$$\underset{-}{\begin{array}{ll} AgBr(s) + e^- \rightleftharpoons Ag(s) + Br^- & E_+° = 0.071 \text{ V} \\ Ag^+ + e^- \rightleftharpoons Ag(s) & E_-° = 0.799 \text{ V} \\ \hline AgBr(s) \rightleftharpoons Ag^+ + Br^- & E° = 0.071 - 0.799 \\ & = -0.728 \text{ V} \\ & K_{sp} = 10^{1E°/0.059\ 16} = 4._9 \times 10^{-13} \end{array}}$$

13-F. **(a)** $E = E_+ - E_-$

$$= \left\{ \underbrace{E_+^\circ}_{0.771\ V} - (0.059\ 16) \log\underbrace{\left(\frac{[Fe^{2+}]}{[Fe^{3+}]}\right)}_{?} \right\} - \underbrace{0.197}_{\substack{\text{Reference} \\ \text{electrode} \\ \text{voltage}}}$$

$$0.703 = \left\{ 0.771 - (0.059\ 16) \log\left(\frac{[Fe^{2+}]}{[Fe^{3+}]}\right)\right\} - 0.197$$

$$(0.059\ 16) \log\left(\frac{[Fe^{2+}]}{[Fe^{3+}]}\right) = -0.129$$

$$\Rightarrow \log\left(\frac{[Fe^{2+}]}{[Fe^{3+}]}\right) = -2.181$$

$$\Rightarrow \frac{[Fe^{2+}]}{[Fe^{3+}]} = 10^{-2.181} = 6.6 \times 10^{-3}$$

(b) **(i)** 0.326 V **(ii)** 0.463 V

Chapter 14

14-A. **(a)** At 0.10 mL: Initial $Cl^- = 2.00$ mmol; added $Ag^+ = 0.020$ mmol

$$[Cl^-] = \frac{(2.00 - 0.020)\ \text{mmol}}{(40.0 + 0.10)\ \text{mL}} = 0.049\ 4\ M$$

$$[Ag^+] = K_{sp}/[Cl^-] = (1.8 \times 10^{-10})/(0.049\ 4)$$
$$= 3.6 \times 10^{-9}\ M$$

In a similar manner, we find

2.50 mL:	$[Cl^-] = 0.035\ 3\ M$	$[Ag^+] = 5.1 \times 10^{-9}\ M$
5.00 mL:	$[Cl^-] = 0.022\ 2\ M$	$[Ag^+] = 8.1 \times 10^{-9}\ M$
7.50 mL:	$[Cl^-] = 0.010\ 5\ M$	$[Ag^+] = 1.7 \times 10^{-8}\ M$
9.90 mL:	$[Cl^-] = 0.000\ 401\ M$	$[Ag^+] = 4.5 \times 10^{-7}\ M$

(b) At V_e: $[Ag^+][Cl^-] = x^2 = K_{sp}$
$$\Rightarrow [Cl^-] = [Ag^+] = \sqrt{K_{sp}} = 1.3 \times 10^{-5}\ M$$

(c) At 10.10 mL: $[Ag^+] = \dfrac{(0.10\ \text{mL})(0.200\ M)}{50.1\ \text{mL}}$
$$= 4.0 \times 10^{-4}\ M$$

At 12.00 mL: $[Ag^+] = \dfrac{(2.00\ \text{mL})(0.200\ M)}{52.0\ \text{mL}}$
$$= 7.7 \times 10^{-3}\ M$$

(d) At 0.10 mL: $E = 0.558 + (0.059\ 16) \log [Ag^+]$
$E = 0.558 + (0.059\ 16) \log (3.6 \times 10^{-9}) = 0.059\ V$
In a similar manner, we find the results below:

mL $AgNO_3$	E (V)
0.10	0.059
2.50	0.067
5.00	0.079
7.50	0.099
9.90	0.182
10.00	0.270
10.10	0.357
12.00	0.433

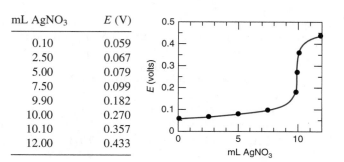

14-B. Cl^- diffuses into the $NaNO_3$, and NO_3^- diffuses into the NaCl. The mobility of Cl^- is greater than that of NO_3^-, so the NaCl region is depleted of Cl^- faster than the $NaNO_3$ region is depleted of NO_3^-. The $NaNO_3$ side becomes negative, and the NaCl side becomes positive.

14-C. As Ca^{2+} diffuses from left to right, the solution on the left side becomes more positive and the right side becomes more negative. The positive charge at the left repels Ca^{2+} and opposes further diffusion of Ca^{2+} across the membrane.

14-D. **(a)** Uncertainty in pH of standard buffers, junction potential, alkaline or acid errors at extreme pH values, equilibration time for electrode.

(b) $(4.63)(0.059\ 16\ V) = 0.274\ V$

(c) If the strongly alkaline solution has a high concentration of Na^+ (as in NaOH), the Na^+ cation competes with H^+ for cation exchange sites on the glass surface. The glass responds as if some H^+ were present, and the apparent pH is lower than the actual pH.

14-E. **(a)**

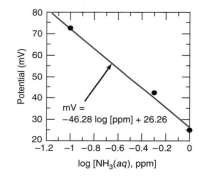

(b) Plugging the experimental values of potential into the equation of the calibration curve above gives
$106 = -46.28 \log [ppm] + 26.26$
$\Rightarrow [ppm] = 0.019$ ppm
$115 = -46.28 \log [ppm] + 26.26$
$\Rightarrow [ppm] = 0.012$ ppm
(Note that the unknown values lie beyond the calibration points, which is not the best practice. It would be a good idea to obtain calibration points at lower concentration to include the full range of unknown points.)

(c) $56 = -46.28 \log [ppm] + 26.26$
$\Rightarrow [ppm] = 0.23$ ppm

Chapter 15

15-A. A monodentate ligand binds to a metal ion through one ligand atom. A multidentate ligand binds through more than one ligand atom. A chelating ligand is a multidentate ligand.

15-B. **(a)** $M^{n+} + Y^{4-} \rightleftharpoons MY^{n-4}$

$K_f = [MY^{n-4}]/([M^{n+}][Y^{4-}])$

(b) At low pH, H^+ competes with M^{n+} in binding to the ligand atoms of EDTA.

(c)

H_6Y^{2+}	H_5Y^+	H_4Y	H_3Y^-	H_2Y^{2-}	HY^{3-}	Y^{4-}

$\uparrow\uparrow\quad\uparrow\uparrow\quad\uparrow\uparrow\quad\uparrow\uparrow\quad\uparrow\uparrow\quad\uparrow$

0.0 | 1.5 | 2.0 | 2.66 | 6.16 | 10.24

pH: 0.75 1.75 2.33 4.41 8.20

15-C. **(a)** Only a small amount of indicator is employed. Most of the Mg^{2+} is not bound to indicator. The free Mg^{2+} reacts with EDTA before MgIn reacts. Therefore the concentration of MgIn is constant until all of the Mg^{2+} has been consumed. Only when MgIn begins to react does the color change.

(b) **(i)** H_3In^{3-} **(ii)** yellow **(iii)** red **(iv)** violet → red

15-D. **(a)** $(25.0 \text{ mL})(0.050\ 0 \text{ M}) = 1.25$ mmol EDTA

(b) $(5.00 \text{ mL})(0.050\ 0 \text{ M}) = 0.25$ mmol Zn^{2+}

(c) mmol Ni^{2+} = mmol EDTA − mmol Zn^{2+}
$= 1.25 - 0.25 = 1.00$ mmol Ni^{2+}

$[Ni^{2+}] = (1.00 \text{ mmol})/(50.0 \text{ mL}) = 0.020\ 0$ M

15-E. **(a)** 50.00 mL contains exactly 1/10 of the $KIO_3 = 0.102\ 2$ g $= 0.477\ 5_7$ mmol KIO_3. Each mol of iodate makes 3 mol of triiodide, so $I_3^- = 3(0.477\ 5_7) = 1.432_7$ mmol.

(b) Two moles of thiosulfate react with one mole of triiodide. Therefore there must have been $2(1.432_7) = 2.865_4$ mmol of thiosulfate in 37.66 mL, so the concentration is $(2.865_4 \text{ mmol})/(37.66 \text{ mL}) = 0.076\ 08_7$ M.

(c) 50.00 mL of KIO_3 makes 1.432_7 mmol I_3^-. The unreacted I_3^- requires 14.22 mL of sodium thiosulfate $= (14.22 \text{ mL})(0.076\ 08_7 \text{ M}) = 1.082_0$ mmol, which reacts with $\frac{1}{2}(1.082_0 \text{ mmol}) = 0.541_0$ mmol I_3^-. The ascorbic acid must have consumed the difference $= 1.432_7 - 0.541_0 = 0.891_7$ mmol I_3^-. Each mole of ascorbic acid consumes one mole of triiodide, so mol ascorbic acid $= 0.891_7$ mmol, which has a mass of $(0.891_7 \times 10^{-3} \text{ mol})(176.13 \text{ g/mol}) = 0.157_1$ g. The weight percent of ascorbic acid in the unknown is $100 \times (0.157_1 \text{ g})/(1.223 \text{ g}) = 12.8$ wt %.

Chapter 16

16-A. 1-C, 2-D, 3-A, 4-E, 5-B

16-B. Your answer to part **(a)** will be different from what is shown here, which was measured on a larger figure than the one in your textbook. However, your answers to parts **(b)**–**(e)** should be the same as those below.

(a) t_r: octane, 200 units; nonane, 260 units
w: octane, 40 units; nonane, 52 units

(b) octane: $N = \dfrac{5.55(200)^2}{(40)^2} = 139$

nonane: $N = \dfrac{5.55(260)^2}{(52)^2} = 139$

(c) octane: $H = L/N = (1.00 \text{ m}/139) = 7.2$ mm
nonane: $H = (1.00 \text{ m}/139) = 7.2$ mm

(d) resolution $= \dfrac{\Delta t_r}{w_{av}} = \dfrac{260 - 200}{\frac{1}{2}(40 + 52)} = 1.30$

(e) $\dfrac{\text{large load}}{\text{small load}} = \left(\dfrac{\text{large column radius}}{\text{small column radius}}\right)^2$

$\dfrac{27.0 \text{ mg}}{3.0 \text{ mg}} = \left(\dfrac{\text{large column radius}}{2.0 \text{ mm}}\right)^2$

$\Rightarrow$ large column radius = 6.0 mm
large column diameter = 12.0 mm

Flow rate should be proportional to the cross-sectional area of the column, which is proportional to the square of the radius.

$\dfrac{\text{large column flow rate}}{\text{small column flow rate}} = \left(\dfrac{\text{large column radius}}{\text{small column radius}}\right)^2$

$= \left(\dfrac{6.0 \text{ mm}}{2.0 \text{ mm}}\right)^2 = 9.0$

If the small column flow rate is 7.0 mL/min, the large column flow rate should be $(9.0)(7.0 \text{ mL/min}) = 63$ mL/min.

16-C. **(a)** Molecules in the gas phase move faster than molecules in liquid. Therefore, longitudinal diffusion in the gas phase is much faster than longitudinal diffusion in the liquid phase, so band broadening by this mechanism is more rapid in gas chromatography than in liquid chromatography.

(b) **(i)** Optimum flow rate gives minimum plate height (23 cm/s for He). **(ii)** When flow is too fast, there is not adequate time for solute to equilibrate between the phases as the mobile phase streams past the stationary phase. This also broadens the band. **(iii)** When flow is too slow, plate height increases (i.e., band broadening gets worse) because solute spends a long time on the column and is broadened by longitudinal diffusion.

(c) There is no broadening by multiple flow paths in an open tubular column.

(d) The longer the column, the better the resolution. We can use a longer open tubular column than a packed column because the particles in the packed column resist flow and require high pressures for high flow rate.

16-D. The peak at 136 mass units is $C_4H_9{}^{79}Br^+$ and the peak at 138 is $C_4H_9{}^{81}Br^+$. Natural bromine consists of 50.5% ^{79}Br and 49.5% ^{81}Br, so the intensities of the peaks are almost equal. Peaks at 107 and 109 mass units are $C_2H_4{}^{79}Br^+$ and $C_2H_4{}^{81}Br^+$. The peak at 57 mass units is $C_4H_9^+$, which has no Br and therefore has no isotopic partner at 59 mass units. The peak at 41 mass units is $C_3H_5^+$.

16-E. $\dfrac{\text{[X]/[S] in unknown}}{\text{[X]/[S] in standard}} = \dfrac{\text{area ratio in unknown}}{\text{area ratio in standard}}$

$\dfrac{\text{[X]/(742 nM)}}{(52.4 \text{ nM}/38.9 \text{ nM})} = \dfrac{1.093}{0.644}$

$\Rightarrow \text{[X]} = 1.70 \times 10^3 \text{ nM} = 1.70 \ \mu\text{M}$

Chapter 17

17-A.   (a) Low boiling solutes are separated well at low temperature, and the retention of high boiling solutes is reduced to a reasonable time at high temperature.

(b) An open tubular column gives higher resolution than a packed column. A packed column can handle much more sample than an open tubular column, which is vital for preparative separations in which we are trying to isolate some quantity of the separated components.

(c) Diffusion of solute in H_2 and He is more rapid than in N_2. Therefore equilibration of solute between mobile phase and stationary phase is faster. The column can be run faster without excessive broadening because of the finite rate of mass transfer between the mobile and stationary phases.

(d) Split injection is the ordinary mode for open tubular columns. Splitless injection is useful for trace and quantitative analysis. On-column injection is useful for thermally sensitive solutes that might decompose during a high-temperature injection.

(e) (i) all analytes; (ii) carbon atoms bearing hydrogen atoms; (iii) molecules with halogens, conjugated $C=O$, CN, NO_2; (iv) P and S; (v) P and N; (vi) S; (vii) all analytes

17-B.   Solvent is competing with solute for adsorption sites. The strength of the solvent-adsorbent interaction is independent of solute.

17-C.   (a) In normal-phase chromatography, the stationary phase is more polar than the mobile phase. In reverse-phase chromatography, the stationary phase is less polar than the mobile phase.

(b) In isocratic elution, the composition of eluent is constant. In gradient elution, the composition of the eluent is changed—usually continuously—from low eluent strength to high eluent strength.

(c) In normal-phase chromatography, polar solvent must compete with analyte for polar sites on the stationary phase. The more polar the solvent, the better it binds to the stationary phase, and the greater is its ability to displace analyte from the stationary phase.

(d) In reverse-phase chromatography, nonpolar analyte adheres to the nonpolar stationary phase. A polar solvent does not compete with analyte for nonpolar sites on the stationary phase. Making the solvent less polar gives it greater ability to displace analyte from the stationary phase.

(e) A guard column is a small, disposable column containing the same stationary phase as the main column. Sample and solvent pass through the guard column first. Any irreversibly bound impurities stick to the guard column, which is eventually thrown away. The guard column prevents junk from being irreversibly bound to the expensive main column and eventually ruining it.

(f) Solid-phase extraction uses a short column containing any of the stationary phases used in HPLC. The column carries out gross separations of one type of analyte from other types of analyte (e.g., separating nonpolar from polar analytes).

Chapter 18

18-A.   (a) Deionized water has been passed through ion-exchange columns to convert cations to H^+ and anions to OH^-, making H_2O. Nonionic impurities (such as neutral organic compounds) are not removed by this process.

(b) Gradient elution with increasing concentration of H^+ is required to displace more and more strongly bound cations from the column.

(c) Smaller hydrated radius: Ca^{2+}. Ca^{2+} is eluted after Mg^{2+}, so Ca^{2+} must be bound to the column more tightly than Mg^{2+}. To be bound more tightly to the cation exchanger, Ca^{2+} must have a smaller hydrated radius.

18-B.   The separator column separates ions by ion exchange, whereas the suppressor exchanges the counterion to reduce the conductivity of eluent.

18-C.   (a) total volume $= \pi(0.80 \text{ cm})^2(20.0 \text{ cm}) = 40.2 \text{ mL}$

(b) If the void volume ($=$ volume of mobile phase excluded from gel) is 18.2 mL, the stationary phase plus its included solvent must occupy $40.2 - 18.2 = 22.0$ mL.

(c) If pores occupy 60.0% of the stationary-phase volume, the pore volume is $(0.600)(22.0 \text{ mL}) = 13.2$ mL. Large molecules excluded from the pores are eluted in the void volume of $x = 18.2$ mL. The smallest molecules (which can enter all of the pores) are eluted in a volume of $y = 18.2 + 13.2 = 31.4$ mL. (In fact, there is usually some adsorption of solutes on the stationary phase, so retention volumes in real columns can be greater than 31.4 mL.)

18-D.   (a) 25 μm diameter: volume $= \pi r^2 \times$ length
$= \pi(12.5 \times 10^{-6} \text{ m})^2(5 \times 10^{-3} \text{ m}) = 2.5 \times 10^{-12} \text{ m}^3$

$(2.5 \times 10^{-12} \text{ m}^3)\left(\dfrac{1 \text{ L}}{10^{-3} \text{ m}^3}\right)$

$= 2.5 \times 10^{-9} \text{ L} = 2.5 \text{ nL}$

20 μm diameter: volume $=$
$\pi(25 \times 10^{-6} \text{ m})^2(5 \times 10^{-3} \text{ m}) = 9.8 \text{ nL}$

(b) Capillary electrophoresis eliminates peak broadening from (1) the finite rate of mass transfer between the mobile and stationary phases and (2) multiple flow paths around stationary phase particles.

18-E. **(a)**

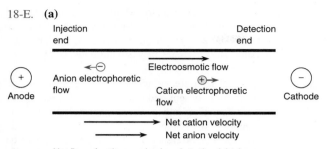

Net flow of cations and anions is to the right, because electroosmotic flow is stronger than electrophoretic flow at high pH

(b) At pH 3, the net flow of cations is to the *right* and the net flow of anions is to the *left*.

(c) When there is no analyte present, the constant concentration of chromate in the background buffer gives a steady ultraviolet absorbance at 254 nm. When Cl^- emerges, it displaces some of the chromate anion (to maintain electroneutrality). Because Cl^- does not absorb at 254 nm, the absorbance *decreases*.

18-F. **(a)** In the absence of micelles, all neutral molecules would move along with the electroosmotic speed of the bulk solvent and arrive at the detector at time t_0. Negative micelles migrate upstream and arrive at time $t_{mc} > t_0$. Neutral molecules partition between the bulk solvent and the micelles, so they arrive between t_0 and t_{mc}. The more time a neutral molecule spends in the micelles, the closer is its migration time to t_{mc}.

(b) (i) pK_a for $C_6H_5NH_2$ is 4.60. At pH 10, the predominant form is neutral.

(ii) Because anthracene arrives at the detector last, it must spend more time inside the micelles, and therefore it is more soluble in the micelles.

Chapter 19

19-A. **(a)** $\dfrac{0.214\ 6\ \text{g AgBr}}{187.772\ \text{g AgBr/mol}} = 1.142\ 9 \times 10^{-3}$ mol AgBr

(b) $[\text{NaBr}] = \dfrac{1.142\ 9 \times 10^{-3}\ \text{mol}}{50.00 \times 10^{-3}\ \text{L}} = 0.022\ 86$ M

19-B. **(a)** Absorbed impurities are taken into a substance. Adsorbed impurities are found on the surface of a substance.

(b) Included impurities occupy lattice sites of the host crystal. Occluded impurities are trapped in a pocket inside the host.

(c) An ideal gravimetric precipitate should be insoluble, easily filtered, possess a known, constant composition, and be stable to heat.

(d) High supersaturation often leads to formation of colloidal product with a large amount of impurities.

(e) Supersaturation can be decreased by increasing temperature (for most solutions), mixing well during addition of precipitant and using dilute reagents. Homogeneous precipitation also reduces supersaturation.

(f) Washing with electrolyte preserves the electric double layer and prevents peptization.

(g) The volatile HNO_3 evaporates during drying. $NaNO_3$ is nonvolatile and will lead to a high mass for the precipitate.

(h) During the first precipitation, the concentration of impurities in the solution is high, giving a relatively high concentration of impurities in the precipitate. In the reprecipitation, the level of solution impurities is reduced, thus giving a purer precipitate.

(i) In thermogravimetric analysis, the mass of a sample is measured as the sample is heated. The mass lost during decomposition provides information about the composition of the sample.

19-C. **(a)** $\dfrac{0.104\ \text{g CeO}_2}{172.114\ \text{g CeO}_2/\text{mol}} = 6.043 \times 10^{-4}$ mol CeO_2

6.043×10^{-4} mol $CeO_2 \times \dfrac{1\ \text{mol Ce}}{1\ \text{mol CeO}_2} =$
6.043×10^{-4} mol Ce

$(6.043 \times 10^{-4}\ \text{mol Ce})(140.115\ \text{g Ce/mol Ce}) =$
$0.084\ 66$ g Ce

(b) wt % Ce $= \dfrac{0.084\ 66\ \text{g Ce}}{4.37\ \text{g unknown}} \times 100$
$= 1.94$ wt %

19-D. **(a)** In *combustion,* a substance is heated in the presence of excess O_2 to convert carbon to CO_2 and hydrogen to H_2O. In *pyrolysis,* the substance is decomposed by heating in the absence of added O_2. All oxygen in the sample is converted to CO by passage through a suitable catalyst.

(b) WO_3 catalyzes the complete combustion of C to CO_2 in the presence of excess O_2. Cu reduces SO_3 to SO_2 and removes excess O_2 from the gas stream.

(c) The tin capsule melts and is oxidized to SnO_2 to liberate heat and crack the sample. Tin uses the available oxygen immediately, ensures that sample oxidation occurs in the gas phase, and acts as an oxidation catalyst.

(d) By dropping the sample in before very much O_2 is present, pyrolysis of the sample to give gaseous products occurs prior to oxidation. This minimizes the formation of nitrogen oxides.

(e) $C_8H_7NO_2SBrCl + 9\frac{1}{4}O_2$
$\rightarrow 8CO_2 + \frac{5}{2}H_2O + \frac{1}{2}N_2 + SO_2 + HBr + HCl$

Chapter 20

20-A. In atomic absorption spectroscopy, light of a specific frequency is passed through a flame containing free atoms. The absorbance of light is measured and is proportional to the concentration of atoms. In atomic emission spectroscopy, no lamp is used. The intensity of light emitted by excited atoms in the flame is measured and is proportional to the concentration of atoms.

20-B. Atoms in a furnace are confined in a small volume for a relatively long time. The detection limit for an element in the unknown is small because the concentration of atoms in the gas phase is relatively high. A large sample volume is not required because gaseous atoms do not rapidly escape from the furnace. In a flame or plasma, the volume of gas phase is relatively large, so the concentration of atoms is relatively low. A great deal of sample is required because the atoms are rapidly moving through and escaping from the flame or plasma.

20-C. **(a)** First find the frequency of light:
$$\nu = c/\lambda = (2.998 \times 10^8 \text{ m/s})/(400 \times 10^{-9} \text{ m})$$
$$= 7.495 \times 10^{14} \text{ s}^{-1}$$
Energy $= h\nu$
$$= (6.626 \times 10^{-34} \text{ J} \cdot \text{s})(7.495 \times 10^{14} \text{ s}^{-1})$$
$$= 4.966 \times 10^{-19} \text{ J}$$

(b) $\dfrac{N^*}{N_0} = \left(\dfrac{g^*}{g_0}\right)e^{-\Delta E/kT}$
$$= \left(\frac{1}{1}\right)e^{-(4.966 \times 10^{-19} \text{ J})/[(1.381 \times 10^{-23} \text{ J/K})(2\,500 \text{ K})]}$$
$$= 5.7 \times 10^{-7}$$

20-D. Signal reaching the detector while the beam is blocked must be from flame emission. Signal reaching the detector when the beam is not blocked is from the lamp and the flame. The difference between these two signals is the desired analytical signal.

20-E. **(a)** Spectral interference arises from overlap of analyte absorption or emission lines with absorptions or emissions from other elements or molecules in the sample or the flame.

(b) Chemical interference is caused by any substance that decreases the extent of atomization of analyte.

(c) Ionization interference is a reduction of the concentration of free atoms by ionization of the atoms.

20-F. **(a)** mol added standard $= (0.001\,00 \text{ L})(0.030\,0 \text{ M}) =$
$$3.00 \times 10^{-5} \text{ mol}$$
$$[S]_f = (3.00 \times 10^{-5} \text{ mol})/(0.010\,0 \text{ L}) = 3.00 \times 10^{-3} \text{ M}$$

(b) $\dfrac{[K^+]_i}{[K^+]_f + [S]_f} = \dfrac{[K^+]_i}{\frac{1}{2}[K^+]_i + [3.00 \times 10^{-3} \text{ M}]}$
$$= \frac{3.00 \text{ mV}}{4.00 \text{ mV}} \Rightarrow [K^+]_i = 3.60 \times 10^{-3} \text{ M}$$

Answers to Problems

Complete solutions to all problems can be found in the Solutions Manual for Exploring Chemical Analysis.

Chapter 1

1-5. **(a)** milliwatt = 10^{-3} watt
(b) picometer = 10^{-12} meter
(c) kiloohm = 10^3 ohm
(d) microcoulomb = 10^{-6} coulomb
(e) terajoule = 10^{12} joule
(f) nanosecond = 10^{-9} second
(g) femtogram = 10^{-15} gram
(h) decipascal = 10^{-1} pascal

1-6. **(a)** 100 fJ or 0.1 pJ   **(b)** 43.172 8 nC
(c) 299.79 THz   **(d)** 0.1 nm or 100 pm
(e) 21 TW   **(f)** 0.483 amol or 483 zmol

1-7. **(a)** 7.457×10^4 W   **(b)** 7.457×10^4 J/s
(c) 1.782×10^4 cal/s   **(d)** 6.416×10^7 cal/h

1-8. **(a)** 0.025 4 m, 39.37 inches
(b) 0.214 mile/s, 770 mile/h
(c) 1.04×10^3 m, 1.04 km, 0.643 mile

1-10. 1.10 M

1-11. 0.054 8 ppm, 54.8 ppb

1-12. 4.4×10^{-3} M, 6.7×10^{-3} M

1-13. **(a)** 70.5 g   **(b)** 29.5 g   **(c)** 0.702 mol

1-14. 6.18 g

1-15. **(a)** 6.6×10^3 L   **(b)** 5.8×10^5 g

1-16. 8.0 g

1-17. **(a)** 55.6 mL   **(b)** 1.80 g/mL

1-18. 5.48 g

1-19. 10^{-3} g/L, 10^3 μg/L, 1 μg/L, 1 mg/L

1-20. 7×10^{-10} M

1-21. **(a)** 764 g ethanol   **(b)** 16.6 M

1-22. **(a)** 0.228 g Ni   **(b)** 1.06 g/mL

1-23. 1.235 M

1-24. dilute 8.26 mL of 12.1 M HCl to 100.0 mL

1-25. **(a)** 3.40 M   **(b)** 14.7 mL

1-26. shredded wheat: 3.6 Cal/g, 102 Cal/oz; doughnut: 3.9, 111; hamburger: 2.8, 79; apple: 0.48, 14

Chapter 2

2-6. 5.403 1 g

2-7. 14.85 g

2-8. 0.296 1 g

2-9. 9.980 mL

2-10. 5.022 mL

2-11. 15.631 mL

Chapter 3

3-1. **(a)** 1.237   **(b)** 1.238   **(c)** 0.135
(d) 2.1   **(e)** 2.00

3-2. **(a)** 0.217   **(b)** 0.216   **(c)** 0.217

3-3. **(a)** 4   **(b)** 4   **(c)** 4

3-4. **(a)** 12.3   **(b)** 75.5   **(c)** 5.520×10^3
(d) 3.04   **(e)** 3.04×10^{-10} **(f)** 11.9
(g) 4.600   **(h)** 4.9×10^{-7}

3-5. **(a)** 12.01   **(b)** 10.9   **(c)** 14   **(d)** 14.3
(e) -17.66 **(f)** 5.97×10^{-3} **(g)** 2.79×10^{-5}

3-6. **(a)** 208.232   **(b)** 560.604

3-9. **(a)** 3.124 ($\pm$0.005) or 3.123_6 ($\pm0.005_2$)
(b) 3.124 ($\pm$0.2%) or 3.123_6 ($\pm0.1_7$%)

3-10. **(a)** 2.1 $\pm$0.2 (or 2.1 $\pm$11%)
(b) 0.151 $\pm$0.009 (or 0.151 $\pm$6%)
(c) 0.22_3 $\pm0.02_4$ ($\pm$11%)
(d) 0.097_1 $\pm0.002_2$ ($\pm2._2$%)

3-11. **(a)** 21.0_9 ($\pm0.1_6$) or 21.1 ($\pm$0.2); relative uncertainty = 0.8%
(b) 27.4_3 ($\pm0.8_6$); relative uncertainty = $3._2$%

(c) $(14._9 \pm 1._3) \times 10^4$ or $(15 \pm 1) \times 10^4$; relative uncertainty = 9%

3-12. (a) $10.18 \ (\pm 0.07) \ (\pm 0.7\%)$
(b) $174 \ (\pm 3) \ (\pm 2\%)$
(c) $0.147 \ (\pm 0.003) \ (\pm 2\%)$
(d) $7.86 \ (\pm 0.01) \ (\pm 0.1\%)$
(e) $2\,185.8 \ (\pm 0.8) \ (\pm 0.04\%)$

3-13. (a) $6.0 \pm 0.2 \ (\pm 3._7\%)$ (b) $1.30_8 \pm 0.09_2 \ (\pm 7._0\%)$
(c) $1.30_8 \ (\pm 0.09_2) \times 10^{-11} \ (\pm 7._0\%)$
(d) $2.7_2 \pm 0.7_8 \ (\pm 29\%)$

3-14. 78.114 ± 0.006

3-15. MW = $95.980 \ (\pm 0.008)$

3-16. (a) $58.442\,5 \pm 0.000\,9$ g/mol
(b) $0.450\,7 \ (\pm 0.000\,5)$ M

3-17. (a) $0.020\,76_6 \pm 0.000\,02_8$ M   (b) yes

3-18. (a) 16.66 mL   (b) $0.169 \ (\pm 0.002)$ M

3-19. 100 J = 23.90 cal = 0.094 78 BTU = 6.241×10^{20} eV

Chapter 4

4-2. 0.683, 0.955, 0.997

4-3. (a) 1.527 67   (b) 0.001 26, 0.082 5%
(c) 1.527 93 $\pm 0.000\,10$

4-5. 108.6_4, 7.1_4, $108.6_4 \pm 6.8_1$

4-6. yes

4-7. 2.299 47 $\pm 0.001\,15$, 2.299 47 $\pm 0.001\,71$

4-8. no

4-9. yes, no

4-10. yes

4-11. discard 0.195

4-12. $y = -2x + 15$

4-13. $-1.299 \ (\pm 0.001) \times 10^4$, $3 \ (\pm 3) \times 10^2$

4-14. (a) $2.0_0 \pm 0.3_8$ (b) $2.0_0 \pm 0.2_6$

4-15. (a) $y \ (\pm 0.005_7) = 0.021\,7_7 \ (\pm 0.000\,1_9)x + 0.004_6$
      $(\pm 0.004_4)$
(c) protein = $23.0_7 \pm 0.2_9$ μg

Chapter 5

5-2. (a) $K = 1/[Ag^+]^3 \ [PO_4^{3-}]$ (b) $K = P_{CO_2}{}^6/P_{O_2}{}^{15/2}$

5-3. (a) $P_A = 0.027_6$ atm, $P_E = 47._4$ atm
(b) 1.2×10^{10}

5-4. unchanged

5-5. 4.5×10^3

5-6. (a) 3.6×10^{-7} (b) 3.6×10^{-7} M
(c) 3.0×10^4

5-7. (a) $7._1 \times 10^{-5}$ M   (b) $1._0 \times 10^{-3}$ g/100 mL

5-8. (a) $6.6_9 \times 10^{-5}$ M   (b) $0.002\,2_2$ g/100 mL
(c) 14.4 ppm

5-9. 1 400 ppb, 76 ppb, 0.98 ppb

5-10. (a) $[Hg_2^{2+}] = 6.8_7 \times 10^{-7}$ M $[IO_3^-] = 1.3_7 \times 10^{-6}$ M
(b) $[Hg_2^{2+}] = 1.3 \times 10^{-14}$ M

5-11. I^- before Br^- before Cl^- before CrO_4^{2-}

5-12. (a) HCN/CN^- HCO_2H/HCO_2^-
(b) H_2O/OH^- HPO_4^{2-}/PO_4^{3-}
(c) H_2O/OH^- HSO_3^-/SO_3^{2-}

5-13. $[H^+] > [OH^-]$ $[OH^-] > [H^+]$

5-14. (a) 4   (b) 9   (c) 3.24   (d) 9.76

5-15. (a) 7.46   (b) 2.9×10^{-7} M

5-16. 2.5×10^{-5} M

5-17. 7.8

5-18. see Table 5-1

5-19. carboxylic acids and ammonium salts, carboxylate anions and amines

5-20. (a) 0.010 M, 2.00   (b) $2.8_6 \times 10^{-13}$ M, pH 12.54
(c) 0.030 M, 1.52
(d) 3.0 M, -0.48 (e) 1.0×10^{-12} M, 12.00

5-21. (b) trichloroacetic acid

5-22. (b) sodium 2-mercaptoethanol

5-23. $2H_2SO_4 \rightleftharpoons H_3SO_4^+ + HSO_4^-$

5-24.

5-25. $OCl^- + H_2O \rightleftharpoons HOCl + OH^-$ $K_b = 3.3 \times 10^{-7}$

Chapter 6

6-2. 43.2 mL, 270.0 mL

6-3. 0.149 M

6-4. 32.0 mL

6-5. (a) 0.045 00 M   (b) 36.42 mg/mL

6-6. 947 mg

6-7. 1.72 mg

6-8. (a) 0.020 34 M   (b) 0.125 7 g   (c) 0.019 83 M

6-9. 0.092 54 M

6-10. (a) 0.105 3 mol/kg solution   (b) 0.305 M

6-11. 9.066 mM

6-12. 3.555 mM

6-13. 30.5 wt %

6-14. 0.020 6 $(\pm 0.000\,7)$ M

Chapter 7

7-1. (a) double   (b) halve   (c) double

7-2. (a) 3.06×10^{-19} J/photon, 184 kJ/mol
(b) 4.97×10^{-19} J/photon, 299 kJ/mol

7-3. (a) orange   (b) violet-blue or violet
(c) blue-green

7-5. absorbance or molar absorptivity versus wavelength

7-7. violet-blue

7-8. (a) 1.20×10^{15} Hz, 4.00×10^4 cm^{-1}, 7.95×10^{-19},
      479 kJ/mol

(b) 1.20×10^{14} Hz, 4 000 cm^{-1}, 7.95×10^{-20} J/photon, 47.9 kJ/mol

7-9. 0.004 4, 0.046, 0.30, 1.00, 2.00, 3.00, 4.00
7-10. 3.56×10^4 M^{-1} cm^{-1}
7-11. **(a)** 2.76×10^{-5} M   **(b)** 2.24 g/L
7-12. **(a)** 7.80×10^{-3} M   **(b)** 7.80×10^{-4} M
     (c) 1.63×10^3 M^{-1} cm^{-1}
7-13. **(a)** 0.52   **(b)** 30%
7-14. **(a)** 0.347   **(b)** 20.2%
7-15. 2.30×10^3 M^{-1} cm^{-1}

Chapter 8

8-1. **(a)** 1.50×10^3 M^{-1} cm^{-1} **(b)** 2.31×10^{-4} M
     (c) 4.69×10^{-4} M   **(d)** 5.87×10^{-3} M
8-2. **(a)** 6.97×10^{-5} M   **(b)** 6.97×10^{-4} M
     (c) 1.02 mg
8-3. **(a)** 1.735 ppm   **(b)** 1.239×10^{-4} M
8-4. ±0.017 ppm
8-6. **(a)** $4.49_3 \times 10^3$ M^{-1} cm^{-1} **(b)** $1.00_6 \times 10^{-4}$ M
     (c) $5.03_0 \times 10^{-4}$ M   **(d)** 16.1%
8-7. ~2 and 0
8-8. **(a)** 1.57×10^{-5} M   **(b)** 0.180   **(c)** 6.60 mg
8-9. **(a)** $2.42_5 \times 10^4$ M^{-1} cm^{-1} **(b)** 1.26%
8-11. 1.73×10^{-5} M
8-12. 2.19×10^{-4} M

Chapter 9

9-1. 4.30
9-2. 9.78
9-3. 2.2×10^{-12} M
9-4. 3.02, 9.51×10^{-2}
9-5. 5.40, 2.65×10^{-5}
9-6. 3.15, 7.05×10^{-4} M, 0.084 3 M, 0.085 0 M
9-7. 3.70
9-8. 7.30
9-9. 5.51, 3.1×10^{-6} M, 0.060 M
9-10. **(a)** 3.03, 0.094   **(b)** 7.00, 0.999
9-11. 5.50
9-12. $K_a = 2.3 \times 10^{-11}$, $K_b = 4.3 \times 10^{-4}$
9-14. 11.28, 0.058 M, $1.9_2 \times 10^{-3}$ M
9-15. 0.007 56%, 0.023 9%, 0.568%
9-16. 10.95
9-17. 9.97, 0.003 6
9-18. $3.3_6 \times 10^{-6}$
9-19. 2.2×10^{-7}
9-20. **(a)** 5.57×10^{-2}, 0.94_4 **(b)** 0.37_1, 0.62_9
     (c) 0.85_5, 0.14_5
9-21. **(a)** 0.98_0, 0.019_6 **(b)** 0.83_4, 0.16_6
     (c) 0.61_3, 0.38_7
9-22. **(a)** 0.090_9, 0.90_9 **(b)** 0.50_0, 0.50_0
     (c) 0.66_6, 0.33_4
9-24. 2.93, 0.118

Chapter 10

10-1. 4.13
10-2. **(b)** 1/1 000   **(c)** $pK_a - 4$
10-3. **(a)** 1.5×10^{-7} **(b)** 0.15
10-4. **(a)** 14   **(b)** 1.4×10^{-7}
10-5. **(a)** 2.3×10^{-7} **(b)** 1.00   **(c)** 23
10-7. 3.59
10-8. **(a)** 8.37   **(b)** 0.423 g   **(c)** 8.33   **(d)** 8.41
10-9. **(b)** 7.18   **(c)** 7.00   **(d)** 6.86 mL
10-10. **(a)** 2.56   **(b)** 2.86
10-11. 3.27 mL
10-12. 13.7 mL
10-13. 4.68 mL
10-14. ii
10-15. **(b)** NaOH
10-16. **(a)** 90.9 mL HCl
10-17. **(a)** $p = 0.940\ 6$ mol, $q = 0.059\ 35$ mol
     (b) Δ(pH) $= -0.077$
10-18. **(a)** red   **(b)** orange   **(c)** yellow   **(d)** red
10-19. **(a)** $A = 2\ 080[\text{HIn}] + 14\ 200[\text{In}^-]$ **(b)** 6.79

Chapter 11

11-1. $V_e = 10.0$ mL; pH = 13.00, 12.95, 12.68, 11.96, 10.96, 7.00, 3.04, 1.75
11-2. $V_e = 12.5$ mL; pH = 1.30, 1.35, 1.60, 2.15, 3.57, 7.00, 10.43, 11.11
11-3. $V_e = 5.00$ mL; pH = 2.66, 3.40, 4.00, 4.60, 5.69, 8.33, 10.96, 11.95
11-4. $V_e = 5.00$ mL; pH = 6.32, 10.64, 11.23, 11.85
11-5. 2.80, 3.65, 4.60, 5.56, 8.65, 11.96
11-6. 8.18
11-7. 3.72
11-8. 0.091 8 M
11-9. **(a)** 0.025 92 M   **(b)** 0.020 31 M   **(c)** 9.69
     (d) 10.07
11-10. $V_e = 10.0$ mL; pH = 11.00, 9.95, 9.00, 8.05, 7.00, 5.02, 3.04, 1.75
11-11. $V_e = 47.79$ mL; pH = 8.74, 5.35, 4.87, 4.40, 3.22, 2.58
11-12. **(a)** 2.2×10^9
     (b) 10.92, 9.57, 9.35, 8.15, 5.53, 2.74
11-13. **(a)** 9.44   **(b)** 2.55   **(c)** 5.15
11-14. no
11-15. yellow, green, blue
11-17. **(a)** colorless → pink
     (b) systematically requires too much NaOH
11-18. cresol red (orange → red) or phenophthalein (colorless → red)
11-19. 10.727 mL
11-20. 0.063 56 M
11-21. **(a)** 0.087 99   **(b)** 25.74 mg   **(c)** 2.860 wt %
11-22. **(a)** 5.62   **(b)** methyl red

Chapter 12

12-1. $K_{a2} = 1.02 \times 10^{-2}$, $K_{b2} = 1.79 \times 10^{-13}$

12-2. 7.09×10^{-3}, 6.33×10^{-8}, 7.1×10^{-13}

12-4. piperazine: $K_{b1} = 5.38 \times 10^{-5}$, $K_{b2} = 2.15 \times 10^{-9}$
phthalate: $K_{b1} = 2.56 \times 10^{-9}$, $K_{b2} = 8.93 \times 10^{-12}$

12-6. 2.49×10^{-8}, 5.78×10^{-10}, 1.34×10^{-11}

12-7. 1.62×10^{-5}, 1.54×10^{-12}

12-8. **(a)** 1.95, 0.089 M, 1.12×10^{-2} M, 2.01×10^{-6} M
(b) 4.27, 3.8×10^{-3} M, 0.100 M, 3.8×10^{-3} M
(c) 9.35, 7.04×10^{-12} M, 2.23×10^{-5} M, 0.100 M

12-9. **(a)** 11.00, 0.099 0 M, 9.95×10^{-4} M, 1.00×10^{-9} M
(b) 7.00, 1.0×10^{-3} M, 0.100 M, 1.0×10^{-3} M
(c) 3.00, 1.00×10^{-9} M, 9.95×10^{-4} M, 0.099 0 M

12-10. 11.60, [B] = 0.296 M, [BH$^+$] = 3.99×10^{-3} M,
[BH$_2^{2+}$] = 2.15×10^{-9} M

12-11. 7.53, [BH$_2^{2+}$] = 9.48×10^{-4} M, [BH$^+$] $\approx$ 0.150 M,
[B] = 9.49×10^{-4} M

12-12. 5.59

12-13. **(a)** HA   **(b)** A$^-$ **(c)** 1.0, 0.10

12-14. **(a)** 4.00   **(b)** 8.00   **(c)** H$_2$A   **(d)** HA$^-$
(e) A^{2-}

12-15. **(a)** 9.00   **(b)** 9.00   **(c)** BH$^+$ **(d)** 1.0×10^3

12-17. **(a)** 5.59   **(b)** 10.74

12-19. 5.41, [H$_3$Arg^{2+}] = 1.3×10^{-5} M, [H$_2$Arg$^+$] = 0.050 M,
[HArg] = 1.3×10^{-5} M, [Arg$^-$] = 1.1×10^{-12} M

12-20. 2$-$

12-21. V_{e1} = 10.0 mL, V_{e2} = 20.0 mL; pH = 2.51, 4.00, 6.00,
8.00, 10.46, 12.21

12-22. V_{e1} = 10.0 mL, V_{e2} = 20.0 mL; pH = 11.49, 10.00,
8.00, 6.00, 3.54, 1.79

12-23. **(a)** phenolphthalein (colorless → red)
(b) *p*-nitrophenol (colorless → yellow)
(c) bromothymol blue (yellow → green)
or bromocresol purple (yellow → yellow + purple)
(d) thymolphthalein (colorless → blue)

12-24. V_{e1} = 40.0 mL, V_{e2} = 80.0 mL; pH = 11.36, 9.73,
7.53, 5.33, 3.41, 1.85

12-25. V_{e1} = 25.0 mL, V_{e2} = 50.0 mL; pH = 7.55, 6.02, 3.86,
1.97

12-26. **(a)** 3.46, 8.42   **(b)** second
(c) thymolphthalein; first trace of blue

Chapter 13

13-1. **(b)** 6.242×10^{18} e$^-$/C   **(c)** 9.649×10^4 C/mol

13-2. **(a)** oxidant: TeO$_3^{2-}$; reductant: S$_2$O$_4^{2-}$
(b) 3.02×10^3 C   **(c)** 0.840 A

13-3. 0.015 9 J

13-4. **(a)** 71.$_5$ A   **(b)** 6.8×10^6 J

13-6. **(a)** 0.572 V   **(b)** 0.568 V

13-7. **(a)** Pt(*s*) I Cr^{2+}(*aq*), Cr^{3+}(*aq*) II Tl$^+$(*aq*) I Tl(*s*)
(b) 0.08$_4$ V   **(c)** Tl$^+$ + Cr^{2+} ⇌ Tl(*s*) + Cr^{3+}
(d) Pt

13-8. **(b)** 0.430 V; reduction occurs at right-hand electrode

13-9. **(a)** electrons flow right to left; $\frac{3}{2}$Br$_2$(*l*) + Al(*s*) ⇌
3Br$^-$ + Al^{3+}
(b) Br$_2$ **(c)** 1.31 kJ   **(d)** 2.69×10^{-8} g/s

13-10. **(a)** -0.357 V; 9.2×10^{-7} **(b)** 0.722 V; 7×10^{48}

13-11. 7×10^{-12}

13-12. **(a)** 1.9×10^{-6} **(b)** -0.386 V; oxidized

13-13. 34 g/L

13-14. **(b)** 0.675 V; 10^{114} **(c)** 0.178 V   **(d)** 8.5

13-15. **(b)** 0.044 V

13-16. **(a)** 0.086 V   **(b)** -0.021 V   **(c)** 0.021 V

13-17. 0.684 V

Chapter 14

14-1. **(a)** Cu^{2+} + 2e$^-$ ⇌ Cu(*s*)   **(c)** 0.112 V

14-2. **(a)** Br$_2$(*aq*) + 2e$^-$ ⇌ 2Br$^-$, E_+ = 1.057 V
(b) 0.816 V

14-3. 0.1 mL: [Ag$^+$] = 1.1×10^{-11} M, $E = -0.090$ V
10.0 mL: [Ag$^+$] = 2.2×10^{-11} M; $E = -0.073$ V
25.0 mL: [Ag$^+$] = $1.0_5 \times 10^{-6}$ M; $E = 0.204$ V
30.0 mL: [Ag$^+$] = 0.012 5 M; $E = 0.445$ V

14-4. (0.1 mL, $E = 0.481$ V), (10.0 mL, 0.445 V), (20.0 mL,
0.194 V), (30.0 mL, -0.039 V)

14-5. left side

14-6. 10.67

14-7. **(a)** $+0.10$ **(b)** 13%

14-9. small

14-10. $+0.029$ 6 V

14-11. **(a)** -0.407 V   **(b)** $1.5_5 \times 10^{-2}$ M

14-12. -0.332 V

14-13. **(a)** 1.60×10^{-4} M   **(b)** 3.59 wt %

14-14. K$^+$

14-15. **(a)** $E = \text{constant} + \beta \left(\dfrac{0.059\ 16}{3} \right) \log ([\text{La}^{3+}]_{\text{outside}})$
(b) 19.7 mV   **(c)** $+25.1$ mV   **(d)** $+100.7$ mV

14-16. **(a)** $E = 51.10\ (\pm 0.24) + 28.14\ (\pm 0.08_5) \log [\text{Ca}^{2+}]$
$(s_y = 0.2_7)$
(b), (c) 2.43 $(\pm 0.06) \times 10^{-3}$ M

Chapter 15

15-1. 10.0 mL, 10.0 mL

15-6. 0.010 3 M

15-7. [Ni^{2+}] = 0.012 4 M, [Zn^{2+}] = 0.007 18 M

15-8. 0.024 30 M

15-9. 1.256 (± 0.003) mM

15-10. 0.014 68 M

15-11. 0.092 6 M.

15-12. 5.150 mg Mg, 20.89 mg Zn, 69.64 mg Mn

15-14. **(a)** 0.029 14 M   **(b)** no

15-15. **(a)** 26   **(b)** 4.02 mg

15-16. mol NH$_3$ = 2 (initial mol H$_2$SO$_4$ $-\frac{1}{2}$ × mol thiosulfate)

15-17. **(a)** 700 **(b)** 1.0 **(c)** 0.34 g I_2/L
15-18. **(a)** 0.125 **(b)** $6.87_5 \pm 0.03_8$

Chapter 16

16-2. 0.1 mm
16-7. **(a)** $4.3_5 \times 10^4$ **(b)** 3.6 μm
16-8. **(a)** $w_{1/2} = 0.8$ mm for ethyl acetate and 2.6 mm for toluene
   (b) $N = 1.1 \times 10^3$ for ethyl acetate and 1.1×10^3 for toluene
16-9. 6.8 cm diameter $\times$ 25 cm long, 17 mL/min
16-10. **(a)** $w_{1/2} = 0.172$ min, $N = 3.58 \times 10^4$
   (b) 0.838 mm
   (c) w(measured) = 0.311 min; $w/w_{1/2}$(measured) = 1.81
16-11. **(a)** heptane: 7.41×10^4 plates, 0.404 mm plate height
   $C_6H_4F_2$: 8.55×10^4 plates, 0.351 mm plate height
   (b) resolution = 1.01
16-12. **(a)** 1.11 cm diameter $\times$ 32.6 cm long
   (b) 0.069 1 mL **(c)** 0.256 mL/min
16-13. 1.47 mL/min
16-14. **(a)** 0.847 mM **(b)** 6.16 mM **(c)** 12.3 mM
16-15. 161 μg/mL
16-16. 0.47 mmol
16-17. $31 = CH_2OH^+$; $41 = C_3H_5^+$; $43 = C_3H_7^+$; $56 = C_4H_8^+$

Chapter 17

17-3. retention time decreases
17-4. retention time increases
17-7. amine will be eluted first
17-8. 0.418 mg/mL
17-9. **(a)** 8.68×10^8 **(b)** 0.273 m^2

Chapter 18

18-1. **(a)** mixture contains RCO_2^-; RNH_2, Na^+, and OH^-; all pass directly through the cation-exchange column
   (b) mixture contains RCO_2H, RNH_3^+, H^+, and Cl^-; RNH_3^+ is retained and the others are eluted
18-3. $VOSO_4 = 80.9\%$, $H_2SO_4 = 10._7\%$, $H_2O = 8._4\%$
18-5. $NH_3 < (CH_3)_3N < CH_3NH_2 < (CH_3)_2NH$

18-6. 38.0%
18-8. 4.9×10^4
18-9. **(a)** 5.7 mL **(b)** 11.5 mL
   (c) adsorption occurs
18-10. cations < neutrals < anions
18-12. **(a)** longitudinal diffusion
18-14. **(a)** $w_{1/2} = 0.75$ min, 1.6×10^4 plates
   (b) plate height = 25 μm
18-15. thiamine < (niacinamide + riboflavin) < niacin; thiamine is most soluble

Chapter 19

19-1. 2.03
19-2. 0.085 38 g
19-3. 50.79% Ni
19-4. 7.22 mL
19-5. 8.665 wt %
19-6. 0.339 g
19-7. **(a)** 5.5_1 mg/100 mL **(b)** 5.834 mg, yes
19-8. Ba, 47.35%; K, 8.279%; Cl, 31.95%
19-9. 11.69 mg CO_2, 2.051 mg H_2O
19-10. **(a)** 51.36% C, 3.639% H **(b)** C:H $\approx$ 6:5
19-11. $C_4H_9NO_2$
19-12. 104.1 ppm
19-13. $C_8H_{9.06 \pm 0.17}N_{0.997 \pm 0.010}$

Chapter 20

20-6. **(a)** 4.699×10^{-19} J **(b)** 3.67×10^{-6}
   (c) +8.4% **(d)** 0.010 3
20-7. **(a)** 6.07×10^{-19} J **(b)** 3.3×10^{-8}
   (c) +12% **(d)** 0.002 0
20-8. 0.005 7 μg/mL
20-9. 17.4 μg/mL
20-10. **(a)** 0.950 $[Cu^{2+}]_i$ **(b)** 1.00 ppm
   (c) 1.04 ppm
20-11. 1.64 μg/mL
20-12. 8.33×10^{-5} M
20-13. **(a)** 0, 10.0, 20.0, 30.0, 40.0 μg/mL
   (b) 204 μg/mL
20-14. **(a)** 7.49 μg/mL **(b)** 25.6 μg/mL

Credits

Chapter 1 **Opener:** A. G. Ewing, Pennsylvania State University. See T. K. Chen, G. Luo, and A. G. Ewing, *Anal. Chem.* 1994, *66*, 3031. This notation refers to page 3031 of volume *66* of the journal *Analytical Chemistry,* published in 1994. **Fig. 1-2:** Data cited in K. S. Johnson, K. H. Coale, and H. W. Jannasch, *Anal. Chem.* 1992, *64*, 1065A. **Fig. 1-3:** K. Levsen, S. Behnert, and H. D. Winkeler, *Fresenius J. Anal. Chem.* 1991, *340*, 665. **Fig. 1-4:** J. Hu and D. H. Stedman, *Anal. Chem.* 1994, *66*, 3384. **Box 1-2:** T. K. Chen, G. Luo, and A. G. Ewing, *Anal. Chem.* 1994, *66*, 3031.

Chapter 2 **p. 18:** Y. Okahata, Y. Matsunobu, K. Ijiro, M. Mukae, A. Murakami, and K. Makino, *J. Am. Chem. Soc.* 1992, *114*, 8299. **Fig. 2-2:** Fisher Scientific, Pittsburgh, PA. **Fig. 2-4:** (*a*) Fisher Scientific, Pittsburgh, PA. (*b*) Hach Co., Loveland, CO. **Fig. 2-7:** Cole-Parmer Co., Niles, IL. **Fig. 2-8:** (*a*) A. H. Thomas Co., Philadelphia, PA. (*b*) Fisher Scientific, Pittsburgh, PA. **Fig. 2-10:** A. H. Thomas Co., Philadelphia, PA. **Fig. 2-11:** Brinkman Instruments, Westbury, NY. **Fig. 2-12:** Hamilton Co., Reno, NV. **Fig. 2-17:** A. H. Thomas Co., Philadelphia, PA. **Fig. 2-18:** Thomas Scientific, Swedesboro, NJ. **Fig. 2-19:** Parr Instrument Co., Moline, IL.

Chapter 3 **p. 36:** 3M Company, St. Paul, MN.

Chapter 4 **p. 54:** R. H. Kardon, *Tissues and Organs* (San Francisco: W. H. Freeman, 1978), p. 39. **Fig. 4-2:** Data from Nobel Lecture of B. Sakmann, *Angew. Chem. Int. Ed. Engl.* 1992, *31*, 830. **Table 4-3:** R. D. Larsen, *J. Chem. Ed.* 1990, *67*, 925. **Table 4-4:** R. B. Dean and W. J. Dixon, *Anal. Chem.* 1951, *23*, 636; D. R. Rorabacher, *Anal. Chem.* 1991, *63*, 139. **Box 4-1:** L. H. Keith, W. Crummett, J. Deegan, Jr., R. A. Libby, J. K. Taylor, and G. Wentler, *Anal. Chem.* 1983, *55*, 2210.

Chapter 5 **p. 72:** C. Dalpra, Potomac River Basin Commission. **Fig. 5-1:** Chip Clark.

Chapter 6 **p. 94:** E. R. Madsen, *The Development of Titrimetric Analysis 'till 1806* (Copenhagen: E. E. C. Gad Publishers, 1958). **Box 6-1:** K. D. Hughes, *Anal. Chem.* 1993, *65*, 883A.

Chapter 7 **p. 116:** R. P. Wayne, *Chemistry of Atmospheres* (Oxford: Clarendon Press, 1991). **Box 7-1:** R. W. Ricci, M. A. Ditzler, and L. P. Nestor, *J. Chem. Ed.* 1994, *71*, 983. **Demo. 7-1:** D. H. Alman and F. W. Billmeyer, Jr., *J. Chem. Ed.* 1976, *53*, 166. **Fig. 7-5:** D. W. Daniel, *J. Chem. Ed.* 1994, *71*, 83. **Fig. 7-6:** A. H. Thomas Co., Philadelphia, PA. **Fig. 7-8:** Varian Australia Pty. Ltd., Victoria, Australia.

Chapter 8 **p. 132:** J. S. Schultz, *Scientific American,* August 1991, p. 65. **Fig. 8-1:** Kenneth Hughes, Georgia Institute of Technology. **Fig. 8-3:** Adapted from E. N. Baker, B. F. Anderson, H. M. Baker, M. Haridas, G. E. Norris, S. V. Rumball, and C. A. Smith, *Pure Appl. Chem.* 1990, *62*, 1067. **Fig. 8-6:** C. M. Byron and T. C. Werner, *J. Chem. Ed.* 1991, *68*, 433. **Fig. 8-7:** Data from B. T. Jones, B. W. Smith, M. B. Leong, M. A. Mignardi, and J. D. Winefordner, *J. Chem. Ed.* 1989, *66*, 357. **Box 8-1:** J. M. Van Emon and V. Lopez-Avila, *Anal. Chem.* 1992, *64*, 79A. **Demo. 8-1:** J. A. DeLuca, *J. Chem. Ed.* 1980, *57*, 541.

Chapter 9 **p. 148:** Sarah Leen, *Matrix* (from *Scientific American,* June 1994, p. 32). **Fig. 9-1:** R. Semonin, *Study of Atmospheric Pollution Scavenging* (Washington, DC: U. S. Department of Energy, 1981), pp. 56–119.

Chapter 10 **p. 164:** R. Kopelman and W. Tan, University of Michigan. See W. Tan, S.-Y. Shi, S. Smith, D. Birnbaum, and R. Kopelman, *Science* 1992, *258*, 778; W. Tan, S.-Y. Shi, and R. Kopelman, *Anal. Chem.* 1992, *64*, 2985. **Fig. 10-1:** M. L. Bender, G. E. Clement, F. J. Kézdy, and H. A. Heck, *J. Am. Chem. Soc.* 1964, *86*, 3680. **Demo. 10-1:** R. L. Barrett, *J. Chem. Ed.* 1955, *32*, 78.

Chapter 11   **p. 182:** M. Gratzl and C. Yi, *Anal. Chem.* 1993, *65*, 2085. Photos courtesy M. Gratzl, Case Western Reserve University. **Fig. 11-6:** Kyoto Electronics Manufacturing USA, Springfield, NJ. **Fig. 11-7:** Fisher Scientific, Pittsburgh, PA. **Fig. 11-8:** Fisher Scientific, Pittsburgh, PA.

Chapter 12   **p. 206:** P. A. Baedecker, U.S. Geological Survey. **Fig. 12-4:** M. F. Perutz, "The Hemoglobin Molecule." Copyright ©1964 by Scientific American, Inc. **Box 12-1:** Bio-Rad Laboratories, Hercules, CA.

Chapter 13   **p. 228:** R. G. Kessel and R. H. Kardon, *Tissues and Organs* (San Francisco: W. H. Freeman & Co., 1978), p. 14; D. Lagunoff, *J. Invest. Dermatol.* 1972, *58*, 296. **C. M. Hall:** Alcoa Co., Pittsburgh, PA.

Chapter 14   **p. 252:** V. V. Cosofret, M. Erdösy, T. A. Johnson, R. P. Buck, R. B. Ash, and M. R. Neuman, *Anal. Chem.* 1995, *67*, 1647. Photos courtesy R. P. Buck, University of North Carolina. **Demo. 14-1:** George Rossman, California Institute of Technology. **Fig. 14-3:** Fisher Scientific, Pittsburgh, PA. **Fig. 14-7:** Fisher Scientific, Pittsburgh, PA. **Fig. 14-9:** R. G. Bates, *Determination of pH: Theory and Practice,* 2nd ed. (New York: Wiley, 1973). **Fig. 14-10:** Sentron, Gig Harbor, WA. **Fig. 14-14:** M. Umemoto, W. Tani, K. Kuwa, and Y. Ujihira, *Anal. Chem.* 1994, *66*, 352A.

Chapter 15   **p. 272:** T. J. Marrone and K. M. Merz, Jr., *J. Am. Chem. Soc.* 1992, *114*, 7542. **Fig. 15-1:** J. Stein, J. P. Fackler, Jr., G. J. McClune, J. A. Fee, and L. T. Chan, *Inorg. Chem.* 1979, *18*, 3511. **Fig. 15-5:** Don Cornelius and Terrell Vanderah, Michelson Laboratory, China Lake, CA. **Fig. 15-6:** R. C. Teitelbaum, S. L. Ruby, and T. J. Marks, *J. Am. Chem. Soc.* 1980, *102*, 3332. **Box 15-1:** P. S. Dobbin and R. C. Hider, *Chem. Br.* 1990, *26*, 565. **Box 15-3:** G. F. Holland and A. M. Stacy, *Acc. Chem. Res.* 1988, *21*, 8.

Chapter 16   **p. 292:** (a) Photo by Maurice Krafft from F. Press and R. Siever, *Understanding Earth* (New York: W. H. Freeman, 1994), p. 107. (b) Philip R. Kyle, New Mexico Tech. **Fig. 16-8:** R. R. Freeman, ed., *High Resolution Gas Chromatography* (Palo Alto, CA: Hewlett Packard Co., 1981). **Fig. 16-12:** A. Illies, P. B. Shevlin, G. Childers, M. Peschke, and J. Tsai, *J. Chem. Ed.* 1995, *72*, 717. **Box 16-3:** G. Eglinton, S. A. Bradshaw, A. Rosell, M. Sarnthein, U. Pflaumann, and R. Tiedemann, *Nature* 1992, *356*, 423.

Chapter 17   **p. 312:** K. L. Steele, D. O. Scott, and C. E. Lunte, *Anal. Chim. Acta* 1991, *246*, 181. **Fig. 17-3:** Alltech Associates, State College, PA. **Fig. 17-4:** Restek Co., Bellefonte, PA. **Fig. 17-5:** J. Madabushi, H. Cai, S. Steams, and W. Wentworth, *Am. Lab.* October 1995, 21. **Fig. 17-6:** H. M. McNair and E. J. Bonelli, *Basic Gas Chromatography* (Palo Alto, CA:

Varian Instrument Division, 1968). **Fig. 17-7:** Varian Associates, Palo Alto, CA. **Fig. 17-8:** Varian Associates, Palo Alto, CA. **Fig. 17-9:** Rainin Instrument Co., Emeryville, CA. **Fig. 17-10:** Chiral Technologies, Exton, PA. **Fig. 17-11:** R. E. Majors, *J. Chromatogr. Sci.* 1973, *11*, 88. **Fig. 17-12:** Perkin-Elmer Corp., Norwalk, CT. **Fig. 17-13:** Phenomenex, Torrance, CA. **Fig. 17-14:** Upchurch Scientific, Oak Harbor, WA. **Fig. 17-16:** Anspec Co., Ann Arbor, MI. **Fig. 17-18:** H. Joshua, *J. Chromatogr.* 1993, *654*, 247.

Chapter 18   **p. 330:** *Upper:* T. M. Olefirowicz and A. G. Ewing, *Anal. Chem.* 1990, *62*, 1872; *Lower:* H.-T. Chang and E. S. Yeung, *Anal. Chem.* 1995, *67*, 1079. **Fig. 18-3:** F. W. E. Strelow, *Anal. Chem.* 1984, *56*, 1053. **Fig. 18-4:** R. J. Joyce, *Am. Environ. Lab.* May 1994, 1. **Fig. 18-7:** Varian Associates, Palo Alto, CA. **Fig. 18-9:** B. J. Compton and L. Kreilgaard, *Anal. Chem.* 1994, *66*, 1175A. **Fig. 18-11:** J. C. Transfiguracion, C. Dolman, D. H. Eidelman, and D. K. Lloyd, *Anal. Chem.* 1995, *67*, 2937. **Fig. 18-14:** C. A. Lucy and T. L. McDonald, *Anal. Chem.* 1995, *67*, 1074. **Fig. 18-15:** Bio-Rad Laboratories, Richmond, CA. **Fig. 18-16:** J. T. Smith, W. Nashabeh, and Z. E. Rassi, *Anal. Chem.* 1994, *66*, 1119.

Chapter 19   **p. 348:** M. Ferretti, R. Udisti, and E. Barbolani, *Fresenius J. Anal. Chem.* 1992, *343*, 607. **Box 19-1:** T. M. Harris, *J. Chem. Ed.* 1995, *72*, 355. **Fig. 19-3:** G. Liptay, ed., *Atlas of Thermoanalytical Curves* (London: Heyden and Son, 1976). **Fig. 19-5:** E. Pella, *Am. Lab.* August 1990, 28. **Fig. 19-6:** E. Pella, *Am. Lab.* August 1990, 28. **Fig. 19-7:** Leco Corp., St. Joseph, MI. **Fig. 19-8:** E. Pella, *Am. Lab.* February 1990, 116. **Table 19-4:** E. M. Hodge, H. P. Patterson, M. C. Williams, and E. S. Gladney, *Am. Lab.* June 1991, 34.

Chapter 20   **p. 366:** T. Nomizu, S. Kaneco, T. Tanaka, D. Ito, H. Kawaguchi, and B. L. Vallee, *Anal. Chem.* 1994, *66*, 3000. **Fig. 20-1:** Instrumentation Laboratory, Wilmington, MA. **Fig. 20-3:** A. P. Thorne, *Anal. Chem.* 1991, *63*, 57A. **Fig. 20-5:** Instrumentation Laboratory, Wilmington, MA. **Fig. 20-6:** Perkin-Elmer Corp., Norwalk, CT. **Fig. 20-7:** V. A. Fassel, *Anal. Chem.* 1979, *51*, 1290A. **Fig. 20-12:** B. T. Jones, B. W. Smith, and J. D. Winefordner, *Anal. Chem.* 1989, *61*, 1670. **Fig. 20-16:** R. J. Gill, *Am. Lab.* November 1993, 24F. **Fig. 20-17:** W. A. Weimer, *Appl. Phys. Lett.* 1988, *52*, 2171. **Fig. 20-18:** R. N. Savage and G. M. Hieftje, *Anal. Chem.* 1979, *51*, 408. **Box 20-1:** Cole-Parmer Instrument Co., Chicago, IL.

Chapter 21   **Fig. 21-10:** P. L. Weber and D. R. Buck, *J. Chem. Ed.* 1994, *71*, 609.

Color Plates   **Plate 6:** M. D. Seltzer, M. Johnson, and D. O'Connor, Michelson Laboratory, China Lake, CA. **Plate 17:** M. C. Ruiz-Martinez, J. Berka, A. Belenkii, F. Foret, A. W. Miller, and B. L. Karger, *Anal. Chem.* 1993, *65*, 2851.

Index

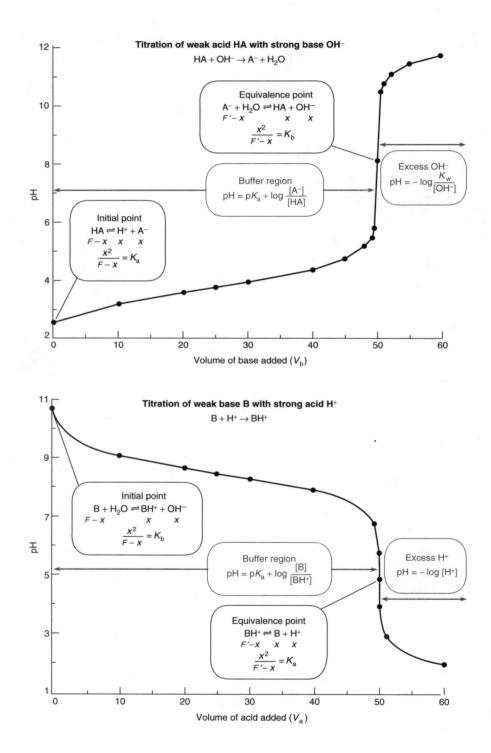